REVIEWS in M
and GEOCH

Volume 44 2001

NANOPARTICLES AND THE ENVIRONMENT

Editors:

JILLIAN F. BANFIELD *Department of Geology & Geophysics*
University of Wisconsin
Madison, Wisconsin

ALEXANDRA NAVROTSKY *Department of Chemical*
Engineering & Materials Science
University of California-Davis
Davis, California

COVER: Colorized, high-resolution transmission electron microscope lattice-fringe image of seven nanoparticles of UO_2 produced by the activity of sulfate-reducing bacteria. Note that the largest uraninite particle is <3 nm in diameter (Suzuki and Banfield, in preparation).
Graphic provided by Jill Banfield, Yohey Suzuki, and Mary Diman.

Series Editor for MSA: ***Paul H. Ribbe***
Virginia Polytechnic Institute and State University
Blacksburg, Virginia

MINERALOGICAL SOCIETY of AMERICA

REVIEWS IN MINERALOGY AND GEOCHEMISTRY

(Formerly: REVIEWS IN MINERALOGY)

ISSN 1529-6466

Volume 44

Nanoparticles and the Environment

ISBN 0-939950-56-1

** This volume is the sixth of a series of review volumes published jointly under the banner of the Mineralogical Society of America and the Geochemical Society. The newly titled *Reviews in Mineralogy and Geochemistry* has been numbered contiguously with the previous series, *Reviews in Mineralogy*.

Additional copies of this volume as well as others in this series may be obtained at moderate cost from:

THE MINERALOGICAL SOCIETY OF AMERICA
1015 EIGHTEENTH STREET, NW, SUITE 601
WASHINGTON, DC 20036 U.S.A.

Dedication

Dr. William C. Luth has had a long and distinguished career in research, education and in the government. He was a leader in experimental petrology and in training graduate students at Stanford University. His efforts at Sandia National Laboratory and at the Department of Energy's headquarters resulted in the initiation and long-term support of many of the cutting edge research projects whose results form the foundations of these short courses. Bill's broad interest in understanding fundamental geochemical processes and their applications to national problems is a continuous thread through both his university and government career. He retired in 1996, but his efforts to foster excellent basic research, and to promote the development of advanced analytical capabilities gave a unique focus to the basic research portfolio in Geosciences at the Department of Energy. He has been, and continues to be, a friend and mentor to many of us. It is appropriate to celebrate his career in education and government service with this series of courses in cutting-edge geochemistry that have particular focus on Department of Energy-related science, at a time when he can still enjoy the recognition of his contributions.

Reviews in Mineralogy and Geochemistry Volume 44

FOREWORD

This volume was prepared in conjunction with a short course, "Nanoparticles in the Environment and Technology," convened on the campus of the University of California-Davis on December 8 and 9, 2001, by the editors, Jill Banfield and Alex Navrotsky, both of whom have been previous short-course conveners and editors of *RiM* volumes (35 and 14, respectively). The U.S. Department of Energy, Office of Basic Energy Sciences—Chemical Sciences, Geosciences, and Biosciences Division, generously supported both the course (scholarships for participating graduate students) and the publication of this book (see *Dedication* on p. iii). This is the sixth volume in the new *Reviews in Mineralogy and Geochemistry* series, begun in the year 2000 under the joint sponsorship of the Mineralogical Society of America and the Geochemical Society.

Paul H. Ribbe, Series Editor for the
Mineralogical Society of America
Blacksburg, Virginia
October 15, 2001

Nanoparticles and the Environment – An Introduction

Jillian F. Banfield
University of Wisconsin–Madison
and
Alexandra Navrotsky
University of California–Davis

Over the years, volumes in this series have taken a variety of forms. Many have focused on mature fields of investigation to draw together a comprehensive body of work and provide a definitive, up to date reference. A few, however, have sought to provide enough coverage of an emerging or re-emerging field to allow the reader to identify important and exciting gaps in current knowledge and opportunities for new research. This volume falls into the later category. Our primary goal in convening the short course and assembling this text is to invigorate future research.

Early *Reviews in Mineralogy* dealt with specific groups of minerals, one (or two) volumes at a time. In contrast, this volume deals explicitly with the topic of crystal size in many different systems. Until recently, the special and complicated nature of the very smallest particles rendered them nearly impossible to study by conventional methods. Even today, the challenges associated with evaluating the size-dependence of a mineral's bulk and surface structures, properties, and reactivity are significant. However, ongoing improvements in sophisticated characterization, theory, and data analysis make particles previously described (often inaccurately) as "amorphous" (or even more mysteriously as "X-ray amorphous") amenable to quantitative evaluation. Thermochemical, crystal chemical, and computational chemical approaches must be combined to understand particles with diameters of 1 to 100 nanometers. Determination of the variation of structure, properties, and reaction kinetics with crystal size requires careful synthesis of size- and perhaps morphology-specific samples. These problems demand integration of mineralogical and geochemical approaches. Thus, it is appropriate that the current issue belongs to the era of *Reviews in Mineralogy and Geochemistry.*

"Nanoparticles and the Environment" targets naturally occurring, finely particulate minerals, many of which form at low temperature. Thus, many of the compounds of interest are those of the "clay fraction". Of course, there have been decades of critical work on the structures, microstructures, and reactivity of finely crystalline or amorphous minerals, especially oxides, oxyhydroxides, hydroxides, and clays. We will not summarize what is known in general about these (for this, the reader is referred to earlier Reviews in Mineralogy volumes). Rather, our goal is to focus on the features of these materials that stem directly or indirectly from their size.

The term "nanoparticles" is much more than a re-labeling designed to align "clay" (sized) minerals with nanotechnology and its goals. The term signifies that the substance has physical dimensions that are small enough to ensure that the structure and/or properties and/or reactivity are measurably particle size dependent, yet the particle is large enough to warrant its distinction from aqueous ions, complexes, or clusters. The chemistry, physics, and geology of particles at this intermediate scale are unique, fascinating, and important. Of particular interest are those properties that emerge only after a cluster of atoms has grown beyond some specific size, and disappear once the particle passes out of the "nanoparticle" size regime.

There are some compelling examples of size-dependent phenomena. It is well known that the melting temperature of nanocrystals (defined as crystals having properties intermediate between molecular and crystalline) decreases dramatically as the radius of the cluster decreases. Buffat and Borel (1976)[1] showed that the ratio of the melting temperature of the cluster (T) with a radius r to that of the bulk (T_0) is given by

$$\frac{T}{T_0} = 1 - \frac{2}{\rho_S \lambda}\left[\frac{\gamma_{SL}}{r-\delta} + \frac{\gamma_l}{r}\left(1 - \frac{\rho_S}{\rho_L}\right)\right]$$

where γ_{SL} is the solid-liquid surface tension and γ_L is the surface tension of the liquid, δ is the liquid layer thickness, ρ_S and ρ_L are the densities of the bulk solid and liquid, respectively, and λ is the heat of fusion. Qi et al. (2001)[2] used molecular dynamics calculations for nickel to further investigate the size-dependence of melting behavior. Their results showed that nanocrystals will melt via surface processes, with the melting point for particles containing N atoms given by $T_N = T_0 - a\,N^{-1/3}$, where a is a constant. For example, for N = 336 the melting temperature is decreased by 44%. Phase stability can be size dependent (e.g., Gravie 1978)[3]. This phenomenon may explain nucleation of what are normally considered metastable phases. Absorption and luminescence spectra for small crystals are determined by the quantum-size effect. Decreasing nanocrystal size correlates with increased total energy of band edge optical transitions. As a consequence,

[1] Buffat P, Borel JP (1976) Size effect on the melting temperature of gold particles. Phys Rev. A 13: 2287-2298

[2] Qi Y, Tahir C, Johnson WL Goddard WA III (2001) Melting and crystallization in Ni nanoclusters: The mesoscale regime. J Chem Phys 115: 385-394

[3] Gravie RC (1978) Stabilization of the tetragonal structure in zirconia microcrystals. J Phys Chem 82: 218-224

the color of some nanocrystals correlates strongly with their particle size. For example, the energy gap of CdSe increases from its bulk value of 1.8 eV up to 3 eV for very small crystals, passing through most of the visible part of the optical spectrum! (Murray et al. 1993; Efros and Rosen 2000).[4,5]

Current world-wide interest in "nanotechnology" and "nanomaterials" offers a unique opportunity for the Earth sciences. Both the level of visibility and the explosion of synthesis and characterization techniques in physics, chemistry, and materials science provide mineralogy and geochemistry with new opportunities. It is important for us to show that the "nano" field consists of more than micromachines and electronic devices, and that nanoscale phenomena permeate and often control natural processes.

Why all the fuss about nanoparticles now? As increasing attention in engineering is focused on making smaller and smaller machines, questions about the fundamental processes that govern nanoparticle form, stability, and reactivity emerge. The geoscience community is well equipped to tackle the basic science concepts associated with these questions. However, we have our own reasons to study size-dependent phenomena.

Size-dependent structure and properties of Earth materials impact the geological processes they participate in. This topic has not been fully explored to date. Chapters in this volume contain descriptions of the inorganic and biological processes by which nanoparticles form, information about the distribution of nanoparticles in the atmosphere, aqueous environments, and soils, discussion of the impact of size on nanoparticle structure, thermodynamics, and reaction kinetics, consideration of the nature of the smallest nanoparticles and molecular clusters, pathways for crystal growth and colloid formation, analysis of the size-dependence of phase stability and magnetic properties, and descriptions of methods for the study of nanoparticles. These questions are explored through both theoretical and experimental approaches.

Nanoparticles participate in every crystallization reaction and they constitute a major source of surface area in environments where virtually every important reaction takes place on a surface. They are components of enzymes and key biomolecules and their presence may record the early existence of life. How can we not be fascinated by these remarkable, and special, forms of matter?

ACKNOWLEDGMENTS

The editors thank their co-authors for their contributions to this enterprise, Dr. Nick Woodward (US Department of Energy, Basic Energy Sciences Program) for financial support for the project, and Dr. Paul Ribbe, Series Editor, for bringing the volume together.

[4,5] Murray CB, Norris DJ, Bawendi MG (1993) Synthesis and characterization of nearly monodisperse CdE (E = S, Se, Te) Semiconductor nanocrystallites. J Am Chem Soc 115:8706-8715. Efros AL, Rosen M (2000) The electronic structure of semiconductor nanocrystals. Ann Rev Mater Sci 30:475-521

TABLE OF CONTENTS

1 Nanoparticles in the Environment

Jillian F. Banfield, Hengzhong Zhang

2 Nanocrystals as Model Systems for Pressure-Induced Structural Phase Transitions

Keren Jacobs, A. Paul Alivisatos

3 Thermochemistry of Nanomaterials

Alexandra Navrotsky

4 Structure, Aggregation and Characterization of Nanoparticles

Glenn A. Waychunas

5 Aqueous Aluminum Polynuclear Complexes and Nanoclusters: A Review

William H. Casey, Brian L. Phillips, Gerhard Furrer

6 Computational Approaches to Nanomineralogy

James R. Rustad, Witold Dzwinel, David A. Yuen

7 Magnetism of Earth, Planetary, and Environmental Nanomaterials

Denis G. Rancourt

8 Atmospheric Nanoparticles

C. Anastasio, S. T. Martin

1

Nanoparticles in the Environment

Jillian F. Banfield* and Hengzhong Zhang

Department of Geology and Geophysics
University of Wisconsin-Madison
1215 West Dayton Street
Madison, Wisconsin 53706

* Current Address: *Department of Earth and Planetary Science*
University of California-Berkeley
Berkeley, California 94720

INTRODUCTION

Nanoparticles are discrete nanometer (10^{-9} m)-scale assemblies of atoms. Thus, they have dimensions between those characteristic of ions (10^{-10} m) and those of macroscopic materials. They are interesting because the number of atoms in the particles is small enough, and a large enough fraction of them are at, or near surfaces, to significantly modify the particle's atomic, electronic, and magnetic structures, physical and chemical properties, and reactivity relative to the bulk material. Nanoparticle surfaces themselves may be distinctive. Particles may be terminated by atomic planes or clusters that are not common, or not found, at surfaces of the bulk mineral. These, and other size-related effects will lead to modified phase stability and changes in reaction kinetics.

What makes a nanoparticle a nanoparticle? Definitions of the size ranges for molecules, nanoparticles, and macroscopic solids must be compound specific. However, a useful upper limit for nanoparticles is the size at which one of its properties deviates from the value for the equivalent bulk material by an amount that is significantly larger than the error of the method used to make the measurement (a few percent). In practice, some characteristic will probably be different enough to warrant description as a "nanoparticle" if it is less than a few tens of nanometers in diameter, and perhaps less than a fraction of a micron in diameter.

Because of the importance of size-dependent property changes to the materials sciences, size-property relationships have been studied in detail for some systems. For example, for semiconductors, size effects become important when the particle diameter is close to the Bohr diameter of excitons in the bulk phase. Generally, semiconductor size quantization effects (relevant for naturally occurring metal sulfides, for example) appear when particles are less than 10 nm in diameter (Vogel and Urban 1997).

Definition of the lower size limit for nanoparticles is fairly subjective because it is difficult to distinguish the smallest nanoparticles from multinuclear clusters and dissolved chemical species. However, the changes in physical and chemical properties of a cluster as it converts to a nanoparticle makes the smaller end of the nanoparticle size range very interesting!

Compared to other planetary materials, nanoparticles have received comparatively little attention. In large part, this is because their structures are difficult to analyze via conventional methods. Analysis of nanoparticle reactivity requires careful synthesis of homogeneous materials, detailed experimental studies, and advanced computational modeling. Recent advances in methods and approaches have yielded many essentially intractable problems solvable. Because of growing awareness of their novel and important physical, chemical, magnetic, and optical properties (a subset of which only "emerge" at the nanoscale), nanoparticles are attracting increasing attention from the

1529-6466/00/0044-0001$05.00

Earth Science community. The field, sometimes referred to as "nanogeoscience," covers topics such as size-dependent structure, stability, and reactivity. Although these topics are not new, advances in methods and new combinations of these evolving methods, ensure that the field is poised for considerable growth. New discoveries may change how we view processes at, and near, the Earth's surface.

Where do nanoparticles occur in natural systems? Where do they come from? Why should we distinguish them as a special state of matter deserving a volume in the RIMG series? This chapter will (1) provide examples of the types of solids that are commonly encountered as nanoparticles in natural systems, (2) detail a subset of the inorganic and biological processes that generate nanoparticles in the environment, (3) review the ways in which nanoparticle stability and reactivity are modified as a consequence of their particle size in order to lay the foundation for consideration of the role of nanoparticles in natural processes; and (4) consider some opportunities for future work on nanoparticles in geological systems.

NANOPARTICLES IN NATURAL SYSTEMS: WHERE DO THEY COME FROM?

Nanoparticles are everywhere … or almost everywhere! To mineralogists who work with environmental samples this is no surprise. In fact, the field of clay mineralogy, where "clays" are defined not as layer silicates but as particles smaller than 2 ∝ μm, is quite mature (e.g., Parker and Rae 1998; Occelli and Kessler 1998; Moore and Reynolds 1989). However, the challenges associated with characterization of the smallest of these particles are considerable, and much remains to be learned about size-related effects. Similarly, biochemists are highly familiar with nanoparticles as components of important biological molecules (e.g., ferritin; Harrison and Dutriazac 1998), cells, and at the smallest size, as active centers in enzymes (e.g., $[Fe_4S_4](S^\gamma\ Cys)_4$ clusters). New knowledge about cluster formation and size-dependent properties of small particles may throw new light on biological function, evolution, and perhaps even the origin of molecules essential to life (e.g., proteins).

From a nanocrystal's perspective, low-temperature environments are special because conditions are conducive to their birth and survival, at least over short to medium time scales. Consequently, planetary surfaces are places where nanoparticles are commonly encountered, and where they play special geochemical and mineralogical roles. However, nanoparticles are relevant in environments other than those where clays are found. In fact, all crystals pass through the "nanocrystal" stage early in their existence. In many geological settings, this stage is highly transitory, and nanoparticle lifetimes may be so short that we barely notice them or find them deserving of special attention. Given sufficient time, energy, and supply of appropriate ions, even at low temperatures nanocrystals grow into microcrystals and macroscopic crystals, thereby losing many of the special features that make them so interesting.

Many pathways operating under a broad diversity of conditions generate a few nanoparticles. Although these processes are deserving of attention, we will not provide a comprehensive review. Instead, we will focus our discussion on the subset of natural phenomena that generate sufficiently large concentrations of nanoparticles that these products become a major source of surface area in the environment. These are also the conditions where nanoparticle-nanoparticle interactions become important, giving rise to interesting and complicated reaction kinetics. Because most chemical reactions take place at interfaces, the nature of the boundaries that separate nanoparticles, and between nanoparticles and their surroundings (e.g., air, solution) is of special interest.

Pathways that produce abundant nanoparticles tend to involve geochemical processes that generate high degrees of supersaturation, leading to production of very many crystal nuclei. Supersaturation can occur as the result of inorganic or biological processes.

NANOPARTICLE FORMATION VIA INORGANIC PATHWAYS

Examples of environments where local high degrees of supersaturation are encountered include those associated with discharge of hydrothermal vent fluids into cold ocean water, regions where streams of highly acidic solutions mix with neutral pH water, zones of mixing between groundwater fluids, and sites of evaporation of soil water solutions.

Consider chemical weathering processes. Silicate, oxide, phosphate, and other minerals that are unstable under Earth surface conditions dissolve or react to form other phases. When solutions formed by dissolution become concentrated due to evaporation, abundant nanoparticle products form, either by direct precipitation from solution or via restructuring of a gel-like precursor (e.g., Combes et al. 1989). Examples of nanoscale products include amorphous or opaline silica (e.g., Jones 1966; Guthrie et al. 1995; Banfield and Barker 1998), hydrous aluminosilicates such as allophane (Hemni and Wada 1976; Wada and Wada 1977) with particle sizes of a few to a few tens of nanometers, and the tubular aluminosilicate imogolite (Gustafsson 2001), aluminosilicates clays such as halloysite, and oxides (e.g., magnetite and hematite formed by subsolidus alteration and weathering of olivine; Banfield et al. 1990) and oxyhydroxides (e.g., goethite formed by weathering of amphibole; Fig. 1). Other minerals, such as biotite, can undergo essentially solid state conversion to form chlorite, vermiculite, or smectite (Banfield and Murakami 1998). This may be accompanied by expulsion of ions such as Ti from the structure, and precipitation of nanocrystalline anatase (TiO_2). Similar reactions are encountered when rocks are slowly cooled and hydrated (e.g., during retrograde metamorphism).

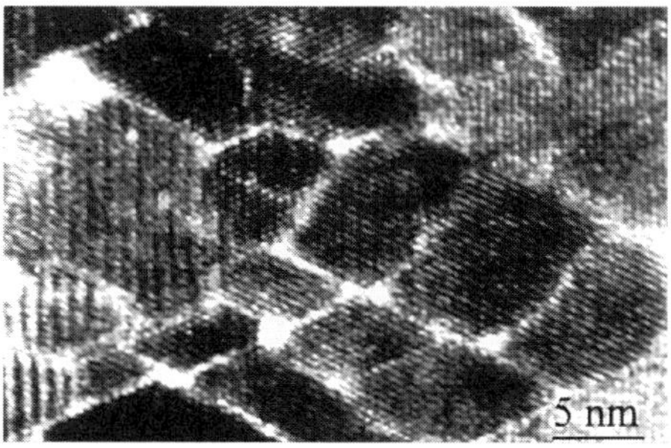

Figure 1. High-resolution transmission electron microscope image of goethite from weathered amphibole. Note the nanometer-scale porosity that separates oriented nanocrystals. Similar aggregates were reported by Smith et al. (1983, 1987) in botryoidal goethite (Banfield and Barker, unpublished data).

Acid mine drainage environments are weathering-dominated systems that generate very abundant nanoparticles. Consequently, we review the geochemistry and mineralogy of acid systems to provide specific examples of relevant processes.

Example: predominantly inorganic nanoparticle formation in acid drainage

Very low pH solutions derived by weathering of metal sulfide-rich rocks (sulfuric acid-rich solutions) carry high concentrations of dissolved ferrous and ferric iron, aluminum, and other metals into the surrounding environment. Solutions at, and near acid generation sites are often hot (>40°C), in part due to the exothermic nature of metal sulfide oxidation reactions, and usually very iron-rich, due to the predominance of iron sulfide minerals (often pyrite, FeS_2) in most ore deposits. The low solubility of oxygen in these fluids and the slow rate of inorganic oxidation of ferrous iron by oxygen at low pH (especially below pH 2) can lead to formation of solutions containing tens of g/L Fe^{2+}. In rocks containing other sulfide phases, concentrations of other elements (e.g., Cu, Zn, Cd, As, Se) can also be considerable.

In addition to ferrous iron, ferric iron can be very abundant in acid drainage solutions, especially where the activity of ferrous iron-oxidizing prokaryotic microorganisms is high (see below for details). In fact, the observed iron speciation may be largely determined by the balance between microbial ferrous iron oxidation rates and the rates at which ferric iron is reduced by oxidation of sulfide, sulfur, and sulfoxy species.

Conditions that sustain high metal loads are typically spatially confined to metal-sulfide-rich regions. Fluid draining from ore bodies (for example) often mix with cool, dilute, oxygenated water in streams or lakes. The higher solubility of oxygen, lower solubility of iron, and faster oxidation kinetics due to the lower temperature and higher pH leads to rapid precipitation of nanoscale iron hydroxides/oxides/oxyhydroxides, and related phases. However, if acid solutions are diluted under low oxygen conditions (e.g., due to intersection of subsurface groundwater flow paths), metals may remain in solution and be transported considerable distances away from the source. Under some conditions, the microorganisms adapted to grow at neutral pH can populate niches in these systems and induce precipitation of abundant nanoparticles (see below).

Schwertmannite, is a common nanoparticle-product of neutralization of sulfuric acid-rich solutions (Bigham et al. 1994). The original structural analysis indicated that sulfate was contained within tunnels similar to those found in akaganeite (FeOOH). However, recent work by Waychunas et al. (2001) suggests that this is a defective, nanoporous phase and that sulfate occupies inner and outer sphere positions on the surface, and probably on the internal surfaces of defect regions within the structure.

Ferrihydrite (often subdivided into 2- and 6-line variants based on whether two or six lines appear in diffraction patterns) is also produced by neutralization of Fe-rich, acid solutions. Six line-ferrihydrite is similar to the phase found in ferritin (e.g., Cowley et al. 2000). The structural state of ferrihydrite has been a matter of considerable debate (e.g., Drits et al. 1993; Janney et al. 2000a; Janney et al. 2001), probably because it has no single form and, when examined by X-ray methods, is inferred to be a mixture of phases. For example, samples characterized by Drits et al. (1993) were reported to contain ferrihydrite with a double hexagonal closed packed ABAC-type stacking, 2-layer material in which ABA and ACA stacking are randomly intermixed, and few-nanometer-diameter hematite (α-Fe_2O_3) particles. The distinction between 2- and 6-line ferrihydrite has been attributed to the degree of crystallinity (Drits et al. 1993). However, based on transmission electron microscope nano-electron diffraction pattern analysis, Janney et al. (2000b) were unable to detect ABAC stacking in 2-layer ferrihydrite. Instead, they suggest that 2-line ferrihydrite contains an essentially randomly stacked succession of closest packed anions sheets. Janney et al. (2000b) propose that some 6-layer ferrihydrite contains regions that are similar to those found in some 2-line ferrihydrite (Janney et al. 2000b).

Goethite (α-FeOOH) is also a common nanocrystalline phase in acid drainage systems. Goethite crystals often have nanoscale-scale porosity (see Fig. 1) and a distinct subgrain structure. This could arise if the large goethites grow via assembly of goethite nanoparticles, as described below, or may arise for other reasons. Nanoporosity is also a

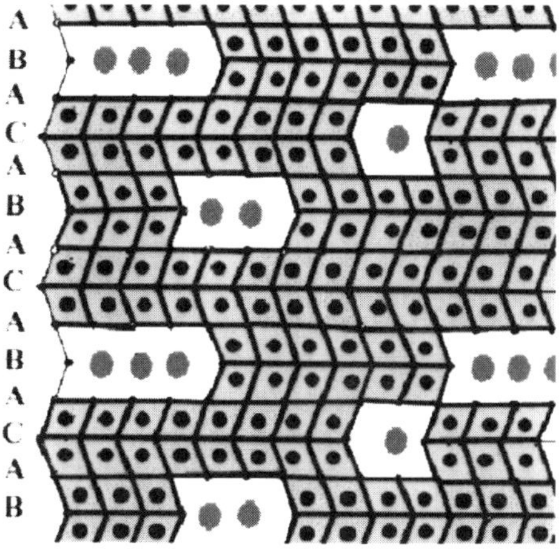

Figure 2. Cartoon illustrating features of the structure of ferrihydrite proposed by Drits et al. (1993). Note the several octahedra-wide spaces. Although not discussed by Drits et al., these might arise if the solid forms by aggregation of smaller iron oxyhydroxide subunits.

feature of one proposed structure for ferrihydrite (Fig. 2, after Drits et al. 1993).

It has been known for a very long time that hydr/oxides can exert important controls on the distributions of ions in the environment. This is especially important where nanoparticulate materials provide sufficient surface area to sorb a large fraction of an ion of interest from solution. For example, fracture and grain coatings, especially those that consist of iron and manganese oxides, can partition high concentrations of metals such as Ni, Co, Pb, Cr, Se, Cd, U, and Zn (Fuller et al. 1996; Charlet and Manceau 1992; Manceau and Charlet 1992, 1994; Spadini et al. 1994; Waite et al. 1994; Barger et al. 1999; Brown et al. 1999; Fowle et al., in prep.). Iron oxides can also strongly adsorb As (Waychunas et al. 1993; Manceau 1995), and P, thus can impact both metal toxicity and ecosystem productivity. In other cases, metals are not adsorbed but coexist as fine-scale intergrowths. For example, 10- to 50-nm crystals of uranium-rich saleeite [$Mg(UO_2)_2(PO_4)_2{\cdot}10H_2O$] and (meta)torbernite are associated with iron oxyhydroxide phases (Murakami et al. 1997). The coordination geometries of surface-adsorbed ions have been characterized in detail in numerous studies. However, questions that relate to the size-dependence of adsorption remain. This topic is considered in more detail below.

Iron-oxide nanoparticles have interesting electromagnetic properties (see Rancourt, this volume), thus are studied for potential materials applications. It is not the purpose of this chapter to review this aspect of nanoscience. However, as an aside we report an intriguing use of γ-Fe_2O_3 (maghemite) nanoparticles in medical applications. Evidently, ferrogels of Fe_2O_3 are suitable for applications in human mobile prostheses because they are able to answer mental messages (Matxain et al. 2000)!

Other inorganic pathways for formation of nanoparticles

Many inorganic processes, in addition to those described above, generate small particles. Nanoparticles will be encountered during the early stages of any crystallization event, regardless of the temperatures. The are also common in some metamorphic rocks. For example, the stability of submicron-diameter sillimanite (Al_2SiO_5) in metamorphic rocks is highly dependent upon its particle size (e.g., fibrolite: Kerrick 1990; Penn et al. 1999). Examples of higher-temperature geological reactions that generate small crystals include nucleation from melts (precipitation from melts, recrystallization of glasses), precipitation from vapors, unmixing reactions within minerals (exsolution), and chemical reactions between minerals. In these cases, the effects of particle size on phase stability, morphology, reactivity, and properties will depend on the nature of the interface between the melt/vapor/crystal and the nuclei. The general topic of nucleation within solids is covered in mineralogical texts, such as Putnis (1992). Nanoparticles can also be produced by physical processes, including impacts, electrical discharges, and condensation reactions within the solar nebula. Carbon-based "buckyballs" investigated for photovoltaic and other applications have been detected in geological settings (e.g., Daly et al. 1993). Nanoparticles probably constitute much of the surface dust on Mars. Nanoparticles that form in, and are reactive components of the Earth's atmosphere are discussed in detail by Anastasio and Martin (this volume).

Nanoparticles may also be important within planetary interiors. For example, phase transitions within the deep Earth may generate materials that are composites of nanoparticles (e.g., within the spinel phase at the olivine-spinel transition at the 400-km discontinuity). These grain sizes may affect both kinetics and rheology (e.g., of ice in planetary interiors; Stern et al. 1997). Chemical reactions in the deep Earth, perhaps between metal and silicate near the core-mantle boundary, may be impacted by nanocrystals.

NANOPARTICLE FORMATION VIA BIOLOGICAL PATHWAYS

The <150°C environments conducive to survival of very small particles are also often populated by microorganisms, some of which are directly responsible for formation (and dissolution) of nanoparticles. Of particular importance in this context are those microorganisms that generate metabolic energy by pathways that involve inorganic ions that participate in redox reactions.

Some microorganisms are able to couple oxidation of inorganic compounds to reduction of inorganic compounds, whereas others pass electrons from organics to inorganic ions. These redox reactions commonly lead to changes in ion solubility and thus, generate nanoscale mineral products. Examples of microbially-induced precipitates include ferric iron minerals produced when oxidation of ferrous iron or manganese is coupled to reduction of oxygen or nitrate and sulfides that form when sulfate is reduced in the presence of dissolved metal ions (additional details are provided below). Microorganisms can also generate nanoparticles when they sequester ions into precipitates that form within the cytoplasm, on surfaces of cells, or on cell-associated polymers. Some internally-precipitated minerals may confer navigational ability (e.g., magnetite or greigite; for review see Bazylinski and Moskowitz 1997), others represent storage devices for valuable resources (e.g., sulfur, phosphate). In other cases, microbes produce minerals in order to reduce the toxicity of a chemical species. For example, intracellular uranium-rich crystals probably develop as part of a detoxification response (Suzuki and Banfield 2001).

Example: Iron-oxidizing microbes and nanoparticle formation in AMD systems

Above we described chemical reactions between sulfide minerals and oxidized aqueous solutions that generate metal-rich fluids that rapidly precipitate nanocrystals via inorganic pathways when cooled and/or diluted. If the processes of dilution and cooling are gradual, environments will develop where iron oxidation kinetics are inhibited sufficiently to allow microorganisms to utilize iron oxidation as a source of energy. Examples include the pH 2-4 habitats of iron-oxidizing chemoautotrophs. Despite the diversity of known iron-oxidizing organisms (e.g., see Bond et al. 2000 and references therein), the vast majority of research has been conducted on *Acidithiobacillus ferrooxidans* (previous *Thiobacillus ferrooxidans*). This organism has the only Fe-based respiratory chain to be studied in any detail and was the first iron-oxidizing microorganism to have its genome completed. A reason for focus on this bacterium was the popular (but probably erroneous) belief that the activity of *A. ferrooxidans* was responsible for production of a major fraction of acid mine drainage. Ironically, the role of this microbe in the environment often may be positive because it populates less extreme environments where its activity promotes rapid precipitation of iron oxyhydroxides and removal of toxic metals through co-precipitation and sorption. In fact, *A. ferrooxidans* is utilized in some acid mine drainage treatment plants to enhance metal removal.

The coupling between iron oxidation, oxygen reduction, and generation of ATP (the primary energy currency of the cell) in *A. ferrooxidans* is illustrated in Figure 3. Ferrous iron released by pyrite dissolution diffuses through a porin in the outer membrane (OM) into the periplasm (P) where it binds to a component of the electron transport chain that spans the inner (or cytoplasmic) membrane (IM). One model has Fe^{2+} oxidoreductase that contains several components with an unknown number of Fe_4S_4 clusters (Yamanaka and Fukumori 1995). One model has Fe^{2+} oxidation occurring at this soluble cytochrome c (rawlings 2001). Electrons are then transferred between rusticyanin and possibly other c-type cytochromes to a terminal oxidase in the inner (cytoplasmic) membrane (IM).

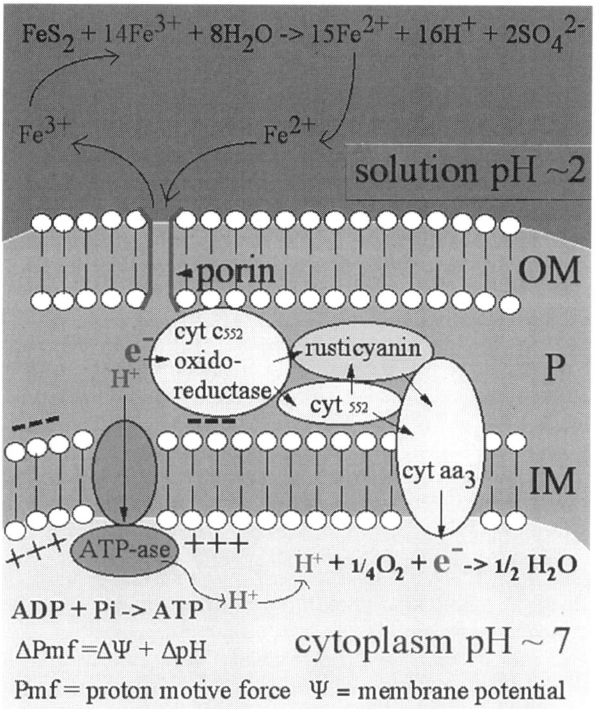

Figure 3. Diagram of a section through the cell wall of *Acidithiobacillus ferrooxidans* modified from Blake et al. (1992) showing the relationship between iron oxidation and pyrite dissolution. OM =outer membrane, P = periplasm, IM = inner or (cytoplasmic) membrane, cty = cytochrome, pmf = proton motive force. Passage of a proton (driven by proton motive force) into the cell catalyzes the conversion of ADP to ATP. Ferrous iron binds to a component of the electron transport chain, probably a cytochrome c, and is oxidized. The electrons are passed to a terminal reductase where they are combined with O_2 and H^+ to form water, preventing acidification of the cytoplasm. Ferric iron can either oxidize pyrite (e.g. within the ore body) or form nanocrystalline iron oxyhydroxide minerals (often in surrounding groundwater or streams).

Rusticyanin is an abundant, highly stable periplasmic blue copper protein (e.g., Cobley and Haddock 1975; Jedlicki et al. 1986; Ronk et al. 1991; Nunzi et al. 1993; Blake et al. 1993). Studies of rusticyanin include kinetic competence in the iron oxidation reaction (Blake and Shute 1994), identification of a *His* ligand to the copper center (Casimiro et al. 1995), a solution NMR structure (Botuyan et al. 1996), and a high-resolution X-ray structure (Walter et al. 1996). Rusticyanin forms a complex with new c-type heme cytochrome in the *A. ferrooxidans* electron transport chain (Giudici-Orticoni et al. 2000).

Preliminary spectroscopic studies indicate that respiratory chains found in different classes of Fe-oxidizing organisms are different (Blake et al. 1993). Research on three strains of *"Leptospirillum ferrooxidans"* showed that iron oxidation involves a molecule that is spectroscopically distinct from rusticyanin (Barr et al. 1990). The molecule was inferred to contain about one Fe and one Zn per cytochrome molecule (Hart et al. 1991). Blake et al. (1993) further described two distinct novel red, redox-active cytochromes from two strains of *"L. ferrooxidans"* (strains DSM 2705 and P3A), and a yellow chromophore (absorption band suggestive of a flavin) from an organism called BC1 that were reducible by Fe^{2+}. To our knowledge, no further characterization of these molecules (and certainly none from the new *L. ferrooxidans* group III; Bond et al. 2000) has been carried out, despite their considerable importance as catalysts for iron-precipitation reactions.

Spectral analyses of cell extracts from the archaea *Acidianus* (formerly *Sulfolobus*) *brierleyi, Acidianus infernus,* and *Metallosphaera sedula* (Barr et al. 1990; Blake et al. 1993) revealed a yellow cytochrome that was reducible in the presence of Fe^{2+}. This similarity led to the proposal for a unique respiratory chain in archaea (Barr et al. 1990; Blake et al. 1993). Recent genome-based analysis of *Sulfolobus acidocaldarius* has led to identification of sequences that code for a cytochrome c oxidase, an iron-sulfur protein, and a blue copper protein (Schäfer et al. 1999). The archeal species studied to date are all relatively closely related, falling within the Cranarchaeota. In contrast, the iron-oxidizing genus *Ferroplasma* falls within the Euryarchaeota. Therefore, the conclusion that iron oxidation by all Archaea involves a common pathway requires evaluation of this taxon.

The activity of enzymes allows for rapid accumulation of insoluble ferric iron aqueous species in solutions where inorganic reactions would occur slowly or not at all. The reaction kinetics may impact the nature of nanoparticles produced (composition, structure, defect structure). However, it should be noted that the crystallization reaction itself is not enzyme mediated and the reaction often does not occur within or on the cell wall. Ferric iron ions diffuse out of the periplasm to form molecular clusters and/or nanoparticles in solution. The precipitation mechanisms probably differ only slightly from those involved in inorganic reactions.

In addition to dilution of acid rock/mine drainage under oxidizing conditions, neutralization can occur under mildly or highly anaerobic conditions. This will create distinctive environments in which microorganisms thrive and nanoparticles form as a result of their activity. We describe two examples of such subsurface systems below. However, before turning to these topics, we note that Fe-based microbial ecosystems are not only found in association with metal sulfide deposits, but may be broadly relevant in the subsurface where Fe-rich minerals (biotite, olivine, pyroxenes, etc.) are present in reasonable abundance and dissolve, releasing aqueous ferrous iron.

Consider an abandoned subsurface deposit rich in pyrite and ore minerals (e.g., PbS, ZnS, $CuFeS_2$, FeAsS, etc.) and located close to, or below, the water table. Surface-derived, oxygenated water will oxidize the deposit, releasing Fe-rich, sulfate solutions, becoming oxygen-depleted in the process. If acidic fluids traveling outward from the deposit mix with groundwater containing very little oxygen, iron (and other metals) will remain in solution at concentrations limited by saturation with respect to ferrous iron minerals. Druschel et al. (2002) note that if the aquifer is carbonate-bearing, a ppm-level ferrous iron concentration limit will be set by siderite solubility. If the fluid becomes highly anoxic and sulfide forms, iron concentrations will be limited at extremely low values by formation of phases such as mackinawite. On the other hand, elevated concentrations of oxygen will limit ferric iron solubility due to precipitation of ferrihydrite or related compounds (see above). So long as the fluid chemistry remains in the window bounded by iron sulfide or ferric oxide precipitation, iron may be transported large distances in the subsurface, even though the pH of the fluid is 7 or higher. Two possible fates exist for iron-bearing groundwater fluids in the subsurface; both can lead to formation of spectacular deposits of nanoparticles, as follows.

Firstly, Fe-rich fluids can mix with more oxidized fluids away from the deposit, creating redox gradients that are suitable for colonization by iron-oxidizing bacteria. Microorganisms (mostly bacteria) occupy specific zones where iron oxidation is thermodynamically favored but still kinetically inhibited by low oxygen (or nitrate) concentrations. The diversity of species in such niches is essentially unknown. Furthermore, the organisms themselves are little understood as they are extremely hard to cultivate due to the exacting nature of their requirements. However, they are not rare in the environment, and in fact may play critical roles in iron cycling in microaerophilic zones (groundwater, swamps, ocean floor).

The best known example of neutrophilic iron-oxidizing bacteria are members of the *Gallionella* and *Leptothrix* taxa (e.g., *Gallionella ferruginia* and *Lepthothrix ochracea*, e.g., Pederson 1997, 1999; Hallbeck and Pedersen 1991). Iron-oxidizing bacteria form part of the red-orange mineral-loaded biomass that often clogs water wells (for review, see Tuhela 1997) and develops along swamp margins (in some cases, producing oil-like films on standing water). In addition, a number of newFe-oxidizing taxa have been isolated recently (Emerson and Moyer 1997; Emerson 2000). *Gallionella ferrugina* has been shown to be an autotroph (fixes CO_2; Hallbeck and Pederson 1991), and this trait has been inferred for the new isolates (e.g., Emerson and Moyer 1997). Because it can sustain autotrophic growth, microbial iron oxidation could potentially underpin a substantial subsurface biosphere.

The products of the enzymatic oxidation of iron by microaerophilic bacteria are characteristically nanocrystalline (Fig. 4). The very small particle size reflects rapid nucleation kinetics due to the high degree of solution supersaturation that accompanies enzyme-mediated iron oxidation. Simultaneously, the low solubility of ferric iron inhibits the nuclei from coarsening via dissolution and reprecipitation pathways. In some cases, nanoparticles of ferrihydrite, goethite, hematite, and feroxyhite (see above) aggregate to form micron-scale colloids (Banfield et al. 2000). In other cases, these iron minerals adhere to cell-associated polymers. For example, *Gallionella* spp. excrete ribbons of polymer that strongly sorb nanoparticulate oxides, leading to highly characteristic mineralized filaments of potential significance as biosignatures (Figs. 5, 6). *Leptothrix* spp. cells are elongate, and form chains that are many tens of microns in length. The cells are coated by a polymer sheath that also sorbs nanoparticulate oxides. Following cell death, elongate tubes coated by, or consisting of aggregated nanoparticles (resembling tiny drinking straws) remain (Fig. 6).

Manganese is cycled by microorganisms in similar ways to iron. Mineralization of manganese oxides (e.g., birnessite) is often attributed to a variety of bacteria, including *Gallionella*, *Leptothrix*, *Sphaerotilus*, *Pseudomonas*, and *Pedomicrobium* spp. Enzymes involved in Mn oxidation have been studied in some detail (e.g., Brouwers et al. 2000).

Figure 4 (top, opposite page). High-resolution transmission electron microscope image of iron oxyhydroxides produced by iron-oxidizing bacteria at near-neutral pH. Based on the positions of rings in the selected area electron diffraction pattern (insert on bottom left), the material is 2-line ferrihydrite. The Fourier transform of the image is inset on the lower right. Note that the particles are typically 2 to 4 nm in diameter (see Banfield et al. 2000).

Example: Nanoparticles formed by microbes in anoxic regions of AMD systems

When metal-rich fluids derived as the result of fluid-rock interactions (e.g., sulfide mineral dissolution) become anoxic, a series of mineral precipitation events are likely. In many cases, onset of reducing conditions is due to the presence of organic carbon. However, despite large thermodynamic disequilibrium, organic carbon is relatively unreactive in the presence of most inorganic environmental oxidants. Some organisms catalyze oxidation of organic carbon coupled to reduction of ions such as manganese, ferric iron, uranium, or sulfate. Multiple niches develop in geochemically stratified systems. For a review of the well known sequence in which microorganisms utilize these, and other (e.g., oxygen, nitrate) oxidants, see Nealson and Stahl (1997). Iron and manganese-reducing microbes dissolve oxide nanoparticles in the environment (e.g., for review, see Lovley 1987), whereas organisms that couple respiration of organics (or hydrogen) to reduction of sulfate or uranyl ions lead to production of abundant precipitates.

Many Bacteria and some Archaea couple sulfate-reduction to oxidation of organic carbon or hydrogen. Consequently, active microbial communities can generate large volumes of sulfide at appreciable rates. For example, as a measure of activity of sulfate-reducing bacteria (SRB) in a sediment, Sahm et al. (1999) determined the proportion of bacterial rRNA derived from SRB and determined that it accounted for ~18% to 25% to the prokaryotic rRNA pool. They estimated that 1 cm^3 of wet sediment contained ~2.4 to 6.1×10^8 cells and calculated cellular sulfate reduction rates of ~0.01-0.09 fmol SO_4^{2-} $cell^{-1}$ day^{-1}. Sulfide minerals are extremely insoluble and sulfide precipitation kinetics are fast. Thus, in the presence of sub-ppm concentrations of metals, the majority of the sulfide will be captured in sulfide minerals. Thus, although a femtomole is a perishingly small quantity (10^{-15} mol), sulfide production rates reported by Sahm et al. (1999) correspond to production of fraction of a milligram quantities of nanocrystalline metal sulfides per cubic centimeter per year (significant, especially over geologic time scales). Because microbial sulfate reduction is closely associated with significant mineralization, we will examine the metabolism of SRB in more detail.

SRB are essentially ubiquitous in aqueous environments that contain organic carbon and sulfate (e.g., subsurface aquifers and lake sediments). Moreover, analysis of a key gene associated with sulfate reduction (dissimilatory sulfite reductase) indicates that microbial sulfate reduction is an ancient trait, suggesting that organisms may have contributed to sulfide mineral formation throughout much of Earth history (Wagner et al. 1998). SRB are tolerant to environmental extremes of heat (some are hyperthermophiles) and salinity (some are halophiles).

The SRB imports both the organic compounds and sulfate into the cytoplasm. The organic is oxidized to CO_2 through a modified citric acid cycle and reducing molecules such as NADH and $FADH_2$ are generated. These coenzymes transport 2 electrons and 1 H^+ and 2 electrons and 2 H^+, respectively, to the cytoplasmic membrane where deydro-genases separate electrons and H^+. Protons are transported outside the membrane to

Figure 5 (bottom, opposite page). Transmission electron microscope of stalks produced by *Gallionella* sp. Coated by few-nanometer-diameter iron oxyhydroxide particles (see Fig. 2). The iron oxyhydroxides also form semi-spherical colloids, seen adhering to the stalks (Banfield, unpublished).

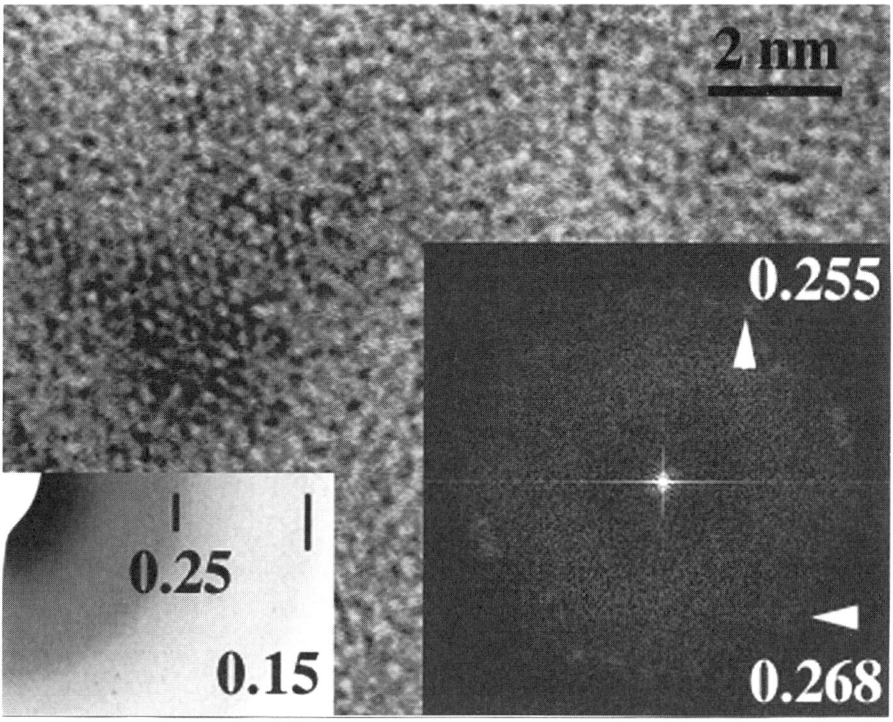
2 nm
0.255
0.25
0.15
0.268

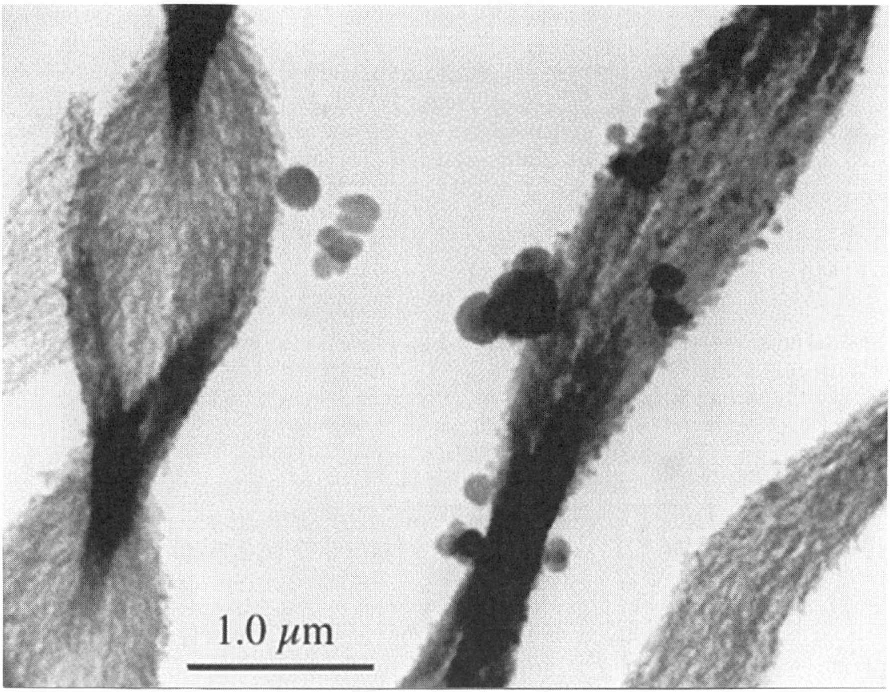
1.0 μm

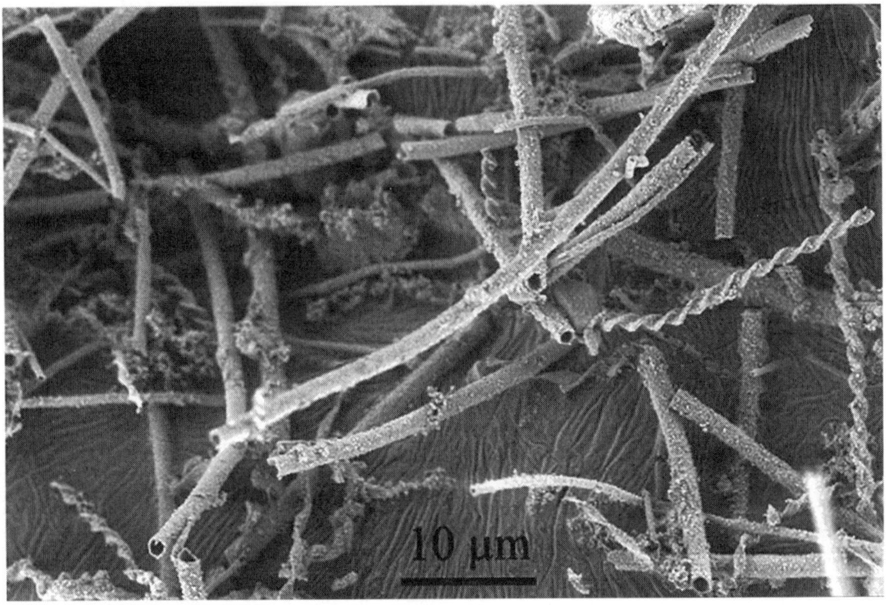

Figure 6. Low-magnification scanning electron microscope image of products of microbial iron oxidation at near-neutral pH. The elongate tubes produced by *Leptothrix* sp. and twisted stalks produced by *Gallionella* sp. (see Fig. 3) are coated in nano-scale iron oxyhydroxides (unpublished data reproduced with permission of Clara Chan).

generate a proton gradient across the membrane. Diffusion of the proton into the cytoplasm through an enzyme complex leads to the formation of ATP, the principal source of energy for cells. Electrons are passed to other membrane-bound enzymes with higher redox potentials to adenine phosphatosulfate (APS), a sulfur-bearing molecule that is generated from sulfate. Sulfite is reduced to hydrogen sulfide via a series of intermediates. Finally, HS^- is expelled through the cell wall and into solution where it may combine with metal ions to form metal-sulfides. Due to high local supersaturation, the metal sulfide minerals that form are typically nanocrystalline. Because SRB metabolism involves uptake of sulfate into the cell, isotopically light hydrogen sulfide, thus sulfide nanocrystals are produced. As for Fe-oxidizing microorganisms, precipitation does not occur in close proximity to the enzymes in the cell wall, but in bulk solution.

It is interesting to note that sulfide, the waste product of anaerobic respiration by SRB, is relatively toxic to microorganisms in high concentrations due to its ability to denature certain proteins and bind to metal centered enzymes (Brock and Madigan 1991; Postgate 1965; Trudinger et al. 1972). Although an incidental process that occurs exterior to the cell, precipitation of very insoluble metal sulfides serves to make the environment more hospitable for microbial communities in these habitats. The process also may contribute to the formation of ore bodies (Druschel et al. 2002).

In anaerobic (or microaerophilic) environments, sulfide accumulates until the solution becomes saturated with a mineral phase for which precipitation kinetics are favorable. Labrenz et al. (2000) and Druschel et al. (2002) present a geochemical model that predicts the sequence of precipitation based on the solution composition and some kinetic considerations. They show that formation of monomineralic deposits of nanoparticulate metal sulfide phases can occur if metals are resupplied by fluid flow at rates that outstrip rates of sulfide production. The phenomenon arises because precipitation reactions remove sulfide as fast as SRB produce it, preventing saturation with a second mineral. When the solution is depleted in the metal incorporated into the first formed sulfide, sulfide accumulates until precipitation of a second phase begins. This can generate spatially distinct zones dominated by specific minerals (e.g., CuS, followed by CdS, then ZnS, PbS, and finally FeS). In contrast, static systems or sulfide-dominated systems will be characterized by formation of many sulfide minerals simultaneously. The resulting mixed nanoparticulate semiconductor materials may have very complex and interesting chemistry (see section on galvanic interactions below).

SRB remove sulfate and metals from solutions. Consequently, there has been considerable interest in stimulating SRB in order to remediate environments contaminated by AMD (e.g., constructed wetlands; Berezowsky 1995; Thompson et al. 2000; Kolmert and Johnson 2001). However, the nanoparticulate nature of products means they are potentially fairly mobile and prone to rapid re-oxidation.

Recent high-resolution transmission electron microscopy of the products of SRB (Labrenz et al. 2000, and unpublished data) growing in groundwater associated with an abandoned Pb-Zn ore deposit revealed that the first formed sulfide mineral is ZnS (Fig. 7). The ZnS particles formed in proximity to SRB are mostly <5 nm, and often <3 nm in diameter. A subset are ~1.5 nm in diameter (Fig. 7a), thus close in size to molecular clusters. The nanoparticle shown in Figure 7b is estimated to contain between 90 and 100 ZnS. The extremely small size of the crystals formed by homogeneous nucleation in solution indicates a very high degree of supersaturation (Nielson 1964; Steefel and van Cappellan 1990). Most nanoparticles flocculate to form micron-scale aggregates (Fig. 8). Note that a sphere a few microns in diameter in Figure 8 contains about a billion nanoparticles!

Luther et al. (1999) reported that natural solutions can contain relatively stable tetrameric $Zn_4S_6(H_2O)_4^{4-}$ clusters ~10 to 16 Å in diameter that have structural similarity to sphalerite. Note that this diameter includes water molecules. The tetrameric cluster is much smaller than those imaged by TEM in Figure 7. However, we anticipate that smaller nanoparticles (i.e., a continuum of larger clusters of between 4 and 100 atoms) will be detected with further work. Other results suggest that a variety of other metal sulfide clusters are present in natural solutions (e.g., FeS, CuS; e.g., Rozan et al. 2000; Theberge et al. 2000; see below for further discussion).

Crystal growth is driven by surface free energy, which is effectively proportional to the density of under coordinated surface ions (e.g., Zhang and Banfield 1998, and below). The persistence of 1- to 5-nm diameter nanocrystals, despite the enormous driving force for crystal growth, suggests that subsequent crystal growth is inhibited by some mechanism. One possible explanation is that uncoordinated surface ions are capped by organic ligands. This strategy has been used to limit crystal growth in experimental metal sulfide syntheses (e.g., Torres-Martinez 1999).

Our work on ZnS formed in association with SRB demonstrates wurtzite is often present as a minor component of the assemblage (Fig. 9; Banfield et al., in prep.). As wurtzite is considered to be unstable below 1020°C (Daslakis and Helz 1997), its formation at 8°C is perhaps surprising. However, it is possible that sphalerite is not the stable phase at low temperatures and pressures due to surface energy contributions to the energetics of both sphalerite and wurtzite (see below). Alternatively, its formation may be associated with the presence of excess sulfide (see below), or with surface-modifying organic ligands (see below).

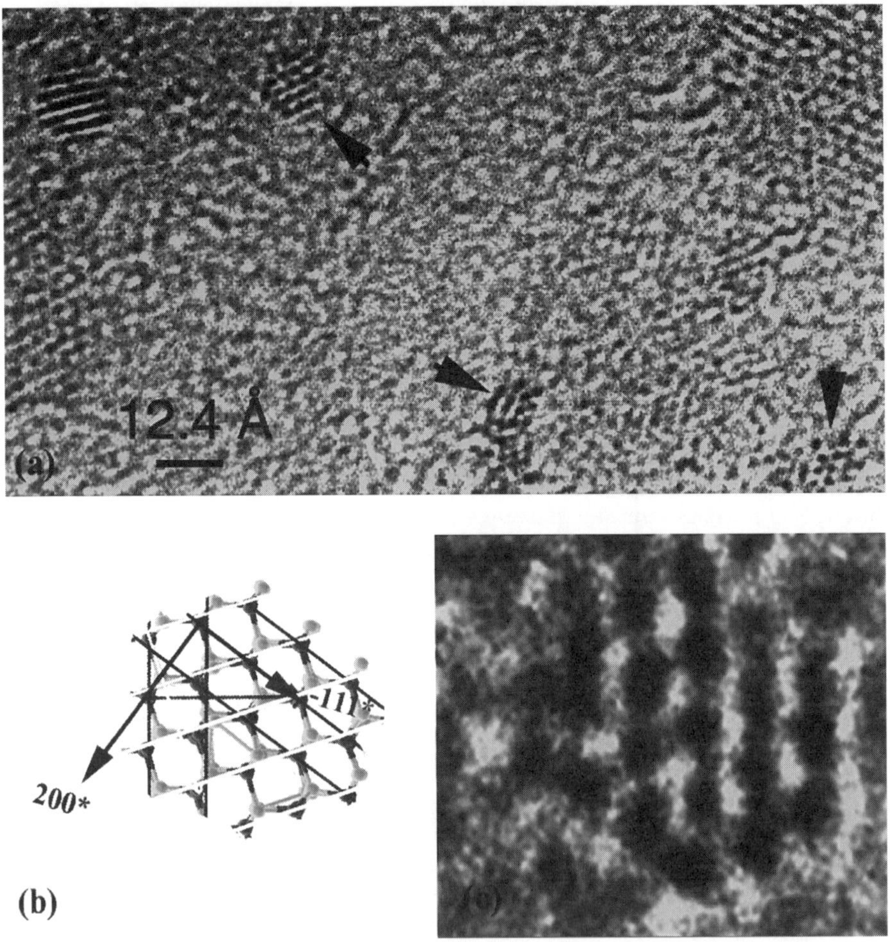

Figure 7. (a) High-resolution transmission electron microscope image of ZnS produced as the result of the activity of sulfate-reducing bacteria (see Labrenz et al. 2000). Note that the small ZnS particles indicated by arrows are sub-2 nm in diameter. (b) Diagram of a sphalerite nanoparticle similar in size to that shown in Figure 7a and 7c (7c is enlarged area of 7a) (Banfield et al., unpublished).

Coarser biogenic ZnS nanoparticles are commonly twinned on very fine scales. We suggest that these microstructures arise as the result of an aggregation-based crystal growth mechanism (see below), probably commencing with clusters analogous to those described by Luther et al. (1999).

Other examples of biological pathways that lead to nanoparticles in the environment

Many minerals have been shown to form as the result of direct or indirect biological processes (e.g., see Ehrlich 1999). Organismal production of a vast diversity of other phases, including phosphates (bones and teeth), actinide minerals (including uraninite and uranium phosphates), elemental sulfur, etc. are well known. In addition to processes described above, high concentrations of nanoparticles are generated in proximity to cells as the result of changes in solution saturation state due to uptake or release of protons or

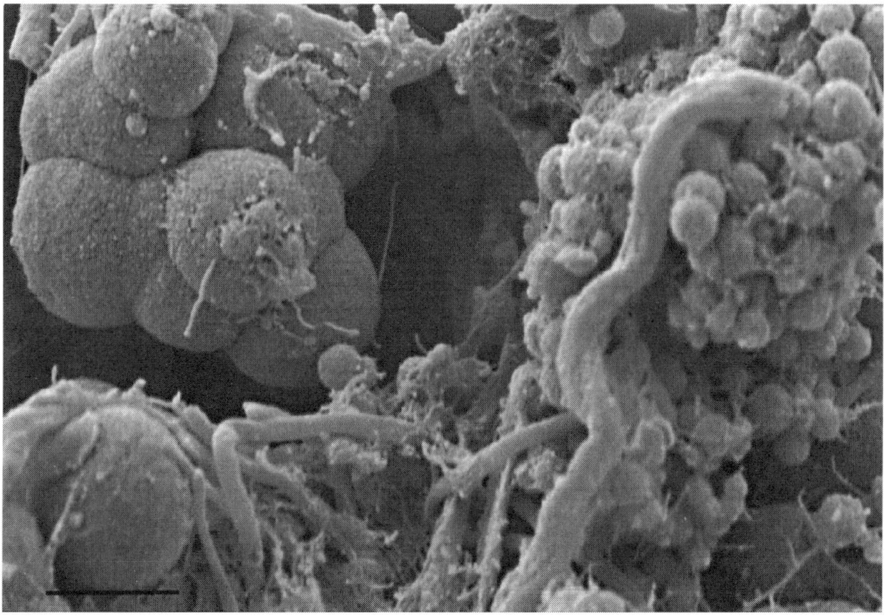

Figure 8. Scanning electron microscope image of a biofilm (elongate cells are evident) dominated by sulfate-reducing bacteria and submicron- to micron-scale spherical aggregates of ZnS nanocrystals. As shown in Figure 7, the ZnS nanocrystals that make up the aggregates are ~1 to 4 nm in diameter. Scale bar = 2 μm (Banfield, unpublished).

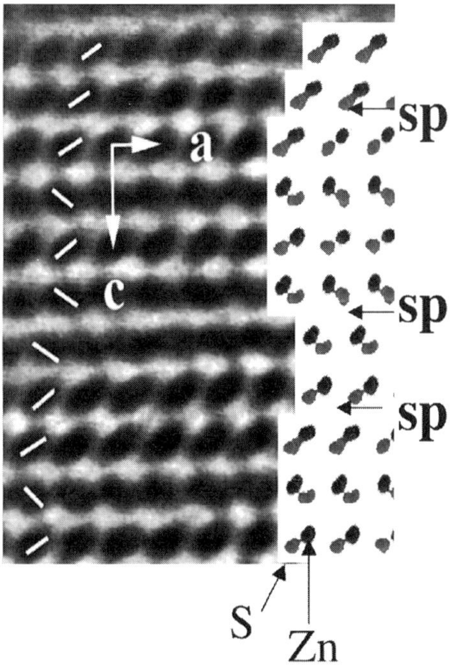

Figure 9. High-resolution transmission electron microscope image of most of the interior of an ~ 4 nm diameter ZnS particle produced as the result of activity of sulfate-reducing bacteria. The image details show that the particle consists of a mixture of wurtzite and sphalerite-like regions. Unit cell axes are shown for the wurtzite region (Banfield et al., unpublished).

organiccompounds from cells or as the result of passive heterogeneous nucleation on cell surface layers. For example, carbonate or sulfate minerals may template on periodic proteinaceous cell surface layers (S-layers) when pH changes (e.g., Schultzlam and Beveridge 1994; Sleytr and Beveridge 1999). Microbial release of byproducts can also lead directly to the precipitation of oxalate minerals. Non-metabolically active cells (e.g., spores) can induce production of nanocrystalline phases due to the presence of extracellular enzymes (e.g., manganese minerals; Bargar et al. 2000). Minerals as disparate as nanoparticle native gold and clays also probably form via biologically impacted routes. Many biomineralizing systems were summarized by Ehrlich (1996). In more detail, the formation of biological carbonates and silicates was reviewed by de Vrind and de Vrind de Jong (1997), oxides by Bazylinski and Moskowitz (1997), Fortin et al. (1997) and Tebo et al. (1997).

MICROBES, NANOPARTICES, AND MINERALOGICAL BIOSIGNATURES

The search for biosignatures is motivated by the search for ways in which geological samples record evidence for the preexistence of life in ancient terrestrial rocks and, if relevant, on Mars and other planets. In the absence of cells or their organic products, attention must turn to minerals. Organisms (especially microorganisms) could impact the form and distribution of minerals in many ways. However, perhaps the most promising category of effects is production of minerals as the result of their metabolic activity. Given that most biominerals generated by microorganisms are nanocrystalline, the topic of mineralogical biosignatures becomes intimately intertwined with nanoparticle formation and fate.

As indicated above, the products of the well-studied iron-oxidizing neutrophiles have high potential as mineralogical biosignatures, and in fact have been used as such (e.g., Alt 1988; Hofmann and Farmer 2000). Cambrian sea-floor silica-iron oxide deposits were described by Duhig et al. (1992), and a Jurassic hydrothermal vent community was described by Little et al. (1999).

Mineral products of anaerobic metabolism may also constitute important biosignatures, especially in low-temperature environments where sulfide mineral development via inorganic routes is virtually impossible.

In addition to the existence of nanocrystalline materials themselves (the particle size of which almost certainly confirms a low-temperature history), the microstructure of nanomaterials or their coarsened counterparts may provide information about their origins. As discussed in more detail below, nanocrystals often grow via aggregation-based pathways and these lead to introduction of point defects, dislocations, fine scale twinning, and other structural intergrowths. These features may serve to distinguish micron-scale metal sulfide crystals grown via inorganic pathways (for example) from crystals formed by coarsening within colloidal aggregates of nanoparticles.

NANOPARTICLES: HOW AND WHY ARE THEY DIFFERENT?

Introduction

Nanoparticles may be solids, liquids or gases. Examples include nanocrystals, nanoscale liquid droplets, and gas bubbles in nanopores in nanostructured solids. In the case of nanoscale liquids, properties such as viscosity deviate from bulk values as the thickness of the water film decreases below a few nanometers. However, recent results suggest that the viscosity change in low ionic strength aqueous solutions is much less than for non-associated liquids (such as organics) and remains within a factor of 3 of its bulk value in films in the range 0.0 ± 0.4 nm to 3.5 ± 1 nm (Raviv et al. 2001). Although we restrict our focus here primarily to the nanoscale solids themselves, unique characteristics of nanoscale volumes of fluids and gases may be of considerable geological significance.

Nanoparticle surfaces. Crystals are always terminated by surfaces. The factors that determine the surface orientation and structure are reviewed by Waychunas (this volume). The structure of a surface on a nanoparticle may be essentially indistinguishable from the equivalent surface on a macroscopic crystal if coordination environments of the atoms that comprise the surface are not disrupted when the surface is created (e.g., by cleaving graphite). However, in general, the surfaces of most commonly encountered minerals (oxides, silicates, etc.) will consist of ions whose environments are modified relative to those in the bulk.

As a particle size shrinks to between tens of nanometers and about a nanometer, a significant and increasing fraction of the atoms are exposed on surfaces (Fig. 10) rather than contained in the bulk (particle interior). Ions with non-optimal coordination geometries on surfaces give rise to excess energy. Consequently, the total energy of the system increases as size decreases because of the enhanced contribution from the energy associated with surface sites. The energy penalty caused by creation of a surface can be minimized by hydration, protonation, surface reconstruction, change in surface site coordination, displacement of atoms, and changes in bond lengths and angles (also see Waychunas, this volume).

In natural environments, particle terminations usually involve both the atoms that constitute the solid surface and those in the surrounding medium (e.g., water, gas molecules, protons, organic molecules, etc.). Because of the predominance of surface sites on nanoparticles, the structure and reactivity of nanoparticles will depend on the nature of the surrounding environment to a much greater degree than for macroscopic materials. The importance of the chemistry of the particle's interface with its surroundings should increase as size decreases.

Nanoparticle structure. The atomic characteristics of surface and near-surface sites, and their energetic consequences, underpin a whole host of characteristics and phenomena that distinguish nanoparticles from macroscopic materials. Consider, for example, the factors that determine which structure a nanocrystal adopts. If at low to moderate pressures, a compound may exist in two polymorphic forms and the total energies of the two structures are not dramatically different, it is possible that the phase "stabilities" will be reversed if the surface energies of the two structures are sufficiently different (Langmuir 1971; McHale et al. 1997; Gribb and Banfield 1997). This topic is elaborated upon below, and considered in detail by Navrotsky (this volume). Of course, the true thermodynamically stable phase will be a macroscopic phase, as nanoparticles (with high surface areas and surface energies) are metastable relative to the bulk material. Other size-dependent phenomena are reviewed below and are considered elsewhere in detail in this volume.

A nanoparticle may be clearly crystalline, i.e., it has a portion of its volume that consists of a periodic array of atoms. In this case, we refer to it as a nanocrystal. Alternatively, the particle may be completely amorphous, i.e., the arrangement of atoms throughout the particle lacks any periodic nature (analogous to that of a glass). The term "nanoparticle" is especially useful for such materials, for cases where the structure is intermediate between that of a nanocrystal and an amorphous solid, or where the degree of crystallinity of the solid is uncertain. A continuum of states may be encountered, ranging from larger particles that are clearly dominated by periodic bulk structure, through smaller particles in which atom periodicities are perturbed due to surface effects, to very small particles (perhaps in a vacuum) in which atomic positions are displaced so far that the structure is no longer clearly periodic. The ways in which bond angles and bond lengths change as molecular clusters grow into nanoparticles are poorly understood, as are the ways in which these changes vary with the environment of the particle (e.g., in solution and as a function of ionic strength, or in vacuum).

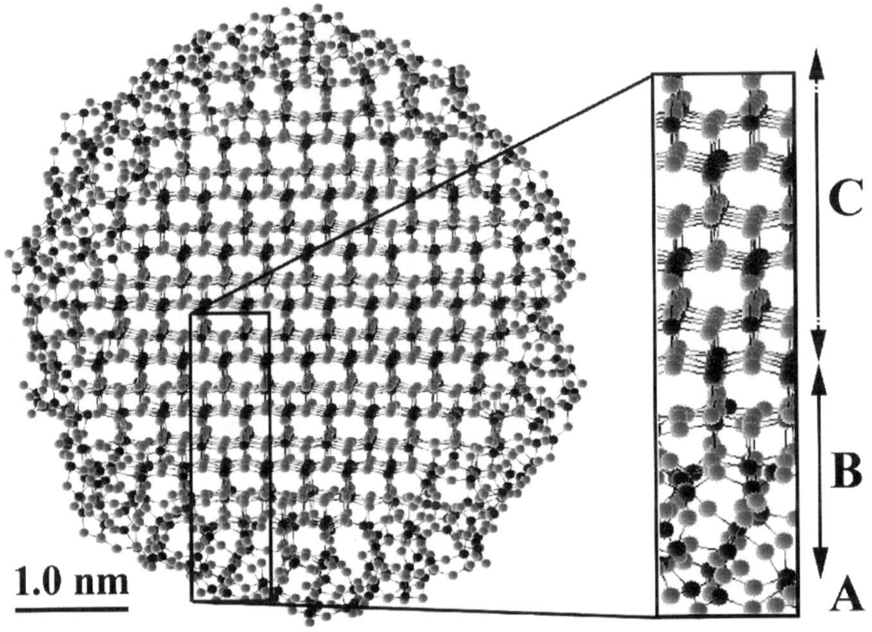

Figure 10. Cross section through the output of a molecular dynamics simulation of a TiO_2 (anatase) particle ~5 nm in diameter. Note that the surface layer (A) and the near-surface layer (B) are highly distorted, but the interior of the nanoparticle (C) retains typical anatase bulk structure. Region C disappears as particle diameter decreases toward ~2 nm (Zhang and Banfield, unpublished).

The impact of the surface atomic layer on adjacent layers within the solid probably decreases smoothly to zero as distance from the surface (into the particle) increases. However, in order to facilitate consideration of size-dependent characteristics and behavior of nanoparticles, it is convenient to subdivide them into three atomic/structural regions (Fig. 10). First, we define a "surface" atomic layer that consists of ions with modified coordination geometries. Second, we specify a "near-surface region," this being a zone of atomic layer(s) near, but not at the surface. These layers are identified by distortion due to proximity to the surface atomic layer (see below for details). Third, we specify "bulk" material with structure equivalent to that within macroscopic crystals. The total energy of the small particle can then be written

$$E(\text{small particle}) = E(\text{bulk + near-surface region}) + E(\text{surface})$$

For comparison, consider a particle that is small enough that its interior consists entirely of "near-surface region." The total energy of such a (tiny) nanoparticle can be written:

$$E(\text{nanoparticle}) = E(\text{near-surface region}) + E(\text{surface})$$

For most materials, $E(\text{surface}) > E(\text{near-surface region}) > E(\text{bulk})$. Thus, the total energy of a mole of nanoparticles is greater than that of a mole of small particles (of course, this is the reason why crystals grow). The energy contribution of the surface is defined as an "excess" quantity. The energy/mole of nanoparticles minus the energy/mole of bulk material is equal to the excess energy/mole due to the surface. Normalized to surface area, this quantity is "surface energy."

Surface free energy and surface stress

Consider an oxide crystal. In the bulk of the crystal, each metal is coordinated by six nearest neighbor oxygen anions. As a whole, the electric charges of anions and metal atoms are balanced in the bulk (in the sense that it is mathematically infinitely large). However, on a clean surface of the crystal, the electric charges are unbalanced due to the disrupted coordination sites. As noted by Waychunas (this volume), surfaces of oxides tend to be terminated by oxygen ions that are associated with unbalanced negative charge. The amount of the unbalanced charge can be estimated from detailed structural analyses (e.g., for TiO_2; Zhang and Banfield 1998). These unbalanced surface charges cause the electronic structure of the surface to deviate from that in the bulk, and as a result the surface carries excess energy, called surface energy, as compared to the bulk. This excess energy is equivalent to the reversible work done when a new surface is created. The term surface free energy (γ_o) is then defined as the reversible work done in creating a new surface of a unit area.

As new surface is created, it is impossible to keep the spatial configuration of the surface atoms exactly the same as in the bulk. Surface atoms will try to rearrange themselves so that the total energy of the crystal is minimized. For instance, dangling oxygen atoms on the surface may form bridged pairs, causing the inter-atomic spacing between their two connected metal atoms on the surface to shorten. The set of relative displacements of surface atoms from their positions prior to energy minimization (or the positions they would take as in the bulk) is called surface strain. The force exerted on any unit length on the surface by the surface atoms is called the surface stress (in liquid, this force is the surface tension). Microscopically, the surface stress (N/m) is the length-integrated difference between the stress tensor (N/m^2) along the normal direction of the surface under consideration, and that of the bulk stress tensor (N/m^2) in the interior. For the above example, the surface stress is tensile, since it causes shorter inter-atomic distances. On the other hand, if the formation of the bridged oxygen pairs on the surface is not favored energetically, the two oxygen atoms repel each other, and a compressive surface stress is generated. This increases inter-atomic distances on the surface. Significant rearrangement of surface atoms that generates a completely different atom layout on the surface as compared to the bulk is called surface reconstruction. It is important to note that although we refer to these phenomena in terms of stress and strain, reconstruction must always lower the overall free energy of the system.

For both a liquid and solid, creating a new surface can be done by dividing the material. In a liquid, suppose there is a boundary within which a certain area (A_o) of the liquid surface is enclosed. If we enlarge the liquid surface area to A by pulling the boundary, the reversible work supplied will be ($A\gamma_A - A_o\gamma_o$). The liquid molecules are mobile and can move from elsewhere when we expand the surface area to provide the necessary surface molecules. In this case, the density of liquid molecules remains unchanged, and thus the surface free energy is constant: $\gamma_A = \gamma_o = \gamma$. Thus, $A\gamma_A - A_o\gamma_o = \gamma(A - A_o)$. During new surface area formation, the external force applied on the boundary per unit length must equate to the force exerted by the liquid on the boundary per unit length. The work done by the external force (see Fig. 11) is $\int Fl dR = \int f dA = f(A - A_o)$. This work is equal to $\gamma(A - A_o)$ in quantity, thus $f = \gamma$ in value. For this reason, γ is also called surface tension for liquid substances. However, for solid

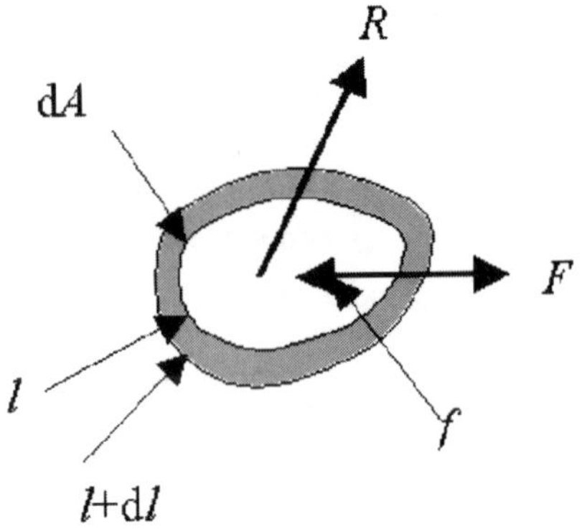

Figure 11. Diagram depicting the stretching of a surface in a polar coordinate system.

substances, due to the low mobility of atoms, the surface density of atoms will differ when we stretch the solid surface elastically. Thus, for a solid, the surface free energy γ after stretching will differ from γ_0 before stretching. As mentioned above, atoms on a solid surface also generate forces on the surface. These forces are generally treated as the surface stress tensors, which can produce surface strain with respect to the state when these forces did not exist. If the extent of surface stretching is measured by the surface strain caused by the surface stress, the energy change caused by the surface stress is $f\,dA$ which must equal the increase in the surface free energy $d(\gamma A)$. Thus, the surface stress $f = d(\gamma A)/dA = \gamma + A d\gamma/dA = \gamma + d\gamma/d\varepsilon$, where ε is the surface strain. From above, it is seen that the surface stress is the reversible work done in elastically stretching a unit area solid surface. The above equation was originally derived by Shuttleworth (1950), and can also be introduced in tensor form (Cammarata and Sieradzki 1994; Ibach 1997).

Obviously, the surface free energy γ must be positive (for clean surfaces), otherwise a solid would spontaneously fragment. However, for solids, the surface stress can be positive or negative, and generally it is not equal to the surface free energy since $d\gamma/d\varepsilon$ is not zero due to the low mobility of atoms in solids. Normally, the magnitude of f is in the same order as γ, being 1-3 times greater (Cammarata 1994; Ibach 1997). If f is positive, a tensile surface stress is present; if it is negative, the surface stress is compressive (Ibach 1997).

For nanocrystalline materials, the surface stress has significant implications for phase stability and surface properties. Surface stress can cause very high excess pressure inside a nanoparticle (see below). Thus, a tensile surface stress should shrink the unit cell in nanoparticles as compared to macroscopic crystals [e.g., in TiO_2 nanoparticles (Cheng et al. 1993)], while a compressive surface stress should produce an expanded unit cell as in $BaTiO_3$ (see Waychunas, this volume) and Fe_2O_3 (Schroeer and Nininger 1967). This excess high pressure also contributes significantly to the free energy that determines the thermodynamic phase stability (Zhang and Banfield 1998). When the difference between the surface stress and the surface free energy $|f-\gamma|$ is large enough, surface atoms undergo structural reconstruction (Ibach 1997; Needs et al. 1991). When there are adsorbates on a surface, the surface stress also changes. The variation of surface stress is a function of the surface coverage; in some cases, the surface stress can change from positive to negative (Ibach 1997). In consequence, it is expected that the surface stress phenomena will be highly related to the phase stability, morphology development, nucleation, crystal growth, etc. of nanocrystalline materials.

Surface energy and particle size

In a more detailed analysis, it must be asked whether surface energies themselves are size dependent. In fact, Tolman (1948 1949) established this to be the case by thermodynamic analysis of liquid droplets. More recently, this question has been examined for solids (Muller et al. 1988, Tomino et al. 1991, Zhang et al. 1999). It can be shown that surface energy can *increase* or *decrease* as particle size decreases. The reasoning is as follows.

The surface energy (per mole) is given by the difference between the total energy of the material and the energy of the interior of the particles. Using the regions defined above (Fig. 10), and a = E("surface")/area, A = surface area/mole, b = E("near-surface-region")/volume, V_b = volume of near-surface material/mole, c = E("bulk")/volume, V_c = volume of bulk/mole particles, the surface energy of a material consisting of small particles is given by:

$$SE_{small}/mole = [a_{small}\cdot A_{small} + b_{small}\cdot V_{b,small} + c\cdot V_{c,small}] - c\cdot(V_{a,small} + V_{b,small} + V_{c,small})$$

and of nanoparticles, for which the interior is the near-surface, is given by:

$$SE_{nano}/mole = [a_{nano}\cdot A_{nano} + b_{nano}\cdot V_{b,nano}] - b_{nano}\cdot(V_{a,nano} + V_{b,nano})$$

It is clear that a mole of structure that is largely disrupted by proximity to the surface will have a higher energy than a mole of bulk structure, so

$$b_{nano}\cdot(V_{a,nano} + V_{b,nano}) > c\cdot(V_{a,small} + V_{b,small} + V_{c,small}).$$

In order to evaluate whether surface energy decreases or increases with decreased particle size, it is necessary to know whether the combined effects of changes in a, b, A, V_b, V_c (bracketed terms above) outweigh the $b_{nano}\cdot(V_{a,nano} + V_{b,nano}) > c\cdot(V_{a,small} + V_{b,small} + V_{c,small})$ effect. If so, then the surface energy will decrease with particle size. However, if the total energy (bracketed terms) of the nanoparticles is greater, due to the possible $a_{nano} > a_{small}$ and/or $b_{nano} > b_{small}$ and larger A and V_b, the surface energy could increase with decreased particle size.

On further consideration of the above construct, it is clear that the value of the surface energy (essentially a macroscopic thermodynamic concept) is dependent upon how the surface, near-surface, and bulk regions are defined (i.e., how the particle is subdivided). For this reason, it is perhaps useful to set aside the surface excess terminology for a moment and focus on the key issues: surface energy is probably often size dependent, and surface sites contribute excess energy to a material.

Surface "pressure" and structural responses in nanoparticles

An important consequence of structural and electronic modification of the surface may be compression of the particle. The surface exerts a pressure (P) on the particle, given by (assuming a spherical particle with radius r)

$$\begin{aligned} P &= \frac{2f}{r} \\ &= \frac{2\gamma}{r} \quad (\text{if } f = \gamma) \\ &= \frac{4\gamma(\text{J/m}^2)}{D(\text{m})} \quad \text{Pa} \\ &= \frac{4\gamma(\text{J/m}^2)}{D(\text{nm})} \quad \text{GPa} \end{aligned}$$

If reasonable values of surface stress f (equated to surface free energy γ as a first approximation, see above) are inserted into this expression, it is apparent that the interiors of many common 2 nm particles could be subjected to forces equivalent to pressures of ~1-4.5 GPa (~10-45 kbar). In large particles, this pressure is dissipated within the first few unit cells, thus the surface energy modifies the total energy and average structure of the material in an insignificant way. However, for a nanoparticle, a significant percentage of the atoms are within a few unit cells of the surface, thus change in structure is expected throughout the particle. The measurable consequence of the surface force is a change in unit cell parameters determined by X-ray diffraction, or a decrease in interatomic distances, as measured by a method such as EXAFS.

How does the effect of surface pressure compare to the effect of external confining pressure? Data published in Cheng et al. (1993; Fig. 2) indicates that *a* is 0.007 Å smaller, and *c* is 0.08 Å smaller than values for a 6 nm compared to macroscopic anatase (TiO_2). Using the bulk modulus for anatase ($B = 2.1 \times 10^{11}$ Pa) we estimate that an external pressure of 3.36 GPa would be required to achieve these changes in cell parameters. For comparison, our estimate of internal pressure calculated assuming that surface stress = surface free energy or surface tension (a first approximation) suggests that the surface pressure caused by surface stress for 6-nm anatase is 0.9 GPa. The use of surface tension to approximate surface stress may largely explain this discrepancy (normally surface stress is equal to, or several times, surface tension). In addition, the bulk modulus of nanocrystalline anatase may be different from that of macroscopic anatase. Nonetheless, the two pressures calculated are of the same order of magnitude, further emphasizing the potentially very substantial magnitude of the particle size effect.

THERMODYNAMICS OF NANOPARTICLE SYSTEMS

Classical thermodynamics deals with macroscopic thermal-mechanical properties and their relationships for massive assemblages of atoms or molecules (i.e., ~10^{23} fundamental particles) in terms of energy conversion and transformation. Studies of phase/chemical changes and equilibria involving nanoparticles are important areas where the classical thermodynamic approach is effective. Because quantum mechanical effects may be marked (e.g., the energy of a nanoparticle may not be continuous) where there are only several hundred (or even only tens) of atoms in a nanoparticle, one may ask, "Is classical thermodynamics still valid for nanoparticle systems?"

A rigorous thermodynamic treatment of nanoparticle systems should at least contain quantum mechanical corrections. However, these treatments are impractical and difficult, considering the vast diversities of thermodynamic systems and the enormous numbers of fundamental particles involved in each. If thermodynamic quantities of a nanoparticle system are determined by conventional methods (such as calorimetry and equilibrium determinations), these quantities bear contributions from quantum mechanical effects and classical thermodynamics may still be applicable, so long as the number of atoms is not too small.

To be general, consider a nanoparticle system comprised by nanoparticles (solid) immersed in an environmental fluid (liquid or gas). There are N_S components in the nanoparticles. The nanoparticles may be present as phase α and/or phase β, depending on their phase stabilities at the equilibrium condition. In the fluid phase (solution), there are N_F species. At thermodynamic equilibrium, the electrochemical potential of the same species in different phases becomes equal, and the total free energy of the nanoparticle system reaches its minimum. Temperature, external pressure, composition, morphology, and size of each phase all affect the free energy of the system. When these variables change, the total free energy *G* also changes in a way determined by the second law of thermodynamics:

$$dG = -(\sum_{i=1}^{N_S+N_F} \bar{S}_i \cdot n_i)dT + (\sum_{i=1}^{N_S+N_F} \bar{V}_i \cdot n_i)d(P+P_{exc}) + \sum_{i=1}^{N_S+N_F} \bar{\mu}_i dn_i + \sum_{j=1}^{J} \gamma_j dA_j + \sum_{k=1}^{K} \sigma_k dl_k + \sum_{l=1}^{L} \lambda_l \quad (1)$$

where T, P, P_{exc}, n_i represents temperature, external pressure, excess pressure induced by surface stress in fine particles (for fluid, this term is zero), as well as the molar number of species i, respectively; S_i, V_i and μ_i are the partial molar entropy, partial molar volume and electrochemical potential (which equals to the chemical potential in the case of neutral species) of species i, respectively; γ_j is the interfacial free energy (J/m^2) of interface j with area A_j (m^2) formed by the fluid phase and one face of the nanoparticle; σ_k is the line energy (J/m) of edge k with length l_k (m) formed by two interfaces; and λ_l is the point energy (J) of vertex l formed by two edges or three faces. If there are grain boundaries inside nanoparticles, contributions to the total free energy from the grain boundaries, junctions of grain boundaries, and vertices of junctions can also be summarized in the last three terms of Equation (1), respectively. In equilibrium at a certain condition (T, P, and parameters determining the geometry of each phase), the partial differential of G with respect to n_i equals zero, producing $N_S + N_F$ coupled differential equations. These $N_S + N_F$ equations can be solved, giving the equilibrium concentrations of the $N_S + N_F$ species in the system.

At constant temperature and external pressure, there are many ways to approach the minimum free energy state (equilibrium state) for the system:

(1) Through phase transformation. Each phase present in the system must be the stable phase at the specified condition (T, P, geometric parameters, and solution compositions). Since geometrical dimensions strongly influence the phase stability for fine powders, at different mean sizes nanoparticles may present as different phases with different structures (discussed further below).

(2) Through coarsening and/or crystal growth of nanoparticles.

(3) Through redistribution of species in different phases. For instance, through dissolution–precipitation, adsorption-desorption, oxidation-reduction, and complexation and dissociation of coordinated species.

Minimization of the total free energy by phase transformation

In a system comprised of a fluid phase and a relatively small quantity of nanoparticulate material, the major contributions to the total free energy come from the two phases themselves. As it is the majority phase, geometrical dimensions have little effect on the thermodynamic properties of the fluid, so the stable phase of the fluid is determined by temperature and external pressure. However, for nanoparticle solids, geometrical factors such as particle size highly affect the thermodynamic properties so that phase stability depends on size in addition to temperature and pressure. Factors that influence surface or interfacial energy may interact in complex ways. In order to concentrate on the effects of size on the phase stability of nanoscale crystals, we idealize our system by assuming that the fluid phase and nanoparticles each contain only one component; there is no chemical interaction between the fluid phase and nanoparticles; the nanoparticles are isotropic and spherical; there are no (or negligible) grain boundaries or other defects within the nanoparticles. These assumptions cannot be equated to a real system. Nevertheless, the approach is useful to capture the basic physics of nanoparticle systems.

Given the above assumptions, the independent thermodynamic variables are temperature T, external pressure P and particle size D (diameter). When the three variables change, the change in the total free energy of the system is only due to the change in the free energy of the nanoparticles. The nanoparticles may undergo a phase transformation from the α-phase to β-phase when T, P and D change in order to reach a minimum energy state. The free energy change from α-phase to β-phase can be described by a modification of the equation by Zhang and Banfield (1998) by introducing an external pressure term [$\Delta V(P-P_o)$]:

$$\begin{aligned}\Delta G(nano-\alpha\rightarrow nano-\beta) &= \Delta G(\infty-\alpha\rightarrow\infty-\beta)\\ &\quad + (\textit{surface free energy \& surface pressure contribution})\\ &= \Delta G^o(\infty,T)+\Delta V\cdot(P-P_o)+2(2t+3)M\left(\frac{\gamma_\beta}{D_\beta\rho_\beta}-\frac{\gamma_\alpha}{D_\alpha\rho_\alpha}\right)\\ &= \Delta G^o(\infty,T)+\Delta V\cdot(P-P_o)+10\cdot M\left(\frac{\gamma_\beta}{D_\beta\rho_\beta}-\frac{\gamma_\alpha}{D_\alpha\rho_\alpha}\right)\end{aligned} \tag{2}$$

where $\Delta G^o(\infty,T)$ is the change of the standard free energy from the macroscopic α-phase to the macroscopic β-phase (i.e., at standard pressure $P_o = 10^5$ Pa and a certain temperature T), ΔV is the change in the molar volume of the two phases (ΔV is considered to be approximately invariant here), M is the molar weight of the nano-particulate material, γ represents the surface or interfacial free energy, and ρ the density of the nanoparticulate material. The last term of Equation (2) represents the contribution of surface free energy and the excess pressure (caused by the surface stress) to the free energy change, where the value of the surface stress is assumed t times that of the surface free energy (in Equation (2), $t = 1$ was assumed; Zhang and Banfield 1998). At a certain condition, if $\Delta G > 0$, then the nano-α-phase is more stable than the nano-β-phase; if $\Delta G < 0$, then the nano-α-phase is less stable than the nano-β-phase; if $\Delta G = 0$, then the nano-α-phase is in equilibrium with the nano-β-phase. Alternatively, the phase stability of a nanoparticulate material can be analyzed with the help of G vs. T, P, and/or D diagrams. Similar to Equation (2), the free energy of a nanophase can be described by

$$\begin{aligned}G(T,P,D) &= G^o(\infty,T)+V\cdot(P-P_o)+\frac{10M\gamma}{D\rho}\\ &= H^o(\infty,T)-TS^o(\infty,T)+V\cdot(P-P_o)+\frac{10M\gamma}{D\rho}\end{aligned} \tag{3}$$

where H^o and S^o represent the standard enthalpy and entropy of the corresponding macroscopic phase. According to this equation, increasing temperature will decrease the free energy; however, either increasing the external pressure or decreasing the particle size will increase the free energy. In the following, we consider four cases of nanoparticulate material stability.

Case 1: At the macroscopic scale, the α-phase is more stable than the β-phase at ambient temperature and pressure (298 K and 1 bar) and the density of the β-phase is higher than that of the α-phase (or the molar volume $V_\beta < V_\alpha$). At high pressure, the β-phase should be more stable than α-phase. Figure 12 demonstrates the variation of the free energies of the two phases with T or P at macroscopic size and nanometer scale. Empirically, a phase with higher density has higher surface energy due to the larger number of atoms with incomplete or distorted coordination sites per unit surface area (Gribb and Banfield 1997, Zhang and Banfield 1998, Blakely 1973). Thus, according to Equation (3), the increase in the free energy at a small size for the nano-β-phase is greater than that for the nano-α-phase at the same particle size D and the same T,P condition (Fig. 12). As a result, the stability region of the α-phase becomes wider for nanometer-scale particles.

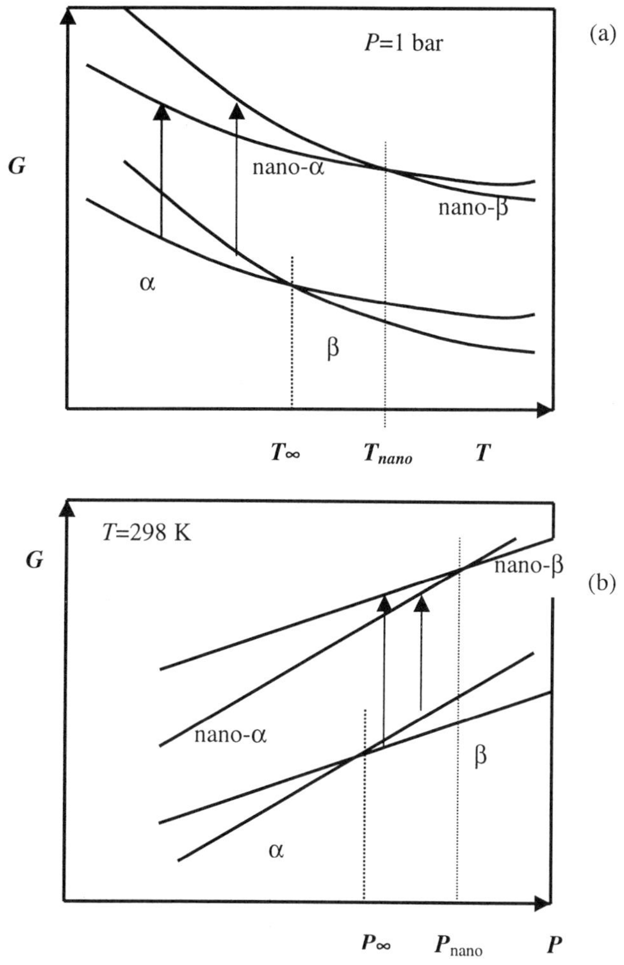

Figure 12. Free energies versus temperature (a) or pressure (b) of phases α and β at the macroscopic scale and nanometer-scale. For this case, the density (and surface free energy) of the β-phase is greater than that of the α-phase, which results in $\gamma_\beta/\rho_\beta > \gamma_\alpha/\rho_\alpha$. At macroscopic scale and under ambient conditions, the α-phase is the stable phase. At nanometer-scale, the stability region of the α-phase becomes even wider (the transition $T_{\text{nano}} > T_\infty$, $P_{\text{nano}} > P_\infty$).

Case 2: At the macroscopic scale, the α-phase is stable compared to the β-phase at ambient temperature and pressure, but the density of the β-phase is lower than that of the α-phase. The variation of the free energies of two such phases with *T* or *P* at macroscopic size and nanometer scale is depicted in Figure 13. In this case, as the increase of the free energy caused by the small particle size is greater for the α-phase than for the β-phase, the stability region of the β-phase becomes wider for nanometer scale particles.

Case 3: At the macroscopic scale, the α-phase is unstable with respect to the β-phase at ambient temperature and pressure (298 K and 1 bar) and the density of the β-phase is greater than that of the α-phase. The variation of the free energies of the two phases with

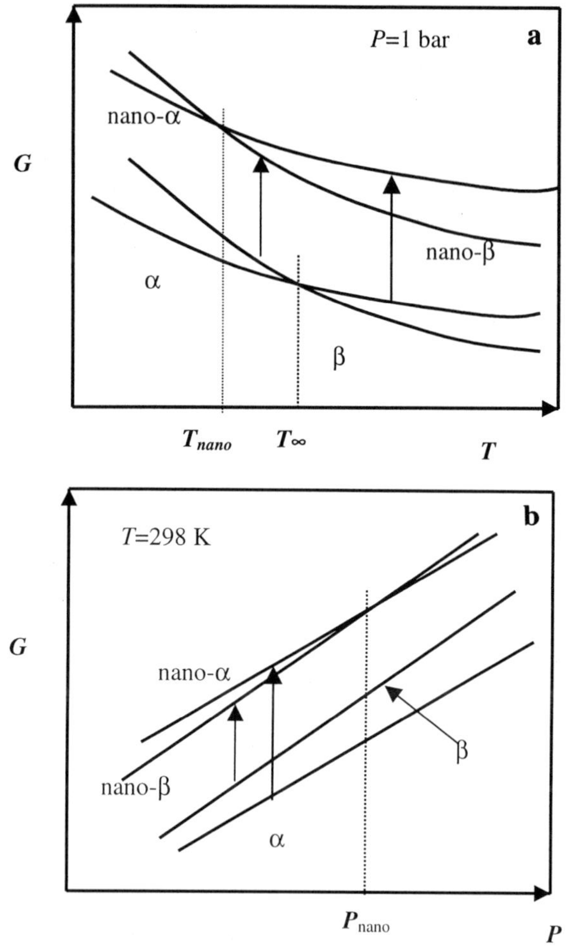

Figure 13. Free energies versus temperature (a) or pressure (b) of phases α and β at the microscopic and nanometer-scales. The density (and surface free energy) of the α-phase is greater than that of the β-phase, which results in $\gamma_\alpha/\rho_\alpha > \gamma_\beta/\rho_\beta$. At macroscopic scale and under ambient conditions, the α-phase is the stable phase. At nanometer-scale, the stability region of the β-phase becomes wider (the transition $T_{nano} < T_\infty$, nano-β-phase which is unstable at ambient temperature now becomes stable at $P < P_{nano}$).

T or P at macroscopic size or nanometer scale is shown in Figure 14. In this case, the α-phase can become stable at $T < T_{nano}$ and $P < P_{nano}$.

Case 4: If the two phases have similar surface free energies and the same particle size, or different surface energies but the last term of Equation (2) is close to zero due to the balanced contributions from the surface free energy γ, the particle diameter D and the density ρ, the modification of the phase stability due to the particle size effect may be hard to detect.

The above considerations are useful when evaluating the stability of two polymorphs

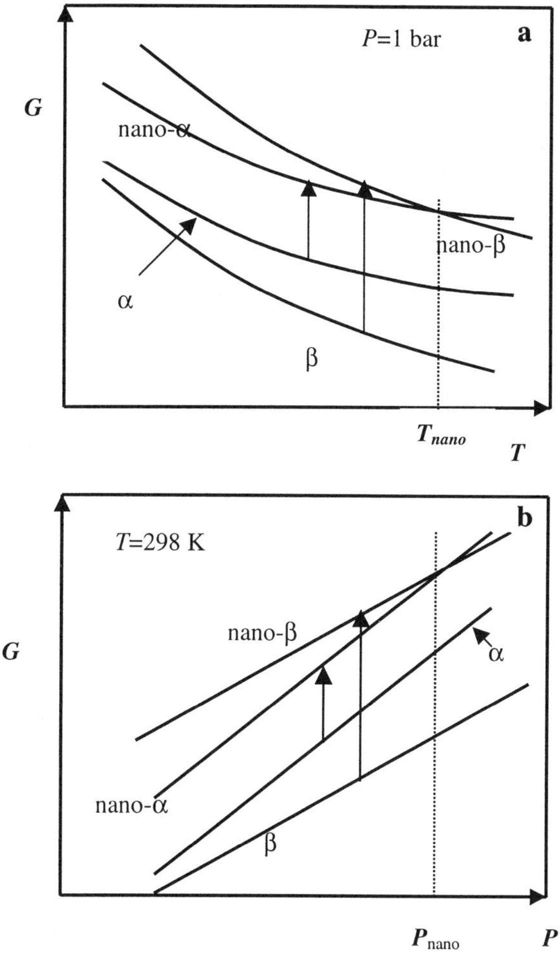

Figure 14. Free energies versus temperature (a) or pressure (b) of phases α and β at the microscopic and nanometer-scales. The density (and surface free energy) of the β-phase is greater than that of the α-phase, which results in $\gamma_\beta/\rho_\beta > \gamma_\alpha/\rho_\alpha$. At the macroscopic scale and under ambient conditions, the α phase is unstable. At the nanometer-scale, the α phase becomes stable at $T < T_{\text{nano}}$ and $P < P_{\text{nano}}$.

that occur as particles of the same size, or when considering the case when an isolated nanoparticle transforms completely to another polymorph (no crystal growth). Complex phase stability relations and transformation sequences may be encountered when a material consists of a mixture of two phases of different particle sizes. For such samples, phase α and phase β may be equally stable at some specific combination of phase α and phase β particle sizes. When the mixture is heated or subjected to pressure, growth of the nanoparticles can lead to stability reversals and re-reversals if the initial size and/or growth rates of the polymorphs differ. Such behavior has been encountered in titania (TiO_2; Zhang and Banfield 2000).

The *T-P* relations at the nanometer scale can also be explored in terms of Equation

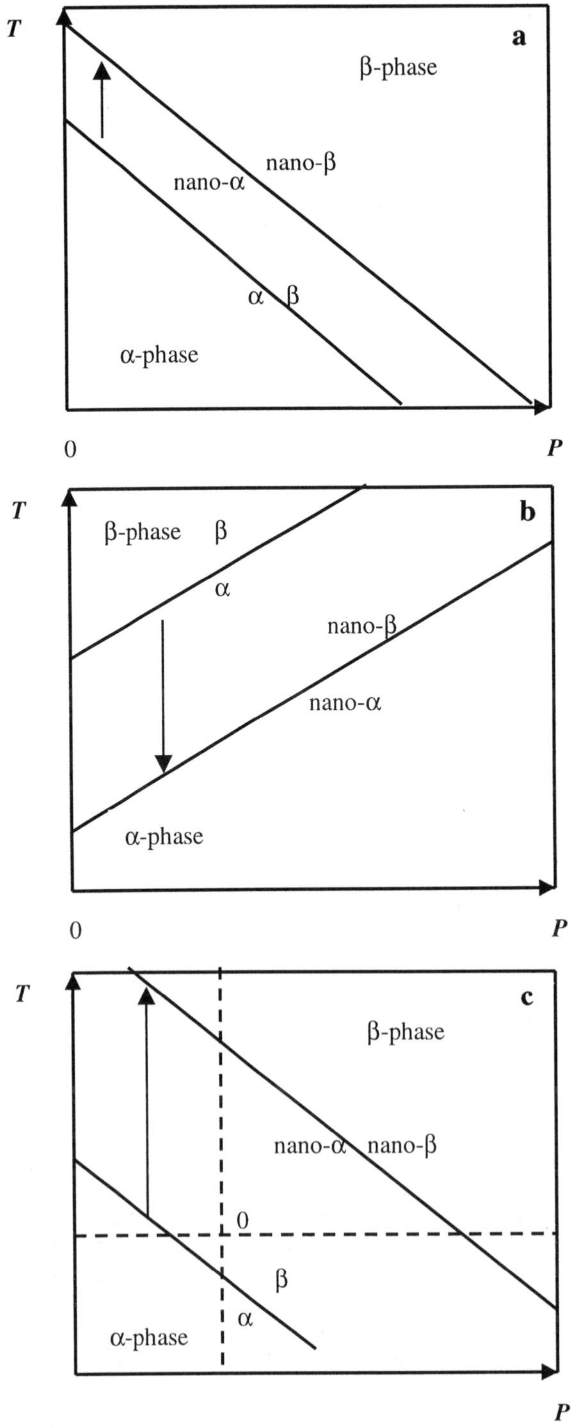

Figure 15. Equilibrium *T-P* relationships for macroscopic phases *vs.* nanophases.

(a) Case 1 (ref. Fig. 12): $A > 0$, $B > 0$, $\gamma_\beta/\rho_\beta > \gamma_\alpha/\rho_\alpha$, $\Delta V < 0$, *term 3 of Equation (5)* > 0. The stability region of nano-α-phase is larger than that of its macroscopic counterpart.

(b) Case 2 (ref. Fig. 13): $A > 0$, $B > 0$, $\gamma_\beta/\rho_\beta < \gamma_\alpha/\rho_\alpha$, $\Delta V > 0$, *term 3 of Equation (5)* < 0. The stability region of nano-α-phase is smaller than that of its macroscopic counterpart.

(c) Case 3 (ref. Fig. 14): $A < 0$, $B > 0$, $\gamma_\beta/\rho_\beta > \gamma_\alpha/\rho_\alpha$, $\Delta V < 0$, *term 3 of Equation (5)* > 0. The metastable α-phase becomes stable at the nanometer scale, i.e. the stability region of the α-phase has been increased.

(2). Usually the variation of $\Delta G^o(\infty,T)$ with temperature can be approximated with a linear relation:

$$\Delta G^o(\infty,T) = A - BT \tag{4}$$

where A and B are regression coefficients. These coefficients may differ from $\Delta H^o(\infty,T)$ [or $\Delta H^o(\infty, 298\ \text{K})$] and $\Delta S^o(\infty,T)$ [or $\Delta S^o(\infty,298\ \text{K})$], respectively, due to their regression nature. Approximate $\Delta G^o(\infty,T)$ in Equation (2) with Equation (4), and setting Equation (2) = 0, we get the *T-P* relationship at a certain particle size D when the α- and β-phases are in equilibrium:

$$T = \frac{A}{B} + \frac{\Delta V}{B}(P - P_o) + \frac{10M}{B \cdot D}\left(\frac{\gamma_\beta}{\rho_\beta} - \frac{\gamma_\alpha}{\rho_\alpha}\right) \tag{5}$$

Here we assume both phases have the same particle size. Figure 15 shows the *T-P* relationship for the above Cases 1-3.

Examples of phase stability in nanoparticle systems

The nano-TiO_2 system. TiO_2 adopts at least 8 structures (Banfield et al. 1993, El Goresy et al. 2001): rutile, anatase, brookite, α-PbO_2 [TiO_2(II)], hollandite [TiO_2(H)], β-VO_2 [TiO_2(B)] and the fluorite and baddeleyite (ZrO_2) structures at extremely high pressures. Anatase, brookite, rutile, TiO_2(B), TiO_2(II), and baddeleyite structures all occur in nature. However, we do not focus on titania for this reason. Rather, we choose to use TiO_2 as a model system because it is an oxide (and oxides are stable phases near the Earth's surface where nanoparticles predominate), it is anhydrous and stable with respect to oxidation under experimental conditions, the low-pressure structures consist only of octahedrally coordinated cations, thus are relatively simple (and rutile is an important structure type), and the phase stabilities of the important polymorphs are close enough at low temperatures and pressures that size can play an important role. Furthermore, the rate of transformation of nanocrystalline anatase to rutile (TiO_2) is clearly sixe-dependent. For example, although 80% of a nanocrystalline anatase sample transformed to rutile after 24 h at 525°C, coarse anatase was completely unreacted after annealing at 700°C for 24 h (Gribb and Banfield 1997).

Zhang and Banfield (1998) have made a detailed thermodynamic analysis of the nanocrystalline anatase and rutile system. Results suggested that at standard pressure, anatase is more stable than rutile when their particle sizes are below ~14.5 nm (curve 1 in Fig. 16). In their calculation, the surface free energies for rutile and anatase were obtained through published data of surface energy calculated by molecule dynamics simulations and experimental data of heat capacity of ultrafine rutile samples:

$\gamma_R(\text{J/m}^2) = 1.98 - 1.48 \times 10^{-4}(T/\text{K-}298)$, and

$\gamma_A(\text{J/m}^2) = 1.32 - 1.48 \times 10^{-4}(T/\text{K-}298)$.

The density data are $\rho_R = 4.249$ g/cm^3 and $\rho_A = 3.893$ g/cm^3. Both the density data and their structure analysis of the unsatisfied surface charges associated with several exposed faces of anatase and rutile support the fact that rutile has higher surface energy than anatase. Unfortunately, direct measurements of surface energies were not available. Based on the available data, the nanocrystalline anatase-rutile system is classified as Case 3 discussed above (α-phase = anatase and β-phase = rutile in Figs. 14 and 15).

If external pressure is applied to this system, the phase boundary should move according to calculations with Equation (2). Calculated results by the present authors (Fig. 16) show that with increased external pressure, anatase becomes less stable (i.e., the

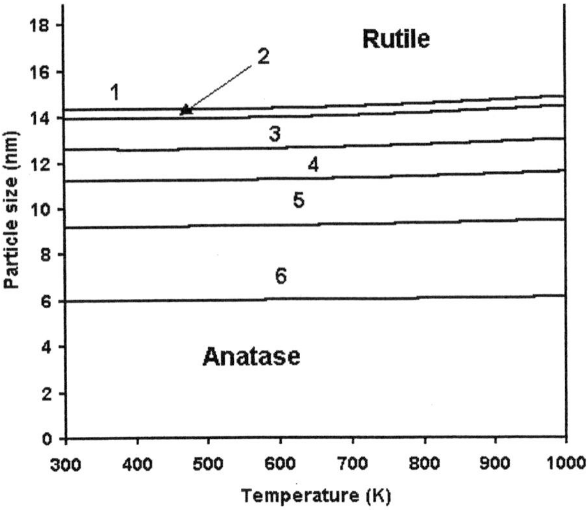

Figure 16. Calculated phase diagram of the nano-TiO_2 system at different external pressures: (1) $P = 10^5$ Pa, (2) $P = 10^8$ Pa, (3) $P = 5 \times 10^8$ Pa, (4) $P = 10^9$ Pa, (5) $P = 2 \times 10^9$ Pa, and (6) $P = 5 \times 10^9$ Pa.

stability region of anatase becomes smaller). At 2 GPa and 5 GPa, respectively, anatase is predicted to be more stable only when the particles sizes are below ~9 nm and ~6 nm, respectively.

Several experimental studies of the phase transformation of nanocrystalline TiO_2 under high pressure have been reported. Based on a Raman spectroscopic study, Wang and Saxena (2001) found that at a pressure higher than ~24 GPa, initially 7- to 11-nm anatase transforms to an amorphous TiO_2 phase. Normally, an amorphous phase is less stable than its crystalline phase and has a lower density and surface energy (see Navrotsky, Table 1, this volume). At nanometer-scale and high pressure, the less stable amorphous phase may be stabilized due to the surface energy contributions. In contrast to this study, Liao et al. (1999) found that at an increased pressure, initially ~38-nm anatase transforms to rutile rather than amorphous titania. At a pressure >4 GPa, the srilankaite ($ZrTi_2O_6$)-like phase may appear. The shape of the reported anatase/rutile boundary in their non-equilibrium *T-P* phase diagram is in agreement with Figure 15c (α-phase = anatase, β-phase = rutile).

Olsen et al. (1999) studied the phase transformation from nanocrystalline rutile to TiO_2(II) (α-PbO_2 structure) at high pressures. The transition pressure from nano-crystalline rutile to TiO_2(II) is lower than that for the corresponding macroscopic phase, which is transformed at 6.0 GPa at ambient temperature (Jamieson and Olinger 1968) and 4.5 GPa at 400°C (Simons and Dachille 1967). Furthermore, the *T-P* relation for nanoparticles is irregular: the slope of the curve (dT/dP) can change from positive to negative values (see Fig. 17b). Olsen et al. (1999) suspected that this could be similar to the unusual curvature in the body-centered cubic (bcc) to face-centered cubic (fcc) transition in the *T-P* phase diagram of iron (Bassett and Weathers 1990). In the iron system, the change of the slope dT/dP of the bcc/fcc boundary from negative to positive as temperature increases is caused by the change in the entropy of the bcc Fe phase from ferromagnetic to paramagnetic (Bassett and Weathers 1990), so $dT/dP = \Delta V/\Delta S$ also changes from negative to positive, resulting the curvature of the boundary.

TiO_2(II) synthesized at higher pressure starts to transform to rutile at 450-600°C over laboratory time scales (Aarik et al. 1996). With respect to TiO_2(II), at standard pressure (1 bar), rutile is considered the stable phase at all temperatures (Jamieson and Olinger

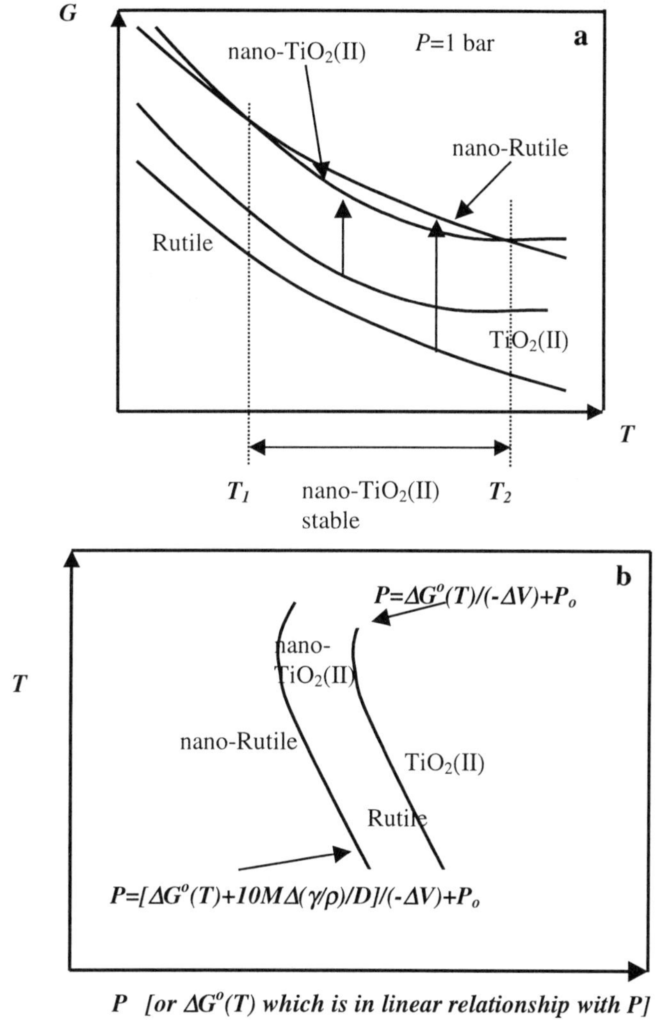

Figure 17. (a) Sketch of the variation of the standard free energies of rutile and $TiO_2(II)$ *vs.* temperature. At nanometer scale, $TiO_2(II)$ becomes stable at $T_1 < T < T_2$. (b) *T-P* (ΔG^o) relationship in the rutile-$TiO_2(II)$ system. The transition pressure from rutile to $TiO_2(II)$ lowers when the particle sizes are at the nanometer scale.

1969). Because $TiO_2(II)$ is only thermodynamically stable at high pressures, it has a higher density (4.33-4.37 g/cm^3; Simons and Dachille 1967), and presumably higher surface energy.

We made the following thermodynamic analysis to interpret the *T-P* relations observed by Olsen et al. (1999). In general, the free energy of a substance decreases with increasing temperature. Consequently, the standard free energy curve of $TiO_2(II)$ should approach that of rutile when temperature increases (Fig. 17a). However, if rutile is the stable phase at all temperatures at standard pressure, the rate of decrease of the standard free energy of $TiO_2(II)$ with temperature should gradually slow down (otherwise the free energy curve for TiO_2(II) would become lower than the free energy curve for rutile at temperatures higher than a certain value). This would make the free energy difference between the two phases even larger at even higher temperatures. This uncommon phenomena may also be related to a change in the sign of the entropy difference $\Delta S_{(rutile \rightarrow II)}$, as in the iron system mentioned above. This gives rise to a minimum in the curve for the change in standard free energy for rutile → TiO_2(II) vs. temperature (Fig. 17b). Accordingly, the change of the transition pressure vs. temperature for macroscopic phases has a similar shape (Fig. 17b) as $\Delta G^o(T)$ vs. T, since the pressure $P = \Delta G^o(T)/(-\Delta V) + P_o$ according to Equation (2) by setting $\Delta G = 0$, where $\Delta V < 0$ (because TiO_2(II) is denser). Although empirically, a higher density phase is likely to have a higher surface energy, it is possible that the increase in the density of TiO_2(II) surpasses the increases in its surface energy or surface free energy so that $\gamma_{II}/\rho_{II} - \gamma_R/\rho_R < 0$, or the third term in Equation (2) is <0 for nanocrystalline phases. In this case, the transition pressure from nanocrystalline rutile to TiO_2(II) becomes lower than that in the macroscopic phase (Fig. 17b). This could account for the irregular *T-P* shape and lowering transition pressure in nanometer scale particles (Fig. 17b) for this system.

Experiments have demonstrated that nanostructured TiO_2(II) thin films (13-70 nm) can be produced by CVD using $TiCl_4 + H_2O$ as reagents under low pressure (250 Pa) conditions (Aarik et al. 1996). At temperatures between 375 and 550°C, highly reproducible TiO_2(II) films can be grown. At 350°C, the product contains a mixture of anatase + rutile/TiO_2(II). At 600°C, the grown film is a mixture of rutile + TiO_2(II). Heating a TiO_2(II) film grown between 400 and 450°C up to 600°C does not induce transformation to rutile. These phenomena suggest that the TiO_2(II) nanophase may be thermodynamically stable at lower pressure only over a certain temperature range (~375-550°C). As assumed in our above thermodynamic analysis, $\gamma_{II}/\rho_{II} < \gamma_R/\rho_R$, thus the curve for the standard free energy of rutile moves up more than that of TiO_2(II) for nanometer scale particles. Thus, according to Equation (3) (Fig. 17a), we expect that the two curves can meet to produce a stable region for TiO_2(II) when $T_1 < T < T_2$ at certain particle diameters. This analysis explains why highly reproducible TiO_2(II) films can only be produced in a certain temperature region.

Hydrothermal conditions are frequently used to synthesize (e.g., Yang et al. 2000; Wang and Ying 1999; Yanagisawa and Ovenstone 1999) and treat (Penn and Banfield 1998, 1999a,b) nanocrystalline titania samples. The surrounding phase now is water or an aqueous solution. Experiments are normally conducted between 100-300°C, and at relatively low pressures (at 300°C the saturated vapor pressure of water is only ~8.5 MPa (*CRC* book), far less than pressures applied in many solid state phase transformation experiments).

There are several factors that can modify phase transformation and crystal growth in nanocrystalline materials under low-temperature and pressure hydrothermal conditions. A solid immersed in water has lower interfacial free energy than surface free energy. For instance, in air we estimate that $\gamma_{(anatase)}$ is ~1.32 J/m^2 (Zhang and Banfield 1998), while in water, $\gamma_{(anatase\text{-}H_2O)}$ is ~1.0 J/m^2 (Zhang and Banfield 1999), being about only 75% of that in air. The surface free energy of rutile in air $\gamma_{(rutile)}$ is estimated to be ~1.91 J/m^2 (Zhang and Banfield 1998), $\gamma_{(rutile\text{-}H_2O)}$ may be ~1.4 J/m^2 if it also decreases by 75%. If we consider only the pressure effect under hydrothermal conditions, anatase might be stable at particle sizes up to ~14 nm (see Fig. 16), considering that the saturated vapor pressure of water is far below several GPa at $T < 300$°C. However, since the interfacial free energies of anatase and rutile also decrease in water, the phase diagram of the nano-anatase and rutile system can be recalculated by inserting the saturated vapor pressure data of water (*CRC* book) into Equation (2) and letting $\Delta G = 0$. Our calculated phase

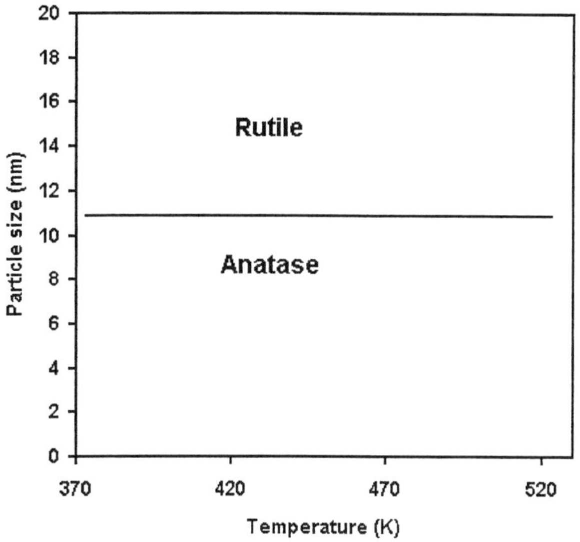

Figure 18. Calculated phase diagram for the nano-TiO_2 system under hydrothermal conditions. The particle size for anatase stability has decreased from ~14 nm at standard pressure and in air to ~11 nm, which signifies that under hydrothermal conditions nanocrystalline anatase is more easily converted to rutile.

diagram (Fig. 18) shows that, under hydrothermal conditions, the stability region of nanocrystalline anatase has been decreased as compared to that in air (Fig. 16), and thus the conversion from anatase to rutile should be thermodynamically favored. Under hydrothermal conditions, the increased pressure also increases nucleation rate (Liao et al. 1998). Thus, hydrothermal conditions should facilitate the phase transformation from nanocrystalline anatase to rutile in the sense of both thermodynamics and kinetics. Experimental data by Penn and Banfield (1999a) support this conclusion: it took ~50 h to convert % 50 anatase to rutile at 465°C in air, yet only ~8 h to achieve the same extent of reaction at 250°C under hydrothermal conditions.

The nanoparticle-fluid interfacial energy values may be substantially different in fluids other than deionized water. Furthermore, interfacial energies may decrease more for one polymorph than another in some solutions. This could significantly perturb size-temperature-pressure phase relations. The solution will also affect reaction kinetics because, under hydrothermal conditions, mass transfer can occur via the liquid, an electric double layer will develop around each nanoparticle, substances can be adsorbed onto surfaces, and nanoparticles may dissolve. The effects of particle size on reaction kinetics are considered below.

Some of these ideas and conclusions are speculative in that they could be refined with more accurate surface energy data. However, the above analyses serves to illustrate phenomena that may be encountered in many systems.

The nano-Al_2O_3 system. There are at least seven polymorphs of Al_2O_3: α-Al_2O_3(corundum), γ-Al_2O_3, δ-Al_2O_3, η-Al_2O_3, θ-Al_2O_3, κ-Al_2O_3, χ-Al_2O_3 (JCPDS cards 4-875/877/878/880, 42-1468; Wen and Yen 2000; FACT database on the web). Boehmite (AlOOH) is often used to prepare corundum via hydrothermal processing. A series of polymorphic phase transformation occurs in the process (Wen and Yen 2000):

$$\text{boehmite} \rightarrow \gamma\text{-}Al_2O_3 \rightarrow \delta\text{-}Al_2O_3 \rightarrow \theta\text{-}Al_2O_3 \rightarrow \alpha\text{-}Al_2O_3$$

Under standard pressure, corundum is the thermodynamically stable phase. However, synthesis of nanocrystalline alumina usually results in γ-Al_2O_3 (Navrotsky, this volume).

α-Al_2O_3 has a higher density (3.99 g/cm^3) than γ-Al_2O_3 (3.68 g/cm^3; Liao et al.1998), which means the former may have a higher surface free energy or surface energy. Elaborate high-temperature solution calorimetry of both nanocrystalline and macroscopic γ-Al_2O_3 and α-Al_2O_3 proved the above deduction: the surface energies of α-Al_2O_3 and γ-Al_2O_3 are, respectively, 2.64 and 1.66 J/m^2 (Navrotsky, this volume; McHale et al. 1997). Therefore, the γ-Al_2O_3–α-Al_2O_3 system can be classified as Case 3 in Figure 14 (where the α-phase = γ-Al_2O_3 and the β-phase = α-Al_2O_3) where nanocrystalline γ-Al_2O_3 becomes more stable when its particle size is below a certain value and $T < T_{nano}$, $P < P_{nano}$.

Liao et al. (1998) have studied the high-pressure–low-temperature phase transition of ~18 nm γ-Al_2O_3 to α-Al_2O_3. A non-equilibrium *T-P* phase diagram (time = 15 min) for the phase transformation was reported. At $P = P_o$, the transition temperature is 1075°C; at 8 GPa, the temperature decreases to 460°C. This *T-P* relationship is in consistent with Figure 15c for Case 3.

Wen and Yen (2000), Chang and Yen et al. (2001) have studied the phase transformation from ultrafine θ-Al_2O_3 to α-Al_2O_3. They discovered and confirmed that only after θ-Al_2O_3 coarsens to above ~20 nm can it start to transform to α-Al_2O_3. This "critical" size phenomenon can be explained with a similar thermodynamic argument as for the nano-TiO_2 (anatase-rutile) system (Fig. 16).

The nano-ZrO_2 system. About ten distinct phases of zirconia (ZrO_2; including monoclinic, tetragonal, cubic, and orthorhombic variants) have been identified to date (Dewhurst and Lowther 1998). At standard pressure, with increasing temperature, ZrO_2 undergoes the following structure transformations (Navrotsky, this volume; Cordfunke and Konings 1990):

$$\text{Monoclinic } (\rho = 5.82 \text{ g/cm}^3) \xrightarrow{1478\text{ K}} \text{tetragonal } (\rho = 5.94 \text{ g/cm}^3) \xrightarrow{2630\text{ K}}$$
$$\text{cubic } (\rho = 6.20 \text{ g/cm}^3) \xrightarrow{>2983\text{ K}} \text{liquid}$$

The density data in parentheses are from Winterer et al. (1995). Zirconia was shown to exhibit phase stability reversal as a function of particle size about two decades ago. Grave (1978) made a detailed thermodynamic analysis of the phase stability of nanometer-sized monoclinic and tetragonal phases of ZrO_2. Surface free energy and internal strain effects on the stability were considered. Surface energies evaluated by Grave (1978) are 1.46 J/m^2 and 1.10 J/m^2 for the monoclinic and tetragonal phases, respectively. In this case, though the densities of the monoclinic and tetragonal phases are not very different, the distinction between their surface energies is obvious. According to Grave's analysis, tetragonal zirconia becomes more stable than monoclinic zirconia at room temperature when the particle size is below ~10 nm. Applying an external pressure, this critical particle size goes up. This is expected because tetragonal phase has a higher density. For the transformation monoclinic → tetragonal, $\Delta V = -4.3 \times 10^{-7}$ m^3/mol < 0, $\gamma/\rho(\text{mono}) = 2.51 \times 10^{-7}$ mJ/g, $\gamma/\rho(\text{tetra}) = 1.85 \times 10^{-7}$ mJ/g. Thus according to Equation (5), the *P-T* phase diagram for nanophase zirconia should be similar to Figure 15a, with α = mono and β = tetra. However, in this case, the phase boundary between the two phases moves down instead of up at nanometer scale because the last term of Equation (5) < 0 as a consequence of $\Delta(\gamma/\rho) < 0$.

The phase transformation from monoclinic to orthorhombic structure under high pressure was studied by Kawasaki et al. (1990). The threshold pressure of the phase transformation shifts toward higher pressures as the crystallite size decreases. For macroscopic material, the pressure is 3.4 GPa. For a 29.3 nm sample, the transformation pressure is 6.1 GPa. The phase stability of monoclinic and orthorhombic zirconia at elevated temperatures and high pressures was studied by Ohtaka et al. (1991). The observed increased transition pressure from the monoclinic to orthorhombic phase in nanocrystalline ZrO_2 is in good agreement with their thermodynamic calculations using the calorimetry data. This case can be described by Case 1, Figure 12b, where $P_{nano} > P_{\infty}$. In this high-pressure transformation, there is a wide pressure region (e.g., at 29.3 nm, ΔP = 2.6 GPa) in which both monoclinic and orthorhombic phases can coexist dynamically. This phenomenon was confirmed by Winterer et al. (1995) and explained as a characteristic of a martensitic transformation. During the phase transformation from monoclinic to orthorhombic structures under high pressure, tetragonal zirconia can form as an intermediate product (Winterer et al. 1995).

Particle size and surface adsorption

A geochemically important property of small particles is their tendency to adsorb ions from solution onto their surfaces. Adsorption events can involve ions that constitute the bulk (i.e., a step in crystal growth) or can involve other ions. The driving force for adsorption is reduction of surface energy (easily appreciated for the special case of crystal growth).

Zhang et al. (1999) suggested that affinity of the surfaces of small particles for ions in solution (at concentrations below surface site saturation) should be size-dependent. Zhang et al. (1999) presented results for adsorption of organic molecules onto TiO_2 surfaces to support the idea that the adsorption coefficient (determined via an approximate Langmuir-type analysis) can increase as size decreases. Other data to support this notion may be the results of Nelson et al. (1999), who found that nanocrystalline microbially-produced manganese oxide sorbed impurities far more strongly than the macroscopic oxide, even when surface area was accounted for.

Given a number of possible alternative explanations for experimental results involving adsorption onto small particles could be advanced (see Navrotsky, this volume), it is perhaps useful to consider the thermodynamic basis for the prediction of a size-dependence of adsorption.

The concept stems of size dependence of the adsorption coefficient stems from structural and energetic considerations. For small particles, an adsorption event decreases the energy of a system composed of small particles of radius r_A by

$$\Delta G_{Ao} = 3V_m \times [\gamma(S) - \gamma(S\text{–}I)]/r_A$$

where V_m is the molar volume, $\gamma(S)$ is the interfacial free energy prior to adsorption, and $\gamma(S\text{-}I)$ is the interfacial free energy after adsorption.

Similarly, for bigger particles of radius r_B,

$$\Delta G_{Bo} = 3V_m \times [\gamma(S) - \gamma(S\text{–}I)]/r_B$$

It follows that $\Delta G_{Ao} > \Delta G_{Bo}$, since $r_A < r_B$. Because an adsorption event on smaller crystals decreases the energy of a system more than adsorption on particles in a system composed of larger particles, the driving force for adsorption is larger, thus adsorption should be favored on small particles.

Another way of viewing this is to consider the effect of particle size on the equilibrium between adsorption and desorption. From the above we deduce that the activation barrier for desorption from smaller particles will be higher than from larger particles, implying a smaller rate constant. Thus, under conditions when adsorption is favored, at equilibrium ions should partition more strongly onto the surfaces of small particles than larger particles. The activation barrier for adsorption is probably size-dependent, especially for very small particles. This may complicate the analysis in as yet unknown ways.

Zhang et al. (1999) considered other factors that might explain the much higher apparent Langmuir-type adsorption coefficient calculated from their experimental data. They evaluated the contribution of additional high-energy *facets* on smaller particles, and concluded that this was unlikely to explain the result. However, it should be noted decreasing particle size will lead to increased importance of high energy *sites*, especially when particles are so small that "facet" has little or no meaning. For such particles, these surface steps may contribute very significantly to the total adsorption.

Nanoparticles and organics

The role of organics and other surface ligands in phase stability control. In the above section we considered the concept that size and phase stability are sometimes intimately interconnected. Because the fundamental link between particle diameter and structure energy is the energy of the surface, it follows that factors that modify surface energy can alter phase relationships. For example, a mineral precipitating from a dilute aqueous solution may form nanocrystals of a polymorph (often the high temperature polymorph due to the lower density) that is stabilized relative to the macroscopic stable phase due to its lower surface energy. However, addition of a compound that binds to surface ions and decreases the surface energy (interfacial energy) of the macroscopic stable phase could lead to its precipitation. Thus, in principal, a nanoparticle could switch from one structure (and set of physical, magnetic, optical properties) to another in response to changes in solution chemistry (if the activation barrier for the transformation is small).

Structural modification due to the presence of surface-bound ligands has been demonstrated experimentally. At room temperature and pressure, the stable phase for macroscopic CdS is hawleyite, which contains tetrahedrally coordinated Cd within a ccp S array (hawleyite is isostructural with sphalerite). Using analysis of EXAFS data, Chemseddine et al. (1997) showed that nanocrystals capped with mixtures of acetate and thiolate exhibited the NaCl structure, the high pressure phase in which Cd is octahedrally coordinated. Heating of the samples at 200°C removed the organic component and resulted in reversion to hawleyite. Similar phenomena may be anticipated in natural environments, especially where nanocrystals (e.g., of ZnS) form due to the activity of anaerobic heterotrophs in environments where both small organic molecules (e.g., acetate) and more complex polymers are present (e.g., within biofilms).

Nanoparticles, colloids and organics. Nanoparticles are often building blocks for colloids encountered in natural systems. Colloidal aggregates consisting of nanoparticles may exhibit variable degrees of internal order. Well known examples of materials in which the components exhibit extremely high degrees of order are opal (regularly packed arrays of opaline silica) and framboidal pyrite (though the pathway for the formation of framboids may involve other factors). Smaller scale arrays of oriented nanoparticles have also been reported. These are described below in the context of aggregation and crystal growth. The kinetics of colloidal flocculation are reviewed by Waychunas (this volume) in some detail.

Nanoparticles can order to form superlattices (see figures in the chapter by Jacobs and Alivisatos, this volume). In all superlattice systems, superstructure is achieved by coating particles with an organic ligand that spans adjacent nanoparticles (and the inorganic particle core is stabilized by the organic surfactant). The organic compound can orient the nanoparticles yet keep them separate and regularly spaced. Both the core and surfactant play roles in determining the superlattice crystallographic symmetry (Collier et al. 1998). Monolayers of ordered and oriented crystallites can be achieved both in solution and at interfaces (e.g., Langmuir-Blodgett films; Murray et al. 1995), and macroscopic crystals of nanoparticles can be obtained (Herron et al. 1993). Manipulation of the surface chemistry through use of specific adsorbed molecules provides some control on crystal structure (see above) as well as allows manipulation of particles into desired superlattice patterns. If a volatile organic is used, it can be subsequently removed to make an inorganic nanoparticle array (Murray et al. 1995). Resulting ordered two- and three-dimensional nanoparticle arrays have attracted considerable attention due to their unique properties and as materials for the study of collective phenomena.

In addition to their technological significance, self-organization in mixed organic-inorganic nanoparticle systems may be relevant to some areas of geoscience. Most probably, the abundant nanoparticles formed by biological pathways (see above) or in solutions rich in organic molecules sorb organics molecules onto their surfaces under some circumstances. In other cases, the organism may generate specific surface-binding ligands as part of the biomineralization process. Thus, nanoparticles may be primed for organic-mediated self-organization phenomena. Superstructure formation may be a key step in formation of minerals by microbes and macroorganisms and, in the presence of appropriate organic confining membranes, may contribute to generation of large(r) single crystals with extraordinary morphologies.

KINETICS IN NANOPARTICLE SYSTEMS

Nanoscale materials can be produced in many ways, including biogenic routes that utilize organic molecules, physical methods such as mechanical ball milling, chemical methods such as chemical vapor deposition (CVD), electrochemical methods, and sol-gel methods. Due to the many pathways, the mechanisms and kinetics of their formation are diverse.

In the processing of nanoparticles, coarsening is common, and may be accompanied by phase transformation to the macroscopic stable structure. Here we will focus on the kinetics of phase transformations and crystal growth in nanocrystalline particles. We will show later that conventional kinetic models that are widely employed for analysis of macroscopic materials behavior may have to be modified prior to their application to nanomaterials.

Brief review of kinetic models for macroscopic solids

In most cases, during the phase transition from one nanocrystal to another, formation of the new phase is accompanied by simultaneous growth of the original crystal. For macroscopic materials, this process is widely described using a formal theory of transformation kinetics, the so-called Johnson-Mehl-Avrami-Kolmogorov (JMAK) theory (described by the JMAK equation). When particles grow in a solution, the theory of Ostwald-ripening is frequently used to describe the growth kinetics. We review these theories as they may be useful for some nanomaterials systems.

Introduction to the JMAK approach. Detailed description of the JMAK formal theory was given by Christian (1975). The basic assumptions of the theory are that the stable phase nucleates as infinitely small droplets; these droplets are rigid, and grow independently; the growing droplets can overlap; and that the dimensions of the parent phase are essentially infinite. Then the volume fraction of transformation, ζ, for a number of nucleation-growth cases can be reduced to the general form of (Christian 1975)

$$\zeta = 1 - \exp(-kt^n) \quad (6)$$

where k and n are empirical model parameters. The parameter k is not a conventional kinetic constant, since it is temperature dependent. The exponent n is determined by the specific reaction mechanism. For instance, for the case of nucleation at a constant rate and diffusion-controlled growth, $n = 5/2$ (Borg and Dienes 1992). Due to the limitations of the assumptions made by this theory, modifications were attempted in order to allow application of the theory to a greater diversity of cases. For example, Weinberg et al. (1997) made several modifications, including consideration of surface nucleation and the effect of finite sample size (Weinberg 1991). The concept of the local Avrami exponent (n) was also developed and applied to some systems (Lu and Wang 1991; Calka and Radlinski1988).

Equation (6) can be written in the form

$$\ln[-\ln(1-\zeta)] = \ln k + n \ln t \tag{7}$$

Thus, the parameters k and n can be evaluated with the so-called log-log plot of the experimental data obtained.

Ostwald ripening in a solution. In a (solid or liquid) solution, coarsening is driven by decrease in surface energy. This process is often referred to as Ostwald ripening. In effect, Ostwald ripening describes the growth of larger particles at the expense of less stable smaller particles via the interdiffusion of adatoms. Smaller particles will eventually disappear. Ostwald ripening can be limited by volume diffusion (diffusion of ions or atoms in the liquid solution or solid matrix), diffusion along the matrix grain boundary, or precipitation / dissolution reactions at the particle / matrix interface. The general kinetic equation for these cases can be written as (Joesten 1991; Glaeser 2001)

$$\overline{D}^n - \overline{D}_o^n = k(t - t_o) \tag{8}$$

where $\overline{D}$ and $\overline{D}_o$ are the mean particle sizes at time t and t_o, respectively, k is a constant at a certain temperature, and n is an exponent relevant to the coarsening mechanisms. For $n = 2$, the coarsening kinetics are inferred to be limited by precipitation /dissolution reactions at the particle /matrix interface. For $n = 3$, the coarsening kinetics are limited by the volume diffusion. When $n = 4$, the coarsening kinetics are considered to be limited by diffusion along the matrix grain boundary, and for $n = 5$, coarsening kinetics are limited by dislocation-pipe-diffusion within a matrix.

Kinetics of amorphous-to-nanocrystalline transformations

Due to the successful applications of the JMAK formal theory to a wide range of nucleation and growth phenomena in macroscopic samples, researchers have tried to employ the JMAK theory to model phase transitions in nanomaterials systems. This has been successful in some cases (Lu and Wang 1991;Luck el al. 1993; Varga et al. 1994; Luck et al. 1996; Damson et al 1996; Missana et al. 1999; Schmidt et al. 2000). However, other studies suggest that applications of the general JMAK theory to nanomaterials systems are only partially satisfactory (Allia et al. 1993; Illekova et al. 1996; Malek et al. 1999), or even inappropriate (Nicolaus et al. 1992; Gloriant et al. 2000).

Metallic amorphous $Ni_{80}P_{20}$ used by Lu and coworkers (Lu and Wang 1991; Luck el al. 1993; Luck et al. 1996) was prepared by rapid quenching. This Ni-P glass alloy can be viewed as a supercooled liquid. During heat treatment, this thermodynamically unstable phase transforms to its stable crystalline state via production of nanocrystals. Luck et al. (1993) used isothermal DSC (differential scanning calorimetry) to monitor the transformation progress, and modeled their data using the JMAK equation. Using the local Avrami exponent concept, they deduced that the transformation process is controlled in different ways in three different stages: by one dimensional surface nucleation and growth ($n = 2.0$) in the early stage, by three dimensional bulk nucleation and growth ($n = 4.0$) in the middle stage, and, by three dimensional growth ($n = 3.0$) in the final stage. These conclusions seem to be supported by their microscopic observations of cross-sections of samples at different annealing times. Varga et al. (1994) and Damson et al. (1996) used a non-isothermal DSC method to investigate the formation of nanocrystals in FINEMET glass alloys ($Fe_{73.5}Cu_1Nb_3Si_{13.5}B_9$). A Kissinger analysis incorporated with the JMAK equation was used to decipher their DSC data. A Kissinger analysis involves a plot of the $\ln(r/T_m^2)$ vs. $1/T$, where r is the DSC scanning rate (K/min), T_m is the DSC peak position in K. The slope of the plot can be used to calculate the activation energy, as in the Arrhenius plot. Results suggested that the kinetics of the phase transformation and particle growth in FINEMET alloys could be described by the JMAK approach. Missana et al. (1999) found the JMAK formulation is a valid approach in analyzing their kinetic data for the amorphous-to-nanocrystalline transformation in SbO_x film grown by dc reactive spluttering. Crystallization kinetics of amorphous $Al_{89}Dy_{11}$, $Al_{84}Dy_{11}Co_5$ and $Al_{84}Dy_6Co_{10}$ alloys prepared by melt spinning were studied by Schmidt et al. (2000). These authors also used the classical JMAK approach to analyze their kinetic data.

Not all researchers have been satisfied with the results of modeling kinetic data for nanocrystal production from an amorphous precursor using the JMAK approach. When researching the crystallization kinetics of amorphous FINEMET alloys, Allia et al. (1993) and Illekova et al. (1996) found that JMAK theory can only partially account for their experimental data because the Avrami exponent adopted an abnormal value ($n \approx 0.9$). The kinetics of the phase transition from amorphous zirconia gel to nanocrystalline tetragonal zirconia had been studied by Malek et al. (1999). They found the JMAK equation can only be applied to cases where the rate of crystallization is slow. For a general treatment, they had to use a two parameter empirical kinetic equation. When studying the crystallization from amorphous $Co_{33}Zr_{67}$ prepared by melt spinning to nanocrystalline state using DSC /TEM/ XRD /VR (Vibrating-Reed), Nicolaus et al. (1992) found the classic JMAK can not be used to fit their kinetic data because the Avrami exponent (n) fails to maintain a constant value. Instead, it varies with temperature (and thus, time). Also, Gloriant et al. (2000) found that it was difficult to apply the JMAK theory to the crystallization process in Al-based amorphous alloys.

We have studied the phase transformation from nanometer-sized amorphous titania (TiO_2) to nanocrystalline anatase at 300 – 400°C (unpublished). Amorphous titania samples were prepared by fast hydrolysis of titanium ethoxide in water at 0°C (Zhang et al. 2001). The extent of transformation was monitored using XRD determination of the phase mixture as a function of time. We also found that the transformation kinetics do not follow the widely employed JMAK equation.

Although the formal JMAK theory has been used to model some nanomaterials systems, serious limitations are imposed due to the fundamental assumptions. In certain instances, the parameters may only be fitting constants with limited or no physical meaning. In many cases, the JMAK theory may not be optimal for analysis of kinetics of reactions involving nanoparticles.

Kinetics of crystalline transformations involving nanoparticles

In a previous section we showed that, for nanoparticles, the "stable" phase can depend upon the crystal size (also see Navrotsky, this volume). Thus, it follows that a crystal could switch from one polymorph to another as it grows. Furthermore, the phase transformation (or reaction) mechanism may be particle size-dependent. This could arise because of novel surface structure/properties or microstructure, modified thermodynamic driving force, or because the probability of surface nucleation is enhanced compared to bulk nucleation by high surface area. In addition, the scale of structural fluctuations within a nanoscale solid may be comparable to, or potentially larger than, the particle itself (see Jacobs and Alivisatos, this volume, for a discussion of related issues for phase transformations induced by pressure). In macroscopic materials, growth of nucleating regions is inhibited by unfavorable interfacial energy until the nuclei is larger than a critical diameter. However, in very small particles, the transformation rate could be enhanced if the interfacial energy component is removed when the interface sweeps out the particle as it is completely transformed. For these, and other reasons, it is important to develop models for reaction kinetics that incorporate particle size explicitly.

Much of the work to date on particle size effects on phase transformation kinetics has involved materials of technological interest (e.g., CdS and related materials, see Jacobs and Alivisatos, this volume) or other model compounds with characteristics that make them amenable to experimental studies. Jacobs and Alivisatos (this volume) tackle the question of pressure driven phase transformations where crystal size is largely invariant. In some ways, analysis of the kinetics of temperature-motivated phase transformations in nanoscale materials is more complex because crystal growth occurs simultaneously with polymorphic reactions. However, temperature is an important geological reality and is also a relevant parameter in design of materials for higher temperature applications. Thus, we consider the complicated problem of temperature-driven reaction kinetics in nanomaterials.

For reasons noted above, titania (TiO_2) has been a popular experimental model system for investigating the fundamental ways in which crystal size alters thermally-driven reactions. Also, temperature leads to rapid coarsening and phase transformations that modify the utility of nanoparticle titania for commercial applications.

Conventional kinetic models that have been used to model the transformation of coarse anatase to rutile, including standard first order reaction (Rao 1961; Suzuki and Kotera 1962; Gennari and Pasquevich 1998), standard second order reaction (Rao 1961), contracting spherical interface model (Shannon and Pask 1965; Heald and Weiss 1972; MacKenzie 1975; Gennari and Pasquevich 1998), model of nucleation and growth of overlapping nuclei (Shannon and Pask 1965; Gennari and Pasquevich 1998), model of one dimensional, linear and branching nuclei and constant growth (Shannon and Pask 1965), model of random nucleation and rapid growth (Shannon and Pask 1965; Gennari and Pasquevich 1998), and the universal JMKA approach (Suzuki and Tukuda 1969; Hishita et al. 1983; Kumar et al. 1993), are incapable of describing the kinetics of the phase transformation from nanocrystalline anatase to rutile. This has been attributed to particle size-related effects (Gribb and Banfield 1997; Zhang and Banfield 1999; Zhang and Banfield 2000). Particle size effects may take many forms.

First we consider the transformation mechanism of anatase to rutile in order to determine the reason for the dependence of the transformation rate on particle size. Penn and Banfield (1998 1999) showed that the oriented assembly of nanoparticles to form larger crystals (see below for details) is accompanied by formation of twins that introduce new atomic arrangements at particle-particle interfaces. In the case of anatase, a {112} twin represents a slab of brookite and thus, a structural state intermediate between anatase and rutile. Penn and Banfield (1999) proposed that the activation barrier for rutile nucleation is lowered by the presence of these twins. Simultaneously, it was noted that the transformation of anatase to rutile in air (Gribb and Banfield 1997) and under hydrothermal conditions (Penn and Banfield 1999) rarely generates partially reacted crystals, suggesting a high activation barrier for rutile nucleation but rapid rutile growth.

Once initiated, a cascade of atomic displacements and distortions occurs, converting anatase to rutile. Thus, it appears that the transformation rate is limited by the nucleation rate, which is increased by an increase in the number particle-particle interfaces, thus a decrease in particle size. Zhang and Banfield (1999) extended this more generally to particle-particle contacts.

New kinetic models were developed to incorporate interface nucleation (Zhang and Banfield 1999) and surface nucleation (Zhang and Banfield 2000), thus to quantitatively interpret the kinetic behavior in the nanocrystalline anatase-rutile system. Surface nucleation and bulk nucleation come into play as temperature increases (Zhang and Banfield 2000). Particle size has been explicitly incorporated into the kinetic equations. The transformation rate scales with the square of the number of anatase nanoparticles in the case of interface nucleation (Zhang and Banfield 1999), or with the number of anatase nanoparticles in the case of surface nucleation (Zhang and Banfield 2000). If the transformation is governed only by interface nucleation, the kinetic equation is:

$$\ln\left[\frac{(D_a / D_o)^3}{(1-\alpha)} - 1\right] = \ln(k_2 N_o) + \ln t \tag{9}$$

where k_2 is the kinetic constant for interface nucleation, N_0 and D_0 are the initial number of anatase particles and average particle size, respectively, D_a is the average particle size of anatase at time t, and α is the fraction of transformation of anatase to rutile at time t. If both interface nucleation and surface nucleation govern the phase transformation, the kinetic equation is:

$$\ln\left[\frac{k_1}{(1-\alpha)} \cdot \left(\frac{D_a}{D_o}\right)^3 + k_2 N_o\right] = k_1 t + \ln(k_1 + k_2 N_o) \tag{10}$$

where k_1 is the kinetic constant for surface nucleation. Nanocrystalline anatase-to-rutile phase transformation kinetics follow Equation (9) at $T < \sim 600°C$. Addition of nanocrystalline γ-Al_2O_3 to the nanocrystalline anatase reduced the anatase-to-rutile phase transformation rate. This supports the interface nucleation model because the admixed alumina reduces the number of anatase-anatase contacts, thus the probability of interface nucleation. At $T > \sim 620°C$ but below where bulk nucleation is favored (~1000°C) reaction kinetics are described by Equation (10) (Zhang and Banfield 2000).

Phase transformations in nanomaterials have been studied in other systems. The phase transformation from nanocrystalline maghemite (γ-Fe_2O_3) to hematite (α-Fe_2O_3) at 385°C obeyed the simple form of the JMAK equation with $n \approx 1.0$ (Ennas et al. 1999). Schimanke and Martin (2000) examined the transition of nanocrystalline γ-to-α-Fe_2O_3 and described it as first order, with an activation energy that increased with increasing crystal size.

Crystal growth of nanocrystalline particles

It is generally believed that crystal growth in solution occurs via an Ostwald-ripening mechanism (see above). Normally, this is assumed to proceed via addition of monomers to a polymer (cluster, nanocrystal, crystal). However, as noted above and described in more detail below and by Waychunas (this volume), this atom-by-atom mode of crystal growth is not unique, especially in nanomaterials.

An equation of the form provided in Equation (8) is often used to fit experimental coarsening data. The exponent, n, is used to infer the rate-controlling step for the growth process. This has proven useful for modeling grain growth in a number of metallic alloy systems (Lu 1996). However, Krill et al. (2001) found that the grain growth rate in nanocrystalline (<150 nm) Fe follows a linear relationship with time. This was attributed to grain boundary migration controlled by redistribution of excess volume localized in the boundary cores. For the grain growth of nanocrystalline Cu, it was not possible to determine a best fit to experimental data using Equation (3) with $n = 2, 3, 4$, making it impossible to deduce the growth mechanism (Ganapathi et al. 1991). For other metals and alloys (Ti, Zr etc.), n falls between 2 and 5, and may be temperature dependent (Malow and Koch 1996). Crystal growth of nanocrystalline ZnO in colloidal suspensions was inferred to occur via Ostwald-ripening controlled by volume diffusion (Wong et al. 1998), though in an organic-bearing system Verges et al. (1990) observed ZnO growth via assembly of nanoparticles into rods.

We found that coarsening of nanocrystalline anatase in air in the temperature range 465-525°C could be described by Equation (8) with $n \approx 3$ (Zhang and Banfield 1999), yielding an activation energy for anatase coarsening of 69 kJ/mol. However, as the coarsening occurred in air, it was unlikely to have been controlled by "volume diffusion" (coarsening of titania in air is probably limited by surface diffusion). This highlights the fact that application of coarsening kinetic models to describe grain growth data so as to infer the growth mechanism is dangerous because growth via different mechanisms may exhibit similar reaction progress. If possible, additional data should be obtained to confirm the deduced reaction pathway.

There is another way in which crystals can be assembled. This involves the solid state combination of large clusters of atoms or nanoparticles, or addition of clusters or nanoparticles to larger crystals. The pieces are assembled in crystallographically specific ways so that interface elimination leads to formation of a larger single crystal. The crystallographic control distinguishes this pathway from simple aggregation, as is often modeled in flocculation studies where the mode of transformation from a random aggregate to a single crystal is not specified (see Waychunas, this volume). Oriented assembly of nanoparticles is an effective form of crystal growth because removal of the interface eliminates the energetic contribution of two surfaces. This pathway occurs in parallel with crystal growth via atom-by-atom diffusion (e.g., removal of necks at particle junctions).

High-resolution transmission electron microscopy (HRTEM) has been critical for documenting oriented aggregation. HRTEM images provide information about particle crystallography, orientation, size, composition, and morphology simultaneously. Electron diffraction data and atomic-resolution images have yielded key insights necessary to determine the structures of nanoscale materials (e.g., Banfield et al. 1991; Banfield and Bailey 1996; Wagner et al. 1999; Cowley et al. 2000; Janney et al. 2001). Although some care must be taken not to over-interpret lattice fringe details (Malm and O'Keefe 1997), HRTEM is invaluable for monitoring the characteristics of particles that have been coarsened to varying degrees. Examples of data used to document crystal growth via oriented aggregation are provided in Figure 19. Crystallographic information (diffraction patterns, lattice fringe details) confirm that the large, irregularly-shaped single crystal (chain) in Figure 19a consists of smaller anatase crystals (with typical near-equilibrium morphologies) that were oriented with respect to each other. Most particle-particle boundaries have been eliminated, but a few remain. This growth mechanism is a form of self organization. Figure 19b illustrates an irregularly shaped ZnS (sphalerite) crystal that contains stacking faults and twins that seem to intersect inflexion points in the particle margins, thus are interpreted to mark the positions of interfaces between the nanoparticles that were combined to form the oriented aggregate (see Fig. 19c; Huang et al., in prep.).

Penn and Banfield (1998a,b) first reported the crystallographically-specific oriented

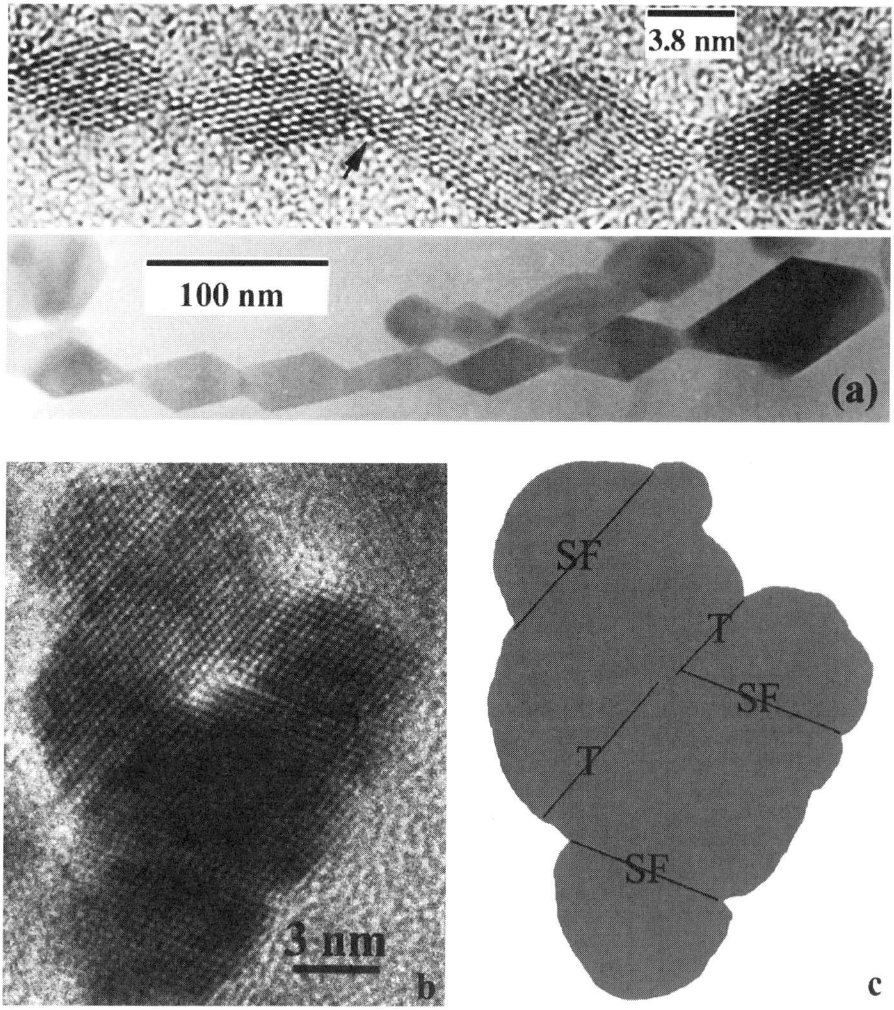

Figure 19. High-resolution transmission elec-tron microscope images of (a) anatase nanoparticles. What appears to be a gain of particles is actually a single crystal because interfaces between adjacent particles have been eliminated (arrow) (Penn and Banfield, after Penn and Banfield 1999). (b) a defective single crystal of ZnS (sphalerite) inferred to have formed by oriented aggregation of nanoparticles. A diagram illustrating the relationship between margins of the crystal in (c) and twins (T) and stacking faults (SF) that probably mark the positions of preexisting interfaces (Huang et al., in prep.) (image provided by F. Huang and J.F. Banfield).

aggregation-based growth of nanocrystalline anatase particles and focused on the implications of this pathway for microstructure development (see below). Other researchers have reported that this mechanism is important for growth of a variety of minerals and inorganic compounds), including anatase (TiO_2, Chemseddine and Moritz 1999), feroxyhite (FeOOH, Banfield et al. 2000; Penn et al. 2001), heterogenite (CoOOH, Penn et al. 2001), zeolites (de Moor et al. 1999; Nikolakis et al. 2000, however the synthesis used organic compounds), TiC (Kuo and Shen 2000), CeO_2 (Lee and Shen

1999) and gold (Privman et al. 1999). These studies differ from those cited above where periodic arrays of nanoparticles separated by (and ordered by) organics were generated. In addition, natural nanometer-sized iron oxyhydroxide products from biomineralization were found to grow via oriented assembly (Banfield et al. 2000; also see Alivisatos 2000).

Although conceptually simple, the process of crystal growth via oriented attachment (in the absence of organics) has not received much attention, except in some very specific circumstances (e.g., fundamental particle theory, which was developed to explain phenomena involving layer silicates; Nadeau et al. 1984). Early theories for crystal growth were sufficiently flexible to accommodate aggregation-based growth. In parallel, considerable experimental work and modeling has been devoted to describing the kinetics of particle aggregation and the resulting (often fractal) nature of aggregate structures. However, flocculation studies (see Waychunas, this volume, for a summary) typically do not constrain aggregate growth to ensure adjacent particles adopt related crystallographic orientations and do not include consideration of how surface protons and water are eliminated or how the aggregate is transformed into a single crystal. Thus, a gap remains between models for aggregate growth and models that describe enlargement of single crystals.

In a solution, nanoparticles interact with each other in a number of ways. Widely separated nanoparticles may be brought into contact by Brownian motion. As they approach each other, electrostatic, van der Waals forces, and hydrogen bonding, in addition to Brownian motion, can cause two nanoparticles to rotate with respect to each other, and collide. Evidently, under some conditions, the collisions that result in fusion are those that involve two nanoparticles in appropriate orientations to form a coherent (or semicoherent) interface. Particle rotation in the absence of a fluid has been modeled computationally (Zhu and Averback 1996), implying that oriented assembly-based crystal growth can also occur in dry systems.

In the crystal growth of nanometer-sized ZnS particles (~2.4 nm) treated under hydrothermal conditions, Huang et al. (in prep.) found that the grain growth of ZnS particles follows a two-stage growth process. Kinetic data for the second stage fit the Ostwald-ripening by volume diffusion ($n = 3$, Eqn. 8) very well. Nevertheless, in order to fit the kinetic data of the first stage with Equation (8) for Ostwald-ripening, we had to use an abnormal and non-physically realistic exponent value of $n \approx 10$. This can not be attributed to the Ostwald-ripening. A new growth equation based on the consideration of oriented aggregation of 2.4 nm ZnS was deduced:

$$D = \frac{D_0(\sqrt[3]{2}k_1 t + 1)}{(k_1 t + 1)} \tag{11}$$

where D_0 and D are the initial average particle size of ZnS (~2.4 nm) and the average particle size at time t, respectively; k_1 is a kinetic constant for oriented attachment of the 2.4-nm ZnS particles. Fitting of the experimental data in the first stage to Equation (11) is excellent. TEM data from the samples support the conclusion that the crystal growth of 2.4-nm ZnS is dominated by orientated aggregation of the ZnS particles in the first stage (see Fig. 19b).

Aggregation and nucleation

Figure 19a illustrates oriented aggregation-based growth involving crystals that are >6 nm to tens of nanometers in diameter. Other data show the same processes involving particles ~3 nm in diameter. Such observations (cited above; also see Anastasio and

Martin, this volume; Waychunas, this volume) lead to the speculation that nanoparticle aggregation is a pathway for crystal growth from molecules to microcrystals in many systems under some conditions. However, processes involving particles with sizes between those of molecules and ~2 nm are very difficult to study with the level of detail needed to clearly resolve reaction mechanisms (though, of course, the existence of particle aggregates can be documented fairly easily).

Formation of multinuclear clusters from aqueous ions defines the initial step in growth of any crystal from solution. Rozan et al. (2000) demonstrated that multinuclear clusters are present and important in the environment (e.g., M_3S_3, M_4S_6, M_2S_4, where M is a metal ion). These may have many fates, one of which is that they aggregate to form even larger clusters. Luther et al. (1999) reported that $Zn_3S_3(H_2O)_6$ rings condense to form a neutral $Zn_6S_6(H_2O)_9$ cluster with a structure analogous to wurtzite. Interestingly, condensation of $Zn_3S_3(H_2O)_6$ rings in the presence of excess bisulfide results in the formation of $Zn_4S_6(H_2O)_4^{4-}$ clusters with structure found in sphalerite (Luther et al. 1999). The inferred cluster aggregation-based pathways are shown in Figure 20.

Luther et al. (1999) suggested that Zn clusters form an amorphous precipitate that develops long-range order with further cross-linking. However, as noted above, biogenic

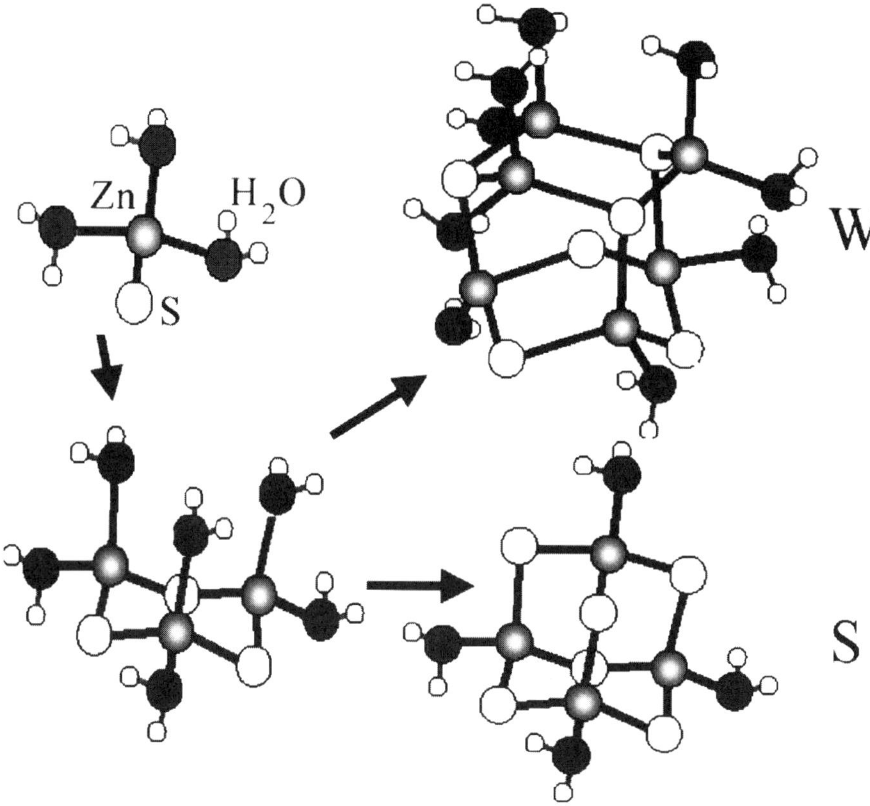

Figure 20. Diagram showing a series of steps that may convert a hydrated ZnS ion into a $Zn_3S_3(H_2O)_3$ cluster and then into either a $Zn_6S_6(H_2O)_9$ cluster that could be assembled into the ZnS wurtzite (W) polymorph or a $Zn_4S_6(H_2O)_4^{4-}$ cluster that could be assembled into the ZnS sphalerite (S) polymorph. After Luther et al. (1999).

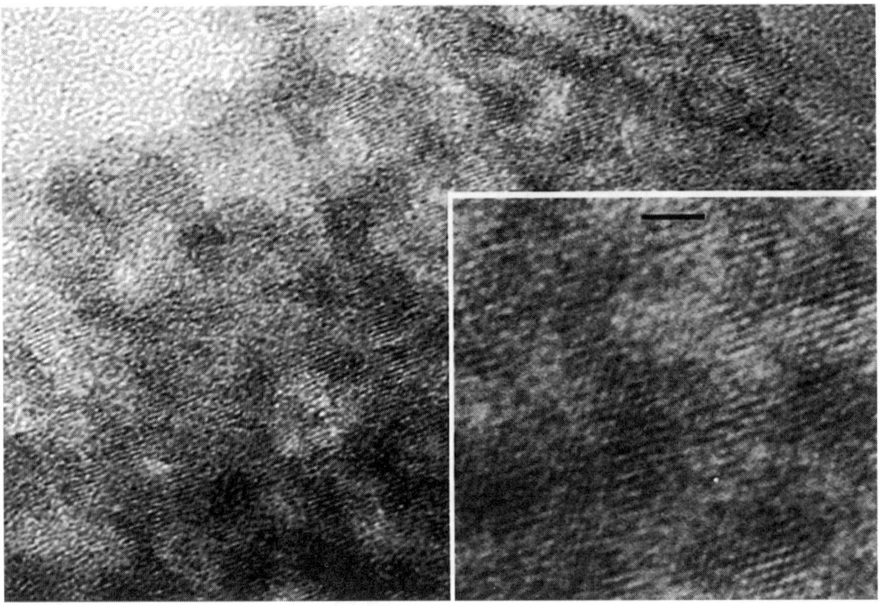

Figure 21. High-resolution transmission electron microscope image of ferrihydrite or feroxyhite formed by the activity of iron-oxidizing bacteria growing in near-neutral pH groundwater (see Banfield et al. 2000). Note that the lattice fringes indicate particles within the material are oriented, but crystalline regions are separated by space. The structure resembles that proposed for ferrihydrite (see Fig. 2), although the size of the pores is slightly different. Insert is a region that has been enlarged and Fourier filtered; scale bar = 2 nm (Banfield, unpublished).

ZnS occurs as very small, crystalline nanoparticles and oriented aggregation-based crystal growth involving few nanometer-diameter ZnS nanoparticles occurs. Thus, we postulate that sphalerite nanoparticles form due to the aggregation of sphalerite-like $Zn_4S_6(H_2O)_4^{4-}$ clusters whereas wurtzite nanoparticles form by oriented aggregation of wurtzite-like $Zn_6S_6(H_2O)_9$ clusters. These consideration may be relevant to many important minerals that are isostructural with sphalerite and wurtzite.

Figure 2 illustrates a proposed defect-bearing structure for ferrihydrite (Drits et al. 1993). Although they were introduced as a way to adjust the model fit to X-ray diffraction patterns, such defects seem reasonable. Given that ferrihydrite and feroxyhite can coarsen via crystallographically-controlled aggregation (see Fig. 21 and Banfield et al. 2000), we speculate that the ferrihydrite particles themselves (with structure analogous to that shown in Fig. 2) could form by imperfect aggregation of nanometer-scale clusters. Although this is mere conjecture, it does serve to illustrate how nanoparticle assembly simultaneously could play a role in structure and microstructure development.

Possible galvanic interactions in nanoparticle mixtures

The particle-size dependence of electrochemical interactions is a potentially very important, but largely unstudied, aspect of geochemically-significant mineral reactivity.

Whether a mineral undergoes oxidation or reduction is determined by the rest potential if there is only one pair of redox process occurring on a mineral surface (the rest potential is the potential an electrochemical system will naturally approach if no external voltage is applied). The redox behavior will be determined by the mixed potential if there are multiple redox processes occurring on the surface (the mixed potential is the electric potential or voltage of a reacting electrochemical system with two or more different redox couples). For example, in the dissolution of copper by ferric ions in strong chloride solutions, the mixed potential is the potential for the reaction $Fe^{3+} + Cu \rightarrow Fe^{2+} + Cu^{+}$. Electrical current flows from the higher potential mineral electrode to the lower potential one. Many factors can affect the behavior of coupled galvanic cells. The chemical compositions of the minerals, solution constituents and pH, and oxygen concentration in the solution, etc., all influence the rest (or mixed) potential, the exchange current density, and the electrical polarization.

Because the internal and surface electronic structures of a material can change with particle size, it is likely that the rest or mixed potentials also vary with particle size. Furthermore, two different sized particles of the same mineral may have slightly different potentials (enough to alter their reactivity). If a mineral assemblage contains a mixture of nano-sized minerals, each with a range of particle sizes, highly complex behavior may result. As the particle size and its distribution vary over time due to nanocrystal growth, the behavior of the galvanic cells may change over time. For instance, the magnitude of the current of a cell, and even the flow direction of the current, may alter if components within the nanoparticle assemblage coarsen at different rates.

Where might this be important? As discussed above, biological activity can result in the simultaneous precipitation of mixtures of nanoscale sulfide minerals under certain conditions. Each mineral will exhibit a particular particle size distribution, dependent on the solution composition, bacterial activity, rate of crystal growth, and the nature of electrochemical interactions between the particles. These electrochemical reactions could lead to oxidation of one type of nanophase sulfide mineral of a certain size, and reduction of another type of nanophase sulfide particle or other species in the solution. In this way, a tremendous number of mineral-solution-mineral galvanic cells could develop, with potentially significant impact on dissolution kinetics, growth kinetics, and the mixture of phases observed. In addition to environmental relevance, these processes may shape the mineralogy of low-temperature ore deposits.

Microstructure development in nanocrystals

Virtually all minerals contain defects. In addition to point defects (e.g., vacancies that exist in a thermodynamically determined equilibrium number, impurities etc.), macroscopic minerals contain line defects (dislocations), and planar defects such as stacking faults, antiphase boundaries and twins. Intergrown layers of different structure or composition, and polytypic disorder also may be present.

Although point defects certainly occur in nanoparticles (and unusual coordination sites are probably common in some very small nanoparticles), it is generally agreed that nanocrystals do not contain dislocations or other extended defects because the energetics of these features are significant and diffusion distances are small. So (in the absence of deformation), given that all big crystals start out small, where do dislocations and planar defects in macroscopic materials come from?

It is probably true that the "as formed" nanoparticles do not contain dislocations (see Waychunas, this volume). However, high-resolution transmission electron microscope studies show that nanocrystals (e.g., ~3 nm size) certainly do contain dislocations, twins, and stacking faults. These may arise due to mistakes during atom-by-atom coarsening of primary nanoparticles. However, a more obvious source is evident.

Nanoparticles, especially those relevant in low-temperature environments, are often rather insoluble compounds (oxides, phosphates, sulfides), thus aggregation-based growth may be favored under many conditions. Aggregation-based growth has been shown to be the source of a diversity of microstructures in minerals. Firstly, Penn and Banfield (1998a) showed that oriented aggregation of nanocrystals with surface steps results in the introduction of dislocations at low angle grain boundaries (subsequent diffusion-based recrystallization removes traces of the individual particles, yielding a single crystal containing a dislocation). This is illustrated in Figure 22. Formation of dislocations at boundaries between coalescing islands in epitaxial films has also been discussed (Cheung et al. 2001).

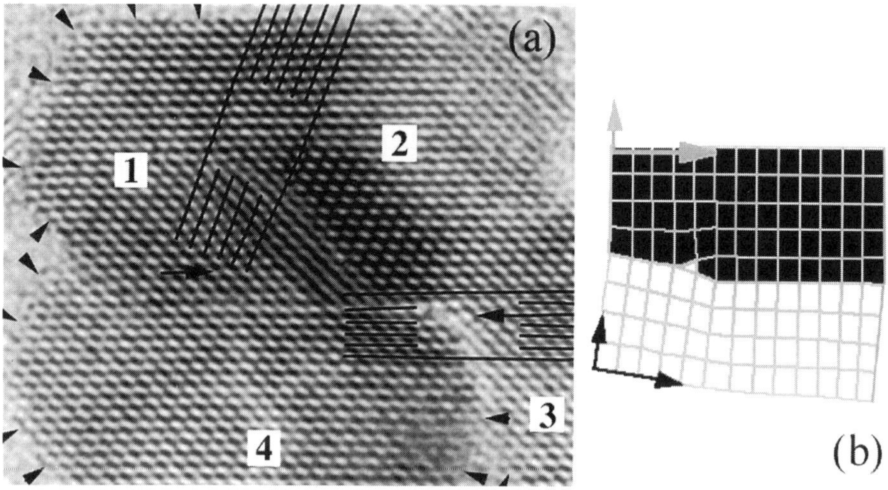

Figure 22. Transmission electron microscope image of an single anatase crystal formed by oriented aggregation. Crystal margins are marked with arrows tips; a subset of the particles that formed the aggregate are numbered. Dislocation positions are indicated by arrows (view at low angle). The diagram on the right illustrates formation of an edge dislocation at an interface where one particle (black) has a surface step (Penn and Banfield, unpublished; see Penn and Banfield 1998).

Penn and Banfield (1998b, 1999) and Banfield et al. (2000) demonstrated that twins and structurally distinct slabs are generated when oriented aggregation involves an interface for which coherence can be obtained in more than one orientation. Multiple aggregation events, perhaps in combination with screw dislocation-based growth, can generate complex polytypes and structural intergrowths (Penn and Banfield 1998).

Formation of point defects as the result of aggregation-based growth is more difficult to establish because of the difficulty in detecting such features in nanoparticles. However, it seems probable that aggregation based growth involving nanoparticles with surface-adsorbed ions will result in impurity sequestration (Banfield et al. 2000; also see Furer 1993). The chemistry of some natural nanomaterials supports this inference (Fowle et al. 2000; Fowle et al., in prep.). However, the validity of this hypothesis awaits experimental confirmation.

SOME RESEARCH OPPORTUNITIES AND CHALLENGES

The geoscience community has been working on nanoparticulate materials for many decades and has made remarkable progress towards understanding the "bulk" structures of finely crystalline silicate clays and oxyhydroxide, oxide, and hydroxide minerals. More recently, the application of spectroscopic methods has revealed a great deal about the nature of surfaces and surface sites on these particles. However, much remains to be learned! We have only scratched the surface in our attempts to understand the size-dependence of structure, physical properties, and reaction kinetics for nanoscale planetary materials. There are many possible challenges and future research. Here are just a few:

Synthesis of materials suitable for quantification of size-related phenomena. In order to employ the characterization tools necessary to make detailed measurements, sufficient quantities of single phases with tightly controlled sizes and dispersion states are required. Partnerships with materials chemists working on materials synthesis problems will be important. Simultaneously, materials chemists should be encouraged to turn their skills to creation of materials relevant to nanogeoscience investigations. Problems that might be investigated using such materials include the size dependence of thermodynamic properties, the size dependence of surface adsorption, size dependent galvanic interactions, and the nature of solution layers surrounding nanoparticles.

Characterization of naturally occurring and synthetic nanoparticles in order to determine the variation in bulk and surface structure and properties with particle size. Most techniques provide valuable insights about some specific characteristic. Some methods (such as transmission electron microscopy) provide information about several of characteristics. Clearly, future studies require integration of as many approaches to characterization and modeling of nanoscale systems as possible. Specifically, direct methods such as high-resolution electron microscope-based imaging and nano-diffraction characterization and nanoscale major element compositional analysis (suitable for consideration of individual nanoparticles) should be combined with methods that provide information about oxidation state, site geometries, and surface structural characteristics (e.g., X-ray-based methods; see Waychunas, this volume; and NMR, Casey et al., this volume). Simultaneously, information about the energetics of nanoparticles is essential (see Navrotsky, this volume). Approaches that provide single particle or near-single particle resolution may have particular significance (X-ray and other synchrotron-based microscopies; see Jacobs and Alivisatos, this volume). However, these data are insufficient. Spectra from small quantities of nanomaterials are difficult to interpret. They often suffer from size-related broadening and are complicated by loss of atomic symmetry due to surface-related stain. In order to extract detailed size-dependent structural parameters, iteration between measurement and data prediction is essential. For example, prediction of atomic coordinates in very small particles via energy-minimization structure modeling (see Rustad et al., this volume); input of resulting coordinates into programs that accurately calculated absorption spectra, and comparison with measured spectra is needed. Better computational methods for predicting structural distortions in small particles will be of great importance.

Nanoparticles in complex, organic-inorganic systems. Organic molecules are sufficiently abundant in many environments that they can impact crystallization and particle aggregation. Of special interest are mixed organic-nanoparticle materials likely to be encountered in biomineralizing systems. Examples include macroorganisms that deposit inorganic-organic composite materials such as bone, and microorganisms that create exquisite single crystals patterned at the nanometer scale. Analysis of the interacting roles of macromolecules and small organic molecules that participate in self organization of nanoparticles to form superperiodic arrays and, perhaps, single crystals may greatly improve our understanding of how detailed control of morphology development is achieved. This knowledge holds additional geological relevance due to the connections between evolution of organisms and the macromolecules they produce and their fossil record.

Self-organization of nanoparticles. Study of the ordering of small crystals to form naturally, the importance of analogous processes in naturals systems has received only patchy attention. Considerable effort has been devoted to the study of flocculation/aggregation, including the generation of structures with fractal geometries (e.g., colloidal gold; Saunders and Shoenly 1995). However, the interactions between nanoparticles during assembly requires detailed analysis. Furthermore, the degree to which crystallographically specific oriented assembly contributes to crystal growth in low temperature and higher temperature systems requires further analysis. Much remains to be learned about consequences of oriented assembly for growth kinetics and microstructure development.

Clusters in the environment and early crystal growth. Molecules with sizes that are intermediate between aqueous ions (few Å) and nanoparticles (> ~13-15 Å) are probably abundant and important in many natural environments. However, they are relatively unstudied because of the challenges associated with their characterization. Recent work by Furer (1993), Luther (1999) and others (see Casey et al. this volume) amplifies suggestions made in earlier studies that large clusters may account for much of what is measured as "dissolved" ions in certain environments.

Size-dependent nanoparticle reactivity. Large clusters/small nanoparticles are probably among the most reactive components in natural systems. How common are they? Are they formed biologically or inorganically? What is their fate? The long-term impact of very small nanoparticles will depend largely upon their fate. Clusters may combine to form nanoparticles or contribute to growth of existing nanoparticles. They may be stabilized by adsorption of organic molecules or other ions, and/or flocculate to form colloids. Because of their small size, the smallest nanoparticles should dissolve readily, potentially leading to rapid release of toxic ions when conditions change. Ultimately, understanding the role of nanoparticles in natural systems requires determination of the form of nanoparticles (e.g., crystalline or amorphous or in between?), their phase stability (e.g., metastable or size-induced stable phase?), surface reactivity (e.g, tendency to adsorb inorganic or organic ions?), aggregation behavior (e.g., reversible flocculation or irreversible aggregation-based crystal growth?), physical properties (e.g, size-dependence of redox potential, thus selective stabilization or dissolution due to biological and geochemical processes?), magnetic properties (e.g., isolated superparamagnetic particles or interacting single domains?), size-dependent optical properties and electronic properties?, size-dependent surface catalytic reactivity (e.g., do nanoparticles adsorb light and catalyze organic degradation?), physical reactivity (unusual defect microstructures?), etc.

Nanoparticle transport in aqueous systems. Nanoparticles are intermediate in size between most clay-sized materials (and colloids) and molecules. Their transport behavior should vary accordingly. Important factors will be nanoparticle size and aggregation state. However, size-dependent surface properties may lead to unanticipated behavior. This could arise, for example, due to modified surface reactivity compared to macroscopic equivalents. Most research on transport of submicron-scale materials has dealt with larger particles or colloidal aggregates. Specific consideration of transport of very small particles may be worthwhile.

In conclusion, nanoparticles are special because their size ensures that they behave in some ways like crystals and in other ways like molecules. Moreover, because they are neither, they possess characteristics not found in either ions or macroscopic materials. Much remains to be learned about nanometer-scale solids and their size-dependent behavior, and the importance of these phenomena in natural systems. We hope that this chapter and volume stimulate research efforts on these and related topics.

Note added in press

Shen and Lee (2001) report formation of {111} twins in ZrO_2 as the result of aggregation-based crystal growth in nanoparticles formed by laser ablation condensation. They directly observed the tetragonal to monoclinic phase transformation induced by the electron beam and showed that reaction is initiated at these twins.

ACKNOWLEDGMENTS

We thank Brian Bischoff and Tye Gribb for their thoughtful contributions to the early development of ideas that seeded a decade (and more!) of research in our lab. We thank Fred Ochs III, Amy Gribb, R. Lee Penn, Michael Finnegan , Ben Gilbert, Glenn Waychunas, William Casey, Alex Navrotsky, and Forrest Huang for their contributions to aspects of the work presented here and many helpful discussions. Research described in this chapter, and associated with preparation of this review, was funded by the Chemical Sciences, Geosciences, and Biosciences Division, Office of Basic Energy Sciences of the Department of Energy, the National Science Foundation Earth Sciences Geochemistry and Petrology and Geology and Palaeontology Programs, and the NASA JPL Astrobiology Institute.

REFERENCES

Aarik J, Aidla A, Uustare T (1996) Atomic-layer growth of TiO_2-II thin films. Philos Mag Letters 73: 115-119

Alivisatos AP (2000) Naturally aligned nanocrystals. Science 289:736-737

Allia P, Baricco M, Tiberto P, Vinai F (1993) Kinetics of the amorphous-to-nanocrystalline transformation in iron-copper-niobium-silicon-boron ($Fe_{73.5}Cu_1Nb_3Si_{13.5}B_9$). J Appl Phys 74:3137-3143

Alt JC (1988) Hydrothermal oxide and nontronite deposits on seamounts in the Eastern Pacific. Marine Geol 81:227-239

Banfield JF, Bailey SW (1996) Evidence for formation of regularly interstratified serpentine-chlorite minerals by tetrahedral inversion in long-period serpentine polytypes. Am Mineral 81:79-91

Banfield JF, Barker WW (1998) Low-temperature alteration in tuffs from Yucca Mountain, Nevada. Clays Clay Minerals 46:27-37

Banfield JF, Bischoff BL, Anderson MA (1993) TiO_2 accessory minerals: Coarsening, and transformation kinetics in pure and doped synthetic nanocrystalline materials. Chem Geol 110:211-231

Banfield JF, Murakami T (1998) Atomic-resolution transmission electron microscope evidence for the mechanism by which chlorite weathers to 1:1 semi-regular chlorite-vermiculite. Am Mineral 83: 348-357

Banfield JF, Veblen DR, Smith DJ (1991) The identification of naturally occurring TiO_2(B) by structure determination using high-resolution electron microscopy, image simulation, and distance-least-squares refinement. Am Mineral 76:343-353

Banfield JF, Veblen DR, Jones BF (1990) Transmission electron microscopy of subsolidus oxidation and weathering of olivine. Contrib Mineral Petrol 106:110-123

Banfield JF, Welch SA, Zhang H, Ebert TT, Penn RL (2000) Aggregation-based crystal growth and microstructure development in natural iron oxyhydroxide biomineralization products. Science 289:751-754

Bargar JR, Persson P, Brown GE (1999) Outer-sphere adsorption of Pb(II)EDTA on goethite. Geochim Cosmochim Acta 63:2957-2969

Bargar JR, Tebo BM, Villinski JE (2000) *In situ* characterization of Mn(II) oxidation by spores of the marine Bacillus sp strain SG-1. Geochim Cosmochim Acta 64:2775-2778

Barr DW, Ingledew WJ, Norris PR (1990). Respiratory chain components of iron-oxidizing acidophilic bacteria. FEMS Micro Lett 70:85-90

Bassett WA, Weathers MS (1990) Stability of the body-centered cubic phase of iron: A thermodynamic analysis. J Geophys Res 95:21709-21711

Bazylinski DA, Moskowitz BM (1997) Microbial biomineralization of magnetic iron minerals: Microbiology, magnetism and environmental significance. Rev Mineral 35:181-223

Berezowsky M (1995) Constructed wetlands for remediation of urban waste waters. Geosci Canada 22:129-141

Bigham JM, Carlson L, Murad E (1994) Swertmannite, a new iron oxyhydroxysulfate from pyhasalmi, Finland, and other localitie. Mineral Mag 58:641-648

Blake RC, Shute EA, Greenwood MM, Spencer GH, Ingledew WJ (1993) Enzymes of aerobic respiration on iron. FEMS Microbiol Rev 11:9-18

Blake RC, Shute EA (1994) Respiratory enzymes of *Thiobacillus ferrooxidans*. Kinetic properties of an acid-stable iron:rusticyanin oxidoreductase. Biochem 33:9220-9228

Blakely JM (1973) Introduction to the Properties of Crystal Surfaces, 1st Edn. Pergamon Press, Oxford.

Bond PL, Smriga SP, Banfield JF (2000) Phylogeny of microorganisms populating a thick, subaerial ,lithotrophic biofilm at an extreme acid mine drainage site. Appl Environ Microbiol 66:3842-3849

Borg RJ, Dienes GJ (1992) The Physical Chemistry of Solids. Academic Press, Boston.

Botuyan MV, Toy-Palmer A, Chung J, Blake RC, Beroza P, Case D, Dyson, H J (1996). NMR solution structure of Cu(I) rusticyanin from *Thiobacillus ferrooxidans*: Structural basis for the extreme acid stability and redox potential. J Molec Biol 263:752-767

Brock TD, Madigan MT (1991) Biology of Microorganisms. Prentice Hall, Englewood Cliffs, New Jersey, 874 p

Brouwers GJ, Vijgenboom E, Corstjens PLAM, De Vrind JPM, de Vrind-de Jong EW (2000) Bacterial Mn^{2+} oxidizing systems and multicopper oxidases: An overview of mechanisms and functions. Geomicrobiol J 17:1-24

Brown GE, Foster AL, Ostergren JD (1999) Mineral surfaces and bioavailability of heavy metals: A molecular-scale perspective. Proc Nat'l Acad Sci USA 96:3388-3395

Calka A, Radlinski AP (1988) Decoupled bulk and surface crystallization in $Pd_{85}Si_{15}$ glassy metallic alloys: Description of isothermal crystallization by a local value of the Avrami exponent. J Mater Res 3:59-66

Cammarata RC (1994) Surface and interface stress effects in thin films. Progr Surface Sci 46:1-38

Cammarata RC, Sieradzki K (1994) Surface and interface stresses. Ann Rev Mater Sci 24:215-234

Casimiro DR, Toy-Palmer A, Blake RC, Dyson HJ (1995) Gene synthesis, high-level expression, and mutagenesis of *Thiobacillus ferrooxidans* rusticyanin: His 85 is a ligand to the blue copper center. Biochem 34:6640-6648.

Chang PL, Yen FS, Cheng KC, Wen HL (2001) Examinations on the critical and primary crystallite sizes during θ-to α-phase transformation of ultrafine alumina powders. Nano Letters 1:253-261

Charlet L, Manceau A (1992) X-ray absorption spectroscopic study of the sorption of Cr(III) at the oxide/water interface. II Adsorption, coprecipitation and surface precipitation on ferric hydrous oxides. J Colloid Interface Sci 148:425-442

Chemseddine A, Fieber-Erdmann M, Holub-Krappe E, Boulmaaz S (1997) EXAFS study of functionalized nanoclusters and nanocluster assemblies Z Phys D 40:566–569

Chemseddine A, Moritz T (1999) Nanostructuring titania: Control over nanocrystal structure, size, shape, and organization. Eur J Inorganic Chem 2:235-245

Cheng BP, Kong J, Luo J, Dong Y (1993) Relation of structure stability and Debye temperature to crystal size of nanometer sized TiO_2 powders. Mater Sci Progress 7:240-243 (in Chinese)

Cheung SH, Zheng LX, Xie MH, Tong SY, Ohtani N (2001) Initial stage of GaN growth and its implication to defect formation in films. Phys Rev B 6403:3304-3344

Christian JW (1975) The Theory of Transformation in Metals and Alloys, 2nd Edn. Pergamon Press, New York

Cobley JG, Haddock BA (1975). The respiratory chain of *Thiobacillus ferrooxidans*: The reduction of cytochromes by Fe^{2+} and the preliminary characterization of rusticyanin, a novel "blue" copper protein. FENS Letters 60:29-33

Collier CP, Vossmeyer T, Heath JR (1998) Nanocrystal superlattices. Ann Rev Phys Chem 49:371-404

Cordfunke EHP, Konings RJM (1990) Thermochemical Data for Reactor Materials and Fission Products. North-Holland, Amsterdam.

Combes JM, Manceau A, Calas G, Bottero JY (1989) Formation of ferric oxides from aqueous solutions: A polyhedral approach by X-ray absorption spectroscopy. I. Hydrolysis and formation of ferric gels. Geochim Cosmochim Acta 53:583-594

Cowley JM, Janney DE, Gerkin RC, Buseck PR (2000) The structure of ferritin cores determined by electron nanodiffraction, J Structural Biol 131:210-216

CRC Book. CRC Handbook of Chemistry and Physics 1985-1986 6th Edition, Eds. R.C. Weast, M.J. Astle, and W.H. Beyer. CRC Press, Boca Raton, Florida

Daly TK, Buseck PR, Williams P, Lewis CF (1993) Fullerenes from a fulgurite. Science 259:1599-1601

Damson B, Wurschum R (1996) Correlation between the kinetics of the amorphous-to-nanocrystalline transformation and the diffusion in alloys. J Appl Phys 80:747-75

de Moor PPEA, Beelen TPM, Komanschek BU, Beck LW, Wagner P, Davis ME, van Santen RA (1999) Imaging the assembly process of the organic-mediated synthesis of a zeolite. Chemistry-Eur J 5: 2083-2088 deVrind-deJong EW, deVrind JPM (1997) Algal deposition of carbonates and silicates. Rev Mineral 35:267-307

Dewhurst JK Lowther JE (1998) Relative stability, structure, and elastic properties of several phases of pure zirconia. Phys Rev B 57:741-747

Drits VA, Sakharov BA, Salyn AL, Manceau A (1993) Structural model for ferrihydrite. Clay Minerals 209:185-208

Drits VA, Sakharov BA, Manceau A (1993) Structure of feroxyhite as determined by simulation of X-ray diffraction curves. Clay Minerals 209-222

Druschel GK, Labrenz M, Thomsen-Ebert T, Fowle DA, Banfield JF (2002) Biogenic precipitation of monomineralic nanocrystalline sulfides: implications of observed and modeled processes to ore deposition. Econ Geol (in review)

Duhig NC, Davidson GJ, Stolz J (1992) Microbial involvement in the formation of Cambrian sea-floor silica-iron oxide deposits, Australia.Geology 20:511-514

Ehrlich HL (1996) Geomicrobiology. Marcel Dekker, New York, 717 p

Ehrlich HL (1999) Microbes as geologic agents: Their role in mineral formation. Geomicrobiol J 16: 135-153

El Goresy A Chen M, Dubrovinsky L, Gillet P, Graup G (2001) An ultradense polymorph of rutile with seven-coordinated titanium from the Ries Crater. Science 293:1467-1470

Elliott P, Ragusa S, Catcheside D (1998) Growth of sulfate-reducing bacteria under acidic conditions in an upflow anaerobic bioreactor as a treatment system for acid mine drainage. Water Res 32:3724-3730

Emerson D, Moyer C (1997) Isolation and characterization of novel iron-oxidizing bacteria that grow at circumneutral pH. Appl Environ Microbiol 63:4784-4799

Ennas G, Marongiu G, Musinu A, Falqui A, Ballirano P, Caminiti R (1999) Characterization of nanocrystalline γ-Fe_2O_3 prepared by wet chemical method. J Mater Res 14:1570-1575

FACT database on Web: Facility for the Analysis of Chemical Thermodynamics, at http://www.crct.polymt1.ca/fact/fact.htm

Fortin D, Ferris FG, Beveridge TJ (1997) Surface-mediated mineral development by bacteria. Rev Mineral 35:161-180

Fowle DA, Druchel GK, Thomsen-Ebert T, Welch SA, Banfield JF (in prep.) Microbial controls on trace metal cycling: Linkages with the iron cycle. *In* The Biogeochemistry of Iron Cycling in Natural Environments. J Coates, C Zhang (eds)

Fowle DA, Thomsen-Ebert T, Welch SA, Banfield JF (in prep.) Role of nanoscale biogenic iron oxyhydroxides on trace metal cycling.

Fowle DA, Kemner K, Kelly S, Thomsen-Ebert T, Banfield JF (in prep.) Novel mechanisms for Zn incorporation by biogenic iron oxyhydroxides.

Fuller CC, Davis JA, Coston JA, Dixon E (1996) Characterization of metal adsorption variability in a sand and gravel aquifer, Cape Cod, Massachusetts, USA. J Contamin Hydrol 22:165-187

Furer G (1993) New aspects on the chemistry of aluminum in soils. Aquatic Sci 55:281-290

Ganapathi SK, Owen DM, Chokshi AH (1991) The kinetics of grain growth in nanocrystalline copper. Scr Metall Mater 25:2699-2704

Gennari FC, Pasquevich DM (1998) Kinetics of the anatase-rutile transformation in TiO_2 in the presence of Fe_2O_3. J Mater Sci 33:1571-1578

Glaeser AM (2001) Grain Growth. *In* Encyclopedia of Materials Science and Technology. KHJ Buschow, RW Cahn, MC Flemings. B Ilschner, EJ Kramer, S Mahajan (eds) Elsevier Science, New York

Gloriant T, Gich M, Surinach S, Baro MD, Greer AL (2000) Evaluation of the volume fraction crystallized during devitrification of Al-based amorphous alloys. Mater Sci Forum 343: 365-370.

Gravie RC (1978) Stabilization of the tetragonal structure in zirconia microcrystals. J Phys Chem 82: 218-224

Gribb AA, Banfield JF (1997) Particle size effects on transformation kinetics and phase stability in nanocrystalline TiO_2. Am Mineral 82:717-728

Giudici-Orticoni MT, Leroy G, Nitschke W, Bruschi M (2000) Characterization of a new dihemic c(4)-type cytochrome isolated from *Thiobacillus ferrooxidans*. Biochem 39:7205-7211

Gustafsson JP (2001) The surface chemistry of imogolite. Clays Clay Minerals 49:73-80

Guthrie GD, Bish DL, Reynolds RC (1995) Modeling the X-ray diffraction pattern of opal-CT. Am Mineral 80:869-872

Hallbeck L, Pedersen K (1991) Autotrophic and mixotrophic growth of *Gallionella ferruginea*. J Gen Microbiol 137:2657-2661

Hart A, Murrell JC, Poole RK, Norris PR (1991) An acid-stable cytochrome in iron-oxidizing *Leptospirillum ferrooxidans*. FEMS Micro Lett 81:89-94

Harrison PM, Dutriazac JE (1998) Ferric oxyhydroxide core of ferritin. Nature 216:1188-1190

Heald EF, Weiss CW (1972) Kinetics and mechanism of the anatase/rutile transformation, as catalyzed by ferric oxide and reducing conditions. Am Mineral 57:10-23

Henmi T, Wada K (1976) Morphology and composition of allophane. Am Mineral 61:379-390

Herron N, Calabrese JC, Farneth WE, Wang Y(1993) Crystal structur eand optical properties of $Cd_{32}S_{14}(SC_6H_5)_{36}.DMF_4$, a cluster with a 15-Å CdS core. Science 259:1426-1428

Hishita S, Mutoh I, Koumoto K, Yanagida H (1983) Inhibition mechanism of the anatase-rutile phase transformation by rare earth oxides. Ceram Int'l 9:61-67

Hofmann BA, Farmer JD (2000) Filamentous fabrics in low-temperature mineral assemblages: Are they fossil biomarkers? Implications for the search for a subsurface fossil record on the early Earth and Mars. Planet Space Sci 48:1077-1086

Huang F, Zhang H, Banfield JF (in prep.) Two-step crystal-growth kinetics observed in hydrothermal coarsening of nanocrystalline ZnS.

Hwang S-L, Shen P, Chu HT, Yui T-F (2000) Nanometer-size α-PbO_2-type TiO_2 in garnet: A thermobarometer for ultrahigh-pressure metamorphism. Science 288:231-324

Ibach H (1997) The role of surface stress in reconstruction, epitaxial growth and stabilization of mesoscopic structures. Surface Sci Report 29:193-263

Illekova E, Czomorova K, Kuhnast F, Fiorani J (1996) Transformation kinetics of the $Fe_{73.5}Cu_1Nb_3Si_{13.5}B_9$ ribbons to the nanocrystalline state. Mater Sci Eng A 205:166-179

Jamieson JC, Olinger B (1968) High-pressure polymorphism of titanium dioxide. Science 161:893-895

Jamieson JC, Olinger B (1969) Pressure-temperature studies of anatase, brookite, rutile, and TiO_2 (II): A discussion. Am Mineral 54:1477-1482

Janney DE, Cowley JM, Buseck PR (2000a) Transmission electron microscopy of synthetic 2-and 6-line ferrihydrite. Clays Clay Minerals 48:111-119

Janney DE, Cowley JM, Buseck PR (2000b) Structure of synthetic 2-line ferrihydrite by electron nanodiffraction. Am Mineral 85:1180-1187

Janney DE, Cowley JM, Buseck PR (2001) Structure of synthetic 6-line ferrihydrite by electron nanodiffraction. Am Mineral 86:327-335

Jedlicki E, Reyes R, Jordana X, Mercereau-Pujalon O, Allende JE (1986) Rusticyanin: Initial studies on the regulation and its synthesis and gene isolation. Biotech Appl Biochem 8:342-350

Joesten RL (1991) Kinetics of coarsening and diffusion-controlled mineral growth. Rev Mineral 26: 507-582

Jones JB, Biddle J, Segnit ER (1966) Opal genesis. Nature 210:1353-1355

Kawasaki S, Yamanaka T, Kume S, Ashida T (1990) Crystallite size effect on the pressur-induced phase transformation of ZrO_2. Solid State Comm 76:527-530

Kerrick DM (1990) Aluminosilicates. Rev Mineral 22, 207 p

Krill CE III, Helfen L, Michels D, Natter H, Fitch A, Masson O, Birringer R (2001) Size-dependent grain-growth kinetics observed in nanocrystalline Fe. Phys Rev Lett 86:842-845

Kolmert A, Johnson DB (2001) Remediation of acidic waste waters using immobilised, acidophilic sulfate-reducing bacteria. J Chem Technol Biotechnol 76:836-843

Kumar KNP, Keizer K, Burggraaf AJ (1993) Textural evolution and phase transformation in titania membranes: Part 1: Unsupported membranes. J Mater Chem 3:1141-1149

Kuo LY, Shen PY (2000) On the condensation and preferred orientation of TiC nanocrystals: Effects of electric field, substrate temperature and second phase. Mater Sci Engin A-Struct Mater 276:99-107

Labrenz M, Druschel GK, Thomsen-Ebert T, Gilbert B, Welch SA, Kemner K, Logan, GA, Summons R, De Stasio G, Bond PL, Lai B, Kelley SD, Banfield JF (2000) Natural formation of sphalerite (ZnS) by sulfate-reducing bacteria. Science 290:1744-1747

Langmuir D (1971) Particle size effect on the reaction goethite = hematite + water. Am J Sci 271:147-156

Lee WH, Shen PY (1999) On the coalescence and twinning of cube-octahedral CeO_2 condensates. J Crystal Growth 205:169-176

Liao S-C, Chen Y-J, Kear BH, Mayo WE (1998) High pressure/low temperature sintering of nano-crystalline alumina. Nanostruct Mater 10:1063-1079

Liao S-C, Chen Y-J, Mayo WE, Kear BH (1999) Transformation-assisted consolidation of bulk nanocrystalline TiO_2. Nanostruct Mater 11:553-557

Little CTS, Herrington RJ, Haymon RM, Danelian T (1999) Early Jurassic hydrothermal vent community from the Franciscan Complex, San Rafael Mountains, California. Geology 27:167-170

Lovley DR (1987) Organic-matter mineralization with the reduction of ferric iron: A review. Geomicrobiol J 5:375-399

Lu K (1996) Nanocrystalline metalscrystallized from amorphous solids: Nanocrystallization, structure, and properties. Mater Sci Engin R 16:161-221

Lu K, Wang JT (1991) Activation energies for crystal nucleation and growth in amorphous alloys. Mater Sci Engin A133:500-503

Luck R, Lu K, Dong ZF (1996) Magnetothermal analysis of the amorphous to nanocrystal transformation. J Non-crystal Solids 205-207, 811-814.

Luck R, Lu K, Frantz W (1993) JMA analysis of the transformation kinetics from the amorphous to the nanocrystalline state. Scr Metall Mater 28:1071-1075

Luther GW, Theberge SM, Rickard DT (1999) Evidence for aqueous clusters as intermediates during zinc sulfide formation. Geochim Cosmochim Acta 64:579-579

MacKenzie KJD (1975) The calcination of titania, V. Kinetics and mechanism of the anatase-rutile transformation in the presence of additives. Trans J British Ceram Soc 74:77-87

Malek J, Mitsuhashi T, Ramirez-Castellanos J, Matsui Y (1999) Calorimetric and high-resolution transmission electron microscopy study of nanocrystallization in zirconia gel. J Mater Res 14: 1834-1843

Malm JO, O'Keefe MA (1996) Deceptive 'lattice spacings' in high-resolution micrographs of metal nanoparticles. Ultramicroscopy 68:13-23

Malow TR, Koch CC (1996) Thermal stability of nanocrystalline materials. Mater Sci Forum 225-227: 595-604

Manceau A, Charlet L (1992) X-ray absorption spectroscopic study of the sorption of Cr(III) at the oxide/water interface. I Molecular mechanism of Cr(III)oxidation on Mn oxides. J Colloid Interface Sci 148:443-458

Manceau A, Charlet L (1994) The Mechanism of Selenate Adsorption on Goethite and Hydrous Ferric Oxide. J Colloid Interface Sci 168:87-93

Manceau A (1995) The mechanism of anion adsorption on Fe oxides: Evidence for the bonding of arsenate tetrahedra on free $Fe(O,OH)_6$ edges. Geochim Cosmochim Acta 59:3647-3653

Matxain JM, Fowler JE, Ugalde JM (2000) Small clusters of II-VI materials: Zn_iO_i. Phys Rev A 6205:53201-1-8

McHale JM, Auroux A, Perrotta AJ, Navrotsky A (1997) Surface energies and thermodynamic phase stability in nanocrystalline aluminas. Science 277:788-791

Missana T, Afonso CN, Petford-Long AK, Doole RC (1999) Amorphous-to-nanocrystalline transformation kinetics in SbO_x films. Philos Mag A 79:2577-2590

Mitsuhashi T, Kleppa OJ (1979) Transformation enthalpies of the TiO_2 polymorphs. J Am Ceram Soc 62:356-357

Moore DM, Reynolds RC Jr (1989) X-ray Diffraction and the Identification and Analysis of Clay Minerals. Oxford University Press, New York, 332 p

Muller E, Vogelsberger W, Fritsche HG (1988) The dependence of the surface energy of regular cluster and small crystallites on the particle size. Cryst Res Technol 23:1153-1159

Murakami T, Ohnuki T, Isobe H, Sato T (1997) Mobility of uranium during weathering. Am Mineral 82:888-899

Murray CB, Kagan CR, Bawendi MG (1995) Self-organization of CdSe nanocrystallites into three-dimensional quantum dot superlatticles. Science 270:1335-1338

Nadeau PH, Wilson MJ, McHardy WJ, Tait JM (1984) Interstratified clays as fundamental particles. Science 225:923-925

Navrotsky A, Kleppa OJ (1967) Enthalpy of the anatase-rutile transformation. J Am Ceram Soc 50:626.

Nealson KH, Stahl DA (1997) Microorganisms and biogeochemical cycles: What can we learn from layered microbial communities? Rev Mineral 35:5-34

Needs RJ, Godfrey MJ, Mansfield M (1991) Theory of surface stress and surface reconstruction. Surface Sci 242:215-221

Nelson YM, Lion LW, Ghiorse WC, Shuler ML (1999) Production of biogenic Mn oxides by *Leptothrix discophora* SS-1 in a chemically defined growth medium and evaluation of their Pb adsorption characteristics. Appl Environ Microbiol 65:175-180

Nicolaus, MM, Sinning HR, Haessner F (1992) Crystallization behavior and generation of a nanocrystalline state from amorphous cobalt-zirconium ($Co_{33}Zr_{67}$). Mater Sci Engin A 150:101-112

Nielson AE (1964) Kinetics of Precipitation. Pergamon Press, New York.151 p.

Nikolakis V, Kokkoli E, Tirrell M, Tsapatsis M, Vlachos DG (2000) Zeolite growth by addition of subcolloidal particles: Modeling and experimental validation. Chem Mater 12:845-853

Nunzi F, Woudstra M, Campese D, Bonicel J, Morin D, Bruschi M (1993) Amino acid sequence of rusticyanin from *Thiobacillus ferrooxidans* and its comparison with other blue copper proteins. Biochim Biophys Acta 1162:28-34

Olsen JS, Gerward L, Jiang JZ (1999) On the rutile/alpha-PbO_2 type phase boundary of TiO_2. J Phys Chem Solids 60:229-233

Occelli ML, Kessler H (1998) Synthesis of Porous Materials: Zeolites, Clays, and Nanostructures. Chemical Industries, Vol 69, 718 p

Ohtaka O, Yamanaka T, Kume S, Ito E, Navrotsky A (1991) Stability of monoclinic and orthorhombic zirconia: Studies by high-pressure equilibria and calorimetry. J Am Ceram Soc 74:505-509
Parker A, Rae JE (1998) Environmental Interactions of Clays. Clays and the Environment. Springer, New York, 271 p
Pedersen K 1997 Microbial life in deep granitic rock. FEMS Microbiol Rev 20:399-414
Pedersen K 1999 Subterranean microorganisms and radioactive waste disposal in Sweden Engin Geol 52:163-176
Penn RL, Banfield JF, Kerrick, DM (1999) TEM investigation of Lewiston, Idaho Fibrolite: Implications for grain boundary energetics and Al_2SiO_5 phase relations. Am Mineral 84:152-159
Penn RL, Banfield JF (1998a) Imperfect oriented attachment: Dislocation generation in defect-free nanocrystals. Science 281:969-971
Penn RL, Banfield JF (1998b) Oriented attachment and growth, twinning, polytypism, and formation of metastable phases: Insights from nanocrystalline TiO_2. Am Mineral 83:1077-1082
Penn RL, Banfield JF (1999) Formation of rutile nuclei at anatase {112} twin interfaces and the phase transformation mechanism in nanocrystalline titania. Am Mineral 84:871-876
Penn RL, Banfield JF (1999) Morphology development and crystal growth in nanocrystalline aggregates under hydrothermal conditions: Insights from titania. Geochim Cosmochim Acta 63:1549-1557
Penn RL, Oskam G, Strathmann GJ, Searson PC, Stone AT, Veblen DR (2001) Epitaxial assembly in aged colloids. J Phys Chem B 105:2177-2182
Postgate JR (1965) Recent advances in the study of sulfate-reducing bacteria. Bacteriol Rev 30:732-738
Privman V, Goia DV, Park J, Matijevic E (1999) Mechanism of formation of monodispersed colloids by aggregation of nanosize precursors. J Colloid Interface Sci 213:36-45
Putnis A (1992) Introduction to mineral sciences. Cambridge University Press, Cambridge, UK, 457 p
Rao CNR (1961) Kinetics and thermodynamics of the crystal structure transformation of spetroscopically pure anatase to rutile. Canadian J Chem 39:498-500
Raviv U, Laurant P, Klein J (2001) Fluidity of water confined to subnanometre films. Nature 413:52-54
Rawlings DE (2001) The molecular genetics of *Thiobacillus ferrooxidans* and other mesophilic, acidophilic, chemolithotropic, iron- or sulfur-oxidizing bacteria. Hydrometall 59:197-201
Ronk M, Shively JE, Shute EA, Blake RC (1991) Amino acid sequence of the blue copper protein rusticyanin from *Thiobacillus ferrooxidans*. Biochem 30:9435-9442
Rozan TF, Lassman ME, Ridge DP, Luther GW (2000) Evidence for iron, copper and zinc complexation as multinuclear sulphide clusters in oxic rivers. Nature 406:879-882
Sahm K, MacGregor BJ, Jorgensen BB, Stahl DA (1999) Sulphate reduction and vertical distribution of sulphate-reducing bacteria quantified by rRNA slot-blot hybridization in a coastal marine sediment. Environ Microbiol 1:65-74
Saunders JA, Schoenly PA (1995) Boiling, colloiddal nucleation and aggregation, and the genesis of Bonanza Au-Ag ores of the Sleeper Deposit, Nevada. Mineral Deposita 30:199-210
Schäfer G, Engelhard M, Müller V (1999) Bioenergetics of the Archea. Microbiol Molec Biol Rev 63: 570-620
Schimanke G, Martin M (2000) *In situ* XRD study of the phase transition of nanocrystalline maghemite (γ-Fe_2O_3) to hematite (α-Fe_2O_3). Solid State Ionics 136:1235-1240
Schmidt U, Eisenschmidt C, Vieweger T, Zahra CY, Zahra A-M (2000) Crystallization of amorphous AlDy- and AlDyCo-alloys. J Non-Crystal Solids 271:29-44
Schroeer D, Nininger RC Jr (1967) Morin transformation in α–Fe_2O_3 microcrystals. Phys Rev Lett 19: 632-634
Schultzlam S, Beveridge TJ (1994) Physicochemical characteristics of mineral-forming S-layer from the cyanobacterium *Synechococcus* strain GL24. Canadian J Microbiol 40:216-223
Shannon RD, Pask JA (1965) Kinetics of the anatase-rutile transformation. J Am Ceram Soc 48:391-398
Shen P, Lee WH (2001) (111)-specific coalescence twinning and martensitic transformation of tetragonal ZrO_2 condensates. Nanoletters (in press) (web release)
Shuttleworth R (1950) The surface tension of solids. Proc Phys Soc London A63:444-457
Simons PY, Dachille F (1967) The structure of TiO_2-II, a high pressure phase of TiO_2. Acta Crystallogr 23:334-336
Sleytr UB, Beveridge TJ (1999) Bacterial S-layers. Trends Microbiol :253-260
Smith KL, Eggleton RA (1983) Botryoidal goethite: A transmission electron microscope study. Clays Clay Minerals 31:392-396
Spadini L, Manceau A, Schindler PW, Charlet L (1994) Structure and Stability of Cd^{2+} Surface Complexes on Ferric Oxides. I. Results from EXAFS Spectroscopy. J Colloid Interface Sci 168:73-86
Steefel CI, van Cappellen P (1990) A new kinetic approach to modeling water-rock interaction: Role of nucleation, precursors, and Ostwald ripening. Geochim Cosmochim Acta 54:2657-2677

Stern LA, Durham WB, Kirby SH (1997) Grain-size-induced weakening of H_2O ices I and II and associated anisotropic recrystallization. J Geophys Res-Solid Earth 102:5313-5325

Suzuki A, Kotera Y (1962) The kinetics of the transition of titanium dioxide. Bull Chem Soc Japan 35:1353-1357

Suzuki A, Tukuda R (1969) Kinetics of the transition of titanium dioxide prepared by sulfate process and chloride process. Bull Chem Soc Japan 42:1853-1957

Suzuki Y, Kelly S, Kemner K, Banfield JF (2001) Microbes make ~2-nanometer diameter crystalline UO_2 particles (abstr) EOS Trans Am Geophys Union, San Francisco, CA, December 2001 (abstr in press)

Suzuki Y, Banfield JF (2001) Aerobic microbial interactions with uranium: Resistance and accumulation (abstr) Proc 11th Goldschmidt Geochemistry Meeting, May 2001

Tebo BM, Ghiorse WC, vanWaasbergen LG, Siering PL, Caspi R (1997) Bacterially mediated mineral formation: Insights into manganese(II) oxidation from molecular genetic and biochemical studies. Rev Mineral 35:225-266

Theberge SM, Luther GW, Rozan TF, Rickard DT (2000) Evidence for aqueous clusters as intermediates during copper sulfide formation. Abstr Am Chem Soc 220:353

Thompson DN, Sayer RL, Noah KS (2000) Sawdust-supported passive bioremediation of western United States acid rock drainage in engineered wetland systems. Minerals Metall Process 17:96-104

Tolman RC (1966) Consideration of Gibbs theory of surface tension. J Chem Phys 16:758-774

Tolman RC (1949) The effect of droplet size on surface tension. J Chem Phys 17:333-337

Tomino H, Kusaka I, Nishioka K, Takai T (1991) Interfacial tension for small nuclei in binary nucleation. J Crystal Growth 113:633-636

Torres-Martinez CL, Nguyen L, Kho R (1999) Biomolecularly capped uniformly sized nanocrystalline materials: Glutathione-capped ZnS nanocrystals. Nanotechnol 10:340-345

Trudinger PA, Lambert IB, Skyring GW (1972) Biogenic sulfide ores: A feasibility study. Econ Geol 67:1114-1127

Tuhela L, Carlson L, Tuovinen OH (1997) Biogeochemical transformations of Fe and Mn in oxic groundwater and well water environments. J Environ Sci Health, A. Environ Sci Engin Toxic Hazardous Substance Control 32:407-426

Varga LK, Bakos E, Kiss LF, Bakonyi I (1994) The kinetics of amorphous-nanocrystalline transformation for a Finemet alloy. Mater Sci Engin A 179:567-571

Verges MA, Mifsud A, Serna CJ (1990) Formation of rod-like zinc-oxide microcrystals in homogenous solutions. J Chem Soc Faraday Trans 86:959-963

Vogel W, Urban J (1997) Sphalerite-Wurtzite Intermediates in Nanocrystalline CdS. Langmuir 13:827-832

Wada SI, Wada K (1977) Density and structure of allophane. Clay Mineral. 12:289-298

Wagner M, Roger AJ, Flax JL, Brusseau GA, Stahl DA (1998) Phylogeny of dissimilatory sulfite reductases supports an early origin of sulfate respiration. J Bacteriol 180:2975-2982

Wagner P, Terasaki O, Ritsch S, Nery JG, Zones SI, Davis ME, Hiraga K (1999) Electron diffraction structure solution of a nanocrystalline zeolite at atomic resolution. J Phys Chem B 103:8245-8250

Waite TD, Davis JA, PayneTE, Waychunas GA, Xu N (1994) Uranium (VI) adsorpton to ferrihydrite—Application of a surface complexation model. Geochim Cosmochim Acta 58:5465-5478

Walter RL, Ealick SE, Friedman AM, Blake RC, Proctor P, Shoham M (1996) Multiple wavelength anomalous diffraction (MAD) crystal structure of rusticyanin: A highly oxidizing cupredoxin with extreme acid stability. J Molec Biol 263:730-751

Wang CC, Ying JY (1999) Sol-gel synthesis and hydrothermal processing of anatase and rutile titania nanocrystals. Chem Mater 11:3113-3120

Wang Z, Saxena SK (2001) Raman spectroscopic study on pressure-induced amorphization in nanocrystalline anatase (TiO_2). Solid State Comm 118:75-78

Waychunas GA, Rea BA, Fuller CC, Davis JA (1993) Surface chemistry of ferrihydrite. 1. EXAFS studies of the geometry of coprecipitated and adsorbed arsenate. Geochim Cosmochim Acta 57:2251-2269

Waychunas GA, Myneni SCB, Traina SC, Bigham JM, Fuller CC, Davis JA (2001) Reanalysis of the schwertmannite structure and the incorporation of SO_4^{2-} groups: An IR, XAS, WAXS and simulation study. (abstr) Proc 11th Goldschmidt Conf, Hot Springs, Virginia, May 2001

Weinberg MC (1991) Surface nucleated transformation kinetics in 2-and 3-demensional finite systems. J Non-crystal Solids 134:116-122

Weinberg MC, Birnie DP III, Shneidman VA (1997) Crystallization kinetics and the JMAK equation. J Non-crystal Solids 219:89-99

Wen HL, Yen SF (2000) Growth characteristics of boehmite-derived ultrafine theta and alpha-alumina particles during phase transformaton. J Crystal Growth 208:696-708

Winterer M, Nitsche R, Redfern SAT, Schmahl WW, Hahn H (1995) Phase stability in nanostructured and coarse grained zirconia at high pressures. Nanostruc Mater 5:679-688

Wong EM, Bonevich JE, Searson PC (1998) Growth kinetics of nanocrystalline ZnO particles from colloidal suspensions. J Phys Chem B 102:7770-7775
Yamanaka T, Fukumori Y (1995) Molecular aspects of the electron transfer system which participates in the oxidation of iron by *Thiobacillus ferrooxidans*. FEMS Microbiol Rev 17:401-413
Yanagisawa K, Ovenstone J (1999) Crystallization of anatase from amorphous titania using the hydrothermal technique: Effects of starting material and temperature. J Phys Chem B 103:7781-87
Yang J, Mei S, Ferreira JMF (2000) Hydrothermal synthesis of nanosized titania powders: Influence of peptization and peptizing agents on the crystalline phases and phase transition. J Am Ceram Soc 83:1361-1368
Zhang H, Banfield JF (1998) Thermodynamic analysis of phase stability of nanocrystalline titania. J Mater Chem 8:2073-2076
Zhang H, Banfield JF (1999) New kinetic model for the nanocrystalline anatase-to-rutile transformation revealing rate dependence on number of particles. Am Mineral 84:528-535
Zhang H, Banfield JF (2000) Understanding polymorphic phase transformation behavior during growth of nanocrystalline aggregates: Insights from TiO_2. J Phys Chem B 104:3481-3487
Zhang H, Banfield JF (2000) Phase transformation of nanocrystalline anatase-to-rutile via combined interface and surface nucleation. J Mater Res 15:437-448
Zhang H, Penn RL, Hamers RJ, Banfield JF (1999) Enhanced adsorption of molecules on surfaces of nanocrystalline particles. J Phys Chem B 103:4656-4662
Zhang H, Finnegan M, Banfield JF (2001) Preparing single-phase nanocrystalline anatase from amorphous titania with particle sizes tailored by temperature. Nano Letters 1:81-85
Zhu HL, Averback RS (1996) Sintering of nano-particle powders: Simulations and experiments. Mater Manufactur Process 11:905-923

Nanocrystals as Model Systems for Pressure-Induced Structural Phase Transitions

Keren Jacobs and A. Paul Alivisatos

Department of Chemistry
University of California–Berkeley
Berkeley, California 94720

INTRODUCTION

An understanding of first-order pressure-induced structural transitions is relevant to many research areas in geological and planetary sciences because it involves the study of materials exposed to high pressures. For example, solid-solid transitions in silicates are responsible for the seismic discontinuities in the earth's mantle (Chudinovskikh and Boehler 2001) and may play a role in plate tectonics and deep earthquakes (Kirby et al. 1991). In geological applications, models of structural transition kinetics simulate rock formation taking place over millions of years (Shekar and Rajan 2001). Despite their importance in earth science applications, the microscopic processes of solid-solid phase transitions are difficult to study in the bulk solid for several reasons (Putnis 1992). In extended solids, the transformation nucleates at defects, which are present at equilibrium even in the highest quality crystals. As a transformed region of the crystal grows larger, mechanical forces generate new defects, which in turn act as new nucleation sites. The difficulties in bulk are compounded by the irreversibility of the kinetics, which depends strongly on the preparation of the sample and its history. The study of first-order phase transitions can be greatly simplified in nanocrystal systems because small crystals can behave as single structural domains and reproducibly cycle through multiple transitions (Wickham et al. 2000). To illustrate the advantages of the simple kinetics in nanoscale systems in exploring fundamental questions of structural phase transitions, this chapter focuses on the CdSe nanocrystal system. Nanocrystalline solid-solid transitions in geologically relevant material such as Fe_2O_3 nanocrystals are also now being studied (Rockenberger et al. 1999), and a comparable understanding of these materials is a current research goal.

The CdSe nanocrystals are a particularly well-controlled and characterized test system (Tolbert et al. 1995). The nanocrystals consist of hundreds to thousands of covalently bonded atoms, synthesized as nearly defect-free crystals with narrow size distribution and controlled shape. At low pressure, the atoms are tetrahedrally bonded and form a hexagonal network of stacked sheets with wurtzite and zinc blende (its mineral name is sphalerite) stacking sequences. At high pressures, the nanocrystals contract by 18% as they transform to a denser six-coordinate rocksalt structure. As in most inorganic solid-solid transitions, there is a substantial barrier to the structural transition at room temperature. To overcome the energetic barrier, the pressure must be raised beyond the thermodynamic transition pressure in the forward transformation, and lowered in the reverse. The result is an experimental hysteresis loop, which reflects the time needed for the system to relax to the equilibrium configuration. In the bulk solid, this barrier is usually related to the fragmentation of the bulk solid into multiple domains during the transition. Nanocrystals are substantially smaller than these domains and are essentially defect free, so that the barrier arises from the microscopic motions of atomic reorganizations. The area of the hysteresis loop provides an initial idea of the energies involved in the transformation process, because it represents the irreversible energy

1529-6466/00/0044-0002$05.00

released in a PV transition cycle. For nanoparticles with a hysteresis width of ~6 GPa, the energy release is estimated to be several orders of magnitude larger than thermal energy at room temperature. This is an indication that in structural transformations, the energy barrier is much larger than the thermal energy accessible in most laboratory experiments.

SINGLE-STRUCTURAL DOMAIN

Experiments on nanocrystal transitions are usually done on ensembles of nearly monodisperse nanocrystals coated with a surfactant allowing them to be dissolved in a solvent (Murray et al. 1993, Peng et al. 1998). The solvent, such as ethylcyclohexane, also acts as the pressure-transmitting medium. The nanocrystals are spatially isolated and separated, and ideally each nanocrystal transforms independently of the others. Pressures are applied using a diamond anvil cell (DAC) and are measured using standard ruby florescence techniques (Barnett et al. 1973) (see Fig. 1). Static temperature can be achieved by placing the cell in a resistively heated ceramic oven, and measured with a thermocouple in contact with the diamond. In the CdSe nanocrystal system, a static temperature limit of 550 K exists because at higher temperatures interparticle diffusion can occur. This limit is close to the synthesis temperature, but should in general depend on the capping group and solvent, and can be determined by the narrowing of X-ray diffraction peaks signifying crystal growth and aggregation.

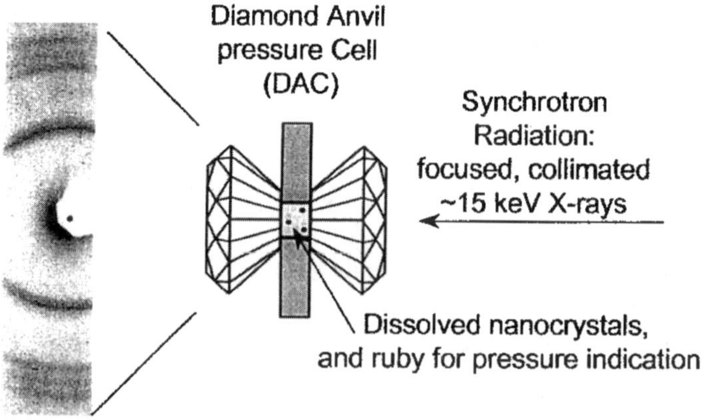

Figure 1. Schematic of an experimental setup for high-pressure X-ray diffraction (XRD) experiments on nanocrystals under pressure of tens of GPa (1 GigaPascal ≅ 10,000 atmospheres). Nanocrystal powder is dissolved in a pressure medium and placed between two opposing diamonds, which are clear to X-rays and visible absorption. The pressure is applied by bringing the diamond closer together, and measured using pieces of ruby chips placed inside the pressure cell. The XRD diffraction peaks are Debye-Scherrer broadened by the finite size of the particles, yielding information on the shape and size of the nanocrystals before and after the transition, such as shown in Figure 3.

The structural transition is directly monitored using X-ray diffraction (XRD) (see Fig. 1), by observing the disappearance and appearance of the characteristic peaks. The limits of this method are mainly from the small sample size of ~100 microns from a solution of a few nanoliters, compressed between two diamonds of ~1-mm thickness. The small physical size of the nanocrystal produces Debye-Scherrer broadening of the diffraction peaks (Cullity 1956), and very concentrated samples are required (optical density of 3-4) for sufficient signal-to-noise. Since XRD requires the focused, bright X-rays available only from a synchrotron light source, a more practical method is often desired. The transitions can often be correlated to optical or vibrational changes in UV-vis or Raman spectroscopy. Samples prepared for these methods are at much lower concentrations (optical density of 0.1-1), allowing nanocrystals to be separated by several lengths.

Several independent observations have led to the conclusion that nanocrystals, at least for the CdSe system, transform as a single structural domain. Single domain behavior produces a macroscopic shape change of the nanocrystal during the transition, discussed in the next section; and is consistent with the observation of simple kinetics behavior, discussed in the Kinetics section. A direct method for observing the microscopic structure of nanocrystals is high-resolution transmission electron microscopy (HRTEM). Electron microscopy is routinely used in viewing samples after synthesis, but cannot currently be done inside the pressure cell. However, recovered samples can be retrieved from the pressure cell after they have undergone phase transition cycles and returned to the initial four-coordinate phase (Jacobs et al. 2001). It must be ensured that the recovered sample is from inside the pressure cell and is not residue from the initial loading.

Figure 2 shows HRTEM images of large representative particles before and after a transition cycle. The nearly isoenergetic wurtzite and zinc blende structures are arranged as stacks in the nanocrystal. The higher fraction of zinc blende stacks in the recovered sample is consistent with interpretations of X-ray diffraction simulations (Wickham et al. 2000). The stacking faults are a particularly simple type of grain boundary, but do not interrupt the connectivity of the atoms in the nanocrystal. Well-defined microscopic structures and volumes are involved in the nanocrystal transition, since the tetrahedral bonding between Cd and Se atoms remains uninterrupted even after pressure cycling.

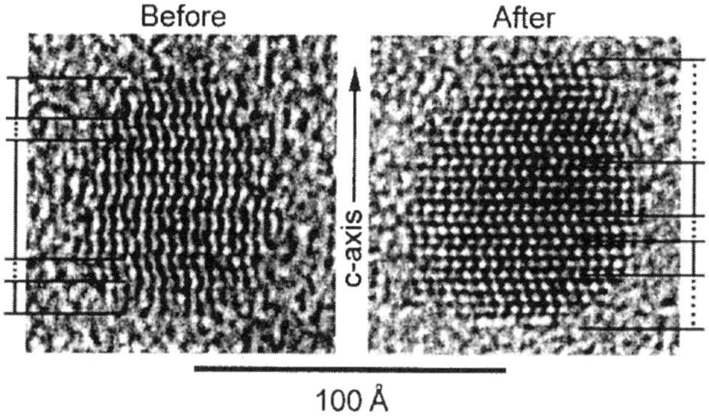

Figure 2. High-resolution transmission electron microscope image of representative CdSe nanocrystals in the four-coordinate structure before and after pressure cycling. The arrow indicates the unique *c*-axis, along which (001) layers are stacked. Two possible stacking sequences, ABAB- and ABCABC-, yield the nearly isoenergetic wurtzite (hexagonally close-packed) and zinc blende (cubic close-packed—f.c.c.) structures within the nanocrystal. Wurtzite stacking is indicated with solid lines; zinc blende stacking with dashed lines. After pressure cycling, the nominally spherical shape recovers and the nanocrystal has an increased fraction of zinc blende layers. The stacking faults do not interrupt the connectivity of the atoms in the nanocrystal [Used by permission of the editor of *Science*, from Jacobs et al (2001), Fig. 1].

SURFACE EFFECTS AND SHAPE CHANGE

The surface energy can play a dominant role in determining the relative thermodynamic stability of two structural nanocrystal solids, as discussed in the chapters by Banfield and Zhang and Navrotsky (this volume). Differences in surface energy between two polymorphs are important in the study of nanocrystalline phase transitions. For example, in melting studies on a wide variety of nanocrystal materials, a depression in melting temperature is observed with decreasing size (Goldstein et al. 1992, Coombes 1972, Buffat and Borel 1976). The data is explained by the notion that the liquid phase is stabilized relative to the solid because surface energy contributions are minimized in liquids. This idea also explains the crossovers in thermodynamic stability at the nanoscale from one structure to another in Navrotsky and Zhang chapter (this volume). Smaller crystallites preferentially favor polymorphs with surfaces of relatively low energy, as illustrated in the nanocrysaline alumina systems (McHale et al. 1997). In solid-solid transitions in nanocrystals, the differences in surface energy between the two crystal structures produces a shift in the thermodynamic transition pressure as a function of nanocrystal size. For example, an observed shift to lower pressure with increasing nanocrystal size in CdSe nanocrystals is explained by higher surface energy in the high-pressure rocksalt structure (Tolbert et al. 1995; Jacobs, in preparation). High-energy surfaces (different from those in an annealed particle) are exposed as a result of the macroscopic shape change accompanying the transformation process, because the experimental temperatures used were too low for surface rearrangements to occur after the transformation. The resulting thermodynamic shift can be significant and should be taken into consideration in interpreting experimental transition pressures as a function of crystal size and in comparison to bulk.

The actual macroscopic shape change has been determined to our knowledge in only CdSe and Si nanoparticles. Experimental determination of shape changes is inferred from XRD patterns because the crystallographic lengths of the crystallite are given by the Debye-Scherrer broadening of diffraction peaks. Si nanoparticles of 500 Å in diameter were found to transform from spherically shaped to an ellipsoidal shape along the (001) axis (Tolbert et al. 1996). In the case of CdSe and InP and CdS, the high-pressure phase is cubic and each diffraction peak is a convolution of the different crystallographic axis. By simulating the XRD patterns, the shape change in CdSe was inferred from the relative widths and intensities in the XRD patterns before and after the transformation. The reason this method works is because the simulated patterns show strong sensitivity towards structural parameters, most importantly to the crystal shape. From anomalously high intensities of the 111 reflection in rocksalt, it was determined that the CdSe nanocrystals covert from a nominally spherically shaped four-coordinate crystal to a slab-shaped rocksalt structure with one set of (111) planes parallel to the long axis (Wickham et al. 2000) (see Fig. 3). The shape change exposes high-energy {111} faces in rocksalt and accounts for the relative instability of the rocksalt structure in small CdSe crystallites.

THERMODYNAMIC SMEARING

It has been implied that the surface of the nanocrystal acts as a perturbation of the total nanocrystal free energy. The assumption has been that, from a thermodynamics viewpoint, the nanocrystal behaves merely as a fragment of the extended solid. However, evidence suggests that structure throughout a nanoparticle in very small particle size may be perturbed relative to the bulk (see chapters by Banfield and Zhang and Waychunas, this volume). Although the thermodynamic properties of the nanocrystal ensemble are well defined, structural phase transitions are rigorously defined only in an infinite medium. A lower limit must exist for the number of atoms required for the crystal in

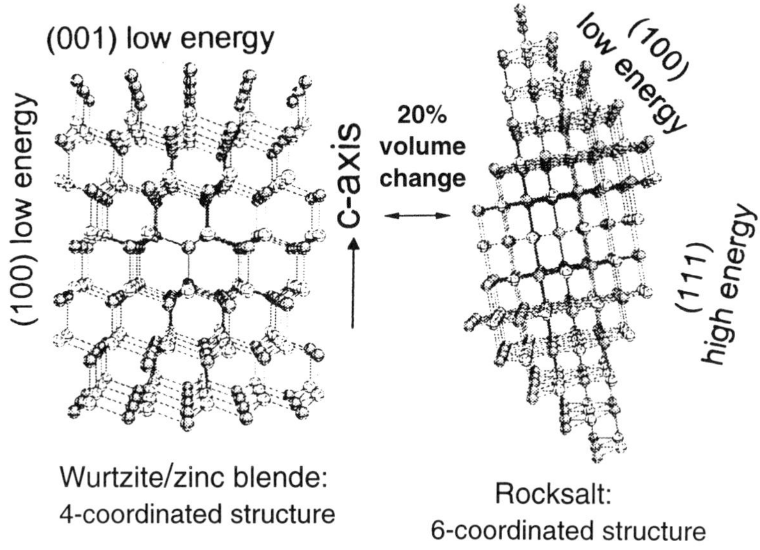

Figure 3. Shape change in CdSe nanocrystals found by comparing simulated powder X-ray diffraction (XRD) patterns to experimental data (Wickham et al. 2000).

practice to undergo a structural phase transition, rather than more closely resembling the limit of a molecular isomerization. A general discussion of this issue is presented from a theoretical standpoint in *Thermodynamics in Small Systems* (Hill 1963). We will outline a parallel interpretation and provide some experimental numbers (Jacobs et al. 2001) showing that CdSe nanocrystal particles contain enough atoms for the study of structural *phase* transition, despite their finite size.

In small systems, smearing of the thermodynamic transition pressure can arise from thermal fluctuations between the structures (see Fig. 4). In the smallest CdSe nanocrystals in the studies (25 Å in diameter, ~300 atoms), the total volume change during the transition is ~ 1500 Å^3. Because the differential of free energy with respect to pressure is volume, the corresponding thermal fluctuations are estimated to smear the nanocrystal transition by less than 0.1 GPa at 500 K (Jacobs et al. 2001). This smearing is insignificant compared to the kinetics-induced hysteresis width in CdSe nanocrystals of ~6 GPa, and implies that the transition is observed "far" from equilibrium even in nanocrystals. The insignificance of the smearing in sizes >20 Å should also apply to geophysically relevant nanoparticles such as Fe_2O_3, which although have smaller volume changes ~7%, have very large hysteresis >30 GPa. The simplest way to view the above calculation is in terms of the Gibbs equilibrium condition, more commonly used for molecular reactions:

$$\Delta G = -k_B T \ln(K) \tag{1}$$

where ΔG is the free energy difference between the two structural phases, k_BT is thermal energy (Boltmann's constant times temperature), and K is the equilibrium constant for the population ratio in the two phases at equilibrium. The transformation of reactants to products in molecular reactions occurs over a range of equilibrium conditions, because substantial thermal population of both phases is possible with finite free energy differences between the phases. Equation 1 is used to estimate the thermal smearing in the nanocrystal transition by substituting the free energy term with its pressure-volume

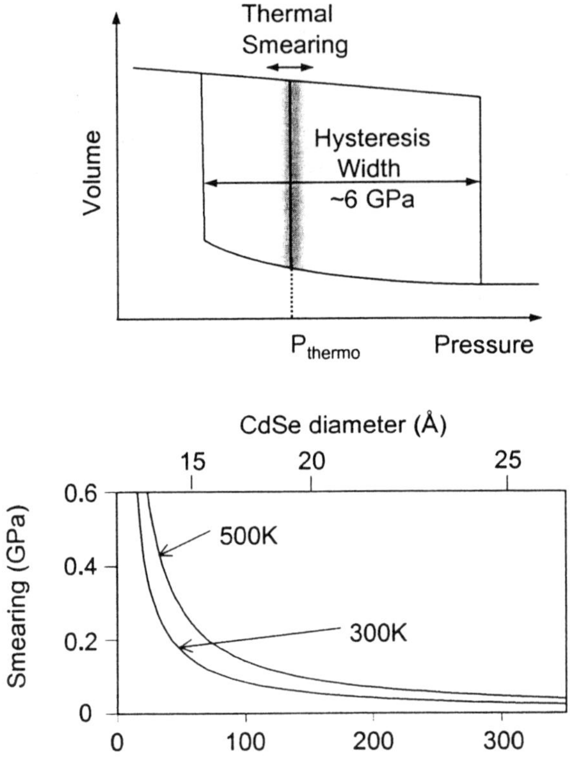

Figure 4. Schematic showing a hysteresis loop for the CdSe nanocrystals with the smearing of the thermodynamic transition pressure caused by the finite nature of the nanocrystal particle. The thermodynamic transition pressure is offset from the hysteresis center to emphasize that in first-order solid-solid transformations, this pressure is unlikely to be precisely centered. The lower plot shows the estimated smearing for CdSe nanocrystals as inversely proportional to the number of atoms in the crystal, at two temperatures, as discussed in the text. Note that nanocrystals are not ordinarily synthesized or studied in sizes smaller than 20 Å in diameter. This figure shows that this thermal smearing is insignificant compared to the large hysteresis width in the CdSe nanocrystals studied (25-130 Å in diameter), such that the transition is bulk-like from this perspective. This means that observed transformations occur at pressures far from equilibrium, where there is little probability of back reaction to the metastable state once a nanocrystal has transformed. In much smaller crystals or with larger temperatures, the smearing could become on the order of the hysteresis width, and the crystals would transform from one structure to the other at thermal equilibrium.

equivalent ($\partial\Delta G = \Delta V\partial\Delta P_t$, ΔV is extensive) and defining smearing as the pressure change relative to the thermodynamic transition pressure (ΔP_t) needed to convert 90% of the nanocrystal population to the thermodynamically stable phase. With this substitution, the smearing scales inversely with the number of atoms in the particle (proportional to the absolute volume change) and increases with thermal energy. In a bulk phase transition, the (extensive) free energy difference diverges except at a single pressure where there is equal population of both phases ($K = 1$). The estimate for minimal smearing in the CdSe nanocrystal transition is consistent with the large energies released in a hysteresis cycle at room temperature, mentioned in the Introduction. If a crystal were still definable for smaller sizes where the smearing is appreciable relative to the hysteresis, the structures would flip back and forth at thermal equilibrium. Structural fluctuations of this type are thought to occur in some gas-phase inorganic cluster species (Berry and Smirnov 2000).

TRANSFORMATION KINETICS: ACTIVATION VOLUME

The hysteresis loop is considered a dynamic property because the kinetics of activated processes should depend on the observation time. The limit of long observation times is important in geophysics for simulating processes occurring over geological time scales. Kinetics experiments are the most important tool in this respect because the study of hysteretic transitions is otherwise limited to measurements taken under non-equilibrium conditions.

Elementary transition state theory provides a general framework for investigating simple chemical reactions and transformations. Transition state theory is based on assigning quasi-equilibrium variables to a transition state with the highest energy along the mechanistic pathway. The most relevant parameter for pressure-induced transitions is the activation volume, or the volume change between the transition state and the reactant. This parameter gives information on the structural changes that occur during a transition, and so is particularly useful for reactions involving significant atomic rearrangements. Experimentally, the activation volume, $\Delta V^{\ddagger}$, is determined from the pressure-dependence of the rate constant:

$$\left(\frac{\partial \ln(k)}{\partial P}\right)_T = \frac{-\Delta V^{\ddagger}}{k_B T} \tag{2}$$

where $k_B T$ is the thermal energy, P is pressure, and k is the rate constant for the relaxation. This expression is the thermodynamic expression for Gibbs free energy with the Arrhenius rate law, and was first used by Evans and Polyani (1935) to explain the intrinsic pressure-dependence of rate constants for activated molecular processes. The activation volume can have an appreciable effect on transition kinetics, as even small microscopic deformations on the order of ~Å^3 can change the rate constant by several orders of magnitude over relatively small pressure changes (<1 GPa). The activation volume is routinely measured for molecular processes, although very little is known about this parameter in structural phase transitions. Since there is only one transformation event per nanocrystal, the activation volume is simpler to interpret in nanosystems than in extended solids (Onodera 1972). (The volume change is for one transformation—or Avagadro's number of transformations, using the gas constant) Nanocrystals also allow for direct comparison of the forward and reverse transformation directions of the same sample because the transition is reproducible.

Extensive measurements of the kinetics to determine rate constants for the nanocrystal transition have been made only on the CdSe system (Chen et al. 1997, Jacobs et al. 2001). Both the forward and reverse transition directions have been studied in spherically shaped crystallites as a function of pressure and temperature. The time-dependence of the transition yields simple transition kinetics that is well described with simple exponential decays (see Fig. 5). This simple rate law describes the transformation process in the nanocrystals even after multiple transformation cycles, and is evidence of the single-domain behavior of the nanocrystal transition. Rate constants for the nanocrystal transition are obtained from the slope of the exponential fits. This is in contrast to the kinetics in the extended solid, which even in the first transformation exhibits complicated time-dependent decays that are usually fit to rate laws such as the Avrami equation.

The measurements of activation volumes and activation energies in both transition directions reveal that the forward and reverse transition kinetics differs. (These

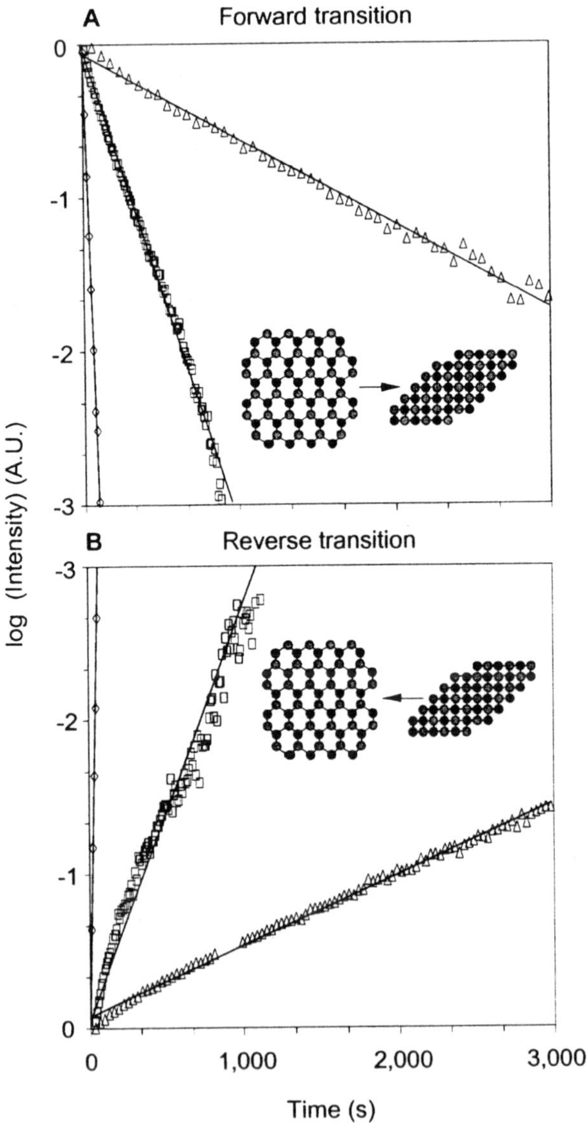

Figure 5. Time-dependence of the transformation in 25-Å CdSe crystals at 463 K at different pressures. (A) The forward transition at Δ 5.2 GPa, □ 5.7 GPa, ◊ 6.9 GPa. (B) The reverse transition at ◊ 0.7 GPa, □ 1.0 GPa, Δ1.2 GPa. (1 GPa ≅ 10,000 atm) The abscissa is the intensity of the four-coordinate electronic absorption feature. Fits are single-exponential decays. Rate constants were obtained from the slope of the fitted lines, and are equivalent to relaxation times (= ln 2/k) in the forward transition of: Δ 21 min, □ 3.6 min, ◊ 20 s. In the reverse transition: ◊ 24 min, □ 3.8 min, and Δ 16 s. Each crystal in the ensemble transforms instantaneously relative to the experiment time such that the relaxation is a measure of the average time required to overcome the kinetic barrier. Once a nanocrystal transforms to the stable structure, it is statistically unlikely that it will fluctuate back unless the pressure is changed accordingly, because the transition is measured far from equilibrium. [Used by permission of the editor of *Science*, from Jacobs et al (2001), Fig. 2.]

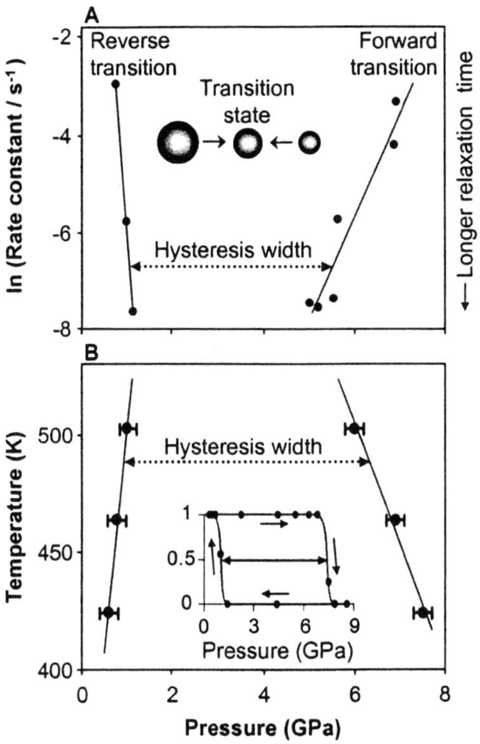

Figure 6. Hysteresis loop dependence on (A) relaxation time and (B) temperature for the transformation in 25-Å diameter CdSe nanocrystals. In (A), the rate constants are determined from time-dependent decays at 463 K, such as those shown in Figure 5. Activation volumes determined from the slopes using Equation (2) are -17 Å^3 and +65 Å^3 in the forward and reverse transitions, respectively. The activation volumes are assumed to be constant over the pressure ranges under study to simplify the analysis. In (B), the temperature data are compiled from measurements of the rate constant for ln (k) = -3.5 at the given temperatures. The inset is a sample hysteresis loop measured for 35 Å nanocrystals with the hysteresis width indicated. The loop starts at low pressure and proceeds in the direction of the arrows, as the normalized ratio of sample transformed is monitored [Used by permission of the editor of *Science*, from Jacobs et al (2001), Fig. 3].

parameters are calculated for experimental pressures, namely, "far" from the thermodynamic transition pressure.) For the case of 25-Å CdSe nanoparticles, the activation volumes are -17 Å^3 and +65 Å^3 in the forward and reverse transition directions, respectively (see Fig. 6A). The activation energies (enthalpic barrier) are 0.9 and 0.2 eV/nanocrystal (88 and 19 kJ/mol of nanocrystals) in the forward and reverse transition directions, respectively. This is in line with the work of Rubie and Brearly (1994), who observed differences in the two transition directions in bulk Mg_2SiO_4 by studying the transformation in samples synthesized as either one of the two structures. From the different activation volumes, a smaller deformation during the forward transition in CdSe nanocrystals implies that the transition state more closely resembles the low-pressure four-coordinate structure than the high-pressure six-coordinate structure. A transition state that is not precisely midway between the two stable minima is reasonable in transformations between phases with differing volumes. The observed asymmetry has profound consequences for the hysteresis loop because it implies that the thermodynamic transition pressure, which can be extrapolated to at very long observation times, is not in the center of the loop. By default, this thermodynamic pressure for structural transitions is frequently taken as the center of the hysteresis (Li and Jeanloz 1987), under the assumption that the two transition directions are symmetric and equivalent (Shekar and Rajan 2001). However, this pressure is observed to be markedly lower than the hysteresis center as the direct result of different activation volumes in the two transition directions. This result was also consistent with the different activation energies in the two transition directions in this CdSe nanocrystal transition, as evident from the greater sensitivity of the forward transition to temperature (see Fig. 6B). The asymmetry is reasonable considering the two structural phases are distinguishable, such that the forward and reverse transition can not be expected to be identical. Nevertheless, it is still difficult to say precisely where this pressure is in systems of different composition or how this relative position within the hysteresis loop depends on crystal size.

Additional information regarding the transition state was determined from opposite signs of the activation volumes in the two transition directions, positive in the reverse transition and negative in the forward transition (from the corresponding slopes in Fig. 6A and Eqn. 2). The opposite signs imply that the transition state in the CdSe nanocrystal transition must be intermediate in volume between that of the two structures, unlike a system that expands in both directions to accommodate the atomic rearrangements (see Fig. 6A schematic). This result is not expected for all solid-solid transitions. For example, the graphite-diamond transition in carbon has a fluid-like transition state (Shekar and Rajan 2001) that hinders the transition with increasing pressure and should therefore produce positive activation volumes in both transition directions. However, the case of an intermediate transition state volume observed for the CdSe nanocrystal transition is probably more common for transitions that exhibit conventional hysteresis loops (the forward transition is facilitated with increasing pressure). It is inherently implied in classical nucleation models for structural transitions (Christian 1965), in which the transition state is an average volume of both structures. In such models, the activation volume is proportional to the critical nucleus size because it is equivalent to the volume change that accompanies its formation. Nucleation theory also predicts that the critical nucleus size should vary with pressure, but is largest at the thermodynamic transition pressure. This premise is consistent with the experimental activation volumes, which are expected to increase in magnitude "near" the thermodynamic pressure in order for the forward and reverse activation volumes to add up suitably to the 18% total volume change across the transition.

The activation volumes in both transition directions is relevant to understanding the dependence of hysteresis loops on observation time, as indicated in Figure 6A. The hysteresis loop narrows with increasing observation time to a half-life corresponding to >10 years at room temperature, at the intercept of the two fitted lines. This is a lower limit because the assumption of constant activation volumes does not hold over larger pressure ranges (see the end of the previous paragraph). The actual relaxation time at the thermodynamic transition pressure may be much longer than 10 years even in the small nanocrystals. Nevertheless, in terms of geologically relevant transitions, this analysis shows that the hysteresis can depend on the observation time, and that this dependence can be understood in terms of magnitudes of the activation volumes.

TRANSFORMATION MECHANISM

The comparison of the CdSe transition to the diamond-graphite transition illustrates that there are several types of transformation mechanisms by which solids transform between polymorphs. How nanocrystals of a particular composition transform can be important by providing insight into the nature of the barrier and the dynamics in the extended solid. For example, if the nanocrystal were small enough to transform between solids in a coherent deformation mechanism, involving coordinated motion of the entire crystal lattice, this would imply a critical nucleus in the bulk larger than several nanometers. The transformation mechanism for CdSe nanocrystals has been the subject of many experimental and theoretical investigations (Moletini et al. 2001, Kodiyalam et al. 2001, Leoni and Nesper 2000). The major issue has been whether the transition is initiated locally in just part of the crystal or involves the entire nanocrystal in a coherent deformation event (Brus et al. 1996).

The microscopic transformation mechanism was recently determined for the CdSe nanocrsytalline transition, based on the kinetics measurements discussed in the previous section and the shape change discussed earlier in this chapter (Jacobs et al. 2001;

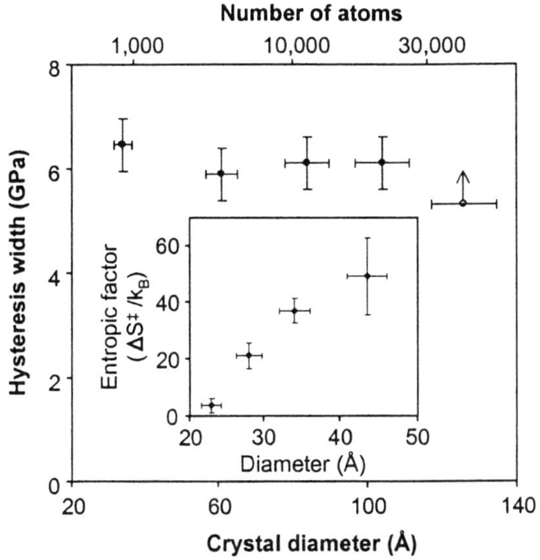

Figure 7. Hysteresis width versus nanocrystal size at room temperature and a relaxation time of ~3 minutes. The width for the largest sample is a lower limit because the 126-Å diameter nanocrystals remain trapped in the high-pressure rocksalt structure as the pressure is released (Jacobs, in preparation). In the inset, $\Delta S^{\ddagger}$ is the entropic barrier and k_B is Boltzmann's constant. [Used by permission of the editor of *Science*, from Jacobs et al. (2001), Fig. 4.]

Wickham et al. 2000). The nanocrystal was determined to transform by a local nucleation event—a sliding-planes model involving the shearing motion of the (001) crystal planes (shown in Fig. 2) relative to one another. This model explains the variation in the stacking sequence along the *c*-axis before and after the transition in the TEM in Figure 2, because the transition is directionally dependent and not sensitive to the stacking along the *c*-axis (Osugi et al. 1966). The transition nucleates without preference at any of the planes along the *c*-axis, as inferred from the strongly increasing frequency factor (entropic factor) with size (see Fig. 7). Although none of the experimental observations alone is conclusive, the sliding planes model, established in the study of martensitic transformation (Porter and Easterling 1992), is the only model under consideration that is consistent with the entire set of data. The following is an outline of the kinetics data supporting the sliding planes mechanism.

Activation volume

The microscopic mechanism in CdSe nanocrystals from which the hysteresis originates is constrained by the small magnitude of the activation volume of ~0.2%, in comparison to the 18% total volume change. The simplest model is a coherent deformation mechanism, as described above, in which the entire nanocrystal transforms in a single coordinated motion. However, this model is inconsistent with the activation volume, because 0.2% distributed throughout the nanocrystal is even smaller than the volume change caused by thermal vibrations of the crystal lattice. A nucleation mechanism involving just part of the crystal is more plausible because it requires only a modest structural deformation. A spherically shaped nucleus is excluded because the activation volume would correspond to a nucleus smaller than the size of a unit cell. (This size was calculated directly from the activation volume magnitude, given that the activation volume is proportional to the critical nucleus size and the overall 18% volume change between the two phases.) In addition, this case is inconsistent with the observation of an increasing activation volume magnitude from 17 to 34 and 124 Å^3 in the forward direction for 25-, 30-, and 35-Å diameter nanocrystals, respectively. The sliding planes mechanism is consistent with the small magnitude of the activation volume, as well as with the increasing activation volumes with size, as the shearing of planes with larger diameter produces larger structural deformations during the transition.

Activation energy (enthalpic barrier)

Increasing activation energies of 0.5, 0.9, 1.2, 1.7, and 2.4 eV/nanocrystal have been reported for CdSe nanocrystal diameters of 20, 25, 27, 34, and 43 Å, for the forward direction (Chen et al. 1997, Jacobs et al. 2001). This increasing trend can be understood as arising from additional bonds that must break as the planes in larger particles slide. The measured activation energies are in remarkable agreement with estimates of the energy necessary to shear the (001) plane, using the stacking fault energy of 14 meV/atom (Takeuchi et al. 1984). For example, the activation energy of 0.9 eV in 25 Å nanocrystals (~50 atoms per plane) compares well with an estimate of 0.7 eV based on the transformation model. Although martensitic transformations are generally considered to be athermal, the shearing appears to be activated in this nanocrystal transition, perhaps because the role of defects has been eliminated.

Entropic factor (entropic barrier)

The size dependence of the hysteresis width at room temperature has been measured in nanocrystal sizes up to 130 Å in diameter (see Fig. 7). The nearly constant hysteresis width of ~6 GPa over this size range suggests the corresponding free energy barrier responsible for the hysteresis is insensitive to nanocrystal size. The increasing entropic factor as a function of size accounts for the observed behavior because it offsets the increasing activation energy with size. This is uncharacteristic of a coherent deformation mechanism, for which the hysteresis width should strongly broaden with size without increasing entropic factors. If present, the entropic factors in a coherent deformation are expected to decrease with size because, the likelihood that there is enough thermal energy in a "breathing" mode should decrease with the number of atoms in the system. Instead, the data are indicative of a nucleation mechanism with multiple sites from which the transition can initiate. In the sliding planes model, this corresponds to greater probability of nucleation in nanocrystals with larger numbers of planes since only one nucleation event occurs per nanocrystal transformation. If high enough temperatures were experimentally feasible in this system, the hysteresis should ultimately narrow with increasing crystal size due to the greater influence of the entropic factor.

The nearly constant hysteresis width is important because it means that, at least in CdSe, the size-dependence of the hysteresis width is not sensitive to the kinetics. The upstroke transition pressure can nevertheless shift with nanocrystal size due to thermodynamic effects, as discussed earlier in this chapter (see SURFACE EFFECTS… above). Some geophysically-relevant systems may not exhibit full hysteresis loops because the high-pressure phase remains metastable as the pressure is released. Thus, special care should be taken to consider both the thermodynamics and kinetics in interpreting hysteresis measurements on such systems.

TEMPERATURE DEPENDENCE

Temperature is an important parameter that has been studied but not fully explored. Figure 6B shows temperature dependent data obtained for 25-Å diameter CdSe nanoparticles. The hysteresis loop systematically narrows with increasing temperature, as expected for an activated process. The larger activation energy discussed earlier is apparent in this figure, as the forward transition is significantly more sensitive to temperature compared to the reverse transition. A considerable shift of the hysteresis loop to lower pressure accompanies the narrowing with increasing temperature. Hysteresis shifts are usually assumed to directly correspond to shifts of the thermodynamic transition pressure (although the pressure is not expected to be precisely in the center of the loop). This assumption was used earlier in this chapter for the shift of the thermodynamics with size attributed to surface effects (see SURFACE EFFECTS... above). Based on this assumption, the shift of the hysteresis with temperature in Figure 6B can be interpreted using the Clausius-Clapeyron equation for the slope of the phase boundary. The hysteresis shift of -0.07 kbar/K corresponds to an entropy change of +65 eV/nanocrystal in the 25-Å nanocrystal, roughly two factors larger than the -0.001 kbar/K shift in bulk CdSe (Jayaraman et al. 1963). A similarly large shift was also observed for 35-Å diameter nanocrystals. An explanation for such a large shift is an additional entropic contribution in the rocksalt phase in nanocrystal relative to the four-coordinate structure. Additional disorder of the surface in the rocksalt nanocrystal may reflect the shape change the nanocrystal undergoes during the transition, and the high-energy faces that are exposed (see Fig. 3). This interpretation suggests that the entropic differences between the phases may be large and are thus important in understanding the nanocrystal structural transition. The shift of the hysteresis to lower pressure with increased temperature is also experimentally related to the observation of different activation energies in the two transition directions. The larger sensitivity to temperature in the forward transition is responsible for the hysteresis shift to lower pressure with temperature (see Fig. 6B). This suggests that one could predict relative difference in activation energies in the forward and reverse transitions in bulk transitions from measurements for the Clausius-Clapeyron slope.

Extending the current state of knowledge could involve measurements at high temperature with shock experiments to access transition pressures closer to the thermodynamic limit. Progress is currently being made to study the transition in individual nanocrystal particles, to both eliminate the ensemble statistics and allow for individual transitions to be observed on ~femtosecond time scales. The study of nanocrystal solid-solid phase transitions to oxide nanosystems should also prove to be useful in understanding the microscopic process of solid-solid transitions relevant to geophysically important systems.

ACKNOWLEDGMENTS

This work was supported by the Director, Office of Energy Research, Office of Science, Division of Materials Sciences, of the U. S. Department of Energy, and the Air Force Office of Scientific Research, Air Force Material Command, USAF.

REFERENCES

Barnett JD, Block S, Piermari GJ (1973) Optical fluorescence system for quantitative pressure measurement in diamond-anvil cell. Rev Sci Instru 44:1-9

Berry RS, Smirnov BM (2000) Phase stability of solid clusters. J Chem Phys 113:728-737

Brus LE, Harkless JAW, Stillinger FH (1996) Theoretical metastability of semiconductor crystallites in high-pressure phases, with application to beta-tin structure silicon. J Am Chem Soc118:4834-4838

Buffat P, Borel JP (1976) Size effect on melting temperature of gold particles. Phys Rev A 13:2287-2298

Bundy FP (1969) Behavior of carbon at very high pressures and temperatures. Proc K Ned Akad B-Ph 72:302

Chen CC, Herhold AB, Johnson CS, Alivisatos AP (1997) Size-dependence of structural metastability in semiconductor nanocrystals. Science 276:398-401

Christian JW (1965) The Theory of Transformations in Metals and Alloys. New York, Pergamon Press

Chudinovskikh L, Boehler R (2001) High-pressure polymorphs of olivine and the 660-km seismic discontinuity. Nature 411:574-577

Coombes CJ (1972) Melting of small particles of lead and indium. J Phys F–Metal Physics 2:441

Cullity BD (1956) Elements of X-ray Diffraction. Addison-Wesley Metallurgy Series. Addison-Wesley, Reading, Massachusetts

Evans MG, Polanyi M (1935) Some applications of the transition state method to the calculation of reaction velocities, especially in solution. Trans Faraday Soc, June 21, 1935

Goldstein AN, Echer CM, Alivisatos AP (1992) Melting in semiconductor nanocrystals. Science 256:1425

Hill TL (1963) Thermodynamics of Small Systems. New York, Dover Publications
Jacobs K, Zaziski D, Scher EC, Herhold AB, Alivisatos AP (2001) Activation volumes for solid-solid transitions in nanocrystals. Science (accepted)
Jayaraman AG, Kennedy C, Klement W (1963) Melting and polymorphic transitions for some group 2-6 compounds at high pressures. Phys Rev 130:2277
Kirby SH, Durham WB, Stern LA (1991) Mantle phase-changes and deep-earthquake faulting in subducting lithosphere. Science 252:216-225
Kodiyalam S, Kalia RK, Kikuchi H, Nakano A, Shimojo F, Vashishta P (2001) Grain boundaries in gallium arsenide nanocrystals under pressure: A parallel molecular-dynamics study. Phys Rev Letters 86:55-58
Leoni S, Nesper R (2000) Elucidation of simple pathways for reconstructive phase transitions using periodic equi-surface (PES) descriptors. The silica phase system. I. Quartz-tridymite. Acta Crystallogr A 56:383-393
Li XY, Jeanloz R (1987) Measurement of the B1-B2 transition pressure in NaCl at high temperatures. Phys Rev B 36:474-479
McHale JM, Auroux A, Perrotta AJ, Navrotsky A (1997) Surface energies and thermodynamic phase stability in nanocrystalline aluminas. Science 277:788-791
Molteni C, Martonak R, Parrinello M (2001) First principles molecular dynamics simulations of pressure-induced structural transformations in silicon clusters. J Chem Phys 114:5358-5365
Murray CB, Norris DJ, Bawendi MG (1993) Synthesis and characterization of nearly monodisperse CdE (E = S, Se, Te) semiconductor nanocrystallites. J Am Chem Soc 115:8706-8715
Onodera A (1972) Kinetics of polymorphic transitions of cadmium chalcogenides under high pressure. Rev Phys Chem Japan 41:1
Osugi J, Shimizu K, Nakamura T, Onodera A (1966) High pressure transition in cadmium sulfide. Rev Phys Chem Japan 36:59
Peng XG, Wickham J, Alivisatos AP (1998) Kinetics of II-VI and III-V colloidal semiconductor nanocrystal growth: Focusing of size distributions. J Am Chem Soc 120:5343-5344
Porter DA, Easterling KE (1992) Phase Transformations in Metals and Alloys. London, Chapman and Hall
Putnis A (1992) Introduction to Mineral Sciences. Cambridge, Cambridge University Press
Rockenberger J, Scher EC, Alivisatos AP (1999) A new nonhydrolytic single-precursor approach to surfactant- capped nanocrystals of transition metal oxides. J Am Chem Soc 121:11595-11596
Rubie DC, Brearley AJ (1994) Phase-transitions between beta-$(Mg,Fe)_2SiO_4$ and gamma- $(Mg,Fe)_2SiO_4$ in the Earth's mantle—Mechanisms and rheological implications. Science 264:1445-1448
Shekar NVC, Rajan KG (2001) Kinetics of pressure-induced structural phase transitions—A review. Bull Mater Sci 24:1-21
Takéuchi S, Suzuki K, Maeda K, Iwanaga (1984) Stacking-fault energy of Ii-Vi compounds. Philos Mag A 50:171-178
Tolbert SH, Alivisatos AP (1995) The wurtzite to rock-salt structural transformation in cdse nanocrystals under high-pressure. J Chem Phys 102:4642-4656
Tolbert SH, Herhold AB, Brus LE, Alivisatos AP (1996) Pressure-induced structural transformations in Si nanocrystals: Surface and shape effects. Phys Rev Letters 76:4384-4387
Wickham JN, Herhold AB, Alivisatos AP (2000) Shape change as an indicator of mechanism in the high-pressure structural transformations of CdSe nanocrystals. Phys Rev Letters 84:4515-4515

3 Thermochemistry of Nanomaterials

Alexandra Navrotsky

Thermochemistry Facility
Department of Chemical Engineering and Materials Science
University of California-Davis
Davis, California 95616

INTRODUCTION

The term "nanoparticle" or "nanomaterial" is somewhat difficult to define rigorously. A nanoparticle has dimensions somewhere in the nanometer regime, that is, a diameter of 1 to 100 nm. Thus nanoparticles span the range from clusters of atoms in solution at the small end to colloidal particles at the large end. Nanoparticles may be amorphous or consist of only a few unit cells of crystalline material. A very large fraction of their atoms are near the surface, see Figure 1. Nanoparticles may be surrounded by vacuum, a gaseous atmosphere, water, or other fluid. In the natural environment, nanoparticles are generally heavily hydrated.

A nanomaterial can be loosely defined to be any material containing heterogeneity at the nanoscale in one or more dimensions. In the broadest sense, then, the following are nanomaterials: phase-separated glasses or crystals with domains in the nanoregime, zeolites and mesoporous materials with pores of nanometer dimensions, clays with nanometer sized alternations of aluminosilicate layers and interlayer hydrated cations, and nanoscale leach layers at the mineral-water interface.

A broad definition in the sense above emphasizes the commonality of phenomena at the nanoscale. In essence "if it quacks like a nanomaterial, it is one." *A nanomaterial is any state of condensed matter whose properties diverge significantly from those of the bulk or of molecules by the emergence of new phenomena not seen at smaller or larger scales.* Such properties are related to nanoscale heterogeneity created by pervasive surfaces, interfaces, chemical variability, or pores. The exact size at which this happens depends both on the system and the property being considered. From the thermodynamic point of view, a material becomes "nano" when its thermodynamic parameters differ significantly from those of the bulk because of diminution in dimension. Such changes may be accompanied by changes in structure, changes in crystallographic symmetry, polymorphism, or amorphization. This chapter will emphasize the thermochemical causes/consequences of nanoscale behavior in materials of importance to geological and environmental science.

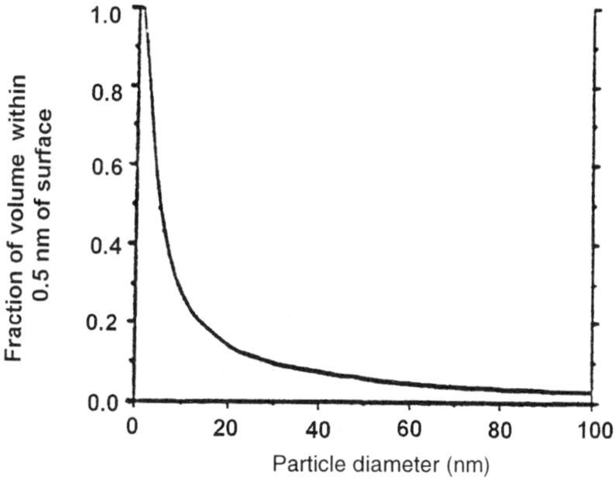

Figure 1. Fraction of atoms within 0.5 nm of the surface of a nanoparticle as a function of its diameter. [Used by permission of the editor of *Materials Research Society Symp Proc*, from Navrotsky (1997), Fig. 3, p. 10.]

1529-6466/00/0044-0003$05.00

Nanomaterials are thermodynamically metastable (in both enthalpy and Gibbs free energy) with respect to bulk macrocrystalline materials. However, they are often only slightly metastable, see below, and the driving force for their coarsening can be very small. That, coupled with slow kinetics of coarsening and/or phase transformation in low temperature geochemical environments, ensures that nanoparticles are prevalent in air, water, soil, and sedimentary settings. Because of their persistence, it is appropriate to investigate their thermodynamic properties.

THERMODYNAMIC ISSUES

A nanoparticle exposed to vacuum or a gas is less stable than a bulk coarse grained material. A surface free energy can be defined in terms of surface enthalpy and entropy contributions.

$$G_{surface}\ (J/m^2) = H_{surface}\ (J/m^2) - TS_{surface}\ (J/(m^2K)) \quad (1)$$

$$\begin{aligned} G_{surface}\ (J/g) &= G_{surface}\ (J/m^2) \times \text{surface area}\ (m^2/g) \\ &= (H_{surface}\ (J/m^2) - TS_{surface}\ (J/(m^2K)) \times \text{surface area}\ (m^2/g) \end{aligned} \quad (2)$$

$$\begin{aligned} G_{surface}\ (J/mol) &= G_{surface}\ (J/m^2) \times \text{surface area}\ (m^2/mol) \\ &= (H_{surface}\ (J/m^2) - TS_{surface}\ (J/(m^2K)) \times \text{surface area}\ (m^2/mol) \end{aligned} \quad (3)$$

The difference between surface energy and surface enthalpy (a PV term) is expected to be small and has, to the best of my knowledge, never been characterized experimentally. One is probably making an error of less than 5% in equating surface energy and enthalpy. The difference between surface enthalpy and surface free energy, the TS term, can in principle be evaluated, see below, but the correction is not expected to exceed 1-20%. One expects the surface entropy to be positive (atoms less tightly bound at the surface leading to higher vibrational entropy), and thus $G_{surface}$ will usually be slightly less positive than $H_{surface}$. The literature commonly reports surface areas in m^2/g and surface energies or free energies in J/m^2. When comparing different oxides, e.g., Fe_2O_3 and Al_2O_3, it is often more useful to think of surface areas in m^2/mol, and this is done in several examples below.

Typical surface energies (enthalpies) are of the order of 0.1 to 3 J/m^2 (see Table 1). For nanoparticles with surface areas of 20,000 m^2/mol (corresponding, for example, to alumina with a surface area of about 200 m^2/g), the enthalpy would be raised by 2 to 60 kJ/mol. Thus the effect on thermochemical properties can be significant.

Different polymorphs have different surface (free) energies (see Table 1). These differences may serve to stabilize, in the nanoregime, polymorphs not stable in the bulk (see Fig. 2). As an example, if the difference in surface free energy between polymorphs is 0.5 J/m^2, a polymorph metastable in the macrocrystalline regime by 10 kJ/mol could be stabilized at surface areas greater than 20,000 m^2/mol. Because many oxides (silica, iron and aluminum oxides and oxyhydroxides, manganese oxides, titania, zirconia, zeolites) show polymorphism with relatively small free energy differences between polymorphs, such crossovers in stability at the nanoscale may be a rather general phenomenon, see below for specific examples.

The surface energy depends on the crystal faces exposed. For example, molecular dynamics simulations (Blonski and Garofalini 1993) give the different surface energies for various orientations of α- and γ-alumina surfaces. However, for mineralogical and ceramic nanoparticles, it is difficult or impossible to control (or measure) which surfaces dominate, and usually only some sort of average (not rigorously defined in terms of

Table 1. Surface thermodynamic properties of oxides

Material	*ΔE (J/m²)*	*ΔH (J/m²)*
α-Al_2O_3 (corundum)	2.04 – 8.04[a]	2.64[d]
	2.57 – 3.27[b]	
	2.03 – 2.50[c]	
γ-Al_2O_3 (defect spinel)	0.87 – 2.54[a]	1.66[d]
$MgAl_2O_4$ (spinel)		1.8[e]
α-Fe_2O_3 (hematite)		1.0[f,h]
γ-Fe_2O_3 (maghemite)		0.6[h]
TiO_2 (rutile)	1.85 – 2.02[j]	1.93[k], 2.2[r]
TiO_2 (anatase)	1.28 – 1.40[j]	1.34[k]
TiO_2 (brookite)		
ZrO_2 (monoclinic)		1.9[q]
ZrO_2 (amorphous)		1.2 > monoclinic[p], 0.7[q]
SiO_2 (quartz)		0.4-1[o], 1.0[n]
SiO_2 (amorphous)		0.09[l], 0.27[n]
SiO_2 (zeolitic)		0.10[n]
AlOOH (boehmite)		0.5±0.1[s]
FeOOH (goethite)		(1.55)[f,i]
FeOOH (lepidocrocite)		(2.08)[g,i]

a. Blonski and Garofalini (1993), molecular dynamics
b. Tasker (1984), static energy calculations
c. Macrodt et al. (1987), molecular dynamics
d. McHale et al. (1997a,b), oxide melt solution calorimetry; all measured quantities in this table are "apparent" quantities in the nomenclature of Diakonov (1998).
e. McHale et al. (1998), derived indirectly from oxide melt solution calorimetry
f. Diakonov et al. (1994), calculated from earlier HF solution calorimetry
g. Diakonov (1998a), calculated for earlier HF solution calorimetry
h. Diakonov (1998b), calculated for earlier HF solution calorimetry
i. Values considered rather uncertain, therefore in ( ).
j. Oliver et al. (1997), simulation, empirical potentials
k. Zhang and Banfield (1999), consistent with Oliver et al. (1997)
l. Moloy et al. (2001), summary of data from various sources
m. Moloy et al. (2001), correlation of measured enthalpies of formation and modeled internal surface areas
n. Iler (1979)
o. Consistent with Brace and Walsh (1962) and Hemingway and Nietkowicz (1993)
p. Molodetsky et al. (2000)
q. Calculated from data of Molodetsky et al. (2000) and Ellsworth et al. (1994); see text
r. Ranade et al. (2001, in preparation), oxide melt calorimetry
s. Majzlan et al. (2000)

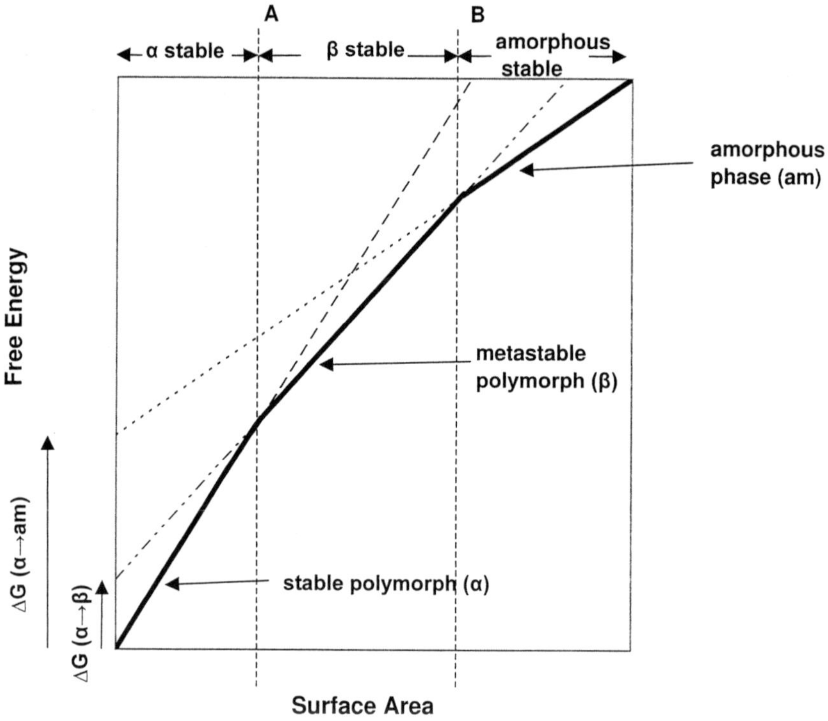

Figure 2. Schematic representation of free energy crossovers in nanoparticle systems. The polymorph stable as a bulk material (α) is stable as a nanoparticle for surface areas between zero and "A." The polymorph β, metastable as a bulk material, is stable as a nanoparticle for surface areas between B and C. For surface areas greater than C, amorphous material is stable. The diagram is drawn such that surface free energy decreases (becomes less positive) in the order of α, β amorphous. All nanoparticles are metastable with respect to coarsening. The heavy line shows the phases that will form at a given particle size if coarsening is precluded. The free energy of the α–β transition and of vitrification (amorphization) for the bulk phases is shown on the y-axis. Amorphization is shown as a first order transition though in fact it may be gradual.

crystal orientations) surface energy is obtained by experiment.

The structure at the surface is generally different from that in the bulk. There may be surface reconstruction (characterized well for single crystal metal surfaces, less well for oxides, and poorly for nanoparticles). This rearrangement may involve the preferred exposure of metal or oxygen as the terminal layer, preferential occupancy by one component in a multicomponent system, and, most importantly for geological systems, hydration. It is likely that all natural nanoparticles have hydrated surfaces, and the surface concentrations of H^+, H_2O, and OH^- depend on the pH. This is related to the zero point of charge, a phenomenon well studied in colloid chemistry (Hunter 1986). Surface hydration, hydroxylation, or protonation, the adsorption of other metal ions, and the formation of leach layers and surface precipitates involving these adsorbates are phenomena which are not distinct and separable. Rather they form a continuum from loosely to tightly bound surface species which interact with the surface. Similarly the structure of the aqueous solution near the surface (within a few molecular distances) is almost certainly different from that of the bulk solution. This becomes even more important when the hydrous region is limited in dimensions, as is true in zeolite pores and clay interlayers. Some thermochemical ramifications of this complexity will be illustrated below. Because of this complexity, it is essential that thermodynamic measurements, structural studies, and modeling proceed hand in hand on the same systems to attain molecular-level understanding of the relations among structure, thermochemistry, and rate processes.

METHODS FOR DETERMINING NANOPARTICLE ENERGETICS

These may be classified into three groups: equilibrium measurements, calorimetric measurements, and computational methods. The first address surface free energy, the second surface enthalpy, and the third (usually) surface energy.

Equilibrium measurements

Contact angle measurements between a liquid of known surface tension and a surface measures surface free energy (Hunter 1986). Such measurements are generally confined to bulk macroporous materials such as silica (see below). This method is usually difficult to apply to nanoparticles. A high temperature application of contact angle measurements is the study of Hodkin and Nicholas (1973). They measured surface and grain boundary energies in stoichiometric UO_2 by equilibration with molten copper, which acted as an inert liquid for interfacial angle measurements at 1400-1700°C.

A metastable material, including one of high surface area, should have a higher solubility (e.g., in an aqueous solution) than the stable phase. If dissolution equilibrium can be demonstrated under conditions where the particle size does not change, solubility measurements offer a means to measure the excess free energy of a fine grained material. Relating this excess free energy to the surface area then provides the surface free energy. There are several difficulties inherent in this approach. One must establish equilibrium (preferably by reversals rather than merely steady state) and assure that the particle size is not changing significantly during the dissolution/reprecipitation process. Furthermore, data interpretation often requires speciation models and activity coefficients for aqueous species. Such models vary among investigators and the thermodynamic data are not always available, complete, or reliable. Peryea and Kittrick (1988) used the aqueous solubility of corundum, gibbsite, boehmite, and diaspore to obtain their thermodynamic properties, but without explicit attention to particle size. Wintsch and Dunning (1985) used aqueous solubility data for quartz to estimate the free energy associated with dislocations. Enüstün and Turkevitch (1960) studied the aqueous solubility of fine particles of $SrSO_4$ to analyze surface free energies.

Calorimetric measurements

Calorimetric studies of surface and nanoparticle energetics fall into several classes: enthalpies of wetting and hydration/dehydration, heat capacity measurements, thermal analysis of coarsening and phase transition, and enthalpy differences by solution calorimetry. These methods measure different quantities, suffer from different potential difficulties, and are generally regarded as complementary.

If a surface is wet by, but not chemically interacting with, a liquid of known surface tension, measurement of the heat of immersion (heat of wetting) can be related to the surface energy (Fowkes 1965).

If there is chemical interaction, as with water, then the heat of immersion gives the integral heat of this interaction, i.e., the integral enthalpy of hydration. The initial state of a nanophase material, clay, or zeolite is difficult to define in the latter case because it is not easy to prepare a totally anhydrous sample. Measurements of immersion enthalpies of partially hydrated samples (partially hydrated → fully hydrated) can give the concentration dependence of the heat of hydration and, in favorable cases, the data can be fit to equations for the partial molar enthalpy of hydration. Combined with measurements of adsorption isotherms (which give free energy), a detailed picture of the ΔH, ΔS, and ΔG of hydration can be obtained. Complicating issues include possible irreversible change of surface structure and surface area during hydration/dehydration cycles, the interpretation of hysteresis in isotherms, and the need for very careful control and analysis of water content. Such calorimetric experiments are generally performed at or near room temperature in very sensitive heat flow calorimeters. Examples of application to zeolites include the work of Barrer and Cram (1971) and Carey and Bish (1996, 1997). The partial molar enthalpy of adsorption of water can also be measured directly by introducing known aliquots of water into a sensitive calorimeter containing the sample. The hydration energetics of nanophase α- and γ-alumina were studied by this method by McHale et al. (1997b). Zeolite hydration enthalpies were measured by Valueva and Goryainov (1992). However, one should realize that physical wetting, on the one hand, and surface hydration/hydroxylation, on the other, represent two ends of a continuum. The strength of chemical interaction at an oxide/water interface will depend on the nature of the oxide, pH, temperature, and other factors. A rule of thumb is that oxides which form stable hydroxides ($MgO/Mg(OH)_2$, $CaO/Ca(OH)_2$, $Al_2O_3/Al(OH)_3$) will show the strongest interactions.

Thermal analysis techniques (differential thermal analysis (DTA), differential scanning calorimetry (DSC), thermogravimetric analysis (TGA), and evolved gas analysis (EGA)) provide qualitative, semiquantitative, and in special cases, quantitative measurements of the energetic evolution of nanophase materials on heating. Variation of the heating rate and the atmosphere surrounding the sample provide additional information. Some examples are given below in the context of specific systems.

Solution calorimetry has proven very useful for studying nanoparticle energetics. In its simplest form, the calorimetric experiment uses the following thermochemical cycle.

Nanomaterial = Dissolved species	(4)
Bulk material = Dissolved species	(5)
Nanomaterial = Bulk material	(6)

$$\Delta H_6 = \Delta H_4 - \Delta H_5$$

The solvent used can be an acid near room temperature or an oxide melt at 700-800°C. Issues to be addressed include the following. The thermochemical cycle must account for any variations in composition between the nano and the bulk material. Variability of oxidation state is a common problem, especially when transition metals (Fe, Mn) are involved, and even a small difference in the extent of oxidation-reduction reactions during calorimetry can introduce errors of 10 kJ or more. Variability in water content must be accounted for; the nanomaterial is generally more hydrated. The small particles may also have adsorbed carbon dioxide (carbonate) and traces of organics which need to be included in the thermochemical cycle. The final dissolved state must indeed be the same for the nano and bulk material and the fate of volatiles (H_2O, CO_2, organics) must be properly characterized. Thus the conceptually simple cycle above must often be modified to include these complications, and the calorimetric study must be carried out in concert with detailed analytical chemistry. Examples are discussed below for specific oxides. An advantage of solution calorimetry is that, since the dissolution process rearranges all the bonds in the material being dissolved, the enthalpy difference between nano and bulk includes the contribution of all sources of energetics—polymorphism, surface energy, strain, defects, disorder, etc. This total enthalpy difference is the appropriate thermodynamic term to put into models of phase stability, dissolution rates, and other processes. The separation of these different effects, at least approximately, is one of the major challenges in interpreting the calorimetric data.

Nanophase materials generally have an excess heat capacity and entropy relative to the bulk. These can be obtained by conventional heat capacity measurements (adiabatic calorimetry, differential scanning calorimetry), although problems with the adsorbed water and other gases are more severe for nanomaterials than for bulk phases. Data at present are fragmentary and it is difficult to evaluate their accuracy. Dugdale et al. (1954) report on excess heat capacity for fine grained rutile. Victor (1962) report data for MgO and BeO, and Sorai et al. (1969) for $Ni(OH)_2$ and $Co(OH)_2$.

Diakonov (1998b) argues that the surface enthalpy derived from the slope of the line relating heat of solution to surface area is "apparent" rather than "true." This is because the number of moles of substance does not increase linearly with surface area. The true thermodynamic surface enthalpy (or other thermodynamic function) is then argued to be 1.5 times the apparent quantity. It is not clear to me whether this argument, as presented initially by Enüstün and Turkevitch (1960) is valid, and this point needs further clarification.

Computation, simulation, and modeling

Modern computational physics provides powerful tools for calculating the energies of nanoparticle surfaces and interfaces. Approaches range from ab initio calculations using density functional theory to lattice energy simulations using empirical potentials to molecular dynamics simulations. It is important that enough atoms be included to obtain a realistic picture of the surface. These theoretical approaches are discussed in detail in the chapter by Rustad in this volume. Issues that can be addressed by theory which are not readily accessible by experiment in nanomaterials include the energetics of specific crystal faces, the structure of the surface or interface, the structure of the aqueous layer directly contiguous to the interface, and the relaxation of structure around impurities and defects. Areas in which theory and experiment can be directly compared include the energetics of various polymorphs, the overall surface energy, hydration energy, and spectroscopic properties. A problem at present is how to assess the accuracy of calculated quantities when no experimental data are available. There is no simple answer at present, but a positive approach is for experimentalists and theorists to work together on the same systems, where experimental values are used to benchmark theoretical predictions, while theory is used to suggest crucial experiments. As an illustration of the use of theoretical predictions, see the section on aluminum oxides below.

SPECIFIC SYSTEMS

Aluminum oxides and oxyhydroxides

Corundum, α-Al_2O_3, is the thermodynamically stable phase of coarsely crystalline aluminum oxide at standard temperature and pressure conditions, but syntheses of nanocrystalline Al_2O_3 usually result in γ-Al_2O_3. Blonski and Garofalini (1993) performed molecular dynamics simulations of various α-Al_2O_3 and γ-Al_2O_3 surfaces (see Table 1). The surface energies for α-Al_2O_3 were significantly greater than those of γ-Al_2O_3. Using their data and assuming preferential exposure of the surfaces with lowest energy, McHale et al. (1997a,b) predicted that γ-Al_2O_3 should become the energetically stable polymorph as specific surface areas exceed ~125 m^2g^{-1}. The thermodynamic stability of γ-Al_2O_3 should be even greater than implied by this energy. Due to the presence of tetrahedral and octahedral sites in its spinel-type structure, and the fairly random distribution of Al^{3+} and vacancies over these sites, γ-Al_2O_3 has a greater entropy than α-Al_2O_3. The entropy change of the α-Al_2O_3 to γ-Al_2O_3 transition, $\Delta S_{\alpha\to\gamma}$, is about +5.7 $J{\cdot}K^{-1}mol^{-1}$ (McHale et al. 1997b). Therefore, at room temperature, γ-Al_2O_3 could be thermodynamically stable with respect to α-Al_2O_3 at specific surface areas >100 $m^2{\cdot}g^{-1}$, and at 800 K (a temperature typical of a hydroxide decomposition) γ-Al_2O_3 might become thermodynamically stable at specific surface areas greater than only 75 $m^2{\cdot}g^{-1}$.

McHale et al. (1997a) measured the enthalpies of drop solution in molten lead borate of several nanosized α- and γ-alumina samples. However, the surfaces of the Al_2O_3 were modified by adsorbed H_2O which could not be completely removed without severe coarsening. The surface energies of the hydrated polymorphs appeared nearly equal, indicating that the heat of chemisorption of H_2O is directly proportional to the surface energy of the anhydrous phase (a relationship predicted by Cerofolini (1975)). Consequently, McHale et al. (1997a) could not determine the anhydrous surface energies without accurate knowledge of the heats of chemisorption of H_2O. These measurements were made on two samples each of α- and γ-Al_2O_3 with a Calvet type microcalorimeter operating near room temperature (McHale et al. 1997b). The differential heat of H_2O adsorption on γ-Al_2O_3 decreases logarithmically with increasing coverage (Freundlich behavior). In contrast, the differential heat of H_2O adsorption on α-Al_2O_3 does not show regular logarithmic decay, and decreases far less rapidly with increasing coverage. This indicates a greater number of high energy sites on α-Al_2O_3 per unit surface area, which are relaxed by the most strongly chemisorbed hydroxyls. This observation is strong evidence that the surface energy of α-Al_2O_3 is higher than that of γ-Al_2O_3.

A quantitative analysis of the heat of adsorption data enables the separation of hydration enthalpies and surface enthalpies for the two alumina polymorphs (McHale et al. 1997b). The resulting variation of enthalpy of the anhydrous material with surface area is shown in Figure 3. The enthalpy (and free energy) crossover postulated above is clearly demonstrated.

Polymorphs of aluminum oxyhydroxide (boehmite, diaspore) and aluminum hydroxide (gibbsite, bayerite, nordstrandite) are common fine-grained constituents of soils, unconsolidated sediments, and sedimentary rocks. Their occurrence in nature is controlled by their thermodynamic properties and the kinetics involved in transformations. The equilibrium state of the system is dictated solely by the free energies of the solid phases and the aqueous phase and can be computed from existing thermodynamic data. However, agreement among published data is rare, rendering the calculation of phase diagrams difficult.

A recent study of the surface enthalpy of boehmite (Majzlan et al. 2000) was prompted by an apparent contradiction between common occurrence of boehmite and gibbsite and the thermodynamic stability of diaspore. It is commonly believed that many mineral assemblages found in sediments and soils owe their occurrence to sluggish rates of transformation to the stable phase assemblage. Indeed, gibbsite and boehmite are the most common minerals of lateritic bauxites (Bardossy and Aleva 1990). Traces of diaspore have been found in many locations, but large quantities of this mineral are reported only from compacted and recrystallized bauxites. In soils, the main aluminum carriers in addition to clays are fine-grained gibbsite and allophane (Buol et al. 1989). Furthermore, fine-grained boehmite is also an important industrial product used because of its relative ease of synthesis and its ability to retain high surface area. Boehmite serves as a catalyst or precursor of transition metal-Al_2O_3 catalysts and catalyst supports. Anovitz et al. (1991) noted that hydromagnesite, anatase, Mg-calcite, and smectites can also persist in metastable equilibrium over geologic time. Although undoubtedly kinetics plays a major role, metastable mineral assemblages can also be stabilized in a thermodynamic sense, i.e., by reducing the magnitude of Gibbs free energy of reaction

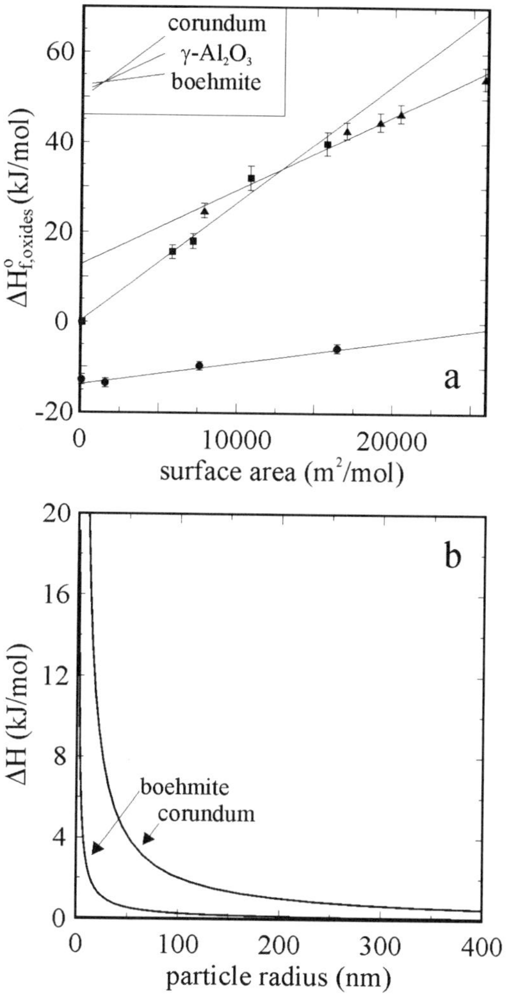

Figure 3. (a) Enthalpy of formation from oxides ($\Delta H^{o}_{f_{oxides}}$) (coarse α-Al_2O_3 or coarse α-Al_2O_3 + H_2O) as a function of surface area for boehmite (circles), γ-Al_2O_3 (triangles), and corundum (squares). Surface enthalpy is equal to the slope of each line, shown for easier comparison in the inset. Data for anhydrous oxides from McHale et al. (1997a). (b) Surface enthalpy as a function of particle size for corundum and boehmite, assuming spherical shape of particles. [Used by permission of the editor of *Clays and Clay Minerals*, from Majzlan et al. (2000), Fig. 1, p. 702.]

between metastable and stable phases (see Fig. 4).

The results of calorimetric studies are shown in Figure 4. This figure combines the data of McHale et al. (1997a,b) on alumina (see above) with the measurements of Majzlan et al. (2000) on boehmite samples of different surface areas. The data clearly show that surface enthalpy is lowest for boehmite, intermediate for γ-alumina, and highest for corundum. In a first approximation, surface energy of an ionic solid will be proportional to the charges of the ions within the structure (Tosi 1964). Thus, replacing the oxygen anions with hydroxyls will lower the surface energy, in agreement with experiment.

One can speculate that other soil aluminum oxyhydroxide phases, diaspore, bayerite, and nordstrandite also exhibit low surface free energies, similar to those for gibbsite and boehmite. More calorimetric studies are needed to quantify these relations.

Apps et al. (1989) estimated the enthalpy (ΔH^{o}_{f}) (-995.3 kJ/mol) and entropy of

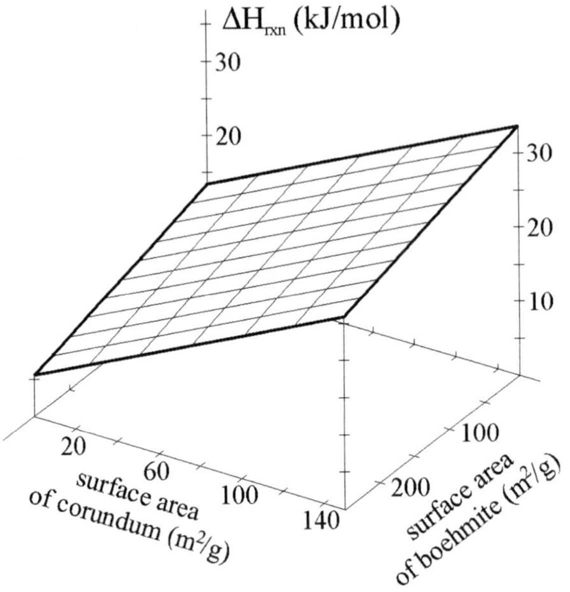

Figure 4. Enthalpy change for the reaction (ΔH_{rxn}) boehmite → corundum + water as a function of surface area of boehmite and corundum. [Used by permission of the editor of *Clays and Clay Minerals*, from Majzlan et al. (2000), Fig. 2, p. 702.]

formation of boehmite at standard temperature and pressure by iterative regression of experimental data obtained from diaspore, boehmite, and gibbsite. Hemingway and Nitkiewicz (1995) estimated ΔH^o_f of boehmite (-996.4±2.2 kJ/mol) from their measurement of third-law entropy (S^o) and an estimate of ΔG^o_f from the solubility measurements of Russell et al. (1955). The first calorimetric determination of ΔH^o_f of boehmite was performed by McHale et al. (1997a) by high-temperature oxide-melt calorimetry. Their measurements, without any excess water correction applied, yielded -995.4±1.6 kJ/mol. The estimate of Majzlan et al. (2000) (-994.0±1.1 kJ/mol) is less negative than ΔH^o_f adopted by Anovitz et al. (1991) (-996.4±0.9 kJ/mol, determined from their ΔG^o_f and S^o) in calculation of the phase diagram of the Al-O-H system. The estimate by Majzlan et al. (2000) makes boehmite more unstable in enthalpy with respect to diaspore (by 2.4 kJ/mol) but does not modify the topology of the proposed phase diagram of Anovitz et al. (1991). Hence, bulk macrocrystalline boehmite is metastable with respect to bulk diaspore at ambient conditions.

The surface enthalpy data emphasize the complexity of assigning thermodynamic stability or instability to low-temperature phase assemblages. One mineral, occurring both in the detrital and authigenic fraction of a sediment, may be simultaneously stable and metastable, depending on particle size. Obviously, a decrease in particle size will not always convert metastable phases into stable phases. Instead, a decrease in particle size may further reduce the driving force for the transformation to a stable macroscopic phase assemblage. In some cases, a change in sign of ΔG of a reaction may occur as a result of variable particle size of reactants and products (e.g., McHale et al. 1997b, Penn et al. 1999).

The available thermodynamic data show that ΔG of the diaspore-boehmite transformation is 2-6 kJ/mol, depending on the reference. A surface energy of diaspore that is larger than that of boehmite by a few tenths of J/m^2 is sufficient to make fine-grained boehmite a stable phase with respect to fine-grained diaspore. The results also suggest that relatively low surface energies (enthalpies) allow hydrated phases to attain high surface areas (small particle size). To further illustrate these effects, Figure 3 shows the effect of increased surface area on the enthalpy of the reaction of boehmite to corundum plus water, calculated from the surface enthalpies above. The enthalpy of reaction becomes markedly more positive (destabilizing the anhydrous assemblage) as the surface area of boehmite increases relative to that of corundum.

Iron oxides and oxyhydroxides

Iron oxides and oxyhydroxides play important technological roles as pigments and magnetic recording materials. In soils and water they regulate the concentrations and dissolution of nutrients (iron) and pollutants (heavy metals). Colloidal and nanophase iron oxides are transported in water and in air. Despite their abundance, they are difficult to characterize because of variable composition, oxidation state, surface area, and crystallinity. The thermochemical database for all but the most stable polymorphs (hematite, magnetite, wustite, goethite) is surprisingly sketchy.

Laberty and Navrotsky (1998) determined the enthalpies of formation of a number of iron oxide and oxyhydroxide polymorphs. Data are listed in Table 2 which also compares the enthalpy relations among aluminum, iron, and manganese. It is evident that the Fe^{3+} oxyhydroxide phases are much less stable relative to the anhydrous ferric phase (hematite) than are the aluminum oxyhydroxides relative to corundum. This is consistent with the much more frequent observation of hematite than of corundum in the field. It is also evident that the iron phases are as rich in polymorphism as the aluminum phases. It is clear that the enthalpy differences for both anhydrous (Al_2O_3, Fe_2O_3, MnO_2) and hydrous (AlOOH, FeOOH, MnOOH) polymorphs are small, setting the stage for nanoscale stability crossovers.

Table 2. Transition enthalpies at 298 K in aluminum, iron, and manganese oxides and oxyhydroxides, ΔH (kJ/mol).

	Metastability	*Dehydration*
α-FeOOH (goethite)[a]	0	10.8±0.6
β-FeOOH (akageneite, 0.032Cl)[a]	4.2±0.4	6.6±0.4
γ-FeOOH (lepidocrocite)[a]	3.1±1	7.7±1
γ-MnOOH (manganite)[b]	0	12.6±0.9
α-MnOOH (groutite)[b]	1.7±1.7	10.9±1.7
AlOOH (diaspore)[c]	0	20.6±2.5
AlOOH (boehmite)[c,d]	4.9±2.4	15.7±2.5
Al_2O_3 (γ-alumina)[d,e]	13.4±2.0	-
Fe_2O_3 (tetragonal maghemite)[a]	14.9±1.5	-
Fe_2O_3 (cubic maghemite)[a]	18.7±0.9	-
MnO_2 (ramsdellite)[b]	5.4±3.6	-

a. Laberty et al. (1998) b. Fritsch et al. (1997) c. Majzlan et al. (2000)
d. McHale et al. (1997a) e. McHale et al. (1997b)

The samples of γ-Fe_2O_3 and FeOOH polymorphs measured by Laberty and Navrotsky (1998) had surface areas of 25-40 m^2/g. Using the surface energy of hematite (Diakonov et al. 1994) to account for the surface energy of these other polymorphs gave correction terms of 4-8 kJ/mol which were included in the tabulation of thermodynamic properties of the bulk phases. Diakonov (1998b) determined the dependence of enthalpy on surface area for hematite and maghemite by analyzing earlier acid solution calorimetric data. He obtained a smaller surface enthalpy for γ-Fe_2O_3 than for α-Fe_2O_3 (see Table 1). The data suggest an enthalpy crossover with the defect spinel becoming more stable in energy at surface areas greater than about 60,000 m^2/mol (535 m^2/g). The results are somewhat questionable because at such large surface areas the samples may be expected to be amorphous and because it is not clear how or whether corrections for adsorbed water were made.

If the surface enthalpy of the oxyhydroxides is a factor of two smaller than that of hematite (in rough analogy to the data for alumina), then the coarse grained iron oxyhydroxides are in fact 2-4 kJ/mol less stable in enthalpy than reported by Laberty and Navrotsky (1998). Diakonov et al. (1994) report surface and bulk thermodynamic properties of goethite (α-FeOOH) and Diakonov (1998a) reports surface and bulk thermodynamic properties of lepidocrocite (γ-FeOOH). They report surface enthalpies of 1.10 J/m^2 for hematite, 1.55 J/m^2 for goethite, and 2.08 J/m^2 for lepidocrocite, based on analysis of very old acid solution calorimetric and heat capacity data (see Table 1). The reliability of these data and analysis is hard to judge. Rather surprisingly, the oxyhydroxides appear to have higher surface enthalpies than hematite. Calorimetric measurements of iron oxides and oxyhydroxides of different surface area are currently underway in the Thermochemistry Facility at Davis to re-evaluate the surface energetics.

Manganese oxides

Oxides based on MnO_2 and having the general formula $M_xMnO_2{\cdot}nH_2O$ (M = alkali, alkaline earth) are common minerals in soils, desert varnish, manganese nodules, and other low temperature environments. They are generally black, ugly, and poorly crystalline and, only relatively recently, have different phases been identified and characterized (Post 1992). They are nanomaterials with variable manganese oxidation state (Mn^{3+}, Mn^{4+}), cation content, water content, and structure.

There are two families of structures: layered and tunnel. Birnessite (with an interlayer spacing of 7 Å) and buserite (with an interlayer spacing of 10 Å) are two relatively well characterized MnO_2-based layered materials. Poorly crystalline layered manganese oxides are sometimes called vernadite. Todorokite, in contrast, is a MnO_2-based open framework phase with a one-dimensional tunnel with edges formed of three octahedra (3 × 3). Hollandite is a 2 × 2 tunnel structure. The structures of all these phases are built by layers or tunnel walls of edge-sharing MnO_6 octahedra with mixed valence. Exchangeable hydrated cations in the tunnel or the interlayer space act as templates for growth and provide charge balance. A variety of other tunnel structures also exist (Post 1992, Brock et al. 1998). These structural characteristics, very different from those of zeolites, result in unique ion-exchange, redox, and semiconducting properties, which enable these layer and framework materials to find application in separations, catalysis, batteries, sensors, and electromagnetic materials.

A fundamental question of both technological and theoretical importance is the stability of these materials in both thermodynamic and kinetic senses. In general, bonding styles (edge-sharing or corner-sharing) of MnO_6 building blocks, interactions of layers and tunnel walls with water and different charge-balancing cations, types of structures (layer or open framework), and defect types and amounts all affect stability. A systematic energetic study of these layer and framework materials is essential to an understanding of all of these effects. The buserite-like phase, which has interlayer space occupied by exchangeable cations and two water layers, is made from the birnessite-like phase, which has interlayer space occupied by exchangeable cations and one water layer, via an ion-exchange reaction in a Mg-rich aqueous solution. The buserite-like phase then can easily transform to the todorokite-like phase by hydrothermal heating, while the birnessite-like phase does so much more reluctantly. Hence, another unanswered fundamental question is how structural, bonding, and energetic factors drive these transformations. A related question is why these materials seldom form large crystals but usually remain as nanomaterials.

Recently, the enthalpies of formation for several natural and synthetic birnessite, buserite, and todorokite, hollandite, and other frameworks have been measured (see Table 3). The energy differences among different structure types can be quite small when their compositions are similar, but the heats of formation span a fairly broad range as the cation, degree of manganese reduction, and water content vary. Assuming no defects on oxygen sites, the intercalation of a mono- or divalent cation reduces tetravalent manganese by the reaction: $nMn^{4+} \rightarrow M^{n+} + nMn^{3+}$. The oxidation of Mn^{3+} to Mn^{4+} is strongly exothermic; for example ΔH = -41 J/mol for the reaction $0.5\ Mn_2O_3 + 0.25\ O_2 = MnO_2$ (Fritsch and Navrotsky 1996). To minimize this effect, Table 3 lists enthalpies of formation from binary oxides for the same average oxidation state, that is, for the reaction.

$$(x/2)M_2O + (x/2)Mn_2O_3 + (1-x)MnO_2\,(\text{pyrolusite}) + nH_2O$$
$$= M_x^{+}Mn_x^{3+}Mn_{1-x}^{4+}O_2 \cdot nH_2O \quad [M = Li,Na,K] \qquad (7)$$

or

$$xMO + xMn_2O_3 + (1-2x)MnO_2\,(\text{pyrolusite}) + nH_2O$$
$$= M_x^{2+}Mn_{2x}^{3+}Mn_{1-2x}^{4+}O_2 \cdot nH_2O \quad [M = Mg,Ca,Sr,Ba,Pb] \qquad (8)$$

This reaction involves no oxidation-reduction and no oxygen gas is evolved or consumed. Nonetheless, its enthalpy varies from near zero to -60 kJ/mol. Part of this variation appears to be related to the hydration state. Tian et al. (2000) measured the energetics of 14 synthetic manganese dioxides, including framework phases with todorokite structure and layer phases with birnessite and buserite structures by drop solution calorimetry in molten $3Na_2O \cdot 4MoO_3$ at 977 K. The composition of these phases with a general formula of $Na_xMg_yMnO_2 \cdot yH_2O$ was systematically varied. Using the measured enthalpies of drop solution, the enthalpies of formation from oxides based on one mole of $Na_xMg_yMnO_2 \cdot yH_2O$ at 298 K were calculated. $\Delta H^o_{f(oxides)}$ shows no obvious correlation with either content of charge-balancing cations or content of water for the layer and framework phases. Using a new parameter, the hydration number of the charge-balancing cations ($N_{(hyd)}$ = the number of water molecules per cation), the $\Delta H^o_{f(oxides)}$ data can be linearly correlated (see Fig. 5). The linear correlations are $\Delta H^o_{f(oxides)} = 16.5N_{(hyd)} - 73.3$ kJ/mol for the layer phases, and $\Delta H^o_{f(oxides)} = 11.7N_{(hyd)} - 64.4$ kJ/mol for the framework phases. The average $N_{(hyd)}$ values for the birnessite, buserite, and todorokite phases are 2, 4, and 2, respectively.

Titanium oxide

Titania is an important accessory oxide phase in the geologic environment. It is also of major technological importance in paints, whiteners, glazes, catalysts, and solar energy conversion. Rutile is the stable high temperature phase, but anatase and brookite are common at low temperature in nanoscale materials in both natural and synthetic samples. All three structures are based on interconnected TiO_6 octahedra, but their linkages and

Table 3. Enthalpies of formation at 298 K of MnO_2-based nanomaterials.

Sample	*Phase*	*Composition*	$\Delta H^o_{f\ (oxides)}$ [f]	$N_{(hyd)}$ [g]
B32[a]	birnessite	$Na_{0.27}Mg_{0.36}MnO_2 \cdot 1.09H_2O$	-45.3±1.7	1.75
B33[a]	birnessite	$Na_{0.27}Mg_{0.34}MnO_2 \cdot 1.19H_2O$	-46.6±1.7	1.95
B34[a]	birnessite	$Na_{0.34}Mg_{0.33}MnO_2 \cdot 1.05H_2O$	-50.6±2.1	1.57
BMg2[a]	birnessite	$Na_{0.46}MgO_{0.08}MnO_2 \cdot 1.42H_2O$	-26.7±2.2	2.61
BMg6[a]	birnessite	$Na_{0.40}Mg_{0.22}MnO_2 \cdot 1.26H_2O$	-28.9±2.2	2.06
BMg10[a]	birnessite	$Na_{0.27}Mg_{0.34}MnO_2 \cdot 1.19H_2O$	-46.6±1.7	1.95
K-OL-1-a[b]	birnessite	$K_{0.125}MnO_2 \cdot 0.19H_2O$	-36.7±1.1	1.50
K-OL-1-b[b]	birnessite	$K_{0.29}MnO_2 \cdot 0.19H_2O$	-69.6±1.0	0.66
Na-OL-1[c]	birnessite	$Na_{0.26}Mg_{0.14}MnO_2 \cdot 0.19H_2O$	-65.6±4.0	0.48
Mg-OL-1[a]	buserite	$Na_{0.02}Mg_{0.24}MnO_2 \cdot 1.19H_2O$	-4.1±2.1	4.51
(Na)Mg-OL-1[a]	buserite	$Na_{0.26}Mg_{0.07}MnO_2 \cdot 1.16H_2O$	-13.3±1.5	3.50
T32[a]	todorokite (3×3 tunnel)	$Na_{0.03}Mg_{0.46}MnO_2 \cdot 0.582H_2O$	-58.4±1.6	1.20
T33[a]	todorokite	$Na_{0.03}Mg_{0.49}MnO_2 \cdot 1.28H_2O$	-38.1±1.7	2.45
T34[a]	todorokite	$Na_{0.03}Mg_{0.47}MnO_2 \cdot 1.04H_2O$	-38.4±2.1	2.06
TMg2[a]	todorokite	$Na_{0.03}Mg_{0.23}MnO_2 \cdot 0.488H_2O$	-37.4±2.0	1.88
TMg6[a]	todorokite	$Na_{0.03}Mg_{0.38}MnO_2 \cdot 1.02H_2O$	-25.8±1.2	2.47
TMg10[a]	todorokite	$Na_{0.03}Mg_{0.49}MnO_2 \cdot 1.28H_2O$	-38.1±1.7	2.45
Mg-OMS-1[c]	todorokite	$Mg_{0.19}MnO_2 \cdot 0.75H_2O$	-22.8±1.9	3.95
large-OL-1[c,d]	birnessite/buserite	$Mg_{0.14}Na_{0.26}MnO_2 \cdot 0.19H_2O$	-65.6±4.0	0.48
small-OL-1[c,d]	birnessite	$Mg_{0.13}Na_{0.04}MnO_2 \cdot 0.19H_2O$	-19.9±2.4	1.11
[Ni]-OL-1[c]	birnessite	$Mg_{0.13}Na_{0.26}Ni_{0.16}MnO_2 \cdot 0.19H_2O$	-60.8±1.7	0.34
[Cu]-OL-1[c]	birnessite	$Mg_{0.14}Na_{0.19}Cu_{0.007}MnO_2 \cdot 0.19H_2O$	-25.3±1.9	0.90
[Ni]-OMS-1[c]	todorokite	$Mg_{0.20}Na_{0.013}Ni_{0.16}MnO_2 \cdot 0.75H_2O$	-10.1±1.7	2.43
[Cu]-OMS-1[c]	todorokite	$Mg_{0.20}Na_{0.013}Cu_{0.16}MnO_2 \cdot 0.75H_2O$	-13.7±1.1	2.43
hollandite[b]	hollandite (2×2 tunnel)	$K_{0.12}Mn_{1.00}O_2 \cdot 1.55H_2O$	-42.5±3.0	12.90
hollandite[b]	hollandite	$K_{0.3}Mn_{1.00}O_2 \cdot 1.55H_2O$	-11.5±3.0	5.17
cryptomelane[b,e]	hollandite	$Na_{0.0125}Al_{0.029}Si_{0.01}K_{0.005}Mn_{0.82}Fe_{0.165}$-$Ba_{0.09}Pb_{0.02}O_2 \cdot 0.099H_2O$	-14.9±3.0	0.59
cryptomelane[b,e]	hollandite	$Na_{0.03}Al_{0.02}K_{0.12}Mn_{0.94}Fe_{0.0375}Sr_{0.016}$-$Ba_{0.012}O_2 \cdot 0.0375H_2O$	-28.9±3.0	0.18
coronadite[b,e]	hollandite	$Pb_{0.175}Mn_{0.99}Al_{0.006}Zn_{0.006}O_2 \cdot 0.19H_2O$	-3.5±3.0	1.01
hollandite[b,e]	hollandite	$Ba_{0.132}Mn_{1.00}O_2 \cdot 0.268H_2O$	-24.1±3.0	2.03
romaneschite[b,e]	2×3 tunnel	$Mg_{0.19}Mn_{1.00}O_2 \cdot 0.75H_2O$	-22.8±3.0	3.94
todorokite[b,e]	todorokite	$K_{0.125}Mn_{1.00}O_2 \cdot 0.19H_2O$	-36.7±3.0	1.52

a. Tian et al. (2000) b. Fritsch et al. (1998) c. Laberty et al. (2000) d. large and small refer to interlayer spacing e. natural sample f. formation from MnO, Mn_2O_3(pyrolusite), plus tunnel cation oxides, no oxidation-reduction involved. g. uncertainty estimated h. Fritsch et al. (1997)

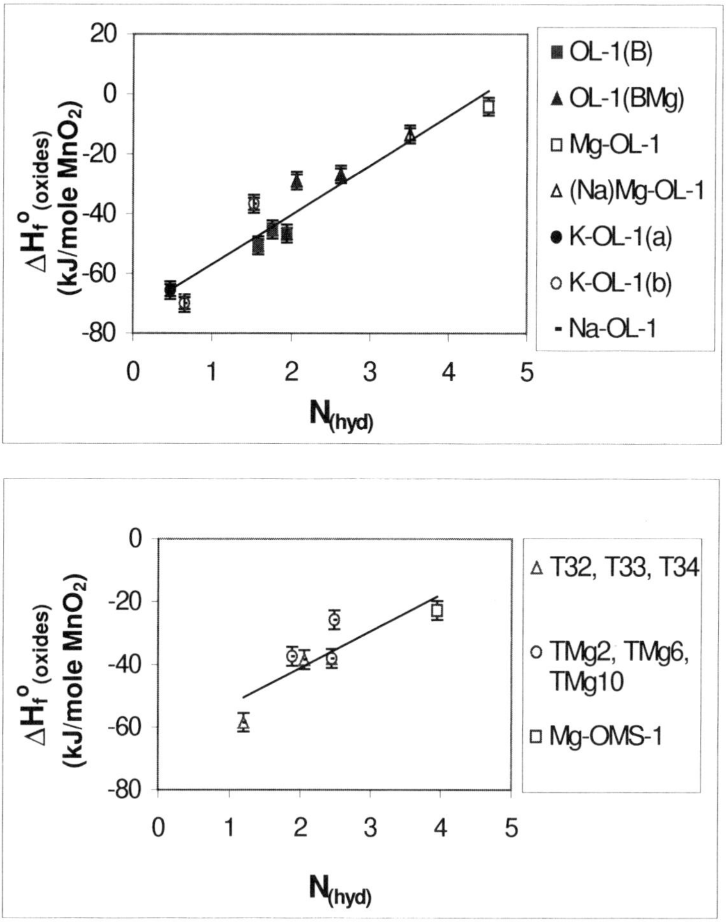

Figure 5. Correlations between enthalpy of formation (Eqn. (7) or (8)) and degree of hydration. [Used by permission of the editor of *J Phys Chem B*, from Tian et al. (2000), Figs. 2c and 3c, p. 5038.]

degree of edge and face sharing differ. Several sequences of phase transformation have been observed concomitant with coarsening. The following transformations are all seen, each under somewhat different conditions of particle size, starting material, temperature, and other parameters (Gribb and Banfield 1997; Zhang and Banfield 1999, 2000a,b).

$$\text{anatase} \rightarrow \text{brookite} \rightarrow \text{rutile} \tag{9}$$

$$\text{brookite} \rightarrow \text{anatase} \rightarrow \text{rutile} \tag{10}$$

$$\text{anatase} \rightarrow \text{rutile} \tag{11}$$

$$\text{brookite} \rightarrow \text{rutile} \tag{12}$$

The observation that nanophase anatase sometimes transforms to nanophase brookite and in other cases nanophase brookite transforms to nanophase anatase implies very closely balanced energetics as a function of particle size. Several competing mechanisms (surface nucleation, interface nucleation, homogeneous nucleation) have been proposed for these reactions as a function of temperature and/or particle size (Zhang and Banfield 1999, 2000a,b).

Table 4. Enthalpies at 298 K of TiO_2 polymorphs.

Phase	*Surface area or particle size*	*Enthalpy relative to bulk rutile (kJ/mol)*
Rutile	coarse	0
Anatase	nd	6.6±0.8[a,h], 5.5±0.9[a]
	nd	3.3±0.4[b,h], 1.9±0.5[b]
	natural, 1-2 mm	2.9±1.3[c,h]
	nd	0.4±0.2[d,h]
	nd	8.4±5.9[e]
	12000 m^2/mol	7.0±1.0[i]
Brookite	nd	0.7±0.4[b]
	natural, coarse single crystals	0.8±0.4[c,h]
	nd	0.4±0.3[f,h]
	6400 m^2/mol	7.8±1.1[i]
α-PbO_2 structure	coarse high pressure phase	3.2±0.7[g]

a. Navrotsky and Kleppa (1967), oxide melt solution calorimetry
b. Mitsuhashi and Kleppa (1979), oxide melt solution calorimetry
c. Mitsuhashi and Kleppa (1979), DSC
d. Rao (1961a), DTA
e. Margrave and Kybett (1965), fluorine bomb calorimetry
f. Rao et al. (1961b), DTA
g. Navrotsky et al. (1967), oxide melt solution calorimetry
h. at 973 to 1173 K
i. Ranade et al. (2001, in preparation)

The energetics of anatase, brookite, and rutile were measured by high temperature oxide melt calorimetry, first by Navrotsky and Kleppa (1967) and later by Mitsuhashi and Kleppa (1979). The results, as well as those of other studies, scatter significantly (see Table 4). Although some other reasons for these discrepancies were proposed by Mitsuhashi and Kleppa (1979), it now seems more likely that much of the difference could result from different particle size and/or water retained in the samples under calorimetric conditions. What is clear is that the enthalpy differences between polymorphs are small enough for complex crossovers in free energy among the three phases to be likely. Also there is no doubt that rutile is the stable polymorph for macrocrystalline materials.

Terwilliger and Chiang (1995) made careful measurements by differential scanning calorimetry of the enthalpy associated with the coarsening of a rutile nanocomposite. They concluded that the specific excess enthalpy at low temperature and small grain size (600-780°C and 30-200 nm) was in the range 0.5 to 1 J/m^2, while that measured over a larger temperature and grain size range (600-1300°C, 30-2000 nm) was higher, 1.3 to 1.7 J/m^2. They concluded, after elegant and careful control experiments to eliminate other factors, that the most likely cause of this size-dependent surface enthalpy is a size-dependent small nonstoichiometry due to the impingement of space charge layers in the grain size and temperature range of their experiments. One wonders whether similar effects could be partially responsible for the complexity of competing nucleation mechanisms inferred by Zhang and Banfield (2000a), and for the particle size dependence of the free energy of adsorption of organic acids (Zhang et al. 1999).

We are completing a more thorough study by high temperature oxide melt solution calorimetry of the effect of particle size and phases present on the energetics of anatase, brookite, and rutile phases (Ranade et al., in preparation). Some preliminary results are summarized here.

We have studied nanorutile samples with surface area ranging from 1758-5833 m^2/mol. The transformation enthalpies versus the surface area gives surface enthalpy of 2.2 J/m^2 (see Fig. 6). A nanoanatase sample (~12000 m^2/mol) and a nanobrookite sample (~6400 m^2/mol) have enthalpies of 7.02±0.96 kJ/mol and 7.75±1.12 kJ/mol, respectively, relative to bulk rutile (see Table 4). Efforts to deconvolute surface energy and transformation enthalpy of anatase and brookite from the energetics of these samples and of nanophase anatase-rutile mixtures and anatase-brookite mixtures are in progress.

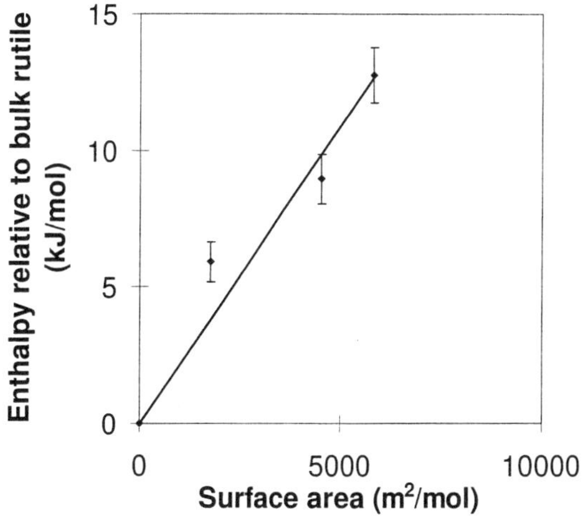

Figure 6. Transformation enthalpy (kJ/mol) against surface area (m^2/mol) of nanorutile samples [Used by permission of the editor of *Chemistry of Materials*, from Ranade et al. (submitted 2001), Fig. 3.]

Zhang et al. (1999) studied the adsorption of organic acids onto anatase nanoparticles. They found a profound enhancement in the adsorption constant (K_{ads}) with decreasing particle size. This was predicted by a thermodynamic model which used a Langmuir isotherm modified to include the dependence of interfacial tension on particle size. This is also discussed in their paper in this volume. In a macroscopic sense, this strong dependence arises from the curvature of the surface. In a microscopic sense, the nanoparticles are faceted rather than spherical. The increased surface free energy arises from the increased contribution of facets having higher surface energy. But as the particle gets smaller, more of its atoms are near the surface (see Fig. 1) and the relaxation of these near-surface atoms plays an increasingly important role. Thus the detailed atomistic picture is probably far more complex. Both atomistic modeling and direct calorimetric measurements would be very interesting.

Zirconium oxide

Zirconia is an important refractory ceramic. It also forms the basis of solid electrolytes in systems such as ZrO_2-Y_2O_3. Both zirconia and hafnia figure prominently in refractory ceramics proposed for the containment of uranium and plutonium radioactive waste. The monoclinic form of ZrO_2, baddeleyite, is a widespread accessory mineral. Zirconia undergoes transitions from monoclinic to tetragonal to cubic, with thermochemical parameters listed in Table 5.

When ZrO_2 is prepared at low temperature by the hydrolysis of $ZrCl_4$ or an

Table 5. Enthalpies at 298 K of ZrO_2 polymorphs.

Phase	*Surface area or particle size*	*Enthalpy relative to bulk baddeleyite (kJ/mol)*
monoclinic (baddeleyite)	coarse	0
tetragonal	coarse	6.0[d]
cubic	coarse	13.5±2.2[c]
		10.6±1.7[e]
		11.5[f]
amorphous	37700 m^2/mol	58.6±3.3[a]
	42000 m^2/mol	50.1±4.9[g]
	coarse	32.6[b]

a. Molodetsky et al. (2000)
b. Estimated using data of Molodetsky et al. (2000) and Ellsworth et al. (1994)
c. Molodetsky et al. (1998)
d. Robie and Hemingway (1995), at 1478 K
e. Lee and Navrotsky (in preparation)
f. Ackermann et. al (1975)
g. Ranade et al. (2001, in preparation)

forms (Clearfield 1964, Garvie 1965, 1978; Wu and Wu 2000, Molodetsky et al. 2000). During dehydration and coarsening on heating, the initial crystalline material is often the tetragonal rather than the monoclinic phase (Garvie 1965, 1978; Wu and Wu 2000, Molodetsky et al. 2000). The small enthalpy and free energy differences among the monoclinic, tetragonal, and cubic polymorphs suggest that this is yet another system in which stability crossovers at the nanoscale are likely. Strong evidence for such a crossover is seen in an experiment by Wu and Wu (2001). They took nanosized (20 nm) tetragonal zirconia, transformed it to the monoclinic phase by grinding (mechanical stress) and then heated it to 900 and 1100°C. The monoclinic material converted back to tetragonal zirconia. This reversal strongly suggests that the tetragonal phase is thermodynamically stable at these grain sizes. Malek et al. (1999) present a detailed study of nanocrystallization kinetics in amorphous zirconia.

Wu and Wu (2000) showed that the apparent stability of tetragonal zirconia made by precipitation synthesis can be enhanced by the replacement of surface hydroxyl groups with surface silanol groups. This may reflect both suppressed grain growth and diminished surface energy. Del Monte et al. (2000) also found an apparent stabilization of tetragonal zirconia by the incorporation of silica into the precipitated zirconia. Whether the major effect is that of constraint and separation of particles by a silica coating or actual substitution of a small amount of silicon in the crystal structure is not clear at present. One may speculate whether natural nanoparticles with adsorbed coatings of other oxides may also have reduced surface energies, and whether this is actually a driving force for adsorption and surface precipitate formation in the environment.

Molodetsky et al. (2000) measured the energetics of amorphous zirconia by high temperature oxide melt solution calorimetry (see Table 5). Using these data and somewhat indirect arguments, they estimated the difference in surface enthalpy between tetragonal and amorphous zirconia to be 1.19±0.08 J/m^2, with the amorphous material having the lower surface enthalpy.

Ellsworth et al. (1994) measured the enthalpy of amorphous (metamict) zircon, $ZrSiO_4$, produced by radiation damage. Assuming that this dense amorphous material has an enthalpy similar to that of a mixture of bulk amorphous silica and amorphous zirconia, the enthalpy of bulk amorphous ZrO_2 is estimated to be about 32.6 kJ/mol above that of baddeleyite. From the particle size (obtained by electron microscopy) of the nanophase amorphous zirconia of Molodetsky et al. (2000) the surface area is estimated as 37700 m^2/mol. Its surface enthalpy can then be estimated as 26000/37700 = 0.7 J/m^2. The surface enthalpy of monoclinic zirconia can then be estimated, using the energy difference of Molodetsky et al. (2000), as 1.2 + 0.7 = 1.9 (±0.5) J/m^2. These values for both amorphous and monoclinic zirconia are physically reasonable in magnitude.

Zirconia also undergoes high pressure phase transitions from the monoclinic to two orthorhombic structures. A thermochemical study of these phases, combined with phase equilibrium observations (Ohtaka et al. 1991) suggests that surface energy changes the apparent position of phase boundaries, as well as enhances the kinetics of transformation. However Ohtaka et al. (1991) were unable to quantify these effects.

Enthalpies of various transitions in zirconia are summarized in Table 5. It is clear that the small enthalpies of these transitions allow free energy crossovers at the nanoscale.

Magnesium aluminate

Nanocrystalline spinel can be synthesized by a variety of routes using nanophase precursors. The formation of $MgAl_2O_4$ from freeze-dried nitrate precursor was studied by thermogravimetric analysis, differential thermal analysis, powder diffraction, transmission electron microscopy, ^{27}Al magic angle spinning NMR, and high-temperature solution calorimetry by McHale et al. (1998). A single phase, slightly alumina-rich spinel of composition $Mg_{0.957}Al_{2.028}O_4$ was obtained from the precursor by calcination at temperatures 1073 K. Transmission electron microscopy revealed that material calcined at 1073 K was nanocrystalline, with grain size on the order of 20 nm. ^{27}Al NMR revealed that this material had an unusually high degree of cation disorder, with an order parameter of 0.59 at room temperature. This degree of disorder, which has previously only been achieved in $MgAl_2O_4$ via neutron bombardment, provides strong thermodynamic evidence that the freeze-dried precursor contained a highly disordered and probably close to random mixture of cations. Significant levels of five-coordinated Al^{3+} were detected in amorphous samples calcined at 973 K. Increasing calcination temperatures resulted in a decrease in the percentage of tetrahedral Al^{3+} and a simultaneous increase in the average particle size of the material. Drop solution calorimetry in $2PbO{\cdot}B_2O_3$ at 975 K revealed an enthalpy difference of 39.9±7.4 kJ mol^{-1} between the disordered nanophase $MgAl_2O_4$ synthesized at 1073 K and the well-crystallized material synthesized at 1773 K. Particle size, cation distribution, and adsorbed H_2O affect the energetics, with the surface energy term dominant. A surface enthalpy of 1.8±0.8 J/m^2 has been estimated. This is similar to the value for γ-Al_2O_3, 1.67 J/m^2 (McHale et al. 1997b).

There are some generalizations possible from this study. (1) In systems where cation or other disorder is possible, low temperature nanophase materials will be extensively disordered. This probably reflects disequilibrium resulting from rapid nanoparticle formation rather than a strong dependence of ordering energy on particle size. (2) There is a gradual evolution on heating from x-ray amorphous, to nanocrystalline and disordered, to macrocrystalline and ordered materials. The lower temperature materials contain a large amount of water, and dehydration, coarsening, and ordering occur together. (3) Calcination of amorphous powder lends to rapid formation of nanocrystalline spinel (or other ternary phase) at temperatures 200-500°C lower than for normal solid state synthesis from mixtures of bulk oxides. This enhancement of reaction rate probably arises mainly from the decreased particle size leading to a shorter diffusion path, but the catalytic effect of water present and the enhanced reactivity at the surface of the nanoparticles may also play a role. The role of nanophases in enhancing the rates of diagenesis and early metamorphism in the geologic environment should be investigated as natural analogues to these synthetic processes. (4) Though the degree of order increases with calcination, this change is probably not the major driver; rather, loss of surface area (intimately tied to water loss) provides the thermodynamic driving force for the formation of a final well crystallized product.

Silica

Quartz does not generally form nanoparticles. Brace and Walsh (1962) measured the energy needed to cleave quartz single crystals. From this work, they obtained surface energies of 0.4, 0.5 and 1.0 J/m^2 for the (1011), (1011), and (1010) faces, respectively (see Table 1), Hemingway and Nietkowicz (1995) reported preliminary data on the enthalpy of solution in hydrofluoric acid of quartz of various size fractions. Their data vary in a nonlinear fashion with both estimated particle size (in the 500-10^5 nm range) and estimated surface area. However each of their samples contained a size fraction sorted to be below a given diameter, so the data can not be easily quantified to obtain a surface enthalpy. Qualitatively, their data appear consistent with a surface enthalpy near the 1 J/m^2 range.

There have been a number of estimates of the surface free energy of amorphous silica, using both contact angle and inverse gas chromatography techniques. These are summarized by Moloy et al. (2001) and yield an average value of 0.104 J/m^2. A solution calorimetric value listed in Iler's compendium on silica (Iler 1979) gives a surface enthalpy for amorphous silica of 0.26 J/m^2. It is not clear to what extent the hydration state of these large surface area, generally hygroscopic materials, was controlled or examined. Iler (1979) also discusses values for the energy of the silica-water interface estimating 0.4 J/m^2 for quartz/water and 0.08 to 0.13 J/m^2 for amorphous silica/water. These values are quite uncertain. Iler (1979) also presents data for increasing solubility of amorphous silica in water with decreasing particle size. The data appear to depend on the temperature of preparation of the amorphous silica. Clearly modern quantitative work, supported by better materials characterization, is needed.

It has been reported (see Moloy et al. (2001) for a summary) that surface free energy terms for amorphous silica have a significant negative temperature dependence, consistent with a positive surface entropy.

Calorimetric studies of the energetics of sol-gel derived amorphous silica have been published (Maniar and Navrotsky 1990, Maniar et al. 1990, Ying et al. 1993a). The energetics of transformation of a hydrous porous gel to a dense anhydrous glass has contributions from dehydration (dehydroxylation) and rearrangement of ring structure which are probably much larger than the surface energy term. Because the initial materials contain a large amount of water in their pores, and dehydration and coarsening are not separable, neither surface areas nor surface enthalpies were determined in those studies.

Ying et al. (1993b) also measured the energetics of a colloidal silica (initial surface area of 157 m^2/g or 9400 m^2/mol) by oxide melt solution calorimetry. They argue that most of its roughly 10 kJ/mol energetic instability relative to bulk glass can be attributed to surface area reduction. This would give an upper bound to the surface enthalpy of about 1 J/m^2, since some of the enthalpy difference seen in calorimetry may be due to structural rearrangements.

The surface energy (surface tension) of supercooled molten silica above 1000°C has been measured as 0.3 J/m^2. This is in the same range as values reported for amorphous silica.

Since the enthalpy difference between quartz and glass is about 9 kJ/mol, and the free energy difference somewhat less than this, a (free) energy crossover at the nanoscale is expected. If one takes the surface enthalpy of quartz as 1 J/m^2 larger than that of amorphous silica (a conservative estimate), the enthalpy crossover will occur at a surface area of 9000 m^2/mol (150 m^2/g) . This crossover at a relatively small surface area is consistent with the observation that all nanosized silica samples, both natural and synthetic, are amorphous.

Zeolites are an increasingly important class of materials currently utilized in a wide variety of industrial applications, including catalysis, ion exchange, gas separation, membrane separation, nuclear waste disposal, chemical sensing, and pollution abatement. These periodic, microporous materials are generally composed of aluminosilicate tetrahedral frameworks that have channel and cage dimensions ranging between 0.2 and 2.0 nm. More than 130 distinct structures have been classified to date according to their unique, specific framework topologies, with about 20 of these crystallized as essentially pure SiO_2. Many significant questions remain about the relationship between structure and energetics. Why, for example, is it possible to form so many structurally distinct, yet energetically similar, structures? Which energetically sensitive structural parameters (bond lengths, bond angles, degree of covalency, etc.) favor, or limit, the creation of a particular structure? Answers to questions such as these are crucial to identifying and understanding the underlying forces and mechanisms responsible for the formation of these remarkable materials, as well as for predicting, and devising synthetic pathways to, new structures.

Petrovic et al. (1993) reported seven high-silica zeolite structures (AFI, EMT, FAU, FER, MEL, MFI, and MTW, topologies in standard International Zeolite Association nomenclature) to be 5.6-14.3 kJ/mol less energetically stable than α-quartz. Navrotsky et al. (1995) reported the enthalpy of MEI was 13.9±0.4 kJ/mol higher than that of α-quartz. Piccione et al. (2000) found eleven high-silica zeolite structures (AST, BEA, CFI, CHA, IFR, ISV, ITE, MEL, MFI, MWW, and SST) to be 6.0-15.5 kJ/mol less stable in enthalpy than α-quartz.

Attempts to correlate these energetic instabilities to specific structural parameters have been partially successful. Ab initio calculations suggested that Si–O–Si bond angles below 135° destabilize silica structures (Geisinger et al. 1985). Kramer et al. (1991) and de vos Burchart et al. (1992), based on a computational approach, identified a linear relationship between energetics and framework density in high-silica zeolites. Petrovic et al. (1993), based on calorimetric measurements, did not find a significant correlation with either framework density or the mean Si–O–Si bond angle, although they did establish a weak correlation between enthalpy and the fraction of Si–O–Si bond angles below 140°. Henson et al. (1994), however, using the Petrovic et al. data and eliminating EMT, claimed a direct correlation of energetics and framework density in both the calorimetric data and lattice energy calculations. Piccione et al. (2000) also identified a linear relationship between measured energetics and both framework density and molar volume. Further, they found no correlation between with silicon (Si–Si) non-bonded distances or loop configurations.

The physical meaning of any energetic correlation with either framework density or molar volume is not obvious. Some of the most important structural characteristics of zeolites are their high porosity and high internal surface area. They are nanomaterials

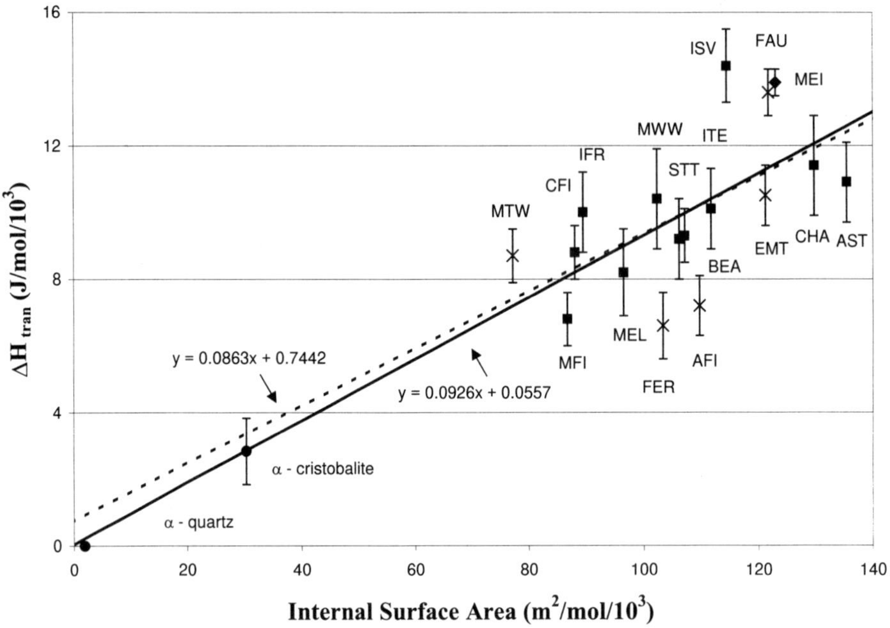

Figure 7. ΔH_{tran} vs. internal surface area for α–quartz, α–cristobalite, and seventeen zeolites using a probe-atom radius of 0.96 Å. The solid regression line includes all 19 data points, while the dashed regression line excludes α–quartz and α–cristobalite. (–■–) Piccione et al.; (–◆–) Navrotsky et al.; (–×–) Petrovic et al.; (–●–) Richet et al.; (—) all data; (- - -) zeolite data only. [Reprinted from *Microporous and Mesoporous Materials*, Moloy EC et al. (submitted 2001), with permission from Elsevier Science.]

because of their internal surfaces. Would it be fruitful, then, to ask whether the energetic destabilization is related to the internal surface area? Would the slope of a correlation between enthalpy and internal surface area provide a reasonable, physically meaningful surface energy? To address this question, Moloy et al. (2001) used Cerius^{2}® simulation software to calculate internal surface area for correlation of energetics. Calculated rather than experimental internal surface areas were used to provide a uniform reference and to avoid the inconsistencies inherent in the very scant experimental data set for silica surface areas, none of which refer to the open zeolitic frameworks. The results are shown in Figure 7. A linear correlation is found, giving an internal surface enthalpy of 0.093 + 0.007 J/m^2. This value is similar to the value of 0.104 J/m^2 for the average external surface free energy of various, mainly amorphous but not microporous or mesoporous, silica phases reported in the literature (see Table 1). Thus it is physically reasonable to consider the metastability of anhydrous silica zeolites to arise from their large internal surface area. The average value of the surface enthalpy (or surface free energy) appears similar for both internal and external surfaces and does not depend significantly on the specific nature of the tetrahedral framework (various zeolite structures or amorphous). It is interesting that the surface enthalpy of the zeolitic silicas appears to be relatively low and much more similar to that of amorphous silica than that of quartz. The bulk energetics also place the zeolitic silicas much closer to amorphous silica than to quartz. The microscopic implications of these general trends, in terms of surface structure, bonding, and reactivity, remain to be explained.

Interfaces, dislocations, and nanocomposites

Not all nanomaterials of geologic, environmental, and technological importance consist of isolated particles. The nanoparticles may be agglomerated either loosely or as a dense nanocompact (for example a dense transparent ceramic consisting of particles too small to scatter visible light). An initially homogeneous macroscopic solid may attain some "nano" character by having a large concentration of planar defects: dislocations, twin planes, or nanoscale phase separated regions. In these cases the important term is interfacial, rather than surface, energy or enthalpy.

There are several studies that suggest that these interfacial enthalpies are an order of magnitude smaller than the surface enthalpies discussed above. Liu et al. (1995) measured the enthalpy of quartz with a dislocation density of about 10^{11} cm^{-2} to be about 0.6 kJ/mol higher than that of undeformed quartz, in reasonable agreement with theoretical values calculated by Hirth and Lothe (1982), Wintsch and Dunning (1985), Green (1972) and Blum et al. (1990). Though such values can not be readily converted to an interfacial energy, they do suggest an overall effect on the thermodynamics that is one to two orders of magnitude smaller than the effect of particle size diminution described above.

A similar conclusion was drawn by Penn et al. (1999). They found an excess energy of about 300 J/mol in fibrolite relative to bulk sillimanite (Al_2SiO_5) due to grain boundaries. Though this effect is small compared to the nanoparticle surface energies discussed above, it is still potentially geologically significant because of the very closely balanced energetics of the kyanite - andalusite - silliminite equilibria.

Moganite is a silica polymorph similar to, but distinct from, quartz. Its structure is related to that of quartz by twinning at the unit cell scale (and accompanying structural distortions) and its enthalpy is 3.4±0.7 kJ/mol higher than that of quartz (Petrovic et al. 1996). If one considers this twining as creating internal interfaces at the unit cell scale, then this relatively small enthalpy difference is consistent with a small interfacial energy. However one should stress that there are no broken bonds at such interfaces.

There are similar findings in phase separated glassy systems. In both lithium silicate and lanthanum-containing potassium silicate systems, there is spectroscopic evidence that the initially prepared optically clear glasses consist of nanodomains of "phase-ordering" or incipient phase separation (Ellison and Navrotsky 1990, Sen et al. 1994). These domains coarsen to form milky white macroscopically phase separated glasses on annealing. Presumably the interfacial areas of the former are much larger than of the latter. The enthalpy release on coarsening is 2.3±2.6 kJ/mol for $Li_2Si_4O_9$ glass (Sen et al. 1994) and about 1 kJ/mol for a glass of composition $(K_2Si_5O_{11})_{0.86}(La_2O_3)_{0.14}$ (Ellison and Navrotsky 1990). These small enthalpies argue for relatively small interfacial energies in these nanocomposite glassy materials.

Another class of nanocomposite materials consists of nanoarchitectured biphasic materials produced by controlled low temperature routes, sometimes involving organic or other templates. Such nano-in-nano composites are technologically interesting because coarsening is hindered by their biphasic nature.

Two recently studied examples are based on TiO_2. One is a TiO_2-ZrO_2 material consisting of a mesoporous anatase scaffold with the pores filled with amorphous zirconia. Another is a TiO_2-MoO_3 material in which anatase cores are surrounded by a shell of amorphous MoO_3. Their energetics have been studied by Ranade et al. (2001). Using the measured drop solution enthalpies, the transformation enthalpies to macroscopic stable crystalline phases have been calculated. The enthalpy of the transformation

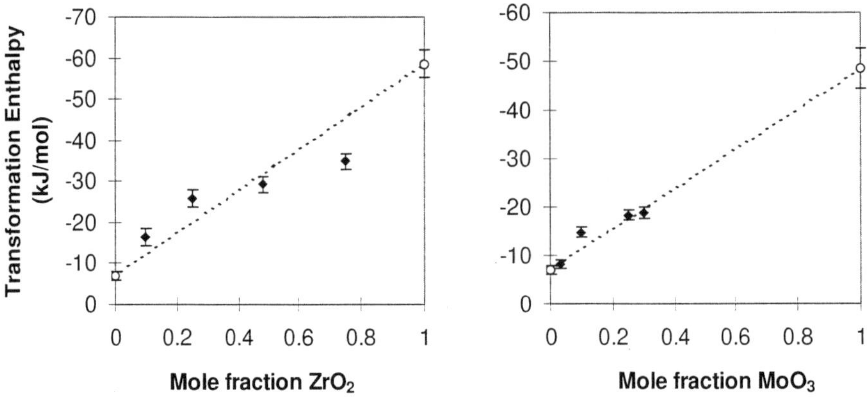

Figure 8. Transformation enthalpies to bulk crystalline phases versus mole fraction for (a) TiO_2-ZrO_2 composites and (b) TiO_2-MoO_3 composites. The solid diamond symbol represents the measured transformation enthalpies and the open symbols and the dashed line joining them represents a mechanical mixture of transformation enthalpies of the nano end-members (Ranade et al. 2001, in preparation).

from amorphous to monoclinic ZrO_2 obtained from this work, -50.08±4.92 kJ/mol, is in reasonable agreement with the value of -58.60±3.30 kJ/mol determined by Molodetsky et al. (2000). The enthalpy of transformation from amorphous to crystalline MoO_3 derived from these details is -51.25±2.01 kJ/mol. The linear variation of energetics with the composition of both series (see Fig. 8) suggests that these materials constitute a nanoscopic two-phase mixture, with their energetics a weighted sum of end-member enthalpies (see Fig. 7). The energetic effects of surface area and of phase transformations could not be separated because the surface area (m^2/mol) varies approximately linearly with mole fraction.

It is interesting that this simple approach using a mixture of nanophase end-members appear to be a good first approximation. The complex morphology of the nanocomposites makes it hard to know whether the BET surface area measurements truly capture the surface areas of both components. What is encouraging in this first study of the energetics of nanocomposites is that a rather simple approach appears to capture the main features of the energetic behavior. This simple thermochemical behavior in turn suggests that the complex nanoarchitecture is kinetically controlled by the synthesis conditions and templating agent(s) and that, other than the effect of polymorphism and surface area, any specific energy terms associated with the nano-meso scale architecture are small and can probably be neglected.

Nanoparticles of iron and aluminum oxides and oxyhydroxides transport both organic and inorganic contaminants in the environment. The systematics developed here may be applied to understanding such natural nanocomposites. For example, it may be possible to treat coating of amorphous uranium or chromium oxides on nanophase (Fe, Al)OOH particles as a mixture of nanophase end-members from the point of view of energetics.

Mixed hydroxides and hydroxycarbonates: from surface adsorbates to nanoparticles to bulk phases

The systems CaO-CO_2-H_2O and MgO-CO_2-H_2O contain a number of ternary phases, see Table 6. What is intriguing in the thermodynamic data is that, when formed from

Table 6. Thermodynamics at 298 K of ternary oxide, hydroxide, and carbonate mixed phases from binary components.

Mineral	*Composition*	*Formation from $M(OH)_2$, MCO_3, H_2O*	
		ΔH°_{298} (kJ/mol)	ΔS°_{298} (J/Kmol)
Nesquehonite	$MgCO_3 \cdot 3H_2O$	-6.6[a]	-79.5 (-26.5 per H_2O)[a]
Hydromagnesite	$4MgCO_3 \cdot Mg(OH)_2 \cdot 4H_2O$	+6.4[b]	-99.5 (-33.1 per H_2O)[a]
Actinite	$Mg(OH)_2 \cdot MgCO_3 \cdot 3H_2O$	-25.4[c]	-105.4 (-35.1 per H_2O)[a]
Monohydrocalcite	$CaCO_3 \cdot H_2O$	-5.1[c]	-30.6 (-30.6 per H_2O)[a]
Ikaite	$CaCO_3 \cdot 6H_2O$	-29.2[c]	-141.7 (-23.6 per H_2O)[a]
Thermonatrite	$Na_2CO_3 \cdot H_2O$	-14.7[a]	-36.9 (-36.9 per H_2O)[a]
Trona	$Na_2CO_3 \cdot NaHCO_3 \cdot 2H_2O$	-32.3[a]	—
Wegscheiderite	$Na_2CO_3 \cdot 3NaHCO_3$	-9.8[a]	—
Cobalt hydrotalcite	$Co_{0.68}Al_{0.32}(OH)_2(CO_3)_{0.16} \cdot 0.779H_2O$	-5.0[b]	—

[a] Robie and Hemingway (1995) [b] Allada et al. (in preparation)

binary carbonates, hydroxides, and liquid water, e.g.,

$$4\ MgCO_3 + Mg(OH)_2 + 4\ H_2O = \text{hydromagnesite} \tag{13}$$

the reactions have small negative enthalpies but surprisingly large negative entropies. The data imply that the incorporation of water in these phases has a very negative entropy (-25 to -35 J/mol K per H_2O) due, presumably, to the structuring of that water in the cation hydration environment within a well-defined crystal structure, although the structural details of these phases are still rather poorly known. For comparison, the enthalpy and entropy of crystallization of ice from liquid water are -40.6 kJ/mol and -109.0 J/molK, respectively. Thus it appears that the ternary compounds have lost amounts of entropy, on a per mole of H_2O basis, about 1/3 of the amount lost when water is immobilized in the solid structure of ice.

These data suggest very negative entropies of hydration for H_2O molecules surrounding cations sandwiched between aluminosilicate layers. If this "structuring" of water is a general phenomenon, it may have significant implications for diffusion and reactivity at layer and nanoparticle surfaces.

The sorption of heavy metal ions on mineral surfaces influences their mobility in the environment. Recent spectroscopic studies of the sorption of nickel on clays and aluminum oxide (Scheidegger et al. 1998) of cobalt on alumina (Towle et al. 1997) and of cobalt on clays (Thompson et al. 1999) argue for the presence of surface precipitate phases. These are probably double hydroxide phases involving transition metals and aluminum, very possibly related to bulk hydrotalcite ($Mg_6Al_2(OH)_{16}CO_3 \cdot H_2O$) and takovite ($Ni_6Al_2(OH)_{16}CO_3) \cdot H_2O$) (Bish and Brindley 1977, Taylor 1984). The surface precipitates appear to form under aqueous conditions which are undersaturated with respect to binary hydroxides, suggesting thermodynamic stability for the mixed-phases.

We are completing an initial solution calorimetric study of a cobalt hydrotalcite

(Allada et al., in preparation). For the reaction

$$0.52\ Co(OH)_2 + 0.16\ CoCO_3 + 0.32\ Al(OH)_3 + 0.779\ H_2O$$
$$= Co_{0.68}Al_{0.32}(OH)_2(CO_3)_{0.16}{\cdot}0.779\ H_2O \qquad (14)$$

ΔH = -5±3 kJ/mol (written on the basis of one mole of cations). Thus this phase is energetically only marginally more stable than a mixture of end-members. Its entropy of formation is unknown. These data suggest that the formation of the surface precipitate, relative to a mixture of cobalt hydroxides and carbonates adsorbed on a fully hydrated alumina surface, probably does not lower the free energy of the system by more than a small amount. Nevertheless, such a surface precipitate may alter further reactivity.

The thermal decomposition of hydrotalcites also is intriguing. They decompose on calcination to nanocrystalline spinel-like phases, and eventually, to crystalline spinels. This pathway initially produces spinels which are hydrated in their large surfaces and poorly ordered in their cation distributions. These materials are very reactive and analogous to "precursors" sought in ceramic synthesis. The thermochemical study of the evolution of $MgAl_2O_4$ from nitrate precursors (McHale et al. 1998) may in fact have encountered hydrotalcite as an intermediate. Further study of the thermal evolution of hydrotalcite nanomaterials may be applicable to both earth and materials sciences.

There have been reports of the easy reduction of takovite (the Ni-hydrotalcite) to metallic nickel on heating (Titulaer et al. 1994) in a reducing atmosphere, with the production of possible Ni/Al_2O_3 catalysts. Can similar reduction processes occur in nature during burial and diagenesis under locally reducing conditions, or during processes similar to serpentinization? If any metallic phases form, these may be nanoparticles, and the distribution of trace elements and heavy metals important to environmental concerns might be seriously affected.

Layered phases similar to hydrotalcites may also contain silicon (Taylor 1984). The cobalt silicate containing material is suggested to decompose, not to Co_2SiO_4 olivine, but to the Co_2SiO_4 spinel (normally stable only above about 6 GPa) on heating. This intriguing observation suggests a route for the preparation of metastable high pressure phases based on the cubic close packing of oxygens attained from the hydroxycarbonate decomposition. Can this occur in nature?

Clearly this field needs more exploration followed by careful structural, thermodynamic, and kinetic studies. There appears to be a continuum starting at adsorbed ionic species, proceeding to reconstructed surface or interfacial structures (surface or interface nanophases) and culminating in new bulk phases. The role of hydration in this process appears critical in the energetics as well as in reaction mechanisms.

FUTURE DIRECTIONS AND UNANSWERED QUESTIONS

Clearly, first and foremost, more data of higher quality are needed for the thermochemistry of nanoparticles and their composites. Measurements of surface enthalpies, hydration enthalpies, excess heat capacities, and other thermodynamic parameters on well defined chemical systems are needed. The question of "apparent" versus "true" surface properties raised by Diakonov (1998b) needs to be resolved and consistent nomenclature adopted. Surface (solid/gas), interface (solid/solid) and "wet" (solid/water) parameters each need to be measured and systematized.

Once more data exist, systematic questions can be formulated. The following questions already appear relevant: Does the crystalline phases of higher symmetry usually or invariably have the lower surface energy? Do amorphous phases always have lower surface energies than crystalline phases? Do hydroxides and oxyhydroxides have lower surface energies than oxides? Are surface entropies always positive? What factors determine whether a calculation/simulation gives reliable results? What is the influence of cation hydration and the confinement of water in pores, channels, and layers on nanophase energetics? To what extent can the energetic properties of nano-composites be predicted from properties of the nanoscale end-members? Many other questions, some of greater importance than these early ones, will arise as more experiments are performed.

The thermochemical parameters then need to be incorporated into geochemical models: Does it make sense to use surface area as another dimension in calculating (metastable) phase equilibria? Is it useful to provide a set of thermodynamic data for "generic" fine particles, say of 100 nm dimensions, for use in geochemical models. What environments, besides the now familiar aqueous low temperature setting, are likely to harbor nanoscale phenomena, and how would thermodynamic modeling be affected? Examples of such environments might include planetary surfaces (e.g., Mars or the Moon), condensation at low pressure from a pre-planetary nebula, dust particles in the upper atmosphere and in deep space, "clouds" in the giant gaseous planets, and rapidly reacting environments in planetary interiors which encounter small grain sizes due to phase transformations. Clearly this short course is being held to mark the onset of a field of endeavor rather than to summarize the well codified results of many years of research. This is especially true in the area of thermodynamics, which itself is undergoing a rebirth as so many new materials are being discovered. This is the beginning, not the middle, nor the end.

ACKNOWLEDGMENTS

Without continuous support from the National Science Foundation and the Department of Energy, much of what is reported here would not have been done. Without the flexibility to explore new directions and interconnections among low temperature geochemistry, high temperature mineral physics, and materials science, this work would have been intellectually poorer. I thank Bill Luth for pointing out to me, nearly 15 years ago, that the thermochemistry of low temperature materials *is* scientifically interesting, societally relevant, and fundable. I thank Mandar Ranade for providing data that will be part of his Ph.D. dissertation.

REFERENCES

Ackermann RJ, Rauh EG, Alexander CA (1975) The thermodynamic properties of gaseous zirconium oxide. High Temp Sci 7:304

Anovitz LM, Perkins D Essene EJ (1991) Metastability in near-surface rocks of minerals in the system Al_2O_3-SiO_2-H_2O. Clays Clay Minerals 39:225-233

Apps JA, Neil JM, Jun CH (1989) Thermochemical properties of gibbsite, bayerite, boehmite, diaspore, and the aluminate ion between 0 and 350°C. Washington, DC: US Nuclear Regulatory Commission (prepared by Lawrence Berkeley Laboratory), p 98

Bardossy G, Aleva GJJ (1990) Lateritic bauxites. Dev Econ Geol 27:624

Barrer RM, Cram PJ (1971) Heats of immersion of outgassed ion-exchanged zeolites. *In* Flanigen EM, Sand LB (eds) Molecular Sieve Zeolites II:150-131, American Chemical Society, Washington, DC

Bish DL, Brindley GW (1977) A reinvestigation of takovite, a nickel aluminum hydroxy-carbonate of the pyroaurite group. Am Mineral 62: 458-464

Blonski S, Garofalini SH (1993) Molecular dynamics simulation of α-alumina and γ-alumina surfaces. Surface Sci 295:263-274

Blum AE, Yund RA, Lasaga A (1990) The effect of dislocation density on the dissolution rate of quartz. Geochim Cosmochim Acta 54:283-297

Brace WF, Walsh JB (1962) Some direct measurement of the surface energy of quartz and orthoclase. Am Mineral 47:1111-1122
Brock SL, Duan N, Tian ZR, Giraldo O, Zhou H, Suib SL (1998) A review of porous manganese oxide materials. Chem Mater 10:2619-2628
Buol SW, Hole FD, McCracken RJ (1989) Soil Genesis and Classification. Iowa State University Press: Ames, Iowa, p 446
Carey JW, Bish DL (1996) Equilibrium in the clinoptilolite-H_2O system. Am Mineral 81:952-962
Carey JW, Bish DL (1997) Calorimetric measurement of the enthalpy of hydration of clinoptilolite. Clays Clay Minerals 45:826-833
Causa M, Dovesi R, Pisani C, Roetti C (1989) *Ab initio* characterization of the (0001) and $(10\bar{1}0)$ crystal faces of α-alumina. Surface Sci 215:259-271
Cerofolini GE (1975) A model which allows for the Freundlich and the Dubinin-Radushkevich adsorption isotherms. Surface Sci 51:333-335
Clearfield A (1964) Crystalline hydrous zirconia. Inorg Chem 3:146-148
de vos Burchart EV, Van Bekkum VA, Van de Graaf B (1992) A consistent molecular mechanics force field for all silica zeolites. Zeolites 12:183-189
del Monte F, Larsen W, Mackenzie JD (2000) Stabilization of tetragonal ZrO_2 in ZrO_2-SiO_2 binary oxides. J Am Ceram Soc 83:628-634
Diakonov I, Khodakovsky I, Schott J, Sergeeva E (1994) Thermodyanmic properties of iron oxides and hydroxides. I. Surface and bulk thermodynamic properties of goethite (α-FeOOH) up to 500 K. Eur J Mineral 6:967-983
Diakonov I (1998a) Thermodynamic properties of iron oxides and hydroxides. III. Surface and bulk thermodynamic properties of lepidocrocite (γ-FeOOH) to 500 K. Eur J Mineral 10:31-41
Diakonov I (1998b) Thermodynamic properties of iron oxides and hydroxides. II. Estimation of the surface and bulk thermodynamic properties of ordered and disordered maghemite (γ-Fe_2O_3). Eur J Mineral 10:17-29
Dugdale JS, Morrison JA, Petterson D (1954). The effect of particle size on the heat capacity of titanium dioxide. Proc R Soc London Ser A 224:228-235
Ellison AJG, Navrotsky A (1990) Thermochemistry and structure of model waste glass compositions. *In* Scientific Basis for Nuclear Waste Management XIII. Oversby VM, Brown PW (eds), Mater Res Soc Symp Proc 176:193-207
Ellsworth S, Navrotsky A, Ewing RC (1994) Energetics of radiation damage in natural zircon ($ZrSiO_4$). Phys Chem Minerals 21:140-149
Enüstün BW, Turkevitch J (1960) Solubility of fine particles of strontium sulfate. J Phys Chem 82: 4503-4509
Fleming S, Rohl A, Lee MY, Gale J, Parkinson G (2000) Atomistic modeling of gibbsite: surface structure and morphology. J Crystal Growth 209:159-166
Fowkes FM (1965) Attractive forces at interfaces. *In* Chemistry and Physics of Interfaces. Washington, DC: American Chemical Societype, p 1-12
Fritsch S, Navrotsky A (1996) Thermodynamic properties of manganese oxides. J Am Ceram Soc 79: 1761-1768
Fritsch S, Post JE, Navrotsky A (1997) Energetics of low temperature polymorphs of manganese dioxide and oxyhydroxide. Geochim Cosmochim Acta 61:2613-2616
Fritsch S, Post JE, Suib SL, Navrotsky A (1998) Thermochemistry of framework and layer manganese dioxide related phases. Chem Mater 10:474-479
Garvie RC (1965) The occurrence of metastable tetragonal zirconia as a crystallite size effect. J Phys Chem 69:1238-1243
Garvie RC (1978) Stabilization of the tetragonal structure in zirconia microcrystals. J Phys Chem 82: 218-224
Geisinger KL, Gibbs GV, Navrotsky A (1985) A molecular orbital study of bond length and angle variation in framework silicates. Phys Chem Minerals 11:266-283
Green HW II (1972) Metastable growth of coesite in highly strained quartz. J Geophys Res 77:2478-2482
Gribb AA, Banfield JF (1997) Particle size effects on transformation kinetics and phase stability in nanocrystalline TiO_2. Am Mineral 82:717-728
Haas KC, Schneider WF, Curioni A, Andreoni W (1998) The chemistry of water on alumina surfaces: reaction dynamics from first principles. Science 282:265-268
Hemingway BS, Robie RA, Apps JA (1991) Revised values for the thermodynamic properties of boehmite, AlO(OH) and related species and phases in the system Al-H-O. Am Mineral 76:445-457
Hemingway BS, Nitkiewicz A (1995) Variation of the enthalpy of solution of quartz in aqueous HF as a function of sample particle size. U S Geological Survey Open File Report 95-510

Henson NJ, Gale JD, Cheetham AK (1994) Theoretical calculations on silica frameworks and their correlation with experiment. Chem Mater 6:1647-1650

Hirth JP, Lothe J (1982) Theory of Dislocations. New York: John Wiley & Sons

Hodkin EN, Nicholas MG (1973) Surface and interfacial properties of stoichiometric uranium dioxide. J Nuclear Mater 47:23-30

Hunter RJ (1986) Foundations of Colloid Science,Volume I. Oxford, UK: Oxford University Press

Kingery WD (1958) Surface tension of some liquid oxides and their temperature coefficients. J Am Ceram Soc 42:6-10

Iler RK (1979) The chemistry of silica: solubility, polymerization, colloid and surface properties, and biochemistry. New York: John Wiley & Sons

Kramer GJ, De Man AJM, Van Santen RA (1991) Zeolites vs. aluminosilicate clusters: the validity of a local description. J Am Chem Soc 113:6435.

Laberty C, Navrotsky A (1998) Energetics of stable and metastable low-temperature iron oxides and oxyhydroxides. Geochim Cosmochim Acta 62:2905-2913

Laberty C, Suib SL, Navrotsky A (2000) Effect of framework and layer substitution in manganese dioxide related phases on the energetics. Chem Mater 12:1660-1665

Langel W, Parrinello M (1995) Ab initio molecular dynamics of H_2O adsorbed on solid MgO. J Chem Phys 103:3240-3252

Lindan PJD, Harison NM, Holender JM, Gillan MJ (1996) First-principles molecular dynamics simulation of water dissociation on TiO_2 (110). Chem Phys Lett 261:246-252

Liu M, Yund RA, Tullis J, Topor L, Navrotsky A (1995) Energy associated with dislocations: a calorimetric study using synthetic quartz. Phys Chem Minerals 22:67-73

Mackrodt WC, Davey RJ, Black SN (1987). J Crystal Growth 80:441

Majzlan J, Navrotsky A, Casey WH (2000) Surface enthalpy of boehmite. Clays Clay Minerals 48:699-707

Malek J, Mitsuhashi T, Ramirez-Castellanos J, Matsui Y (1999) Calorimetric and high-resolution transmission electron microscopy study of nanocrystallization in zirconia gel. J Mater Res 14: 1834-1843

Maniar PD, Navrotsky A (1990) Energetics of High Surface Energy Silicas. J Non-Crystalline Solids 120:20-25

Maniar PD, Navrotsky A, Rabinovich EM, Ying JY, Benziger JB (1990) Energetics and structure of sol-gel silicas. J. Non-Crystalline Solids 124:101-111

Margrave JL, Kybett BD (1965) Thermodynamic and kinetic studies of oxides and other refractory materials at high temperature. Tech Rept AFMO-TR, p 65-123

McHale JM, Navrotsky A, Perrotta AJ (1997a) Effects of increased surface area and chemisorbed H_2O on the relative stability of nanocrystalline α-Al_2O_3 and γ-Al_2O_3. J Phys Chem B 101:603-613

McHale JM, Auroux A, Perrotta AJ, Navrotsky A (1997b) Surface energies and thermodynamic phase stability in nanocrystalline aluminas. Science 277:788-791

McHale JM, Navrotsky A, Kirkpatrick RJ (1998) Nanocrystalline spinel from freeze dried nitrates: synthesis, energetics of product formation, and cation distribution. Chem Mater 10:1083-1090

Mitsuhashi T, Kleppa OJ (1979) Transformation enthalpies of the TiO_2 polymorphs. J Am Ceram Soc 62:356-357

Molodetsky, I, Navrotsky A, Lajavardi M, Brune A (1998) The energetics of cubic zirconia from solution calorimetry of yttria- and calcia-stabilized zirconia. Z Physik Chem 207:59-65

Molodetsky I, Navrotsky A, Paskowitz MJ, Leppert VJ, Risbud SH (2000) Energetics of X-ray-amorphous zirconia and the role of surface energy in its formation. J Non-Crystalline Solids 262:106-113

Moloy EC, Davila LP, Shackelford JF, Navrotsky A (2001) High-silica zeolites: a relationship between energetics and internal surface area. Microporous Mesoporous Materials (submitted)

Navrotsky A, Kleppa OJ (1967) Enthalpy of the anatase-rutile transformation. J Am Ceram Soc 50:626

Navrotsky A, Jamieson JC, Kleppa OJ (1967) Enthalpy of transformation of a high-pressure polymorph of titanium dioxide to the rutile modification. Science 158:388-389

Navrotsky A, Petrovic I, Hu Y, Chen CY, Davis ME (1995) Little energetic limitation to microporous and mesoporous materials. Microporous Mater 4:95-98

Navrotsky A (1997) Progress and new directions in high temperature calorimetry revisited. Phys Chem Minerals 24:222-241

Navrotsky A (2000) Nanomaterials in the environment, agriculture, and technology (NEAT). J Nanoparticle Res 2:321-323

Nitkiewicz A, Kerrick DM, Hemingway BS (1983) The effect of particle size on the enthalpy of solution of quartz: implications for phase equilibria and solution calorimetry. Abstr Progr of the Ann Meeting, Geol Soc Am, p. 653.

Ohtaka O, Yamanaka T, Kume S, Ito E, Navrotsky A (1991) Stability of monoclinic and orthorhombic zirconia: studies by high-pressure phase equilibria and calorimetry. J Am Ceram Soc 74:505-509

Oliver PM, Watson GW, Kelsey ET, Parker SC (1997) Atomistic simulation of the surface structure of the TiO_2 polymorphs rutile and anatase. J Mater Chem 7:563-568

Penn RL, Banfield JF, Kerrick DM (1999) TEM investigation of Lewiston, Idaho, fibrolite: microstructure and grain boundary energetics. Am Mineral 84:152-159

Peryea EJ, Kittrick JA (1988) Relative solubility of corundum, gibbsite, boehmite, and diaspore at standard state conditions. Clays Clay Minerals 36:391-396

Petrovic I, Heaney PJ, Navrotsky A (1996) Thermochemistry of the new silica polymorph moganite. Phys Chem Minerals 23:119-126

Petrovic I, Navrotsky A, Davis ME, Zones SI (1993) Thermochemical study of the stability of frameworks in high silica zeolites. Chem Mater 5:1805-1813

Piccione PM, Laberty C, Yang S, Camblor MA, Navrotsky A, Davis ME (2000) Thermochemistry of pure-silica zeolites. J Phys Chem B 104:10001-10011

Post JE (1992) Crystal structures of manganese oxide minerals. Catena Suppl 21:51-73

Ranade MR, Elder SH, Navrotsky A (2001) Energetics of nanostructured TiO_2-ZrO_2 and TiO_2-MoO_3 composite materials. Chemistry of Materials (submitted)

Rao CNR (1961a) Kinetics and thermodynamics of the crystal structure transformation of spectroscopically pure anatase to rutile. Can J Chem 39:498-500

Rao CNR, Yoganarasimhan SR, Faeth PA (1961b) Studies on the brookite-rutile transformation. Trans Faraday Soc 57:504-510

Robie RA, Hemingway BS (1995) Thermodynamic properties of minerals and related substances at 298.15 K and 1 bar (10^5 Pascals) pressure and at higher temperatures. US Geol Surv Bull 2131.

Russell AS, Edwards JD, Taylor CS (1955) Solubility of hydrated aluminas in NaOH solutions. J Metals 203:1123-1128

Scheidegger AM, Strawn DG, Lamble GM, Sparks DL (1998) The kinetics of mixed Ni-AL hydroxide formation on clay and aluminum oxide minerals: a time-resolved XAFS study. Geochim Cosmochim Acta 62:2233-2245

Schuiling RD, Vink BW (1967) Stability relations of some titanium minerals (sphene, perovskite, rutile, anatase). Geochim Cosmochim Acta 31:2399-2411

Sen S, Gerardin C, Navrotsky A, Dickinson JE (1994) Energetics and structural changes associated with phase separation and crystallization in lithium silicate glasses. J Non-Crystalline Solids 168:64-75

Sorai M, Kosaki A, Suga H, Seki S (1969) Particle size effect on the magnetic properties and surface heat capacities of β-$Co(OH)_2$ and $Ni(OH)_2$ crystals between 1.5 and 300 K. J Chem Thermodyn 1:119-140

Tasker PW (1984) Surfaces of Magnesia and Alumina, Kingery WD (ed). Columbus, Ohio: American Ceramic Society

Taylor RM (1984) The rapid formation of crystalline double hydroxy salts and other compounds by controlled hydrolysis. Clay Minerals 19:591-603

Terwilliger CD, Chiang YM (1995) Measurements of excess enthalpy in ultrafine-grained titanium dioxide. J Am Ceram Soc 78:2045-2055

Thompson HA, Parks GA, Brown Jr, GE (1999) Dynamic interactions of dissolution, surface adsorption, and precipitation in an aging cobalt(II)-clay-water system. Geochim Cosmochim Acta 63:1767-1779

Tian ZR, Xia G, Luo J, Suib SL, Navrotsky A (2000) Effects of water, cations, and structure of energetics of layer and framework phases, $Na_xMg_yMnO_2{\cdot}nH_2O$. J Phys Chem B 104:5035-5039

Titulaer MK, Jansen JBH, Geus JW (1994) The quantity of reduced nickel in synthetic takovite: effects of preparation conditions and calcination temperature. Clays Clay Minerals 42:249-258

Tosi MP (1964) Cohesion of ionic solids in the Born model. Solid State Phys 16:1-120

Towle SN, Bargar JR, Brown, Jr GE, Parks GA (1997) Surface precipitation of Co(II)(aq) on Al_2O_3. J Coll Interf Sci 187:62-82

Valueva GP, Goryainov SV (1992) Chabazite during dehydration. Russian Geol Geophys 33:68-75

Victor AC (1962) Effects of particle size on low temperature heat capacities. J Chem Phys 36:2812-2813

Wintsch RP, Dunning J (1985) The effect of dislocation density on the aqueous solubility of quartz and some geologic implications: a theoretical approach. J Geophys Res 90:3649-3657

Wu NL, Wu TF (2000) Enhanced phase stability for tetragonal zirconia in precipitation synthesis. J Am Ceram Soc 83:3225-3227

Wu NL, Wu TF (2001) Thermodynamic stability of tetragonal zirconia nanocrystallites. J Mater Res 16:666-669

Ying JY, Benziger JB, Navrotsky A (1993a) The structural evolution of alkoxide silica gels to glass: effect of catalyst pH. J Am Ceram Soc 76:2571-2582

Ying JY, Benziger JB, Navrotsky A (1993b) Structural evolution of colloidal silica gels to glass. J Am Ceram Soc 76:2561-2570

Zhang H, Banfield JF (1998) Thermodynamic analysis of phase stability of nanocrystalline titania. J Mater Chem 8:2073-2076

Zhang H, Banfield JF (1999) New kinetic model for the nanocrystalline anatase-to-rutile transformation revealing rate dependence on number of particles. Am Mineral 84:528-353
Zhang H, Penn RL, Hamers RJ, Banfield JF (1999) Enhanced adsorption of molecules on surfaces of nanocrystalline particles. J Phys Chem B 103:4656-4662
Zhang H, Banfield JF (2000a) Phase transformation of nanocrystalline anatase-to-rutile via combined interface and surface nucleation. J Mater Res 15:437-448
Zhang H, Banfield JF (2000b) Understanding polymorphic phase transformation behavior during growth of nanocrystalline aggregates: insights from TiO_2. J Phys Chem B 104:3481-3487

4 Structure, Aggregation and Characterization of Nanoparticles

Glenn A. Waychunas

Geochemistry Department
Earth Sciences Division, MS 70-108B
E.O. Lawrence Berkeley National Laboratory
One Cyclotron Road
Berkeley, California 94720

INTRODUCTION

Structural aspects of natural nanomaterials

A large number of mineral species occur only as micron-sized and smaller crystallites. This includes most of the iron and manganese oxyhydroxide minerals, and other species whose formation processes and growth conditions limit ultimate size. Microscopic investigation of these species generally reveals sub-micron structure down to the nanometer level, including evidence of aggregation, agglomeration and assembly of nanometer units into larger crystals and clots. The bulk of studies in the literature dealing with nanoparticle structure and growth deal with metals, silicon, and other semiconductor materials. A great deal of attention has been given to the electronic properties of such solids, owing to both new commercial applications and new fundamental physics and chemistry tied to this area. Most applicable mineralogical or geochemical studies have not addressed the same issues, instead being more concerned with relatively bulk chemical properties. Very little has been done to understand how natural nanoparticulates (and related types of natural nanomaterials) form, how their microstructure is related to the growth process, and how their structure varies from larger crystallites or bulk material of the same composition. Magnetic and electronic properties of natural nanomaterials are similarly understudied.

In this chapter aspects of nucleation, aggregation and growth processes that give rise to specific microstructures and forms of nanomaterials are considered. Next the way in which the surface structure of nanoparticulates may differ from the interior, and how physical structure may be modified by reduced particle size is examined. The various techniques by which nanoparticle structure, size, microstructure, shape and size distribution are determined are then considered with examples. Finally some of the outstanding problems associated with nanoparticle structure and growth are identified, emphasizing natural processes and compositions.

Definitions

Naturally occurring nanomaterials exist in a variety of complex forms. In this chapter a short set of definitions will be stated for clarity. *Nanocrystals* are single crystals with sizes from a few nm up to about 100 nm. They may be aggregated into larger units with a wide spectrum of microstructures. *Nanoparticles* are units of minerals, mineraloids or solids smaller in size than 100 nm, and composed of aggregated nanocrystals, nanoclusters or other molecular units, and combinations of these. *Nanoclusters* are individual molecular units that have well-defined structure, but too small to be true crystals. Al_{13} and Zn_3S_3 solution complexes are types of nanoclusters with sizes from sub nanometer to a few nm. *Nanoporous* materials are substances with pores or voids of nanoscale dimensions. These materials can be single crystals, such as zeolites or

1529-6466/00/0044-0004$10.00

molecular sieves with cage-like nanopores; amorphous materials with nanoscale voids; poorly crystalline solids with grain-boundary or defect pores; or composite materials. Some of the literature treats clay and similar layer minerals as nanoporous solids.

These categories of nanomaterials are useful due to the rather different physical behavior, growth and structure of each type. For example, nanoclusters are closely akin to molecules in terms of transport and reactive processes. Nanoparticles are somewhat like classical colloids, but in a smaller size regime. Nanoporous solids have unique abilities to encapsulate, bind and react with other nanoparticles and nanoclusters, as well as molecules. Nanocrystals can have relaxed or contracted lattices, and unusual reactivity relative to their larger counterparts.

Colloidal particles can have dimensions from small clusters of atoms up to about 10 μm in diameter. They are classically defined as any suspended fine particles in water. Hence smaller colloidal particles are the same as nanoparticles. In the formation of nanoparticles one often speaks of polymer and monomer units. *Monomers* refer to the smallest growth unit, either a single atom, ion or molecule. *Polymers* are units consisting of two or more monomers.

The *bulk* of a nanomaterial will refer to that portion that is not clearly defined as part of the surface, i.e., not within several atomic layers of the surface. This implies that 1-nm nanomaterials consist only of surface. *Surface volume* will refer to the volume of a nanocrystal or nanoparticle that is assumed to be part of the surface. The *habit* of a crystal is the particular external form that is presented within the options allowed by the point group symmetry. Common descriptive terms are tabular (tablet-like), cubic, acicular, and so forth. Occasionally, habits inconsistent with point group symmetry can occur from kinetic phenomena.

GROWTH AND AGGREGATION PROCESSES

Growth processes

The growth of nanomaterials is a rich field of research that can only be treated in a selective fashion in this chapter. The main concerns are the principle issues that determine formation and growth in the nanoscale range, how growth influences microstructure, the stability of particular sizes, shapes and phases, and structure-property relationships. Unlike most nanomaterials, natural nanomaterials can be formed by either inorganic or biologically-connected (enzymatic) processes. In some cases, the same composition nanomaterials can be formed by either pathway.

Inorganic vs. organic (enzymatic)

Inorganic growth processes generally are treated as near equilibrium, i.e., at low supersaturation in a solution, or at moderate undercooling in a melt-solid system. This situation can give rise to slow growth and a decreased rate of nucleation. Hence it is ideal for the formation of large crystallites. On the other hand, extreme undercooling or sudden generation of high supersaturation levels can lead to very high nucleation rates creating small crystallites or particulates. In non-homogeneous systems, e.g., those where nanomaterials are formed on substrates, the nanomaterial-substrate interface can act to enhance nucleation rate, or influence local concentration (and thus supersaturation) due to surface complexation. In contrast to these processes, enzymatic pathways for nanomaterial growth may take place far from equilibrium, and often under conditions for which nanoparticle growth could otherwise not occur. For example, oxidation of Fe^{2+} in natural waters is much more rapid in the presence of specific bacteria. The organism *Thiobacillus ferrooxidans* can increase this rate by as much as six orders of magnitude

(Singer and Stumm 1970). The oxidized iron can precipitate not only in the surroundings of the bacteria, but also on the bacteria surfaces or exopolymers, so that both catalysis and heterogeneous nucleation effects obtain. Analogous reactions occur for dissolved manganese, oxidizing it into extremely insoluble Mn^{4+}, and for many other metals including uranyl species (Suzuki and Banfield 1999). Further, the process may be controlled by the electron transfer at the cell enzyme/solution interfaces, to restrict growth rate even perhaps to atom by atom. These differences in growth pathway can potentially be utilized to prepare nanomaterials of unique composition, structure and properties.

NUCLEATION AND GROWTH

Classical nucleation theory (CNT)

Classical theory assumes that the formation of a condensed phase commences with the development of a "critical nucleus". This nucleus is assumed to form by random fluctuations from smaller units, generally single atoms or molecules. Once formed this nucleus has a greater probability of growing further than of losing mass, and the system can enter into a growth stage.

The nucleus is assumed to have the same structure, density, etc. as the bulk phase, and have the same interaction energies with adjacent phases (surface tension, etc.). This conception has proved valuable in general, but a few of its assumptions are difficult to prove or incorrect on the molecular level (Talanquer and Oxtoby 1994). Other assumptions are practically impossible to test (ten Wolde and Frenkel 1997), as critical nucleus formation is a fleeting event that may not be amenable to any direct imaging method. However, as nanoparticles are of the same size domain as theoretical critical nuclei, an understanding of the nucleation process is essential to defining and possibly controlling the pathways for nanoparticle formation.

Homogeneous nucleation

The classical model of random homogeneous nucleation is due to Volmer and Weber (1926) and Becker and Doring (1935), who developed a theory for the formation of nuclei of liquid droplets in supersaturated vapors. Figure 1 shows the free energy of formation for a given nucleus as a function of size and temperature. Using spherical particles of radius r, the free energy is given by:

$$\Delta G = 4/3\ \pi r^3\ \Delta g + 4\pi r^2\sigma \tag{1}$$

where Δg is the free energy change per unit volume of the nucleus, and σ is the free energy per unit area of the interface. It is clear that if Δg is negative (the case for supersaturation) then the free energy change is a maximum for some size nucleus depending on the temperature. The lower the temperature the smaller this nucleus, and the smaller the needed change in free energy to stabilize it. If the temperature-dependence of the interface energy is neglected, the critical values $r = -2\sigma/\Delta g$, and $\Delta G = 16\pi\sigma^3/3(\Delta g)^2$ are obtained by solving the derivative $[d(\Delta G)/dr = 0]$ of Equation (1). With a few other assumptions, it is possible to quantify the number of critical nuclei and the rate of nucleation for this model (see e.g., Raghavan and Cohen 1975). We assume that the addition of a single atom defines the boundary between a sub- and super-critical nucleus, and that we can use Maxwell-Boltzmann statistics. If the total number of atoms be N_T, then the number of critical-sized nuclei is approximately:

$$N = N_T \exp(-\Delta G/kT) \tag{2}$$

If the largest subcritical nucleus is surrounded by s atoms in the system, the frequency ν_1

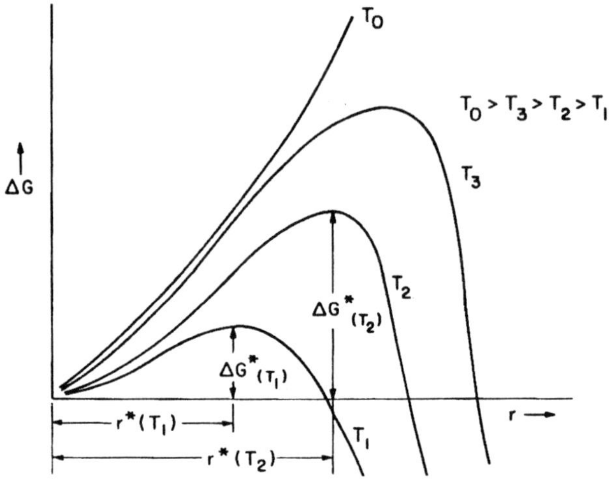

Figure 1. Free energy required for the stability of a critical nucleus. At the lowest temperature both the energy barrier and the critical radius for a nucleus is the smallest, and both increase with increasing temperature. At some temperature T_0, the critical radius is too large for a stable nucleus to form. This corresponds to a system that is not supersaturated. After Raghaven and Cohen (1975).

for which these atoms can join the particle is:

$$\nu_1 = s\nu \exp(-\Delta G_D/kT) \tag{3}$$

where ν is the frequency of lattice vibrations and ΔG_D is the free energy for atomic diffusion across the interface. The nucleation rate is then (events per unit time and unit volume):

$$I = N_T s\nu \exp[-(\Delta G+\Delta G_D)/kT] \tag{4}$$

Heterogeneous nucleation

Clem and Fisher (1958) use a similar treatment as above to derive the solid state nucleation kinetics for new phases at grain boundaries. They neglect orientation of the critical nucleus with respect to the host, strain energy, and coherency effects. Nucleation at the grain boundary interface removes boundary energy. Their treatment yields the following critical values:

$$R = -2\sigma_{\alpha\beta}/\Delta g \tag{5}$$

$$r = -2\sigma_{\alpha\beta}(\sin\theta)/\Delta g \tag{6}$$

$$\Delta G = 8\pi\sigma^3{}_{\alpha\beta}(2\text{-}3\cos\theta+\cos^3\theta)/3(\Delta g)^2 \tag{7}$$

where R is the radius of curvature of the spherical surfaces of the nucleus, 2θ is the included angle of the nucleus edge as controlled by surface tension (Fig. 2), $\sigma_{\alpha\beta}$ is the matrix-nucleus interfacial energy, and $\sigma_{\alpha\beta}$ is the grain boundary interfacial energy.

Using assumptions as before a nucleation rate can be derived:

$$I_{hetero} = \Sigma N_i \nu \sigma_i \exp[-(\Delta G_i+\Delta G_D)/kT] \tag{8}$$

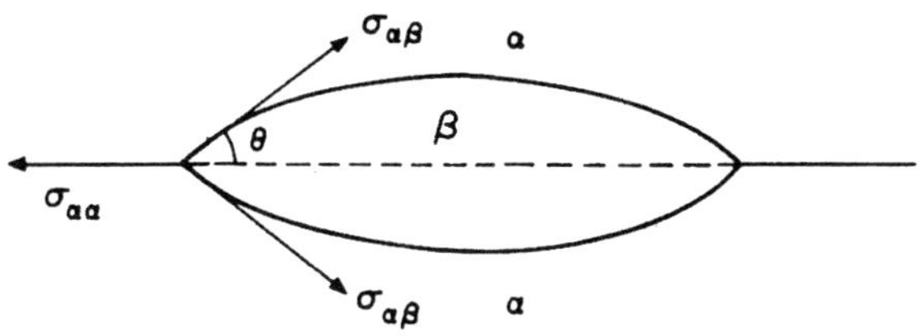

Figure 2. Model geometry for the nucleation of a new phase at a grain boundary. After Raghaven and Cohen (1975).

where N_i is the number of sites per unit volume having nucleation barrier ΔG_i. The main difference between Equations (4) and (8) is that N_i is orders of magnitude smaller than N_T, and ΔG_i is usually significantly smaller than ΔG. In general the exponential dominates so that I_{hetero} is much greater than I.

PROBLEMS WITH CLASSICAL MODELS

These classical approaches (later modified with energy terms including strain, coherency with a substrate, temperature-dependence, interface volume changes, and so on) all assume that random fluctuations somehow create the subcritical nucleus, and that the step from subcritical to supercritical is the addition of a single atom. However no aspect of classical theory takes into account molecular level phenomena. The theory is thus a macroscopic theory, and hence it is not surprising that different types of molecular bonding in physical systems gives rise to large deviations from classically predicted nucleation rates (Talanquer and Oxtoby 1994; Marasli and Hunt 1998; Auer and Frenkel 2001).

Dynamical nucleation processes: Smoluchowski's approach

One of the assumptions of CNT is that formation of the critical nucleus occurs by aggregation of monomer units to the sub-critical nucleus, i.e., the concentration of polymer clusters below critical nucleus size is negligible. This neglects cluster-cluster aggregation as a nucleation process, which seems to go against intuition. However Alexandrowicz (1993) has shown via scaling theory that, given reasonable assumptions, monomer addition/subtraction is by far the most important process. An entirely different way of considering nucleation utilizes a summation over all of the competing processes that can produce critical-size nuclei and further growth. This approach was first suggested by Smoluchowski (1917), and is generally employed today in aggregation processes. This approach yields direct insight into the dynamical aspects of nucleus formation as it considers each discrete type of aggregation/disaggregation rate (from atom and molecule monomers adding to a given size cluster, to larger cluster addition/breakup). It has been termed a "kinetic" approach to nucleation and growth as discrete energies and thermodynamic equilibria are replaced by probabilistic terms for cluster formation (Hettema and McFeaters 1996). Hence the concept of a critical nucleus is replaced by particle size distributions as a function of time.

Smoluchowski's equation in its simplest form describes rates of cluster formation and reduction:

$$\frac{dn_k}{dt} = \frac{1}{2} \sum_{i+j=k} K(i,j)n_i n_j - n_k \sum_{i=1}^{\infty} K(i,k)n_i$$

The first term represents the rate of creation of clusters of a given size from smaller clusters, while the second term represents the loss of such clusters via reactions with other clusters. The equation is thus a description of a set of coupled rate equations. The K terms are reaction rate coefficients and may themselves be complex mathematical expressions. For example, cluster rearrangement, fragmentation into multiple smaller clusters, and sticking probability can all be embedded in a given K term. As fractal theory can be used to describe the results of aggregating clusters (see below), the K terms are often referred to as "kernels" in keeping with fractal notation. The scaling behavior of kernels as a function of cluster size is a focus of such treatments.

Classical growth mechanisms vs. aggregation processes

The contrasting treatments of nucleation have consequences for the initial formation of nanoscale particles. If CNT is accurate on a molecular scale, then nanoparticles below the critical nucleus size would be difficult to preserve. Attachment of molecules to a subcritical nanoparticle surface could be used to alter surface energy and produce stability. Still other technological tricks could be used to reduce the degradation rate of metastable nanoparticles, e.g., fast trapping in noble gas solids. But CNT allows for the formation of a narrow size distribution of nanoparticles if formation conditions can be controlled. On the other hand, aggregation processes would make it difficult to prepare nanoparticles with a narrow size distribution, let alone shape, even through the suitable adjustment of reaction rates. The size distribution is a crucial issue in any description of crystal growth via aggregation, as typical processes of self assembly and superlattice formation require building units of nearly identical size and shape. However, subcolloidal aggregation of nanoclusters to form zeolites (Nikolakis et al. 2000) has been described and modeled, and superlattice formation of metallic clusters has been described in several systems (e.g., Sun and Murray 1999). An important concept is *oriented aggregation* of nanoparticles, which allows for growth without recrystallization on a local scale. This process has been studied in detail by Banfield and co-workers (Banfield et al. 2000; Penn and Banfield 1998; Zhang and Banfield 2000)).

Classical growth mechanisms and laws

Much as CNT assumed monomer attachment, most classical theories of crystal growth assume that material is added to a growing crystal face essentially atom by atom (Fig. 3). The most common type of growth is layer-by-layer growth (also called Frank-Van der Merwe growth in cases of epitaxy (Zangwill 1988)). In this type of growth the energy reduction by the nucleation of another layer is small compared to monomer attachment within a layer, e.g., if forming a new layer forms only a single bond, while intralayer attachment forms 2 or more. Hence new layers are unlikely to start until underlying ones are essentially complete. A modification of this type of growth occurs if there is a screw dislocation exiting at the surface of the growth face. In that case the layer need not be nucleated once growth is initiated. Many crystals grow by this sort of process (Fig. 4).

It is assumed in classical growth models that all added monomers are attached permanently and do not diffuse off into the fluid, solution or gas state once attached. In the layer-by-layer case a newly attached atom or molecule will diffuse along the layer surface until it finds the edge of the currently growing layer. It will then take a position corresponding to the maximum number of interatomic bonds that it can make. The layer will then fill in according to the trade off between surface diffusion rates, bonding energies and availability of multibond attachment sites. Although this type of growth is highly idealized, it is appropriate for many systems, and can predict growth variations produced by ionic strength variations, temperature, surface impurities and other factors.

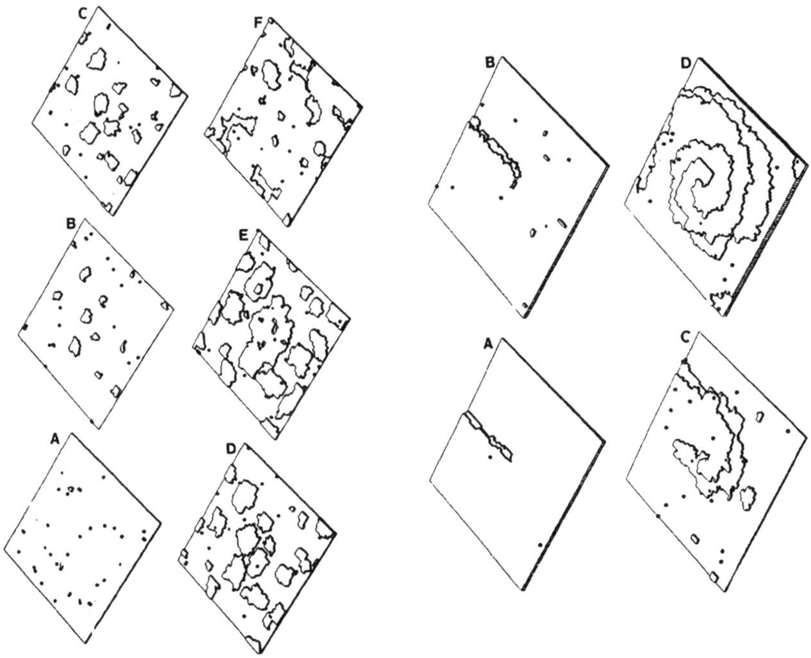

Figure 3. Left side: A to F, atom by atom simulated growth. New growth layer initiation occurs only when most sites at the edges of the layer are filled. Right side: A to D, growth at a surface having a screw dislocation. New layers continuously form. After Kirkpatrick (1981).

Growth topologies

The habit of a euhedral crystal is due to the rate at which particular faces grow, while the equilibrium faces (those nominally appearing initially on a crystal surface) are dictated by surface energy considerations. The greatest reduction in surface free energy will occur when the highest energy faces are reduced in size. This leads in general to the faster growth at high energy faces. Quite unusual crystal habits can be produced by rapid growth from supersaturated systems. In such case both equilibrium and kinetic considerations play a role in determining the type of crystallization. A construction which predicts equilibrium crystal faces based on a knowledge of surface energies and crystal symmetry is due to Wulff (1901), and has been examined in detail by Herring (1951). An example is shown in Figure 5. The basis of this construction is the minimum energy for the surface of a crystal with fixed volume (see Hartman 1973 for derivation), where it can be shown that:

$$\gamma_i = \lambda h_i \tag{10}$$

for each distinct crystallographic face with normal h_i and surface energy γ_i. and λ depending only on absolute volume. An equivalent formulation is:

$$\gamma_1 : \gamma_2 : \gamma_3 : \ldots = h_1 : h_2 : h_3 : \ldots \tag{11}$$

The construction takes the form (in two dimensions) of a plot of surface energies as a function of directions using polar coordinates. For each plane a normal is defined, and then a perpendicular line segment where the normal strikes the energy curve. The distance of a given surface plane from the crystal center of mass turns out to be proportional to the surface energy of that plane. For particular directions corresponding to

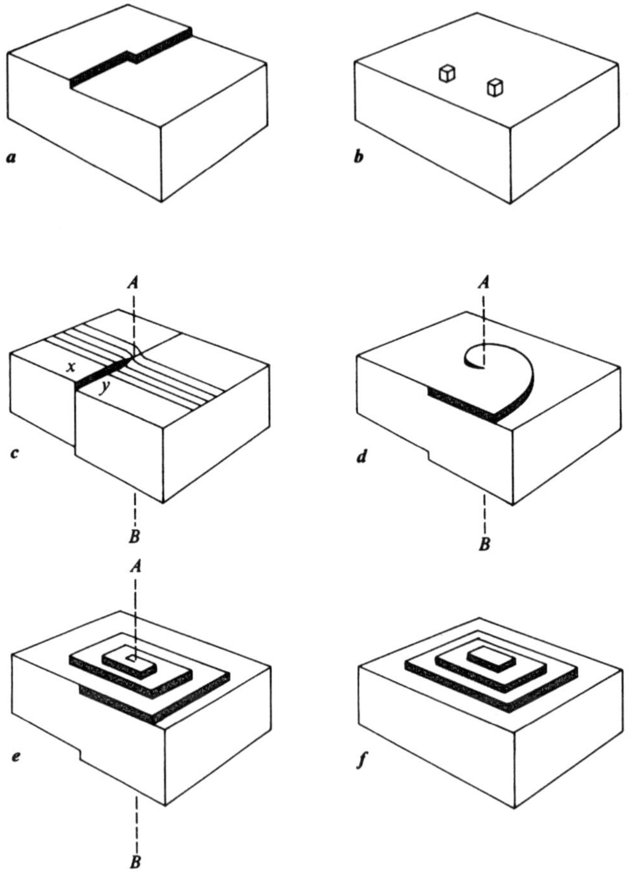

Figure 4. Surface growth features and habits. (a) Single layer growth showing lowest energy attachment site. (b) New layer nucleation via single atoms. (c) Presence of screw dislocation with axis normal to plane of surface. (d) Attachment sites on growing spiral. (e) Truncation of layers in response to anisotropic surface attachment energies. (f) "Pyramidal" habit induced by spiral growth. From Zoltai and Stout (1984) with permission from Burgess Publishing Company.

principal crystal faces there are cusps in this plot. The smallest enclosed area within this line segment construction is the equilibrium crystal form. This process must be done in three dimensions to assuredly predict the equilibrium form for a lower symmetry crystal. The three dimensional Wulff plot can be sectioned to show the relative energy of faces in particular planes.

The competition between equilibrium and kinetic factors are shown in Figure 6 using two dimensional crystal growth. Three orientations of the Wulff construction sections are shown for crystals of CBr_4-8% C_2Cl_6 along with corresponding dendritic growth patterns. Dendritic patterns develop when crystallization is too rapid to allow solvent to flow away from growing faces, or heat of crystallization to dissipate. Hence solvent can be trapped between elements of growth, or local heating raises solubility and decreases supersaturation. Two dimensional growth is markedly prone to complex patterning

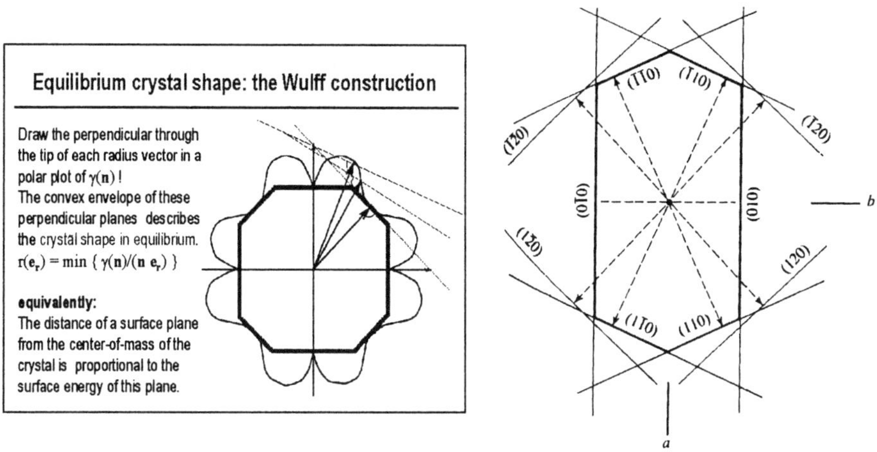

Figure 5. Wulff construction. Left: definition and general form of construction. Right: Wulff construction showing the equilibrium form of a forsterite crystal. After Zoltai and Stout (1984).

caused by various kinds of growth instabilities. For slow growth crystallites closely approximate Wulff construction predictions.

Skeletal crystals are examples of very rapid growth in particular directions most commonly due to crystallographically controlled screw dislocations and diffusion-limited growth conditions. These crystals fill in with lateral growth of dendrites as growth conditions change. Hopper-growth crystals, generally produced by vapor-solid crystallization, are formed when the edges of the crystal grow much faster than new nutrients can diffuse to the interior of the growth faces (Fig. 7). In dramatic contrast, crystals grown from a melt may have little exterior face development. An example are the "boules" produced when pulling single crystals from their melts in the Czochralski crystallization process. In this case the surface energy of different faces of the crystal is similar enough so that no face is favored for growth and the surface form is determined by diffusion rates in the melt. A final issue associated with Wulff constructions is the treatment of vicinal faces, i.e., planes slightly misoriented from a major (hkl) plane such as (1 1 23) adjacent to (001) but tilted toward (1 1 1). A detailed Wulff construction shows cusp structures corresponding to non-equilibrium but metastable suites of vicinal faces (Fig. 8). Cusps exists for all rational ratios of Miller indices. The vicinal faces will tend to reorganize into microfacet-terrace structures if $\gamma(\alpha) > [\gamma_0\sin(\beta-\alpha)+\gamma_1\sin\alpha]/\sin\beta$, where the microfacet inclination angle is β with surface energy γ_1, and the terrace surface energy is γ_0. (Fig. 8).

Inorganic vs. biologically produced crystallites

Hydrothermal, low temperature precipitation, and vapor deposition produce different crystal habits and sizes due to differences in growth rates, diffusion rates, and surface energy. Most Fe-Mn oxide-hydroxide phases are formed at low temperatures (below 100°C), but several are known to form under hydrothermal conditions (above 100°C and up to several kbars pressure). Larger crystals are produced from hydrothermal processes as transport is much more rapid, and solubility of the growing phase in the growth medium (solution) is high. The latter growth process can produce crystals up to 10 cm in size. Ice crystals (snowflakes) are the most commonly observed vapor deposition-produced natural phases. These usually show complex dendritic patterns caused by rapid

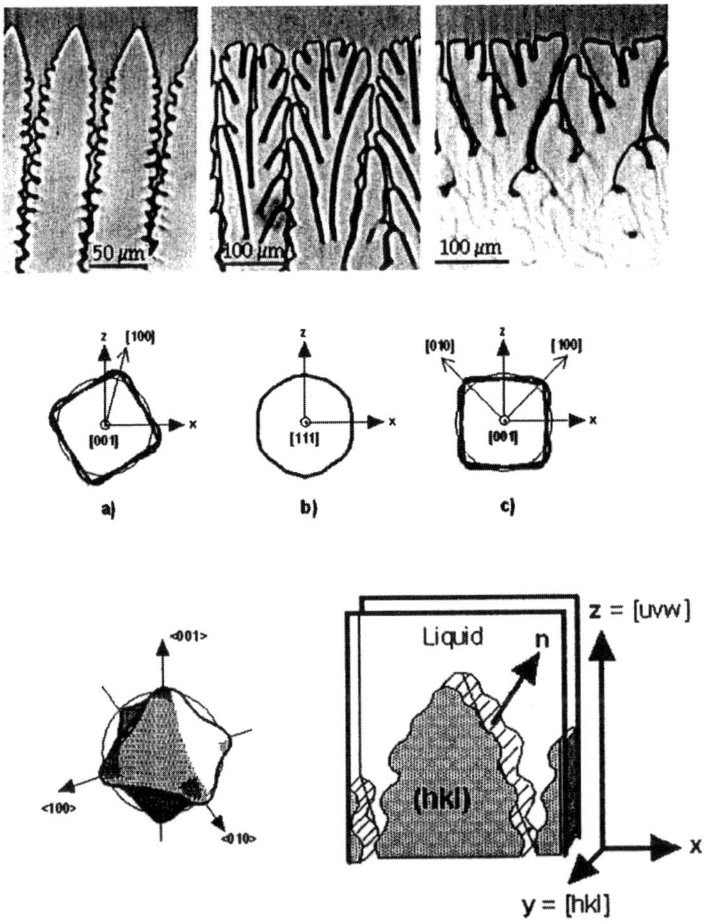

Figure 6. Dendrite formation. Top: Rapid two dimensional crystal growth in halocarbons. Different degrees of supersaturation (or undercooling) lead to fluctuations and limitations in nutrient supply to growing faces. Center: Wulff plots with orientation appropriate for each type of dendrite. Bottom left: three dimensional Wulff plot for this system. Bottom right: diagram of dendrite head structure showing equilibrium face and fluctuations in growth rate. From www.gps.jussieu.fr/engl/cell.htm.

growth as well as rapid changes in growth conditions during formation. Figure 9 shows microscopic images of synthetic vapor grown ice crystals.

In the case of biologically produced crystallites, low temperatures that normally restrict solubility of oxides, hydroxides and sulfides (except in strongly acidic solutions) can be affected by the presence of organic ligands and enzymes that alter the energy of growing crystal surfaces and increase or control solubility. This can result in modified habits, unusual growth patterns, complex microstructure, and dramatically increased crystallite size. An example from nature is the alteration of calcium carbonate growth to form complex microporous (or nanoporous) composite structures in the skeletons of sea organisms. This has been remarkably well modeled by the self assembly of liquid crystal materials, leading to suggestions that suitably surfactant-controlled $CaCO_3$ nanocrystal

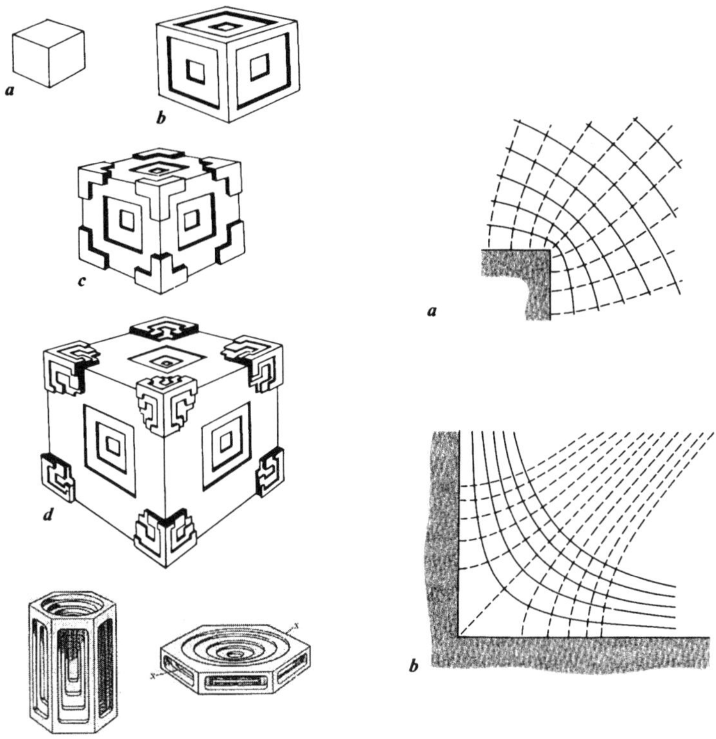

Figure 7. Hopper growth. Left: (a-d) successive growth stages with relatively slow nutrient transport to face centers. This type of growth is very common in halite and fluorite. Left bottom: two examples of hopper growth in ice crystals. Right: Flow density maps for nutrient transport near corners and reentrant angles. Highest flow is for closest grid spacings. From Zoltai and Stout (1984) with permission from Burgess Publishing Company.

growth could produce such structures in the laboratory (McGrath 2001). Another example is the development of iron oxyhydroxide nanoparticle aggregates formed via microorganism-catalyzed iron oxidation (Banfield et al. 2000). In this case nucleation of organized aggregates with well-defined nanocrystallite-nanocrystallite orientation occurs within a mass of randomly oriented nanoparticles. This suggests an initial assembly of nanocrystals by rotation and subsequent surface attachment when favorable orientations are presented. Under certain conditions several stages of such assembly may occur, allowing successively larger crystalline grains to form without a continuous classical growth process.

As inorganic crystal surfaces, growth patterns and growth rates can be strongly affected by organic and trace inorganic surface species, it is not fanciful to imagine the eventual intentional construction of inorganic structures with practically any desired degree and dimensionality of nanoporosity, density, surface area and shape.

How do these growth considerations apply to nanoparticles and nanocrystals?

Examples of the crystal forms of several types of natural and synthetic nanocrystals are shown elsewhere in this volume. Due to their small size, diffusion even through the

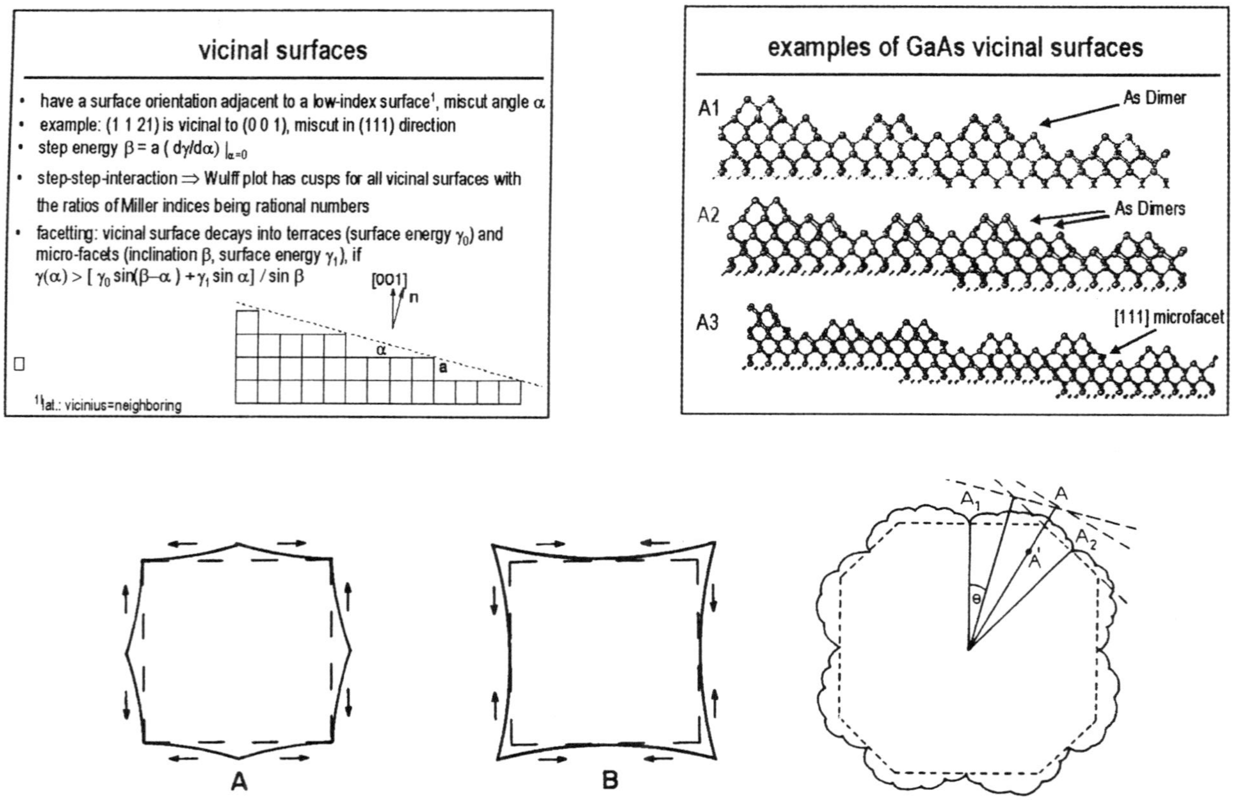

Figure 8. Top right: Definition of vicinal faces. Top left: examples of vicinal faces on GaAs. Bottom right: Development of vicinal faces due to nutrient flow. Bottom left: Wulff construction showing possible metastable low-angle faces near principle directions, A_1 and A_2. From www.fhi-berlin.mpg.de/th/member/kratzer_p.html.

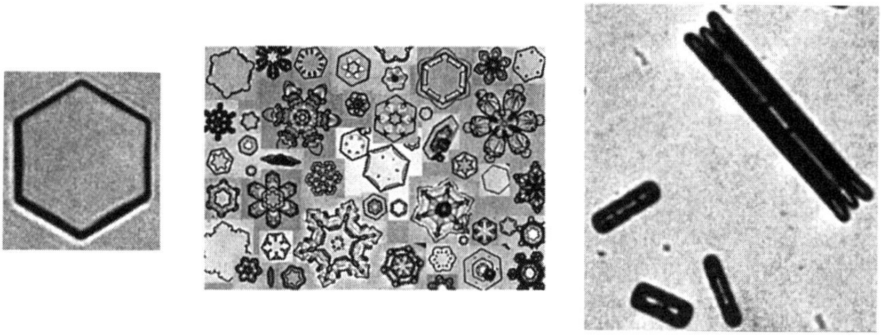

Figure 9. Habits for ice crystals grown under slightly different temperature (supersaturation) conditions. Left: perfect hexagonal crystals grown at low supersaturation. Right: rod-like hopper crystals grown at high supersaturation. Middle: Mosaic of various kinds of ice crystals with varying degrees of supersaturation and changes in conditions during growth. Crystals range from tens to hundreds of microns in diameter. From www.its.caltech.edu/~atomic/snowcrystals/.

interior of a nanocrystal ought to be rapid. Hence for individual nanocrystals, dendritic or other diffusion-limited forms should either not form, or be quickly converted into more stable surfaces. However, alteration of growth conditions to enhance surface energy anisotropy could conceivably be done to dramatically alter nanocrystal habits. On the other hand, the nature of well-defined surface terraces breaks down as nanocrystal diameter decreases. The surface can be considered in the small size limit as consisting only of microfacets. A complicating aspect of natural nanocrystals is that the surfaces will be in contact with an aqueous growth solution, and many types of surface species or impurities will be present. This will tend to affect surface energy and could lead to selective poisoning of particular face or surface feature development. All of these aspects present opportunities and challenges for the controlled development of special nanocrystal habits suited for particular chemical or environmental applications.

AGGREGATION MECHANISMS

The alternative to classic growth mechanisms is growth by aggregation reactions. In this case growth occurs from the attachment of clusters of atoms, critical nuclei, or larger units. A large body of research has been developed that explores the many diverse types of aggregation motifs and mechanisms. We have already noted the Smoluchowski approach to growth kinetics, and this formalism is sufficiently flexible to model many types of aggregation behavior. Before treating this further we now introduce other topics that will be of use in discussing surface structure and nanoparticle properties later in the chapter, and which clarify aspects of aggregation behavior.

Particle-particle interaction forces

Aggregation occurs because of a net attractive forces between particles. Besides varying in type, these forces also have specific ranges of action. A most thorough and approachable summary of all types of intermolecular and surface forces has been compiled by Israelachvili (1992). Here we briefly discuss only the most important forces on mineralogical nanoparticles: Covalent bonding, electrostatic interactions, dipole-charge, dipole-dipole, van der Waals and hydrophobic. Table 1 shows a comparison of these forces. The range of the force is defined as the distance where the force falls to kT. Interaction energies are given in kJ/mol for a vacuum interaction, unless otherwise indicated. *Covalent* forces are the strongest binding forces between atoms, but they have a short range as direct orbital overlap is required. Hence aggregation is only indirectly connected with covalent forces through particle-particle binding via sorbed molecular

Table 1. Interaction forces.

Interaction	*Energy* (vacuum)	*Directionality*	*Range* (nm)	*Example*
Covalent	100-900 kJ/mol	high	0.1-0.2	C-H (CH_4) [430] Si-O (Silicates) [370] C=O (HCHO) [690]
Coulombic	100-1000 kJ/mol	zero	100	NaCl [880] Na^+Cl^- in water [11.2]
Dipole-charge (fixed)	30-400 kJ/mol	high	0.5	Na^+ H_2O [1.2] Mg^{2+} H_2O [3.2]
Dipole-dipole (fixed)	up to ~200 kJ/mol	high	0.4	O^-H^+ O^-H^+ [2.6]
H Bond	10-40 kJ/mol	moderate	0.2	OH···OH [0.5]
Van der Waals[1]				
Dipole-dipole (free rotation/ Keesom or orientation force)				
	0-101 10^{-79}Jm^6	weak	0.4	H_2O-H_2O [96] HCl-HCl [11] CH_3Cl-CH_3Cl [101]
Dipole-induced dipole (free rotation/Debye or induction force)				
	0-32 10^{-79}Jm^6	weak	0.4	H_2O-H_2O [10] HCl-HCl [6] CH_3Cl-CH_3Cl [32]
Induced dipole-induced dipole (London or dispersion force)				
	2-400 10^{-79}Jm^6	weak	0.2-10	CH_3Cl-CH_3Cl [282] H_2O-H_2O [33] HCl-HCl [106]
Total van der Waals	2-415 10^{-79}Jm^6	weak	0.2-10	H_2O-H_2O [139] (50) CH_4-CH_4 [102] (10)
Hydrophobic	0-20 kJ/mol	weak	10	CH_4-CH_4 in H_2O (56)

[1] The contributing energies are scaled by the magnitude of the van der Waals energy coefficients in units of 10^{-79} J m^6. For total van der Waals energies (nn) the units are kJ/mol. Converting between these units requires accurate knowledge of van der Waals radii as the interactions scale as r^{-6}.

units. *Coulombic* forces are almost as strong as covalent forces, and have a long interaction range in water so that they have major effects on particle aggregation. At long distances (> nm) mixed charges on a surface average out. but for very near interactions the alternating charges of ionic solids become significant, and can lead to attractive interactions between similarly net charged surfaces. *Fixed dipole-charge* interactions can be about half as strong as covalent bonding, and are very important in ion solvation. The force varies as r^{-2}. At room temperature and typical ion-water molecule distances the dipole-charge interaction is much larger than kT and thus creates a solvation sphere about the ion that can behave practically like a molecular unit. This is especially true for small highly charged species such as Al^{3+}. *Fixed dipole-dipole* interactions can be as large as

1/5 to 1/10 of covalent bonding forces, but are only significantly larger than kT for molecules with large dipole moments. This is a crucial consideration as water has a relatively large dipole moment and the small proton allows for short interaction distances with other dipoles and thus strong dipole-dipole attraction. This strongly directional type of bond is usually called a *hydrogen bond*, and is responsible for many unique properties of water.

Van der Waals forces are actually composed of three components: freely rotating dipole-dipole, freely rotating dipole-induced-dipole, and induced dipole-induced dipole (dispersion or London) interactions. Net van der Waal forces are almost always attractive, as opposed to Coulombic and dipolar forces which can be strongly repulsive. The fall off with distance is by r^{-6}, so the interaction distance is crucial. It might seem as though freely rotating dipole interactions ought to cancel out, but this does not happen due to Boltzmann weighting factors which favor the lowest energy orientations. The interaction energy of two permanent dipoles is inversely dependent upon the temperature, while the interaction energy of dipole-induced dipole is independent of temperature. The magnitude of induced dipole interactions depends on the polarizability of the molecule. The dispersion component is due to the finite instantaneous dipole moment of any atom's electron distribution. It is always attractive. The effect is called the dispersion effect since the electronic movement that gives rise to it also causes dispersion of light, i.e., the variation of refractive index of a substance with the frequency (color) of the light. Van der Waals forces are quite weak, but are usually equal to a few kT at room temperature. As might be expected, van der Waals forces are difficult to calculate in complex systems, as other nearby atoms affect the polarizability of the electron distribution. A detailed discussion of van der Waals contributions to molecular interactions is available in Israelachvili (1992), and dispersion forces are treated quantitatively in Mahanty and Ninham (1976). *Hydrophobic* forces are the net attraction observed for hydrophobic molecules and surfaces in water. This can be significantly larger than their attractive forces in vacuum, as shown for methane molecules in Table 1. The bases of this force is an entropic phenomenon due to rearrangement of water hydrogen bonding configurations about closely approaching solvated species. The force has much longer range than covalent bonding. Israelachvili and Pashley (1982) found that the force between two macroscopic curved hydrophobic surfaces in water decayed exponentially in the range of 0-10 nm with a decay exponent of 1 nm. The free energy of species dimerization was found to linearly relate to the species diameter. Hydrophobic interactions have important roles in biological surface phenomena, molecular self-assembly and protein folding, and hence are an important aspect of nanoparticulates with hydrophobic components.

Electrical double layer, Derjaguin approximation and DLVO theory

Particle-particle interactions involving the forces above are not simple sums over each type of interaction. Moreover the electrical structure that develops at a solution-solid interface has a large effect on the net attractive or repulsive forces between particulates. This electrical structure is called the electrical double layer in its simplest depiction. The double layer results from the interaction of a charged surface with charged solution species and dipoles. The surface charge occurs due to ion adsorption or ionization of a surface chemical group. Transition metal oxide and hydroxide surfaces equilibrate with aqueous solutions by trading protons and hydroxyls from metal surface sites. At low pH values the surfaces are more protonated and tend to be positively charged, with the polarity reversing as hydroxyls increase with increasing pH. Specific dissociation constants dictate at what pH a surface switches from net positive to net negative, a pH value sometimes referred to as the point of zero charge.

Much has been written in the geochemical literature about electrical double layer theory (e.g., Sposito 1984; Stumm 1987; Davis and Kent 1990), so we show only the main models here. These are the capacitance, diffuse, Guoy-Chapman and Stern-Grahame models (Fig. 10). The capacitance model relates the electrochemical potential in the vicinity of an interface to a parallel plate capacitor. This has a linear dependence of potential with interface position, and is only a fair approximation of an actual interface situation. A more realistic model assumes only counterions in solution, and the application of the Poisson equation in electrostatics to solve for the variation of potential with distance. This is combined with the Boltzmann distribution to yield a Poisson-Boltzmann equation which will give the potential, electric field and counterion density at any point in the double layer (Israelachvili 1992). A linearized version of this equation, applicable to low potentials (using the Debye-Hückel approximation) yields a potential profile with distance that is non-linear, and is attributable to a "diffuse" swarm of counterions balancing the surface charge. A more general calculation is the Guoy-Chapman model derived for a planar double layer without assumptions of low potentials. Although Guoy-Chapman models are closer in form to observations, they still did not agree with experiment, particularly for mercury electrodes. This led to the Stern-Grahame model which uses a compact (Stern or Helmholtz layer) of counterion charges against the surface, and a diffuse layer derived from a Guoy-Chapman type treatment farther from the surface. The potential is assumed to be linear across the Stern layer.

One of the problems in calculating the forces acting on particles in solution is that we can calculate interaction energies for molecules (Table 1) but do not have a way to readily relate these to the forces between interacting real particles. This is especially important as particle-particle and surface-surface forces can now be measured by several methods. Use of a force law with the theoretically simple parallel planar surface model will always yield infinite forces as the planes are themselves infinite. A solution is obtained by use of the Derjaguin approximation (Derjaguin 1934) wherein the force between two interacting spherical particles can be given in terms of the energy per unit area of two flat surfaces of fixed separation. This strategy is applicable to any force law, whether attractive, repulsive or periodic.

Using this approximation Derjaguin and Landau (1941) and Verwey and Overbeek (1948) described a model (DLVO) which incorporated double layer electrostatic forces with van der Waals forces, and which explained much of colloid behavior. Figure 11 shows the basic nature of the DLVO model. As particles of the same type are considered to have the same surface charge, the net double layer forces are repulsive while the van der Waals forces are always attractive. For short distances the van der Waals forces dominate due to the r^{-6} power law, but at distances of a few nm the balance of forces can create local minima and thus stability at a well-defined separation. In this model the van der Waal forces are also assumed to be independent of electrolyte concentration and pH, which is not too bad a first approximation. The general form of the DLVO model is shown in the inset. There is a primary minimum, a secondary minimum, and an energy barrier. Curve a represents highly charged surfaces in a dilute electrolyte. Large planar surfaces will feel a long range repulsive force, and would become attractive if forced together with a great expense of work. Nanoparticles would be stable and resist aggregation. Curve b represents somewhat higher concentration electrolytes and shows the creation of the secondary minimum. Planar surfaces would be at equilibrium at this secondary minimum, and nanoparticles would either remain dispersed in solution or form an association at the same minimum. A colloid formed with such a energy minimum is called *kinetically stable* as opposed to *thermodynamically stable* as would occur if the particles were forced together and reached the primary minimum point. The energy barrier is lowered if the surface charges are lower, and this situation is represented by curve c. Large surfaces will equilibrate at the secondary minimum, but nanoparticles will

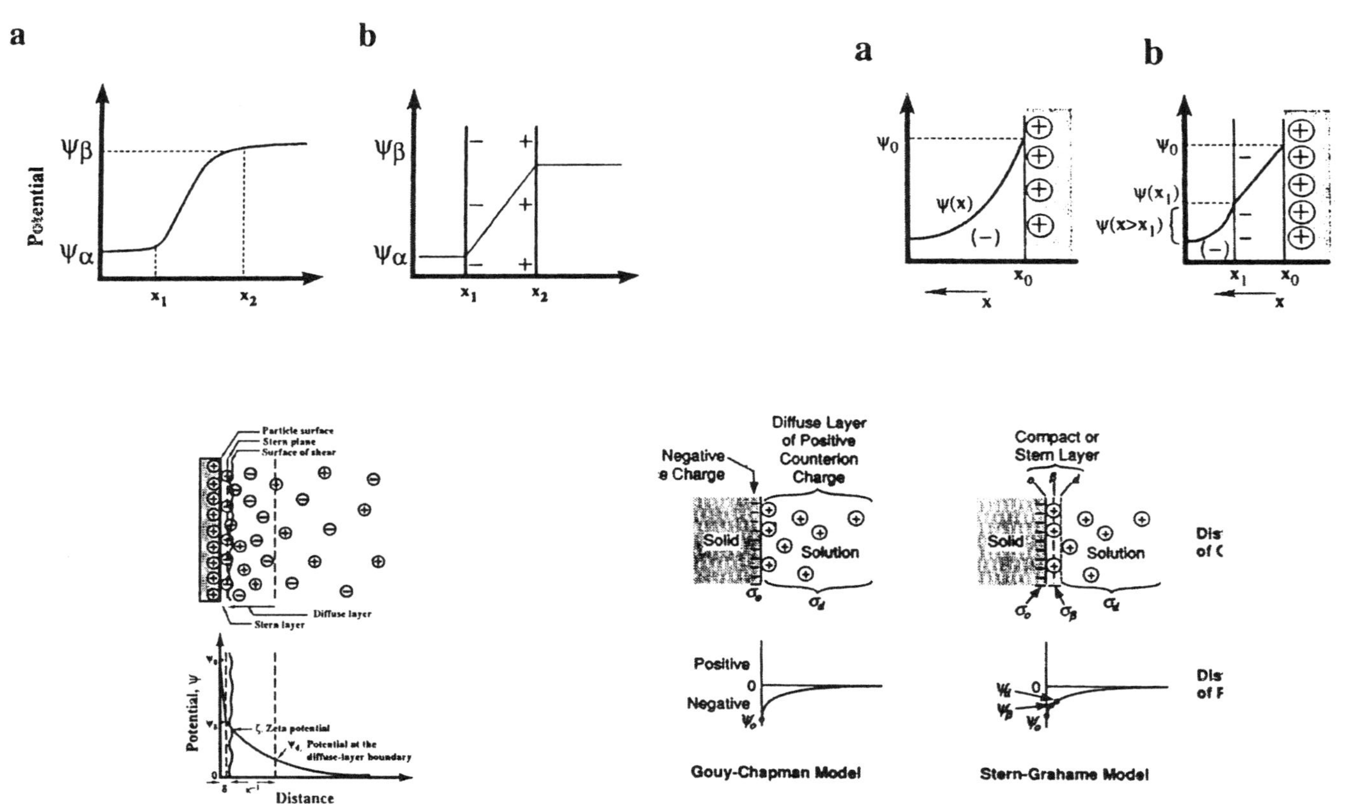

Figure 10. Electrical double layer models. Top right: (a) typical type of potential vs. composition plot for a charged surface compared to (b) constant capacitance model. Top left: Two double-layer models. (a) Diffuse double layer. (b) Part parallel plate capacitor and part diffuse layer. Bottom left: Stern layer model. Incorporation of adsorbed ions to surface. From Hiemenz and Rajagopalan (1997) Bottom right: Comparison of Gouy-Chapman and Stern-Grahame models of the electrical double layer. From Davis and Kent (1990).

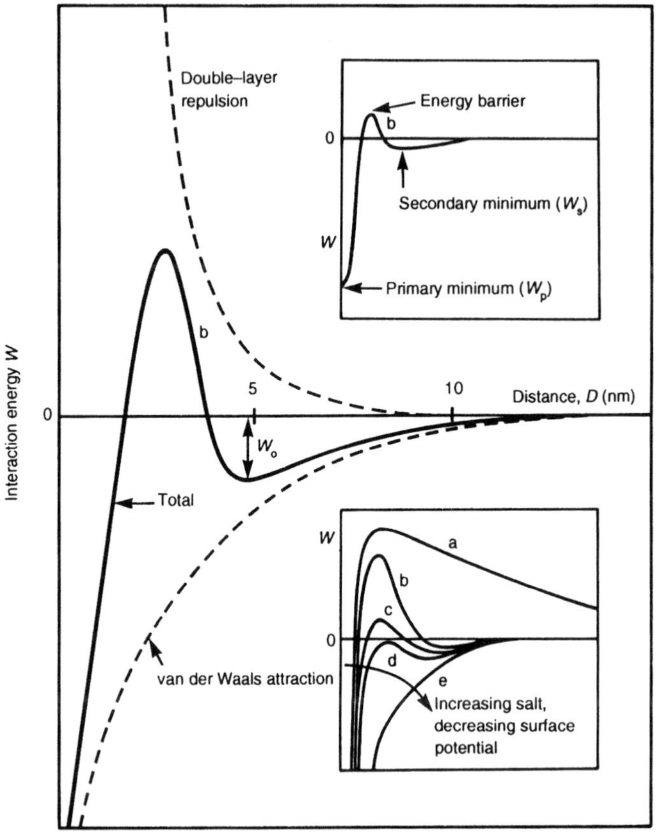

Figure 11. Interaction energy with separation distance for DLVO forces between interacting surfaces or particles. Top inset shows definitions for critical points on energy curve. Bottom inset shows the shape of the curve for different conditions. (a) Highly charged surfaces, dilute electrolyte. (b) Highly charged surfaces and more concentrated electrolyte. (c) Moderate surface charge and more concentrated electrolyte. (d) Moderate surface charge and highly concentrated electrolyte. (e) Weak or nil surface charge. After Israelachvili (1991).

slowly aggregate (called *coagulation* or *flocculation*) due to the lower energy barrier. Above some electrolyte concentration known as the *critical coagulation concentration*, the energy barrier will fall into the attractive(negative) range, and nanoparticles will rapidly aggregate (curve d). Such a situation is referred to as an *unstable colloid.* Finally with near zero surface charges the interaction is mainly just a van der Waals attraction and large planar surfaces or any size particles coalesce rapidly (curve e).

Aggregation kinetics and kernels in the Smoluchowski equation

In contrast to what is clear from the DLVO model, early classical models of aggregation are based on an assumption that interparticle interactions are negligible until the particles contact. This contact then is assumed to result in adhesion 100% of the time. However, as particles are attracted together the fluid in the intervening space must be ejected, creating a hydrodynamic response in the particles, such as rotation. Neglect of this effect in aggregation is termed *rectilinear* modeling, whereas inclusion is termed *curvilinear* modeling. The fact that hydrodynamic effects can combine with Brownian motion to rotate particles as they approach one another, opens the possibility that specific orientations for attachment may be slightly favored due to local strong directional interaction forces. This gives added credence to the possibility of oriented aggregation which is considered in detail in the chapter by Banfield and Zhang in this volume.

The adjustments made to the kinetic aggregation theory of Smoluchowski due to curvilinear effects, as well as to problems caused by the original simplifying assumptions, have been reviewed in detail by Thomas et al. (1999). Due to the form of the Smoluchowski equation, it is possible to make such adjustments by improvements to the kernels. For example, the rotation of particles in the curvilinear model can be taken as a decrease in their probability for attachment. A larger adjustment is the addition of particle-particle interactions through the use of DLVO and related theories. The addition of the kinetic effects of these theories alters the rates of particle collision, probability of particle adhesion, action of particles of different sizes, and the probability of floc breakup. Another important assumption of the original Smoluchowski theory is the that perfectly solid spherical particles were coalescing into perfectly solid spherical aggregates. This is true for types of vapor deposition processes by which the original theory was tested, but not in general. In environmental systems the situation is about as opposite as it can be, with aggregates more properly described as fractal objects (Li and Ganczarczyk 1989). Use of fractal geometry allows for quantification of the geometric scaling properties of rather diverse types of physical aggregates, and enables separation of aggregation models on the basis of kinetic factors and cluster form (Meakin 1992).

Fractal dimensions of an aggregate

If it is assumed that a solid particle aggregates to a solid mass with no interparticle space, the mass of the aggregate as a function of r is given by

$$m(r) = \rho(4/3)\pi r^3 \qquad (12)$$

If there are actually empty spaces between the particles then the mass must be reduced and a relationship like

$$m(r) = \rho(4/3)\pi r^{d_f} \qquad (13)$$

might be used, where $d_f < 3$, representing a reduced spatial dimension. Remarkably, this works for a large proportion of physical cases, and the quantity d_f is known as the fractal dimension. It is possible to measure the fractal dimension by a number of techniques (some of which are discussed below). Fractal analysis and scaling properties are broad subjects that cannot be treated here in detail, but the crucial idea is that aggregates can be characterized in terms of fractal properties, and that kinetic formation models can be similarly derived so that aggregate structure can be related to particular formation processes.

Aggregation topologies

If the probability of particle adhesion controls the rate of formation of an aggregate, the process is known as *reaction-limited aggregation* (RLA). If the rate of formation is controlled by the rate of diffusion of particles to the growing aggregate, then the process is called *diffusion-limited aggregation* (DLA). DLA is a very general phenomena for all kinds of pattern formation (Halsey 2000). In keeping with the original Smoluchowski assumption of primary particle aggregation only, the classical version of DLA was derived by Witten and Sander (1981) and is called a diffusion-limited monomer-cluster aggregation (DLMCA) process. There are many types of variations upon these basic

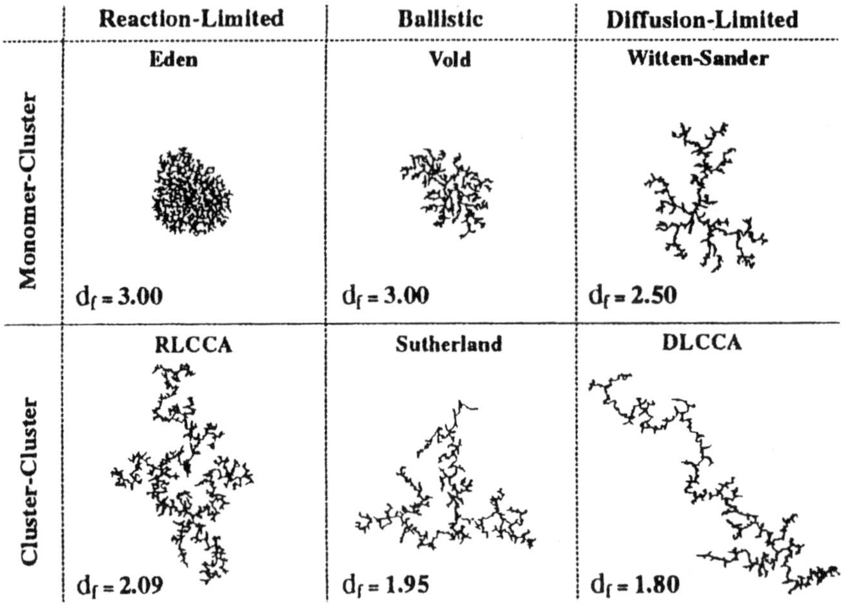

Figure 12. Different types of aggregation geometries. d_f refers to the fractal dimension of the aggregates. From Hiemenz and Rajagopalan (1997) with permission from Marcel-Dekker.

models, including permitting the adhesion of small aggregates to form larger ones, which leads to more expanded aggregates and lower fractal dimensions. This is an example *of diffusion-limited-cluster-cluster aggregation* or DLCCA. Other models, shown in Figure 12 include the reaction-limited cluster-cluster aggregation (RLCCA), rectilinear "ballistic" monomer-cluster (Vold and Vold 1983), rectilinear "ballistic" cluster-cluster (Sutherland 1970), and reaction-limited monomer-cluster (Eden 1961).

Another type of aggregation that differs markedly from these geometries is that imposed by so called "self-assembly" processes. For nanocrystals it has been possible recently to prepare particular habits and sizes, and stabilize these with organic or inorganic secondary coatings. The resulting entities are stable and can pack into super-lattices in two or three dimensions. The nanocrystals are effectively the hard atomic-like cores that preserve lattice ordering, while the coatings perform as interparticle molecular bonds balancing the attractive forces against forcing the core nanocrystals into fractal aggregation (Yin and Wang 1997; Murray et al. 1995; Sun and Murray 1999; Caruso 2001). The strength and type of interparticle bond is mitigated by the type or organic molecules used, e.g., the length of the chains in these molecules or the types of functional groups (see Fig. 13). A somewhat different process is the reorganization of original aggregates into a different type of aggregate, rather than into a single crystal. This has been demonstrated in MnO_2 nanomaterials by Xiao et al. (1998) and Benaissa et al. (1997). In this case an initial low density random aggregate is slowly modified by the nucleation and growth of nanofibers which grow out of the aggregate surface by diffusional processes. The intersection of fibers from dispersed initial aggregates ultimately produces a mass of interlocking fibers (Fig. 14). This scenario suggests that hybrid aggregation/nucleation and growth processes may be important in both technological and environmental nanomaterial formation. Finally, rather different aggregation topologies can be produced by crystallographically oriented aggregation processes, such as the incorporation of particular defects, the introduction of polytypes,

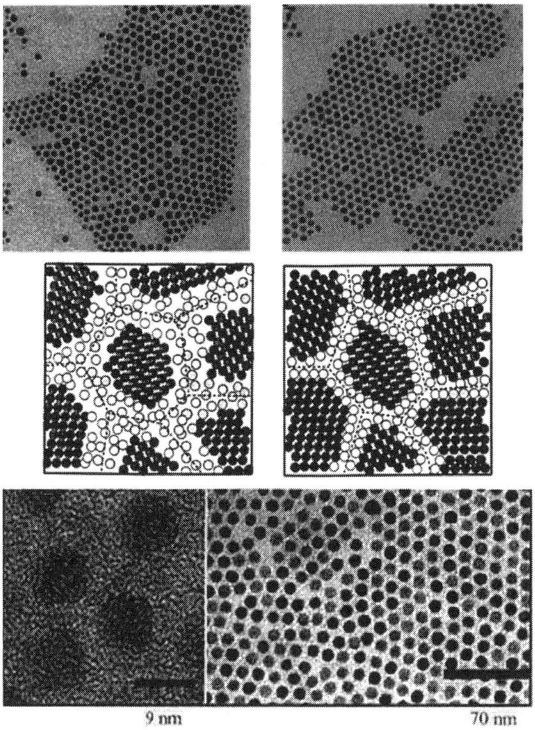

Figure 13. Aggregation/assembly of nanoparticles and nanocrystallites into superlattices and grain structures. Top right HRTEM image: assembly of capped Au nanocrystals into mosaic structures due to polydispersity. Note how sizes are sorted and produce new types of voids and curved registry. Top left HRTEM image: capped Au nanocrystals forming well-defined regular grain structures. Note incorporation of stacking faults and types of void spaces. From Martin et al. (2000), used with permission of the American Chemical Society, publishers of the *Journal of Physical Cemistry*. Middle left: Model for self-assembly of nanosemiconductor crystals having a large disordered region at the grain boundaries. Middle right: Analogous model with well-ordered grain boundary structure (from Weissmuller 1996). Bottom left: capped Pt nanocrystals. Bottom right: Self assembly of capped Pt nanocrystals.

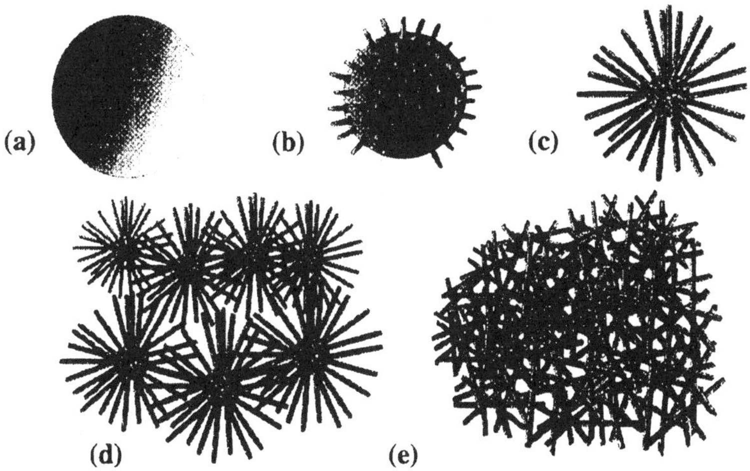

Figure 14. MnO_2 nanoparticle evolution during formation. (a) Initial aggregation produces densely packed clot of nanocystals. (b) Nucleation of elongated nanocrystals and growth normal to surface of resorbing clot. (c) Nearly complete recrystallization of clot into spherulitic shapes. (d) Aggregation of spherulitic forms into loose clot. (e) Interweaving further growth of aggregate. After Benaissa et al. (1997).

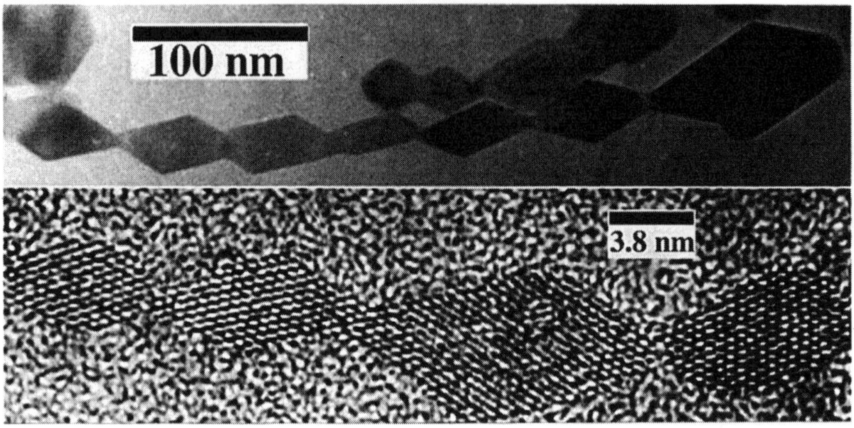

Figure 15. Aggregated nanoparticles of titania forming a single-crystal chain, an example of oriented aggregation. From Penn and Banfield (1999) with permission from Elsevier, publishers of *Geochemica et Cosmochimica Acta.*

and the nucleation of metastable phases (Penn and Banfield 1998a; 1998b). An example of a single crystal chain of nanoparticles formed by oriented aggregation is shown in Figure 15.

Simulations and state of knowledge

Simulations of growth by aggregation using various kinetic models consistent with DLVO theory have been done extensively over the past few decades, generally obtaining good agreement with observation for idealized cases. [For breaking news on many-body interactions that deviate from DLVO, see www.che.utexas.edu/~jeff/research/edl.html.] A thorough review of kinetic simulations is available (Meakin 1992), and numerous examples can be found in the NATO conference volume by Stanley and Ostrowsky (1986). From the point of view of physical theory, there remains much to be done in the area of fractal aggregation on surfaces, an area of obvious importance to geochemistry, and in the area of concentrated systems, such as in situations of low water saturation and increasing concentration. Thomas et al. (1999) make the point that many different aspects of porosity, nonsphericity, interaction forces, hydrodynamics and aggregation kinetics have been dealt with in the available literature, but generally in isolation from one another. Hence work needs to be done to integrate the available theory into a more realistic and applicable form, especially for geochemical systems. Another area which has not received much attention, is the coarsening (or *Ostwald ripening*) of aggregated nanoparticles into larger and more solid aggregates, or into single crystals. This is important for ceramic as well as geochemical processes such as diagenesis, solution encapsulation, impurity capture, and particulate transport. Krill et al. (1997) have reviewed the issues of stabilization of grain size and recrystallization in metallic nanocrystalline materials including the relevant factors of grain boundary motion and grain boundary energy.

STRUCTURE

Scales of structure

The discussions above concerning aggregation provide some background to the problems of characterizing the structure of nanoparticles and nanoparticulate aggregates.

First, there is the problem of *polydispersity* of particles and aggregates. It is one thing to attempt to deduce the structure of a set of objects that are all similar in density and size, but quite another to sort out structure in ensembles where size, density, fractal dimension and other attributes are widely varying. Systems that are usually studied in nanotechnology, such as semiconductor and superconductor materials (see chapter by Jacobs and Alivisatos), and fine droplets or colloidal particles in solution, are much more amenable to structural analysis than environmental aggregates due to their more limited polydispersity. Second, there is the problem of intraparticulate diversity. One of the most important aspects of nanotechnology is the change in physical structure occasioned by reduced particle diameter (see discussion and examples of by Banfield and Zhang, this volume). This can take the form of unusual surface structure, or bulk structure. The degree of structural change may be dependent on particle shape and diameter, thus creating a distribution in structure variation and microstructures. Hence in the general case we need to consider the structure of particle surfaces, interiors, intergrain boundaries and aggregates, all with possible complex dependencies on size distributions. Attempts to sort all of this out, with even the complete arsenal of available structure probes, will require careful experimentation on selected and probably idealized environmental systems. On the other hand, the immense information carried in a complete description of the structural state of environmental nanoscale materials provides ample reason for pursuing this type of investigation.

Surface vs. bulk structure

Surfaces of any crystal have important differences from the interior of the crystal:

1. The long range periodic symmetry is broken, affecting bonding orbital systems, vibrational states, and thus many aspects of spectroscopic signature. Some of the coordination polyhedra of a structure must be different on a surface than in the bulk. Valence states on the surface may be altered by electron transfers associated with adjustments in bonding. This may be especially important in environmental Fe-Mn oxyhydroxides.
2. The chemistry of the surface may differ from the bulk. Except for materials prepared by high temperature anneal under high vacuum conditions, the surfaces of oxides are oxygen terminated. Hence the cations may have complete oxygen coordination, but none of the surface oxygens have complete metal coordination as in the bulk. In the rigorous sense, this means that a small oxide particle has an excess of oxygens and is thus nonstoichiometric. In other types of nanoparticles there may be a chemical gradient near the surface as one type of ion may tend to migrate to the surface. These considerations suggest types of investigations that can be used to specifically detect bulk vs. surface structural differences.
3. The differences is surface bonding can cause surface relaxation and/or rearrangement. Figure 16 shows surface models for α-Al_2O_3 as obtained from crystal truncation rod (CTR) X-ray diffraction measurements. The differences among the vacuum-equilibrated, water-equilibrated and ideally bulk terminated structures are shown. Not only do the surface oxygens have reduced metal coordination, they and other atoms move (or "relax") from their bulk positions. This relaxation process at the surface can be understood using two different concepts for ion interactions: electrostatic and "bond valence".
4. A corollary to (2) is that environmental nanoparticles will be generally surrounded by water, either due to saturated conditions, or adsorbed water. Hence the surfaces of the particles must adjust to the bonding with water molecules, and possibly excess or surplus protons as dictated by the pH. This alone may cause a significant difference in the nature of the particle surface relative to a dry surface, e.g., a cleaved particle exposed to a dry atmosphere of pressure with sufficient oxygen partial pressure to maintain all metal valence states as in the bulk.

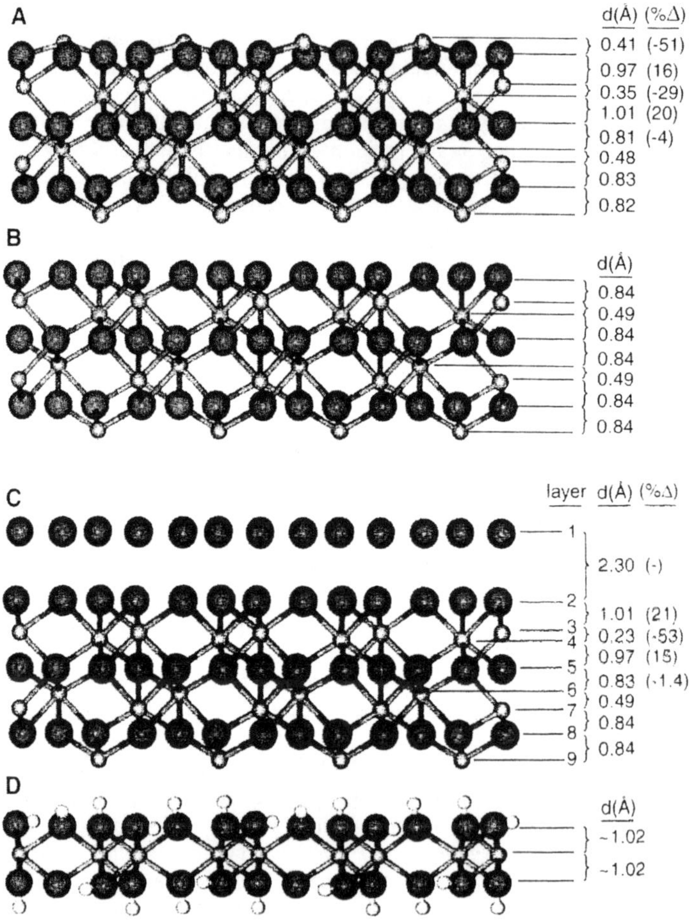

Figure 16. Surface structure the (0001) surface of sapphire (α-Al_2O_3) from Eng et al. (2000) with permission of the editor of Science. (a) Structure of vacuum-equilibrated dry surface. Al metal atoms sit on the surface. (b) Ideal bulk terminated surface with no relaxation or reconstruction. (c) Model for the wet-equilibrated surface at one atmosphere from surface scattering (crystal truncation rod) diffraction measurements. Al metal atoms have shifted, the surface is oxygen terminated, and an organized water monolayer is required to accurately describe the observations. (d) Structure of gibbsite or γ-$Al(OH)_3$. The relaxed wet-equilibrated sapphire surface is intermediate between this structure and that of the bulk terminated structure (B).

Surface features

Symmetry considerations. Even in the case of perfect single crystals of large size that do not differ in atomic arrangement except for face terminations, the surface has a local symmetry that cannot include symmetry elements requiring operations in the z (surface normal) dimension. These operations include some types of glide planes, all screw axes, and center of symmetry, and the possible spatial symmetries are reduced from 230 space groups to 17 plane groups. This alone can have consequences for the type of bonding at surfaces. The symmetry is further broken by the presence of growth features such as steps and terraces (see Figs. 10 and 11) which can be consequences of growth conditions as noted earlier. For the smallest nanocrystals, there are too few atoms to produce stable faces on the crystal, and in many materials the crystallite shape may be a consequence of electronic properties, i.e., covalency or electron delocalization effects involving the entire unit. On the other hand, if the crystallites are prepared near thermal equilibrium and are large enough (but not so large as to have habit altered by kinetic effects), it is expectable that the particle face and shapes will be predictable by the Wulff construction. This is testable provided that reliable calculations or measurements of the surface energies are obtainable.

Besides these basic aspects of surfaces, there are surface expressions of grain boundaries, and defects such as twinning planes and dislocations. Dislocations are expected to be unstable in nanocrystals (Gao and Gleiter 1987; Milligan et al. 1993; Thomas et al. 1990), though observations on this have been limited mainly to metallic and semiconductor nanomaterials. The absence of dislocations can have marked effects on mechanical properties of nanoparticulate phases. As certain types of planar "defects" are also consequences of non-stoichiometry, e.g., the so-called crystallographic shear structures called Wadsley and Magneli phases (Greenwood 1970; Wadsley 1964) common to TiO_x, MoO_x and WO_x systems, it is possible that types of nonstoichiometry could stabilize stacking faults in nanocrystals that would otherwise be energetically unfavorable.

Relaxation of surfaces. A common aspect of surfaces is a relaxation from the interatomic distances observed for the bulk material. The relaxation can be a net contraction or a net expansion depending on the type of material and surface chemistry. For oxide nanocrystals we can invoke the bond valence model to predict a surface contraction. Consider a simple NaCl-structure oxide like MgO. As we noted earlier the surface will be oxygen terminated, so that the surface oxygens do not have full coordination by Mg atoms. In the bond valence method, we sum the 'valence" contributions about an anion to see if its valence needs are achieved. The valence sums are determined by the Mg formal charge and coordination, so that a six-coordinated Mg has a net valence contribution of $2/6 = 1/3$. The oxygen normally has six surrounding Mg atoms that contribute $6 \times 1/3$ valence contributions to add to 2. At the surface of MgO, the oxygen has coordination of 5 or less (depending on its exact position at the surface). Hence its valence is unsatisfied. It can make up this loss by decreasing its bond distance from one or more of the nearby Mg atoms, as it can also be shown that valence contribution is a function of bond length (Brown and Shannon 1973). Hence the overall effect of bond valence analysis suggests a surface contraction, at least in the top oxygen layer. But this situation may be altered by protonation of the surface oxygens. Single protonation may give the hypothetical oxygens too much bond valence (H contributes a full 1.00 valence contribution if not also hydrogen bonded to other atoms.) This bond valence sum would indicate a expansion of the top hydroxyl layer to lengthen the Mg-OH bonds. A mineralogically relevent example is the surface of α-Al_2O_3 (sapphire or corundum). Eng et al. (2000) showed via surface X-ray diffraction that the (0001) surface equilibrated in air (i.e., with a few monolayers of surface water), has a structure that is intermediate between the oxygen terminated surface with no relaxation, and the gibbsite (γ-$Al(OH)_3$) structure (Fig. 16). This means that the nearest surface Al atoms are displaced 0.14 Å down into the crystal, and the next lower Al atoms displaced 0.11 upwards towards the surface relative to the bulk structure. The oxygen layers above and below these Al atoms increase in separation by 0.04 Å, so that the Al-O distances increase relative to the bulk, i.e., a surface expansion.

For a different interpretation of surface relaxation, see Linford (1973) and Onodera

(1991), who consider the role of capillary forces in surface stress and relaxation. Still other mechanisms for surface relaxation involve the changes in vibrational amplitude of surface atoms, which surprisingly can be either larger or smaller than in the bulk phase depending on the vibrational modes active at the surface (Schreyer and Chatelain 1985). Anharmonic vibrations can result in net changes in mean atomic positions if vibrational amplitudes at surfaces are larger than in the bulk.

Reconstructions, distorted or altered coordination environments, incomplete coordination polyhedra and changes in surface bonding. Significant repositioning and ordering of atoms on a surface can lead to surface cells that are quite a bit larger than the projection of the unit cell onto the surface plane. Smaller analogs of the familiar 7×7 reconstruction of the silicon surface, and reconstructions seen in metal single crystals could also occur on mineralogical nanocrystal surfaces, but there has been little observational evidence to date. The small sizes of crystallite faces might disrupt equilibration of reconstructions, as these represent energy reductions that may be smaller than variations in interfacial energy for small particles. It has already been mentioned how oxygen termination of an oxide surface leads to lower coordination for the oxygens. However cation coordination can also be altered due to relaxation effects at a surface, or affected by the presence of surface defects and other features. For example, at the edges and corners of ledges or steps on an oxide surface a metal atom (say in MgO) could have two or three oxygens about it having reduced coordinations. As the symmetry constraints of the structure do not apply to such a site on the surface, the metal atom is free to reconfigure its coordination environment to lower the local energy. The coordination could distort or change in number. Changes such as this have been suggested for the ferrihydrite surface (Zhao et al. 1994), for TiO_2 surfaces, and are consistent with bond valence concepts. For the TiO_2 surfaces, Chen et al. (1999) suggested that the surfaces of 1.9 nm particles have a significant number of highly distorted and probably 5-coordinated Ti sites. XANES analysis indicated that the geometry of these sites was a square pyramid. Binding of other moleules on the surface of the particles restored the coordination geometry to 6-coordination. Current quantum mechanical calculations appear inadequate to predict whether such coordination changes would occur. Hence for both reconstructions and coordination changes we may be forced to rely on characterization of experimental materials for the present.

Bulk nanoparticle features

Lattice contraction/expansion. Just as the surface can experience a relaxation, the bulk of a nanocrystal may expand or contract in lattice dimensions. One of the first clear observations of this was in Cu metal by Montano et al. (1986), who showed that the Cu-Cu distance decreased as nanocrystallites were reduced in size. The Cu-Cu distance appeared to form a continuous trend toward that of Cu-Cu gaseous dimers. Lattice contraction is also seen in nanosized silicon particles having thin surface coverages of amorphous silicon dioxide (Hofmeister et al. 1999). Due to the oxide layer this effect may arise more from compressive stresses than a change in silicon bonding. Lattice contraction is observed in Sn and Bi particles (Nanda et al. 2001), and explained by treating the nanoparticles as if they are droplets with a high effective external pressure, i.e., strain induced by the large surface area. In nanosized Pt particles it has been suggested that the lattice contraction is due to an increase in cohesive energy for particular geometric shapes (Khanna et al. 1983). Aggregation, occupation of zeolite cages and surface stabilizing films all reduce the contraction effect in Pd, Pt and Au nanocrystals (Fritsche and Buttner 1999). In the case of oxide nanoparticle systems both expansions and contractions have been observed. Tsunekawa et al. (2000a) studied nanocrystalline $BaTiO_3$ particles in the size range from 250 to 15 nm. With size reduction to the smallest sizes, a 2.5% expansion in the lattice parameter is observed. At 80 nm there is also a change in the overall space group symmetry, from tetragonal (ferroelectric) to cubic. Tsunekawa et al. (2000b) examined the lattice expansion in both $BaTiO_3$ and CeO_2 by photoelectron spectroscopy (XPS) and simulation methods. They found in the case of CeO_2 that the expansion of the lattice was caused by reduction of some of the Ce^{4+} ions to Ce^{3+}, and in $BaTiO_3$ by the increase of ionicity of the Ti ions. In contrast, Cheng et al. (1993) found a net lattice contraction in nanoparticle TiO_2. These studies suggest a major difference between metal and oxide nanoparticles, which may be extremely important for catalysis processes.

Grain boundaries, stacking faults and grain boundary encapsulation. A large body of literature has dealt with the nature of grain boundaries in metals and other materials (see e.g., Bollman 1970). Stacking faults are similarly well understood crystallographically. The question here is whether these structural aspects will be different in nanoparticles, i.e., will they occur in different geometries or densities, and will they have a large effect on nanoparticle properties? As noted earlier, dislocations should be disfavored in small nanocrystallites. This is not only due to rapid diffusion rates, but also repulsive forces between dislocations that restrict their density (Alivisatos 1997). However, dislocations, stacking faults and grain boundary structures will be generated by particle aggregation. Penn and Banfield (1998) showed that dislocations could be created by small misorientations at the interface between assembled nanoparticles of TiO_2 (anatase). Crystallites of 5-6 nm in diameter were reacted under hydrothermal conditions to produce coarsening. HRTEM photos showed clear evidence for dislocation formation at primary particle attachment zones (Fig. 17). Particularly if surfaces of nanoparticles are chemically altered, e.g., passivated by a thin coating of water or surfactant (i.e., "capped"), or altered in composition or stoichiometry, a large density of defects can be preserved at grain boundaries during aggregation. These defects may dominate the physical properties of the aggregate, and kinetic factors will determine their stability. (See also chapter by Banfield and Zhang for more details of grain boundary and nanoparticle defects.) Grain boundary density and structure are also important factors in aggregate recrystallization. In this case the energy added by these kinds of defects can serve as a driving force for diffusion and recrystallization. On the other hand, particular types of grain boundaries may be stabilized with impurities, and act to frustrate recrystallization processes. An example of the simulation of grain boundaries is that of MgO nanocrystallite films (Sayle and Watson 2000). Although thin films can retain enormous strain energy making them in this sense poor analogs of monodispersed nanoparticles, they are technologically widespread and widely studied. Figure 18 shows the results of Sayle and Watson's simulation of 2 to 3 layers of a nanocrystalline MgO film. Particularly notable are the fairly well-defined type of voids at the grain boundaries, many of which are the cores of dislocations (label A). Other features are the mixed screw-edge dislocations (label B), and regions of hexagonal MgO (label C). The calculated model shows how the dislocation cores could be stabilized by doping with impurities, and also how some of the lattice mismatch between grain boundaries can be made up with an expanded type of local lattice, in this case pseudohexagonal.

Impurities and point defects. Point defects are stable equilibrium species and both impurities and vacancies will be present in any nanoparticles. Due to the expansion/contraction effects we have discussed above, impurities or vacancies may migrate into the bulk or to the surface depending on size and relative charge. Grain boundary "decoration" by impurities are common in metallic alloys, but little studied in nanoparticulates. In mineral systems, aggregation of nanometer particles could provide grain boundary sites for species not readily sorbed onto typical grain surfaces. Hence aggregation of grains could provide a new mechanism for remediation of toxic agents,

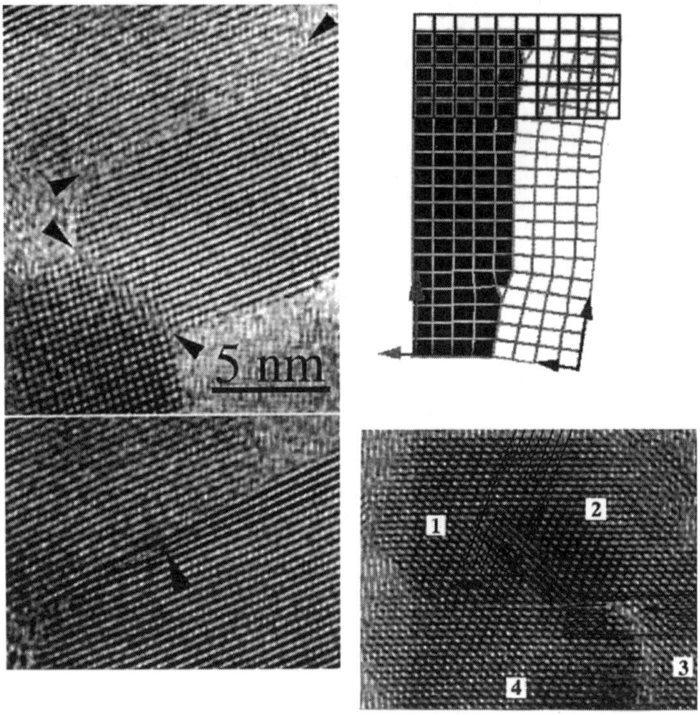

Figure 17. Origin mechanism for dislocations in TiO_2. Left: Three attached nanoparticles (top) showing interface regions. Upper interface is reproduced (bottom) to clarify edge dislocation. Center, top: Diagram illustrating misorientation of idealized grains and the formation of dislocations. Lower grains join initially with surface steps causing a rotation. Third crystal is attached atop the other two oriented with the left crystallite, but rotated with respect to the right crystallite. Both edge and screw dislocations are thus formed. Center, bottom: HRTEM image of a crystal formed by attachment of four primary particles. Arrowheads and lines indicate edge dislocations.

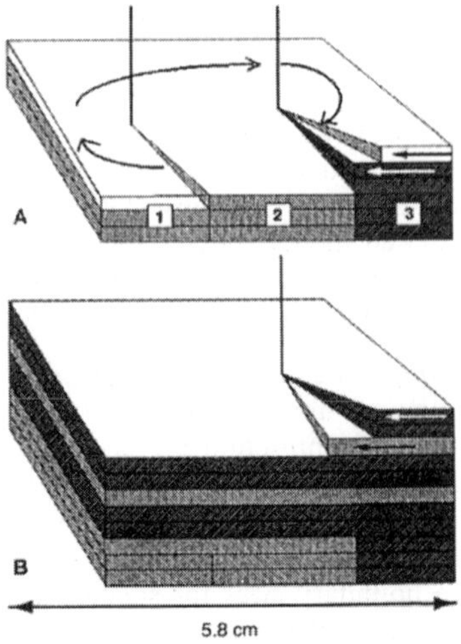

Right (A): Block diagrams showing how growth at two adjacent screw dislocations can occur and give rise to polytypism. Three separate crystals have joined with slight orientational mismatch to create the dislocations. (B) The growth occurs at each leading step and allows intimate stratification of the crystals. From Penn and Banfield (1998a).

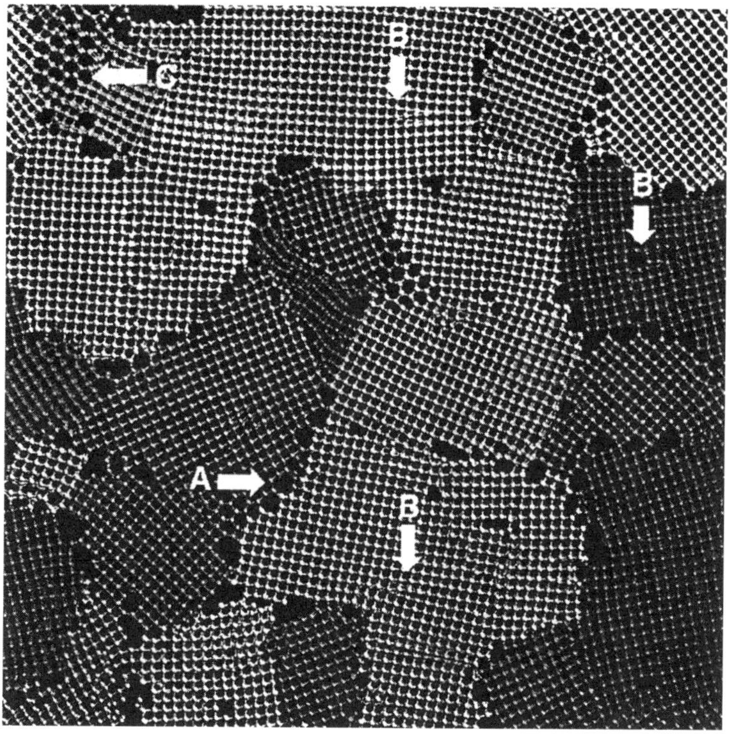

Figure 18. Simulated MgO grain and grain-boundary structure. Note the regular voids at A, defects at B and hexagonal structure at C. From Sayle and Watson (2000), used with permission of the Royal Society of Chemistry, publishers of the *Journal of Materials Chemistry*.

and also explain how seemingly incompatible impurities are incorporated during growth (Banfield et al. 2000).

STRUCTURE AND SHAPE/SIZE DETERMINATION

Atomic structure—TEM, X-ray scattering and diffraction

The details of nanoparticle crystal structure are mainly determined by scattering and imaging techniques, sometimes assisted by spectroscopy that can decisively assign some aspect of the structure. The chapter by Banfield and Zhang gives examples of TEM imaging and image interpretation, hence we will focus here on the scattering techniques.

The most widely used and versatile technique remains X-ray diffraction and scattering. Traditional diffraction involves measurement of the Bragg peaks in either powder or single crystal patterns, usually by comparison with kinematical simulations of the same patterns. This is a mature field and structural analysis is straightforward if a good diffraction pattern can be obtained. General reviews of powder diffraction theory and application can be found in Klug and Alexander (1974) and Azaroff (1968), as well as the RIM volume on powder diffraction (Bish and Post 1989), while single crystal diffraction is treated well by Warren (1969) and Schwartz and Cohen (1987). A good general discussion of nanoparticle X-ray characterization is given by Zanchet et al. (2000).

Bragg peak broadening

The Bragg equation for the scattering from a series of parallel planes of spacing d, is given by

$$n\lambda = 2d \sin \theta \tag{14}$$

where λ is the scattered radiation wavelength, and θ is the scattering angle, equivalent to half of the angular change in direction of the incident and scattered beams, 2θ. This relation simply indicates that the scattered beam has to match in phase the beams scattered from each plane of atoms. Hence the path difference from plane to plane must be a multiple of wavelengths. The integer n indicates the number of wavelengths that add to one path difference.

Small particles present difficulties for Bragg analysis. Particles in the sub-micron range will cause Bragg peaks to be broadened, with the peak widths increasing with decreasing particle size until the Bragg pattern grades into a diffuse non-Bragg or Debye scattering profile (see below). This broadening can be used to determine particle size, provided that the effects of strain on peak broadening can be taken into account. Particle shape also affects the degree of broadening of specific hkl reflections, and results in size uncertainties of 20% or more unless the crystallite shape is independently known. The broadening of Bragg peaks is created by decreasing coherence of scattering as the crystallite size decreases. Because X-rays can penetrate a fraction of a micron even into dense materials, scattering can readily occur from atomic planes from depths such as this within a crystal. For very intense scattering some of the scattered X-rays will be reflected backward toward the depths of the crystal and cancel some of the forward scattering. This is the phenomenon of dynamical diffraction, which is not needed for the evaluation of small particle diffraction as such cancellation effects are small for small crystals. The usual treatment is with the kinematical (or two-beam) theory where only incident and forward scattered X-ray beams are considered. From Figure 19 we can see that a large crystal with many atomic planes contributing to a Bragg reflection can only have finite Bragg intensity over a small angular range. All other directions of scattering are out of phase with one another and completely cancel. As the crystal gets smaller there are fewer and fewer contributing planes, and hence the phase cancellation is progressively relaxed. In the limit of a single scattering pair of atoms the scattering is not cancelled in any direction, but is only modified into a diffuse scattering pattern. Note that this model only applies within individual crystallites. The scattering from separate crystallites is effectively completely independent. A micron-sized particle composed of aggregated nanometer sized crystallites will have broadening associated with the latter size. The basic concept of broadening of X-ray reflections due to particle size was indicated by Scherrer (1918), who derived a semi-empirical formula to determine particle size. Later work gave more quantitative analysis of peak broadening and shape (Wilson 1963; Warren and Averbach 1950).

Bragg analysis and Rietveld method

Crystallites above about 50 nm (and smaller for electron diffraction) are large enough to yield broadened but characteristic diffraction patterns. By characteristic it is meant that the symmetry and geometric arrangement of the structure will be clearly reflected in the diffraction pattern. The normal way in which this is represented is via the crystallographic structure factor:

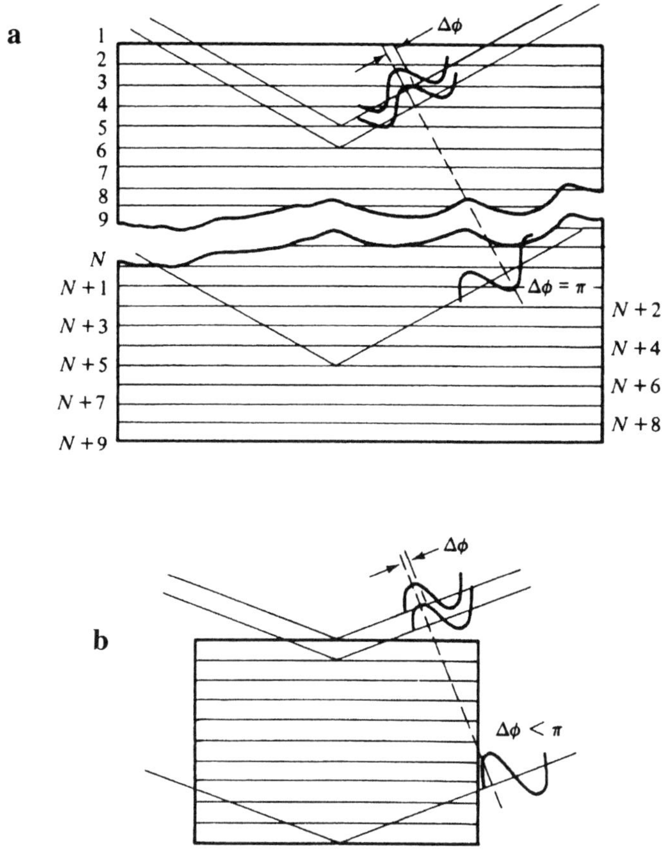

Figure 19. Origin of diffraction peak broadening in the kinematical approximation. (a) Large crystal with orientation just off the Bragg angle. For each different plane there is a slight phase shift in the scattered radiation. For some plane well below the top of the crystal surface there is a plane scattering out of phase with it. Hence for large crystals the reflection peak is quite sharp. (b) For small crystals scattering from the top and bottom-most planes do not create such a large phase shift when the crystal is set off the Bragg angle. Hence the range of angles where the scattered radiation is cancelled is reduced and the reflection broadens. From Schultz (1982) with permission of Prentice-Hall.

$$F_{hkl} = \Sigma f_n e^{2\pi i (hx_n + ky_n + lz_n)} \qquad (15)$$

In this equation f_n is the *atomic scattering factor* or *atomic form factor*, h,k,l are the Miller indices of the reflecting plane, and x_n,y_n,z_n are the coordinates of the scattering atom in decimal fractions of unit cell parameters, *a*,*b*,*c*, respectively. For simple structures the structure factor indicates what types of Bragg planes in a given kind of structure can produce a diffraction peak, i.e., have non-cancelled, coherent scattering, and also indicates the relative intensity of the "allowed" peaks (Warren 1969). A few structure factors for simple crystal structures are shown below.

Face-centered cubic (*F*)lattice	hkl mixed:	$F_{hkl} = 0$
	hkl unmixed:	$F_{hkl} = 4\,\Sigma\, f_n\, e^{2\pi i\,(hx_n + ky_n + lz_n)}$
	MgO hkl all even:	$F_{hkl} = 4\,(f_{Mg} + f_O)$
	MgO hkl all odd:	$F_{hkl} = 4\,(f_{Mg} - f_O)$
	MgO hkl mixed:	$F_{hkl} = 0$
Body-centered cubic (*I*)lattice	h+k+l = even:	$F_{hkl} = 2\,\Sigma\, f_n\, e^{2\pi i\,(hx_n + ky_n + lz_n)}$
	h+k+l = odd:	$F_{hkl} = 0$
Hexagonal lattice:	h+2k = 3n,l even:	$F^2_{hkl} = 4\,f^2$
	h+2k = 3n+1, l odd:	$F^2_{hkl} = 3\,f^2$
	h+2k = 3n+1, l even:	$F^2_{hkl} = f^2$
	h+2k = 3n, l odd:	$F^2_{hkl} = 0$

For structures with complicated unit cells and many symmetry elements the structure factor can be quite complex. However from these relationships it can be seen that there will be characteristic diffraction patterns for FCC, BCC or other atomic packings, and the intensities will have certain relative values (scaled by the scattering power of the individual atoms). With this information, the identity of simple structures can be deduced from the diffraction patterns, usually by comparison with many model calculated patterns. If the patterns are good enough, the presence of more complex structural aspects can be identified, e.g., the space group of the structure which defines the three dimensional set of symmetry elements of the structure can be determined subject to certain limitations, the presence of stacking faults which produce non-Bragg scattered intensity can be detected, and the effects of twinning characterized. In such case the structure can be refined via a variety of techniques to yield cell dimensions and cell angles. Full single crystal analysis of a nanoparticle has not been done, with the current minimum size limit using a synchrotron X-ray source being about 0.5 μm. This type of analysis requires one to be able to mount the single nanocrystal, orient it along known crystallographic directions, and collect diffraction data from a large set of unique diffraction planes. This body of information contains three dimensional structure information, and is needed to refine the positions and thermal vibrations of each atom in a unit cell of the structure. In the future this type of analysis may be possible for 100-nm crystallites extracted from materials, and possibly to similarly sized crystallites at the surface of aggregates. Present focussed X-ray beams at synchrotron sources are as small as a micron in diameter, and sizes to about 100 nm are possible. Single crystal structure analysis is described in detail by Azaroff (1968) and Glusker and Trueblood (1985).

An alternative to traditional single crystal diffraction analysis is the method described by Rietveld (1969), which has been widely developed for use on polycrystalline samples. This is an extremely important structural technique for poorly crystallized minerals (Post and Bish 1989). Rietveld analysis consists essentially in the calculation and least squares fitting of a complete powder diffraction pattern. This includes not only the structure factors which are proportional to the peak areas, but disorder, site occupation, strain, texturing and background intensity. Overlapping peaks due to various similarly spaced atomic planes are easily handled, as no extraction of areas or structure factors is done. Because of its full use of a powder pattern the Rietveld technique can be used to refine atomic positions in relatively complex structures, where use of structure factors alone would not be sufficient. Because the powder data input is essentially from a (ideally) random material and is thus averaged over all directions, the scattering data is one dimensional. Rietveld analysis has the advantage of being applicable to aggregates of nanoparticles as well as powders, but has not yet been applied widely due to the general poor quality of diffraction data from small quantities of aggregates with crystallites below 50-100 nm. Rietveld analysis is regularly used with neutron diffraction as large samples are necessary and generally these must be powders or aggregates of single crystals.

Debye equation and scattering from small crystallites

For small crystallites (less than a few hundred atoms or so) we can calculate the diffraction as the sum over the scattering contributions (and interferences) of each atom pair, i.e., we can neglect the scattering plane treatment as with Bragg diffraction. This is done with the Debye equation:

$$I(S) = D\sum_{m}\sum_{n} f_m f_n \frac{\sin Sr_{mn}}{Sr_{mn}} \tag{16}$$

This is a kinematical relation which is summed over each pair mn of atoms. Note that the sums would get enormous with any sort of macrosized crystal, so the calculation is impractical except for small clusters and molecules. The main feature is that the scattering for any one atom pair is a Fourier series of the form (sin Sr)/Sr, a dependence which is common to all types of scattering where there is interference between the scattering from a given pair of elements. Other parameters are S, the scattering wavevector, f, the atomic scattering factor, D, the Debye-Waller factor which allows for damping of the Fourier oscillations as scattering angle increases, and N is the number of atoms in the cluster. D has a form like

$$D = e^{-(\Delta x)^2 \frac{\sin^2\theta}{\lambda^2}} \tag{17}$$

where Δx is the rms atomic displacement from the equilibrium position. This "blurring" of the atomic position reduces the intensity of scattering in a given direction, and preferentially at higher scattering angles. The Debye-Waller term is a way of introducing random structural or thermal disorder effects into the calculation. Because it is completely general, the Debye scattering equation is an excellent way to incorporate stacking faults, twinning, and other types of structural defects into a nanocrystal scattering pattern. For small metal nanocrystals having particular types of atomic packing and habits the results of the Debye equation calculation are shown in Figure 20. The effects of crystallite size also show in these calculated patterns.

The Fourier nature of the sums in the Debye expression allow the scattering function to be recast via a one-dimensional Fourier transform into a spherically symmetric picture of the interatomic distances r_{mn}. The resulting function is called a radial distribution function (RDF) and in the usual form is a series of peaks on a parabolic background which itself represents the mean atomic density of the sample. Other forms of the RDF subtract this average background to produce oscillations about zero. Different interatomic distances produce different sinusoidal frequencies in the Debye relation, and thus give rise to the discrete peaks (Fig. 21) in the RDF. The size of the region measured in scattering space (also called reciprocal or momentum space) dictates the amount of information in the scattering pattern. Smaller regions will show less of the sinusoidal variations in scattering, and hence will yield poorer estimates of the sinusoidal frequency and interatomic distances. In other words, smaller regions of scattering space will result in broader peaks in the Fourier Transform of the scattering pattern.

The size of scattering space is defined with a quantity like S:

$$S = 4\pi\,(\sin\theta)/\lambda \tag{18}$$

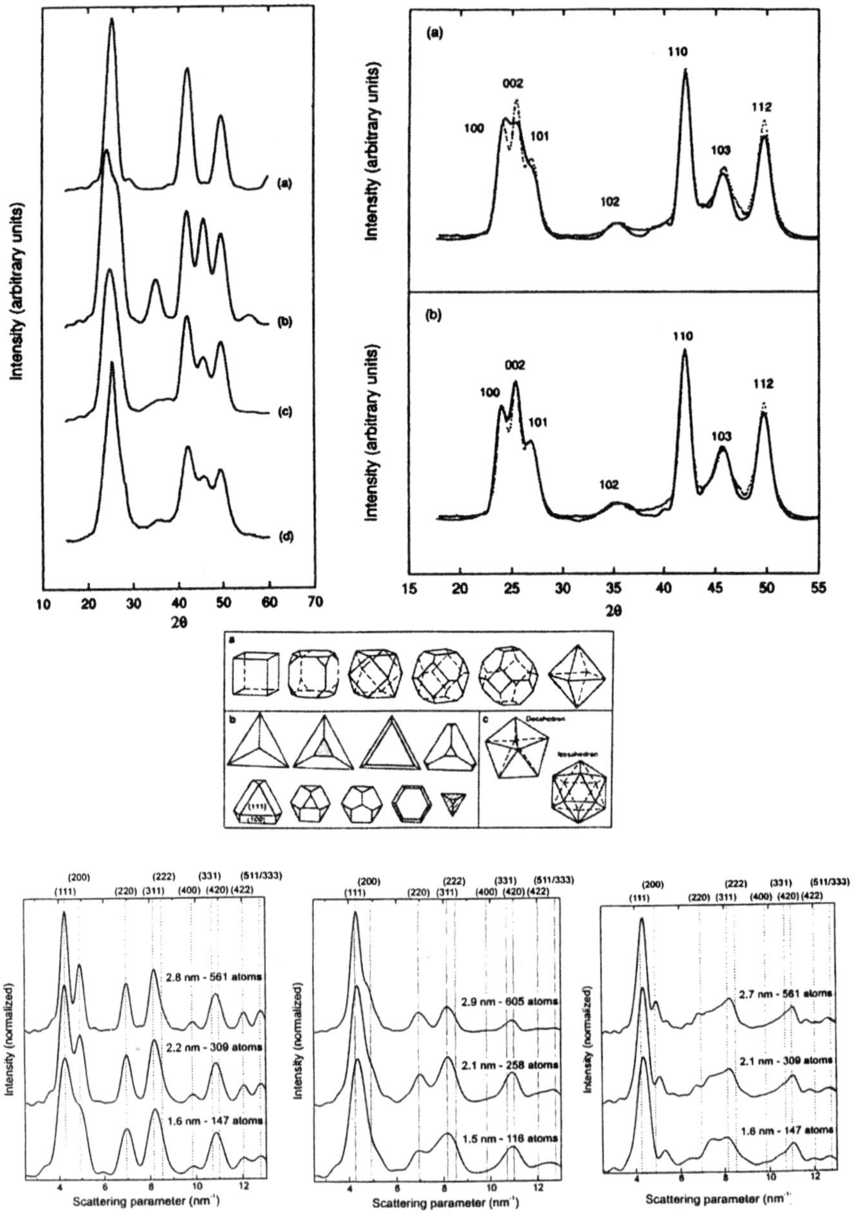

Figure 20. Calculated X-ray scattering patterns for various types and sizes of Pt nanocrystallites. Top left: 3.5 nm (a) sphalerite (b) wurtzite (c) wurtzite with one stacking fault (d) experimental powder spectrum with ca. 3.5.nm avg. crystallites. Top right: Experimental powder diffraction pattern of ca. 8.0 nm crystallites (dotted line) compared to: (a) spherical and (b) prolate particles (solid line) Center: (a) progression of habits of cuboctahedral shapes of nanocrystals. (b) change in shape as {111} faces increase and {100} decrease. (c) decahedron and icosahedron multiply twinned forms. Bottom left to right: three successive sizes of cuboctahedral nanocrystallites; three successive sizes of decahedral nanocrystallites; three successive sizes of icosahedral nanocrystallites. From Zanchet et al. (2000), used with permission of Wiley-VCH.

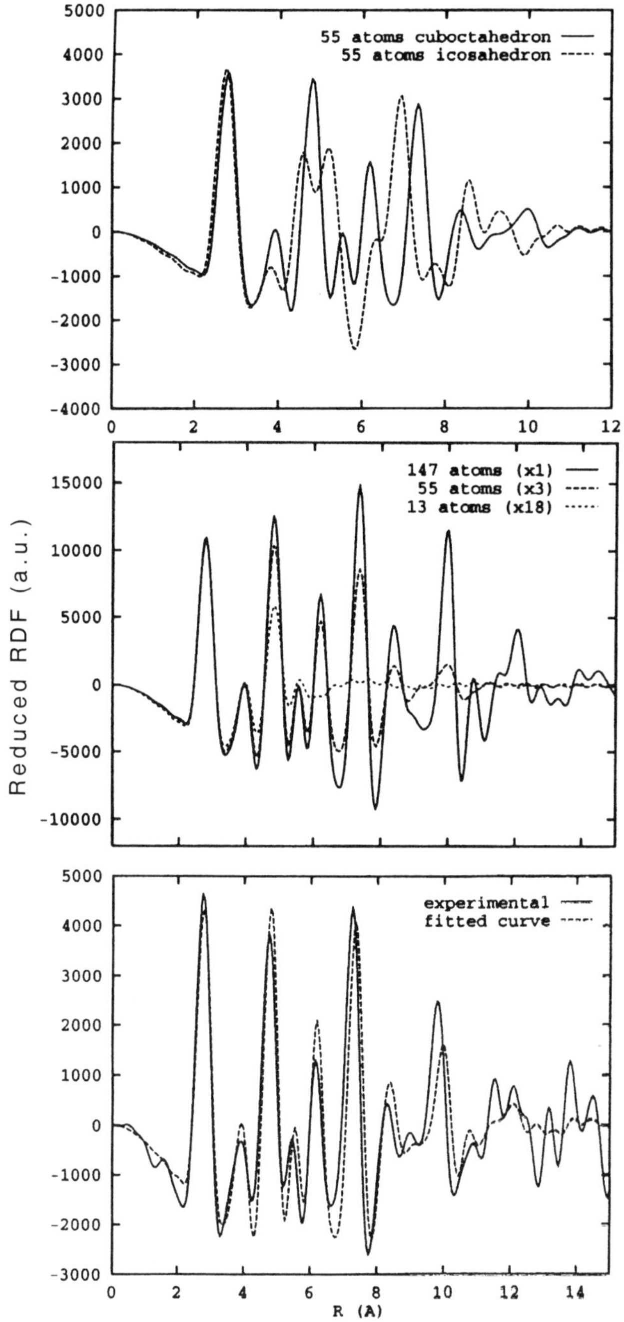

Figure 21. Radial distribution functions calculated using a Fourier transform of scattering patterns produced with a Debye equation. Top: Cuboctahedron (cluster with both octahedral {111} and cube {100} faces) and icosahedron (multiply twinned hcp structure) clusters of the same size. Center: Cuboctahedra of different sizes. Bottom: Experimental and simulated cluster RDF of a Pt colloid. The fit is a 90:10 mixture of 55 and 147 cuboctahedral clusters, respectively. After Casanove et al. (1997).

where θ is the scattering angle and λ is the X-ray wavelength. S has units of reciprocal length, usually Å^{-1}. From Equation (18) the largest amount of scattering space is explored with the largest scattering angles and smallest wavelengths. A powder diffractometer usually uses CuK_α X-radiation (λ = 1.54 Å) and may have a scanning angle (2θ) of 120°. This yields a maximum S value of about 7 Å^{-1}. Using the full 0-7 range of S would yield peaks in the Fourier transform of about 0.45 Å in width, so that it would be difficult to sort out different atom-atom pairwise distances. For standard wide-angle X-ray scattering (WAXS) analysis, MoK_α or AgK_α radiation is used, with scattering angles up to 170°. This yields S values up to about 20 Å^{-1} and resolution in interatomic distances of about 0.16 Å.

In performing the Fourier transform the scattering data must first be corrected for absorption effects, air scattering components, the presence of inelastic scattering components, and effects due to the polarization of the incident X-ray source and any additional diffractive optics in the scattering apparatus (Klug and Alexander 1974). The data are then converted into S-space via Equation (18) and fit with a scattering profile indicative of the average scattering atom in the sample. This function is just the weighted atomic scattering for the sample composition, which is assumed to have the same profile for any type of atom in S space. The fitting is done only by setting a scaling factor for the atomic scattering curve. Once fit, the data will be seen to oscillate about the atomic scattering curve, i.e., the atomic scattering curve represents the total scattering without any interference effects as would be caused by some ordering of the atoms in the sample. The fitted atomic scattering is then subtracted and multiplied by S to give the function S I(S), which is analogous to the output of (16). The Fourier transform is usually done with the expressions:

$$\begin{aligned} D(r) &= 4\pi r^2 \rho_0 + \frac{2r}{\pi}\int_0^\infty SI(S)\,\mathit{sin}\,Sr\ dr \\ G(r) &= D(r) - 4\pi r^2 \rho_0 \end{aligned} \qquad (20)$$

Here the first term in the D(r) function is the average radial density function of the material which has the form of a parabola with ρ_0 equal to the average density in electron^2 units, and the second term oscillates about the parabola indicating distances of high and low atomic (electronic) density. The G(r) function subtracts off the parabola so that the oscillation is about zero. Several representative RDFs of minerals are shown in Figure 22. There are several other types of radial distribution functions used in the literature, as well as the analog in EXAFS analysis below. For a complete review of the various types of analysis see Klug and Alexander (1974) and Warren (1969).

Examples. Nanoparticulates consisting of arsenate sorbed onto ferrihydrite were examined by Waychunas et al. (1996) using WAXS and EXAFS measurements. The crystallite size in the particles was reduced due to coprecipitation with high concentrations of arsenate, which sorbed on particle surfaces and effectively poisoned crystal growth. The SI(S) functions and RDF's are shown in Figure 23, for particles of different mean size and arsenate concentration. It is very clear from the SI(S) functions that as the crystallite size decreases so do the higher frequencies and thus the larger distance peaks in the RDFs. The smallest crystallites (As/Fe = 0.7) averaged about 0.8 nm in diameter.

Another approach to the analysis of a WAXS pattern is called Debye function analysis (DFA) and has been applied by several groups (Reinhard et al. 1997, 1998; Gnutzmann and Vogel 1990). The main difficulty in any diffraction experiment is that a unique structural model cannot usually be extracted from the data. This is obvious with powder diffraction data where for a complex structure there are far more structural

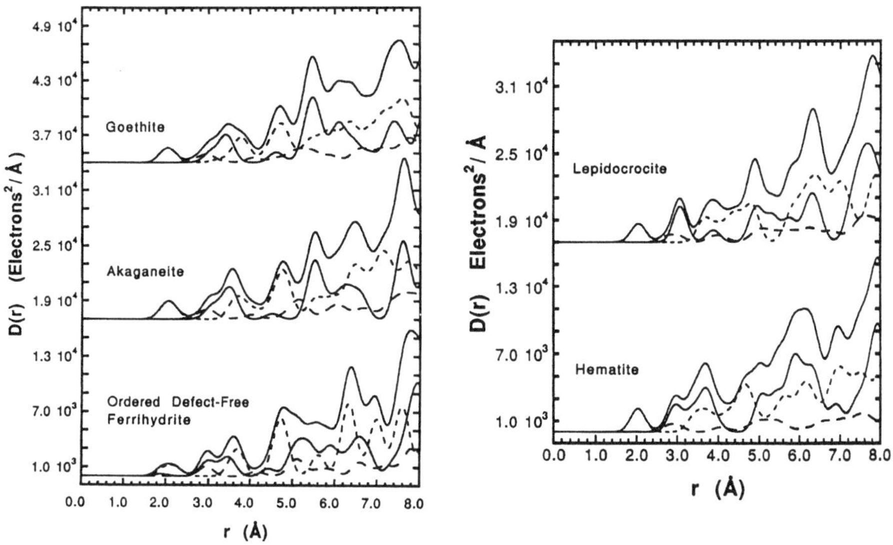

Figure 22. Calculated RDF functions for 2.0-nm diameter nanocrystallites of iron minerals. The overall RDF is a sum over all pair correlations including Fe-Fe, O-O, Fe-O, Fe-H, O-H and H-H, but the proton correlations are neglected here for clarity (they are very small compared to other contributions). Contributions shown are Fe-Fe (solid line), Fe-O (short dash line), and O-O (long dash line). From Waychunas et al. (1996).

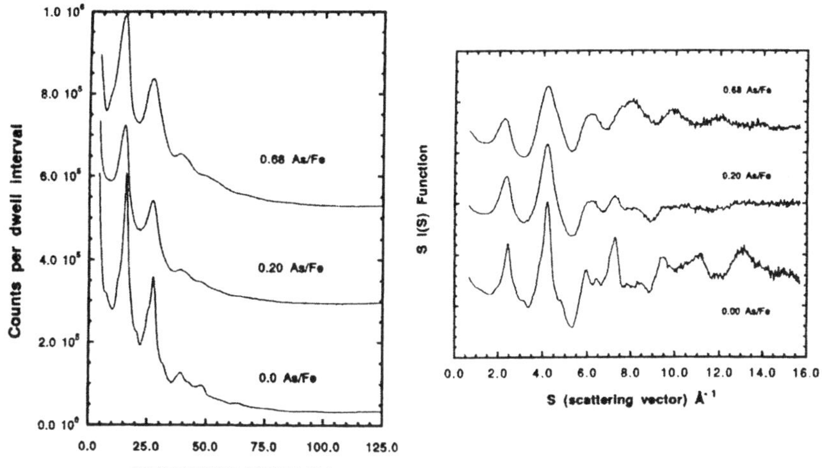

Figure 23. Example of WAXS scattering analysis for ferrihydrite particles with sorbed arsenate. Left: Raw scattering data for ferrihydrite prepared with coprecipitated arsenate in different proportions. Right: SI(S) function extracted from the scattering data. From Waychunas et al. (1996).

parameters than can be measured even with Rietveld analysis. This is also the case for applications of the Debye equation as smaller nanocrystals will have broadened patterns with overlapping features, yielding progressively greater structural ambiguity. With DFA pre-calculated patterns as a function of crystallite size and structure are added to produce the best fit with observations (Fig. 24). The method has been useful in identifying the

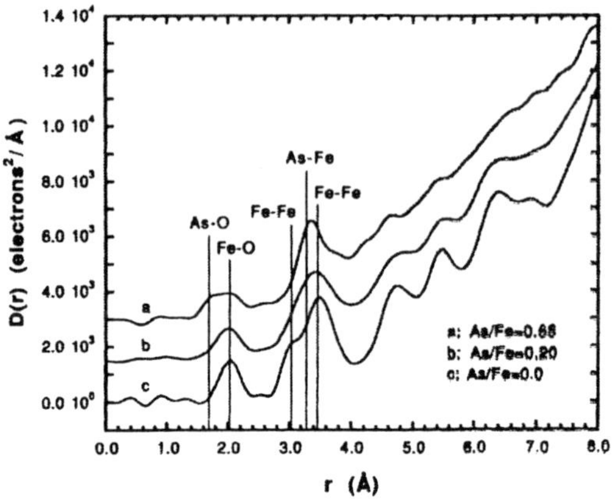

Figure 23, continued. Example of WAXS scattering analysis for ferrihydrite particles with sorbed arsenate.RDFs calculated from the SI(S) functions. The loss of higher frequency oscillations in the SI(S) function indicates loss of longer-range correlations in the RDF. The highest arsenate ferrihydrite particles are about 1 nm in diameter. From Waychunas et al. (1996).

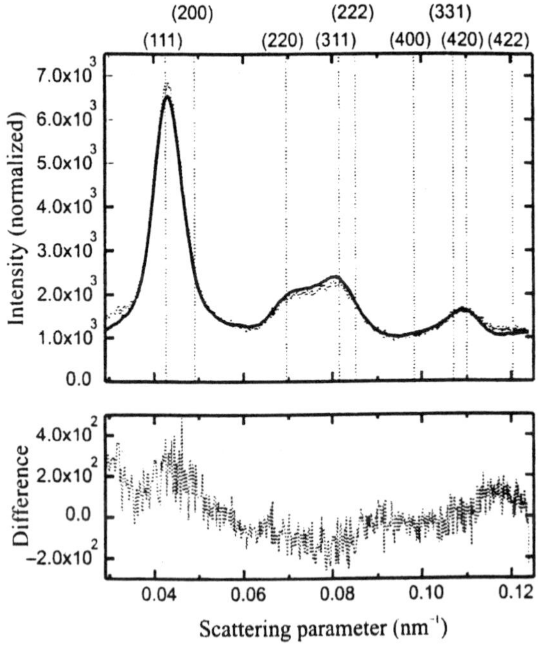

Figure 24. Debye function analysis done on a colloid of 2-nm gold particles. Solid line is the fit using icosahedral and dodecahedral nanoparticles. Dotted vertical lines indicate FCC structure peak positions. Essentially no FCC or cuboctahedral particles are present. The bottom plot shows the difference between calculation and observation. From Zanchet et al. (2000), used with permission of Wiley-VCH.

presence of multiply twinned particles (MTPs) in nanoparticle diffraction patterns (Zanchet et al. 2000).

Short range order: X-ray absorption spectroscopy (XANES and EXAFS)

X-ray absorption spectroscopy is a powerful tool for nanoparticle analysis due to its selectivity and independence of sample physical state. It is limited in range to the region within about 0.5-0.7 nm of a particular (chosen) absorber atom in the structure, but can be applied to amorphous or even liquid samples. The basic theory behind the origins and analysis of the extended X-ray absorption fine structure (EXAFS) has been well described by Sayers et al. (1970, 1971) and Lee et al. (1981). with mineralogical applications detailed by Brown et al. (1988). The crucial aspect of the EXAFS spectrum is that it is formed by an electron backscattering process in the vicinity of the absorber

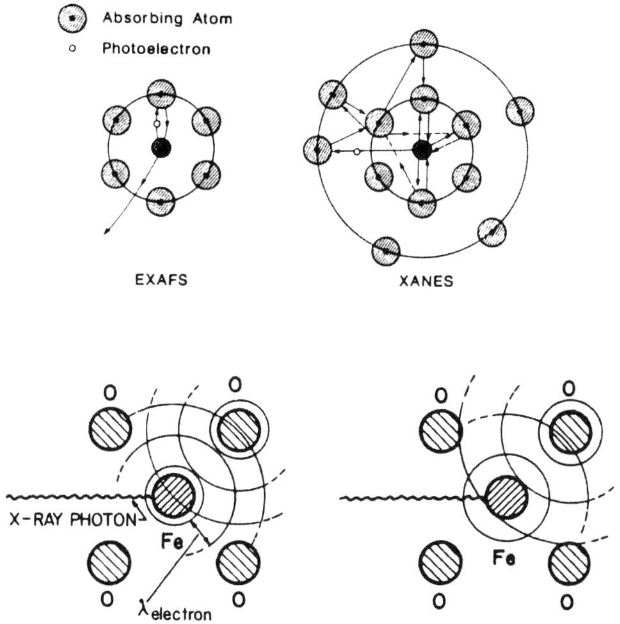

Figure 25. Cartoons showing the fundamental processes giving rise to the EXAFS and XANES spectra. Top left: EXAFS process. The ejected inner shell electron scatters off of a single backscatterer before returning to interfere with the outgoing electric wave at the absorber. Top right: XANES process. At electron kinetic energies close to the edge, the photoelectron has a large mean free path. Hence scattering off of several neighbors and over a large range of pathways is likely. This effect damps out quickly with increasing energy away from the absorption edge. After Calas et al.(1987). Bottom: interference process in EXAFS. The photoelectron leaves the absorber atom as a wave train that interferes with its own reflection off of neighbor atoms. Left: destructive interference. Right: constructive interference. From Brown et al. (1988).

atom. At the absorption edge the energy threshold for ionization of an atomic electron is reached. At energies above the edge the ejected electron leaves the absorber atom as a photoelectric wave with kinetic energy equivalent to the difference in edge and incident X-ray energies. In the EXAFS range of about 30 eV or more above the edge, the photoelectric wave scatters (mainly) only once from neighboring atoms with the backscattering wave interfering with the outgoing wave at the absorber. This interference modulates the absorption spectrum with a sinusoidal oscillation whose frequency depends on the absorber-neighbor distance. Hence embedded in the EXAFS spectrum is the scattering pattern associated with the local structure near that absorber (Fig. 25). The EXAFS spectrum is analyzed in a manner analogous to the X-ray diffuse scattering as per the Debye equation. The spectrum is extracted and suitably normalized to the sample composition, then converted to a scattering function by converting from energy to momentum space. The wavevector in this case is not S, but k: $k = [0.262\ (E{-}E_0)]^{1/2}$, where E is the incident X-ray energy, and E_0 is the energy at the edge corresponding to zero photoelectron kinetic energy. Typically the quantity k^n I(k) is Fourier transformed, with $n = 2$ or 3 depending on the weighting desired. Figure 26 shows a typical sequence of analysis from raw absorption edge spectra to Fourier transform. It is important to note

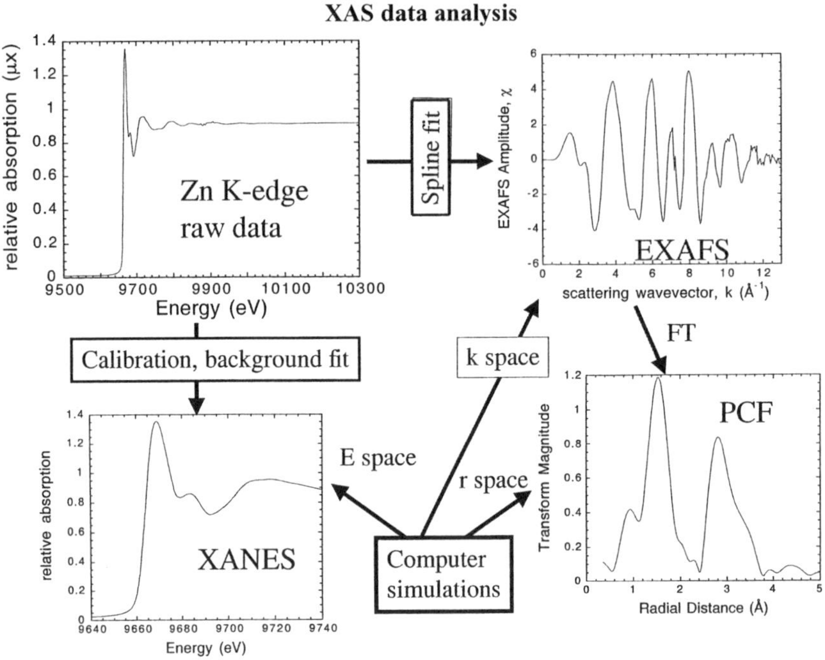

Figure 26. EXAFS and XANES analysis scheme. Top left: Raw data with background removed. Top right: Fit of EXAFS region with cubic spline and subtraction yields EXAFS oscillations; shown is the EXAFS weighted by k^3. Bottom left: XANES spectral region expanded. Bottom right: PCF or pair correlation function (also called Radial Structure Function to distinguish it from the X-ray analog RDF); shown is the Fourier transform of the spectrum at the top left.

with EXAFS analysis that the Fourier transformed function is not a typical RDF, though it is similar. The difference comes from the phasing of the scattering process creating the EXAFS features. This is an electron backscattering process that is affected by local atomic fields, and hence the frequencies of backscattering from a given absorber-neighbor atom pair are shifted from values directly proportional to the distances. This means that the peaks in the Fourier transform do not occur at the actual absorber-neighbor distances. In practice, these distances are accurately obtained by fitting the raw EXAFS data with model functions that closely estimate the phase changes during scattering. For clarity we call the EXAFS-derived "RDF" a radial structure function or RSF. There is no analogous phase shift effect with X-ray scattering.

Examples. One use of EXAFS analysis takes advantage of the total reflection of X-rays off of a sample surface at small incidence angles to examine only the surface of the sample. This method is particularly useful for examination of the structure of nanoprecipitates at the surface. Such a study is one by Waychunas et al. (1999) where nanoparticles of iron oxide were studied on quartz single crystal surfaces. In this study the iron oxide particles were found to be highly textured, with a particular crystallographic direction always normal to the quartz surface. As EXAFS spectra are collected with X-ray synchrotron radiation, which is highly plane polarized, it was therefore possible to collect K-edge EXAFS spectra with X-ray electric vector both in the plane of the particles and normal to this plane (Abruna 1991). This yielded two different but complementary EXAFS spectra that could be analyzed to yield additional structural information. This process is shown in Figure 27, where the RSFs for different types of

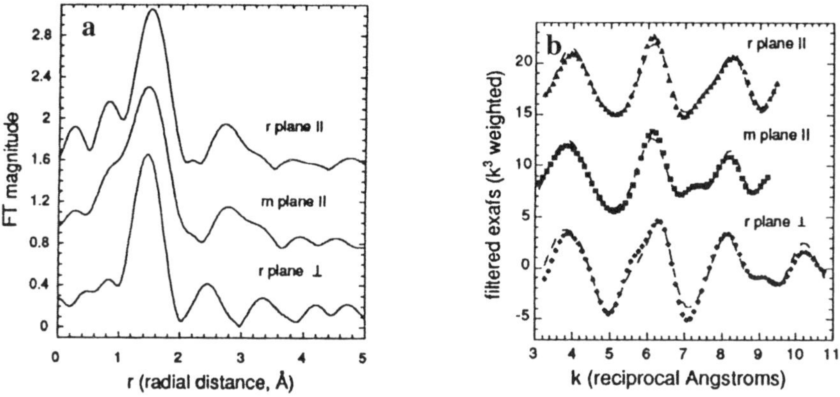

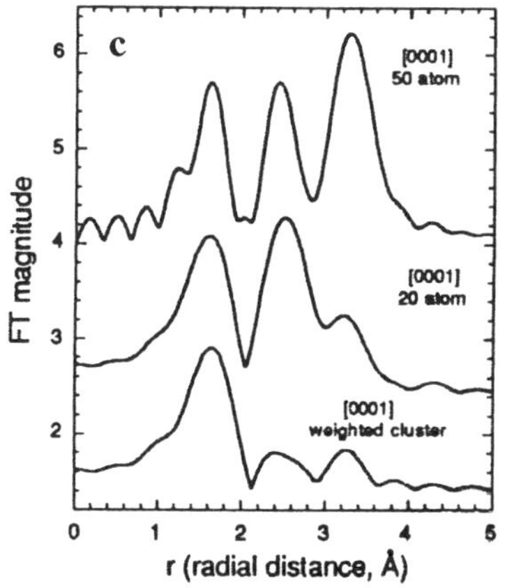

Figure 27. Analysis of polarized EXAFS data. (a) EXAFS radial structure functions for irom oxide precipitates on quartz single crystal surfaces. r and m refer to the $(10\bar{1}1)$and $(11\bar{2}0)$ surface planes of quartz. The parallel and perpendicular refer to the polarization direction of the X-ray beam and thus the probe direction of the EXAFS scattering process. (b) Raw polarized EXAFS and fits for the same samples in (a). (c) Polarized structure function simulations. Top: radial structure function for a single Fe atom within a 50-atom hematite crystal with [0001] orientation. Middle: Same for 20-atom crystal. Bottom: Weighted average of all Fe structure func-tions in the 20-atom crystal. The analysis suggests highly textured hematite-like nanocrystals on the quartz surface but no epitaxial relationship. From Waychunas et al. (1999).

particles had structures similar to α-Fe_2O_3, with their crystallographic [0001] directions normal to the quartz surfaces. Additionally, the amplitudes of the peaks in the RSF could be used to deduce the particle size and shape. This is possible as a bulk crystal of α-Fe_2O_3 has 13 Fe next-nearest-neighbor atoms about a given Fe atom, with different Fe-Fe distances as a function of direction. These correlations give rise to well defined peaks in the RSF. However in small particles, the surfaces will have much smaller numbers of Fe-Fe neighbors, and hence the RSF peak amplitudes will be affected. This type of analysis to deduce particle or crystallite size works well in the 0.5- to 3-nm size range, and in this case showed the precipitates to be about 1 nm in size.

Another example of nanoparticle size analysis via EXAFS is the work of Frenkel (1999) on Pt and Pt-Ru particles supported on carbon. This study utilizes the longer absorber-neighbor distances within nanocrystals, showing that determinations of the relative numbers of these farther neighbors are more sensitive to crystallite size than closer neighbors. In the best case all neighbor coordinations can be used to obtain a good estimate of crystallite size and shape even without the added information provided by polarized studies (Fig. 28).

A large number of EXAFS studies have been done on semiconductor materials. For example, Rockenberger et al. (1997) examined nanocrystals of CdS in the range 1.3 to 12 nm in diameter with different types of surface stabilizing agents. Thiol-capped nanoparticles had an expanded mean Cd-S distance, while polyphosphate capped

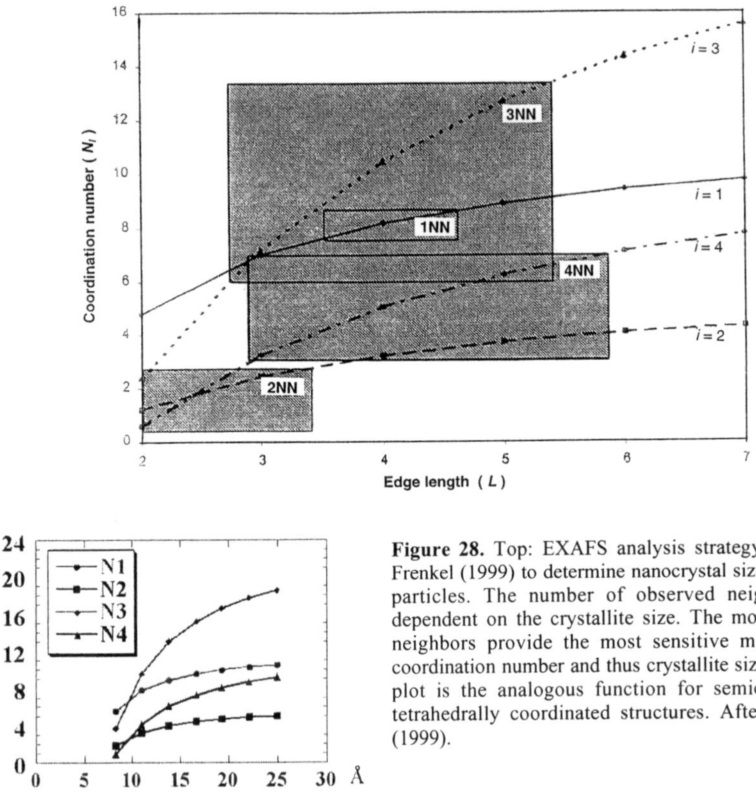

Figure 28. Top: EXAFS analysis strategy used by Frenkel (1999) to determine nanocrystal size in metal particles. The number of observed neighbors is dependent on the crystallite size. The more distant neighbors provide the most sensitive measure of coordination number and thus crystallite size. Bottom plot is the analogous function for semiconductor tetrahedrally coordinated structures. After Frenkel (1999).

nanoparticles had an expanded mean Cd-S distance, while polyphosphate-capped nanoparticles were slightly contracted. As the expected effect is one of contraction, Rockenberger et al. (1997) attributed the differing behavior to steric hindrances among the thiol groups, whereas the polyphosphate groups had no such steric interaction. The EXAFS analysis showed further that static disorder (see next section) increased in the nanoparticles, with a maximum for particles with 3-nm diameters. It was suggested that this size corresponded to the largest thermodynamically-grown nanocrystals, which enlarged further only via Ostwald ripening or aggregation. Fitting of the EXAFS-derived pair distribution function for the Cd-S bonds using a cumulate expansion allowed measurement of the anharmonicity of these distributions. A plot of the third cumulate term vs. decreasing particle diameter showed two distinct trends, one attributed to cubic (sphalerite) structure where the cumulate value increased from zero at large particle size, and a second where the cumulate value increased from a finite value. This latter trend was attributed to hexagonal (wurtzite) structure (Fig. 29). As Cd-X pair correlations are nearly identical between these two structures, the anharmonicity variation with particle diameter is possibly the only clear method for distinguishing these structures in small nanocrystals.

Carter et al. (1997) used EXAFS and other measurements to study the surface structure of CdSe nanocrystals. They found that the Se-Cd coordination number varied with particle size and that Se was only coordinated by Cd. This indicated that surface Se atoms do not have all bonds satisfied, and might lead to special optical properties of these nanocrystals.

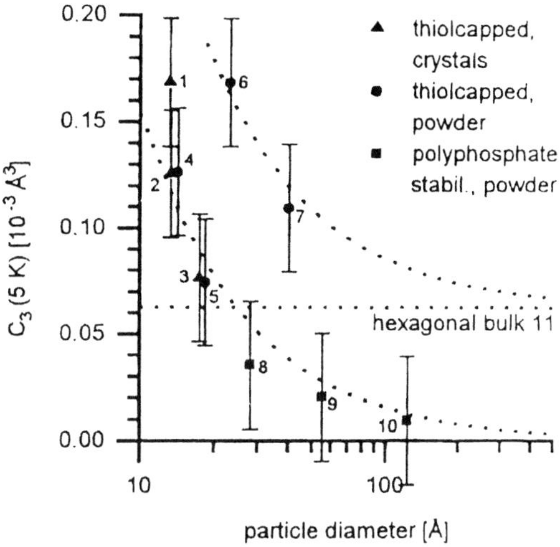

Figure 29. Trends of third cumulate as a function of nanocrystallite diameter for sphalerite and wurtzite structures. The bulk wurtzite-structure material has finite C_3 value which increases with decreasing size, while the sphalerite structure has a zero C_3 value for bulk. These differences allow discrimination of these two structure types in nanoparticles. From Rockenberger et al. (1997), used with permission of the editor of the *Journal of Physical Chemistry*.

Rockenberger et al. (1998) examined 1.8-nm nanocrystals of CdTe capped with mercaptoethanol molecules, and thus having surface CdS bonds. They found that the surface CdS local structure was distorted relative to the bulk CdTe, and attributed this to strain because of the differences in CdTe and CdS bonding. The coordination numbers derived from the EXAFS analysis were consistent with models of individual nanocrystals with discrete stoichiometries, e.g., $Cd_{54}Te_{32}(S\text{-ethanol})_{52}^{8-}$.

Disorder problems in EXAFS analysis

The advantage of EXAFS is that it can often be employed where other structure probes cannot. The standard EXAFS equation has a term for the loss of amplitude with increasing scattering vector. It has the form

$$DW = e^{-2k^2\sigma_i^2} \tag{21}$$

This term essentially uses a Gaussian distribution function, i.e., it holds exactly for a "harmonic" distribution or a "harmonic" set of bond vibrations. For static cases this means that the distribution of N-x bond distances needs to approximate a Gaussian function. For vibrational disorder this means that the vibrations must not be so large that the harmonic part of the potential function (the trough in the potential energy vs. bond length plot) is breached. Such larger vibrations lead to "anharmonic" vibrations where there is a net change in atom mean positions. Figure 30 shows a worst case scenario for extreme anharmonicity. In the example a sharp Gaussian distribution at a shorter interatomic distance is shown to dominate a similar function with an additional Gaussian broad "tail." The entire coordination date contained in the tail would be lost with typical EXAFS analysis. For most cases that are not so severe, the EXAFS can be fit using a cumulate moment expansion to model the distribution function, without any tacit assumption of harmonicity (Gaussian) distributions. With the cumulate expansion, the EXAFS DW term has the form:

$$DW(\text{cumulate}) = e^{-2k^2\sigma_i^2 + \frac{2}{3}\sigma_i^4 k^4 \cdots} \tag{22}$$

where we now have an exponential in even-order terms of the cumulate expansion. This alone indicates a different kind of amplitude decay with increasing k. However the phase is also affected so that one has:

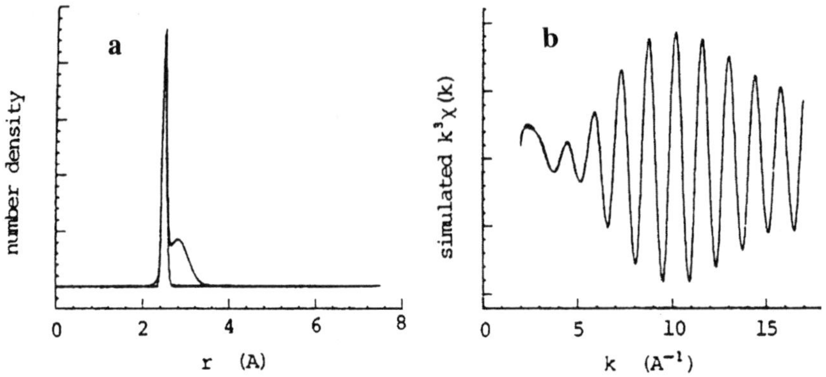

Figure 30. Effect of extreme static disorder in EXAFS analysis. Left: Gaussian distributions of equal area but with much different widths. Both the sum of the two functions and the single narrow distribution function are shown. Right: Simulated EXAFS functions for the functions at left. It is seen that there is no detectable difference in the EXAFS except at low k-values. This difference would overlap with XANES and be extremely difficult to analyze. Hence physical distributions with a broad "tail" will have reduced coordination numbers via standard EXAFS analysis, as well as an artificially produced distance "contraction". For cases not as severe as this, cumulate analysis can quantify the degree of static disorder and allow more correct results. After Kortright et al. (1983).

$$\chi \propto \sum_i F_i(k)\, sin\left[2kr_i - \frac{4\sigma_i^2 k}{r_i}\left(1+\frac{r_i}{\lambda}\right) - \frac{4}{3}\sigma_i^3 k^3 \cdots\right] \tag{23}$$

All this means is that a non-Gaussian distribution function will cause a different damping rate of the EXAFS amplitudes, and a shift in the apparent r values of neighbor distances. The third cumulate is the most reliably calculated and gives a good estimate of anharmonicity for cases without extreme anharmonicity (Tranquada and Ingalls 1983). In favorable cases the EXAFS cumulate moment analysis can predict anharmonicity-based distance effects in first and second neighbor about an absorber atom. The method is difficult to apply to complex many-atom systems. Hence EXAFS analysis of very disordered nanoparticles or mineralogical solids must be done with caution.

Empirical and *ab initio* XANES analysis

The structure very close to the X-ray absorption edges is called the X-ray absorption near edge structure or XANES. This region contains both electronic transitions within the absorber atom (called bound state transitions and just below the E_0 energy), and multiple scattering features. The latter are due to the very long mean free paths of electrons (e.g., hundreds of Å) that are ejected from the atom just above the edge energy, i.e., with little kinetic energy. These electrons can scatter around the local area of the absorber atom producing absorption features on the edge each linked to a particular MS path. As with other types of scattering already discussed, MS also produces a sinusoidal modulation, but this is damped out quickly due to the strong dependence of MS amplitude on k. Thus most significant MS features are observed on the absorption edge proper. Such features may partially cancel one another so that the XANES is not easy to interpret in terms of structure except via *ab initio* XANES model calculations. However, the sensitivity of the MS and electronic absorption features to absorber valence, coordination number and neighbor distances can be used to empirically determine these quantities if closely similar model compound spectra are available for comparison. The great advantage of XANES analysis is that the XANES carry a large amount of structural information over a small energy range, and are often of much larger amplitude than the EXAFS. Hence for small quantities of samples, XANES information is much more easily measured.

An example of an empirical study is that of Chen et al. (1999) where some of the structural details of TiO_2 nanoparticle surfaces can be deduced from progressive changes in the XANES. Past work has shown that the position (energy) and intensity of the electronic absorption features on the low energy side of the Ti K-edge can be interpreted in terms of valence, coordination number and Ti site distortion (Farges 1997; Waychunas 1987). Chen et al. (1999) showed that smaller TiO_2 particles had intensification of these features, consistent with increased Ti-site distortion on the surface of the particles. EXAFS analysis was also done and indicated a higher concentration of 5-coordinated sites on the TiO_2 particle surfaces, which was consistent with the XANES interpretation (Fig. 31).

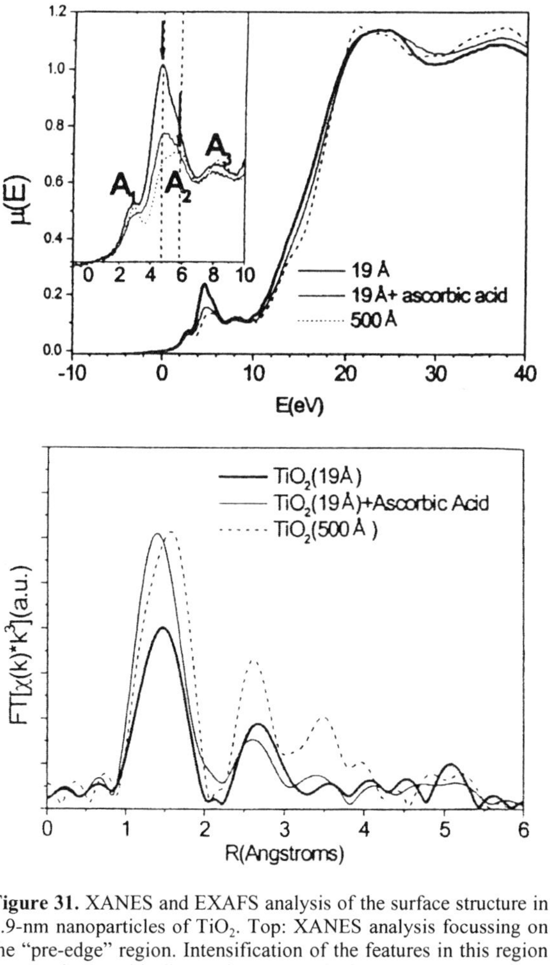

Figure 31. XANES and EXAFS analysis of the surface structure in 1.9-nm nanoparticles of TiO_2. Top: XANES analysis focussing on the "pre-edge" region. Intensification of the features in this region are consistent with distorted or reduced Ti coordination. Bottom: EXAFS analysis shows reduced peak area and coordination in the same samples with XANES intensification. Both analyses point to 5-coordinated Ti on particle surfaces. Capping the particles with ascorbic acid removes the distorted sites. After Chen et al. (1999).

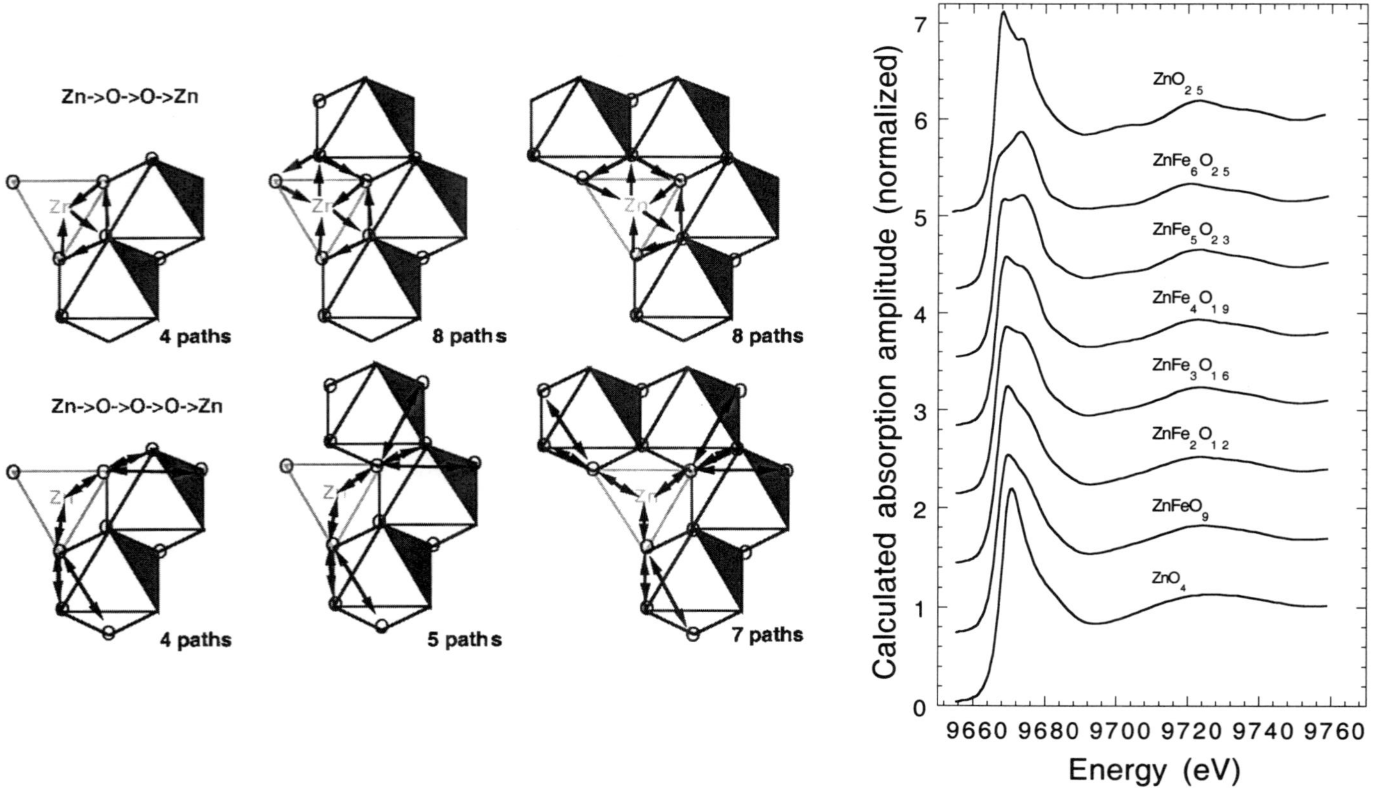

Figure 32. Zn MS calculations and XANES. Left: Small cluster of Fe oxyhydroxide polymers with sorbed tetrahedral Zn^{2+} complex. Two of the multiple scattering paths are shown that vary in number as a function of cluster size. Similar dependencies of other MS paths create non-linear changes in XANES features as a function of cluster size. Right: Calculated XANES for various clusters as depicted at left. All MS paths used in a self-consistent calculation (Feff 8). From Waychunas et al. (2001).

An example of the use of *ab initio* XANES calculations to determine nanoparticle structure is the Zn/ferrihydrite sorption system examined by Waychunas et al. (2001). In the case of sorption complexes the XANES spectrum of the sorbed species will contain information about the local structure of the substrate, and thus the structural nature of the full sorption complex. In the Zn/ferrihydrite system it was observed via EXAFS that the number of Fe^{3+} next nearest neighbors about the sorbed Zn ion decreased as the Zn sorption density increased. Direct calculation of the XANES structure identified MS paths that changed in number as a function of cluster size (and thus number of neighbor Fe atoms), and gave rise to XANES features that changed in intensity (Fig. 32). These changes agreed well with the structural interpretation of the EXAFS and the crystal chemistry of Zn-Fe^{3+} hydroxides.

Size and shape determination—X-ray and neutron small-angle scattering (SAXS and SANS)

The scattering relationships discussed earlier indicated a reciprocal relationship between scattering angle and r, the separation of scattering elements. This can be seen by combining the Bragg relation (Eqn. 14) with Equation (18). Substituting 2d $\sin\theta$ for λ in Equation (16) and cancelling the $\sin\theta$ yields $2\pi/d = S$. Thus the larger the spacing the smaller the value of S, and the smaller the scattering angle for any given λ. If scattering angles down in the millidegree range can be measured, length scales in the micron range can be detected. At scales such as this, atomic scattering centers are not individually sensed, structure factors are not sensitive to order, and only form factors are important. Thus only large scale electron density differences are detected. Hence the size and shape of colloids, nanoparticle aggregates and large molecular units can be determined, though there is little sensitivity to crystallinity.

The two important factors that mitigate the intensity of scattering are the structure factor $F^2(S)$ and the form factor sometimes indicated as P(S). The structure factor describes the external relationships of scattering elements, such as atoms, and hence the phasing and interferences of the net scattering. The form factor describes the internal distribution of electron density within a scattering element, and thus mitigates the intramolecular or intraparticle scattering. In our discussion earlier the only form factor mentioned was the atomic scattering factor, f. To a reasonable approximation all atoms have identical f values as a function of scattering angle, and scaling as the atomic number (number of electrons). However for small angle scattering the structure factor has little importance and we concentrate on detailed evaluation of the form factor.

In the small angle regime there are subsets of specific scattering phenomena that can be identified and analyzed. These are the center-of-mass region, the Guinier region, the Fractal region, and the Porod region (Fig. 33). In the center-of-mass region, at the smallest measurable scattering angles, the size of S is such that very long lengths are being probed by the scattering. These are much larger than the size of the particles or aggregates, and thus only an averaged electron density of a particle can be sensed. There is no interparticle interference phenomena here. Analysis of this region gives a rough average idea of particle size. In the Guinier region at somewhat larger S, there is some degree of intraparticle interference which enables analysis of the particle shape. Specifically what can be extracted is the radius of gyration, R_g, which is the electron density analog of the second moment of inertia. Guinier's (1963) approximate expression for the scattering in this region is:

$$I(S) = N_p n_e^2 \exp(-S^2 R_g^2/3) \quad (24)$$

where N_p is the number of scattering units, and n_e is the number of electrons per unit. A plot of log I versus S^2 can be used to identify R_g, and give information on the mean

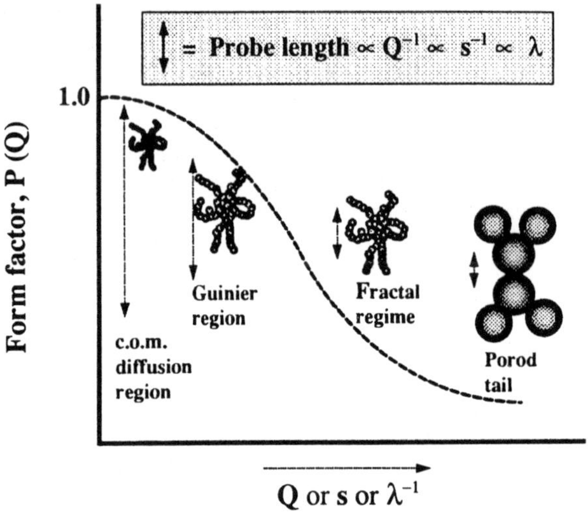

Figure 33. Regions within the small angle scattering regime that are sensitive to differing types of structures and organization. C.O.M. refers to "center of mass" region. From Hiemenz and Rajagopalan (1997), used by permission of Marcel-Dekker.

shapes of the nanoparticles. For example, if Rs is the radius of a spherical particle, the appropriate R_g is equal to $3R_s2/5$. In the case of a thin rod of length R_l, R_g is equal to $R_l^2/12$. The Guinier approximation was derived for spherical particles that are highly dispersed, and is less accurate for needle-like particles and dense solutions. At values of S above the Guinier region, the fractal nature of a particle becomes important in creating interference effects. Here the probe length is consistent with aggregate size, but is much larger than the primary particles in the aggregate. Figure 34 shows a conceptual diagram of the fractal region. It can be shown that the plot of a log I(S) versus log S function is linear in this region if there is self-similarity independent of size. The slope in such a plot yields the exponent of S, which is equivalent to the fractal dimension of the aggregate. The region of largest S or Porod region has the largest falloff in SAXS intensity with S. This tail region can be used to extract the surface area of nanoparticulates using Porod's Law:

$$I(S) = I_e\, 2\pi\, \rho^2\, A/S^4 \qquad (25)$$

where A is the average particle surface area, r is the electron density difference between particle and matrix, and I_e is a constant. From the combined analysis of these regions, along with analysis of scattering from still longer wavelenths such as light, a considerable amount of information can be obtained about particle size, shape, area and size distribution.

Determination of particle size distributions

Particle size distributions are important information for evaluating the synthesis, growth and kinetics of formation of nanoparticles and nanoparticle aggregates. The main contrasting techniques are imaging via TEM, and SAXS analysis. TEM analysis is independent of the type of aggregation, shape and size of nanoparticles, whereas SAXS interpretation requires assumptions about the nature of the sample. On the other hand, SAXS can be done on *in situ* systems, and often in real time during particle formation.

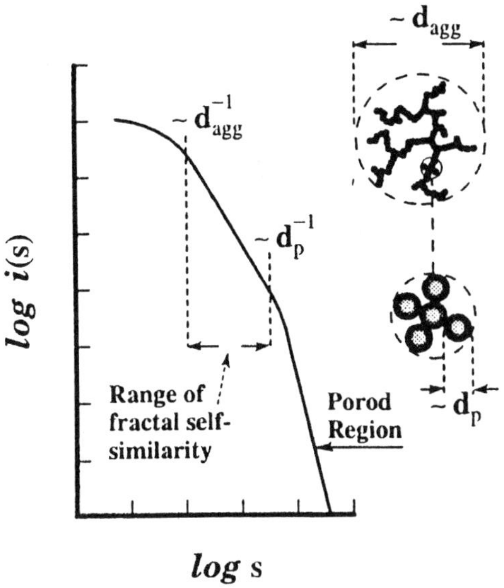

Figure 34. Cartoon showing the SAXS region where self-similarity within aggregates can be detected. Fitting of this region can reveal the fractal dimension of the aggregate, and if combined with analysis of other regions, yield information on the complete distribution function. From Hiemenz and Rajagopalan (1997), used by permission of Marcel-Dekker.

Further, there is no limitation imposed by having to dry and mount the sample, which may occasion aggregation or other physical change. A comparison of these techniques has been made by Rieker et al. (1999) on the same set of idealized hard silica sphere samples. For near-spherical particles that are dispersed and relatively dilute, it is possible to define a unique particle size distribution by fitting the Porod and Guinier regions of the SAXS pattern. Fits to experimental SAXS spectra are shown in Figure 35 compared to histogram distributions determined from TEM images. The SAXS-derived distributions were assumed to be Gaussian. The agreement is quite good, with most of the mismatch in mean particle size due to difficulty in accurately calibrating the TEM size distribution. Other comparisons of SAXS and TEM has been done on a wider set of nanoparticles including $CaCO_3$, Fe_2O_3, TiN and SiO_2 by Xiang et al. (2000) giving good agreement for highly dispersed particles, and by Turkovic et al. (1997) on nanophase TiO_2, also giving good agreement.

Rate of growth/aggregation experiments with SAXS

Stopped or continuous flow reaction methods can be utilized with synchrotron-based SAXS measurements to determine reaction kinetics, including protein folding, nanoparticle growth, breakdown, and aggregation. Using a synchrotron X-ray source, the SAXS spectrum can be collected with a one or two dimensional detector with good quality in periods as short as 0.001 s.

A system such as this has been developed at the SRS at Daresbury, UK, that also allows simultaneous wide angle X-ray scattering and FTIR spectroscopy on the same sample. An example of this type of experiment has been described by Eastoe et al. (1998) where the breakdown kinetics of synthetic micelles was measured. Another study is that of Connolly et al. (1998) where the kinetics of superlattice formation from capped (by dodecanethiol) Ag nanocrystals was studied. The superlattice 111 reflection was monitored to determine growth rate and (111) plane spacing at 1 s intervals. A continuous reduction in (111) superlattice spacing with time suggested gradual solvent evaporation from the growing structure. Real time SAXS experiments of similar type are now also possible at SSRL, ESRF and other synchrotron sources. This type of technique is

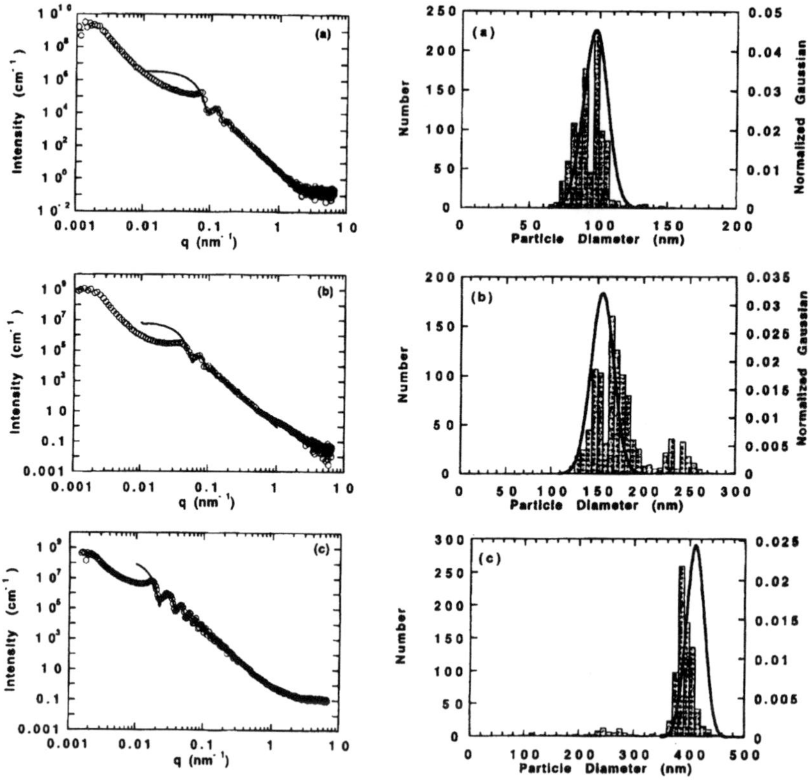

Figure 35. Comparison of distribution functions extracted from SAXS data versus from TEM observations. There are some systematic errors in the TEM analysis. However some larger scale particles appear to be lost in the SAXS analysis. As TEM particle counting favors larger particles this could also be a systematic effect. After Rieker et al. (1999).

especially valuable in studying nanomineral aggregation kinetics in aqueous solutions, or assembly of precipitated nanoparticles in biological systems due to the high electron density differences, and the intractability of other characterization approaches.

Small-angle neutron scattering (SANS)

SANS analysis is essentially the same as SAXS except that the length scales probed are somewhat different, and the intensity of scattering from atoms does not scale as the atomic number as with X-rays, but rather as a nuclear property that is effectively independent of atomic number. Additionally, the scattering factors, which are called neutron scattering lengths, can have negative values. This means that for some specific sample compositions it is possible to have net neutron scattering cross sections that are zero, a situation referred to as a "contrast match". A seemingly unusual aspect of neutron scattering is that protons scatter as effectively as most other atoms, but with a much different (and negative) scattering length than deuterium. Hence polymer/solvent and biological systems can be adjusted with isotopic substitutions to remove scattering due to the surrounding solution, or the effects of substrates can be removed in the study of thin films. For mineralogical nanomaterials SANS can be used to selectively examine the proton distribution within and on the surface of particles, or as a complementary tool to SAXS analysis. A tutorial on SANS is presently available on the internet at www.isis.rl.ac.uk/largescale/log/documents/sans.htm.

SANS is useful in examining pore structures in nanoporous materials, or water in fractal networks (Li et al. 1994). Examples of aggregation-state analyses include studies of silica and titania nanoparticles (Hyeon-Lee et al. 1998), catanionic surfactants (Brasher and Kaler 1996) and TiO_2-SiO_2 and ZrO_2-SiO_2 sol-gel materials (Miranda Salvado et al. 1996). Particle size distributions obtained via SANS, SAXS and TEM on nanophase TiO_2-VO_2 catalyst particles are compared by Albertini et al. (1993).

Light scattering techniques

Scattering of longer wavelength photons in the IR, visible and UV offers a number of advantages over X-ray and neutron scattering. First, the wavelengths are substantially longer so that very long length scales can be probed at technically feasible scattering angles. The range of possible scattering vector, S, overlaps and extends that available with SAXS by two orders of magnitude. Second, these wavelength light sources can be made exceedingly bright, monochromatic and coherent using lasers. Both of these conditions enable static light scattering (SLS), and photon correlation spectroscopy (PCS). A third desirable attribute is that the scattering can readily be done in solutions, i.e., *in situ*, which can be more problematic for SANS and SAXS. [PCS is also known as dynamic light scattering or DLS.] SLS is applicable to the study of nanoparticles in the 10-1000 nm regime, while SAXS mainly is used for the 1-50 nm range. The scattering equations are essentially the same for each technique, except that scattering contrast comes from the electron density differences in SAXS, while it comes from refractive index differences in SLS (Chu and Lu 2000). A comparison of the range for S and scattering lengths of different techniques is shown in Table 2.

Table 2.

Method	*Wavelength (nm)*	*S range* (nm^{-1})	*Length scale (nm)*
SLS/PCS	360–1000	0.001–0.2	1000–5
FPI	400–600	Not applicable	10–1
SAXS	0.05–0.3	0.02–1.0	50–1
WAXS	0.15–0.15	0.6–7.0	1.6–0.14
SANS	0.4	0.007–0.9	150–1
WANS	0.4	10–50	0.1–0.02

PCS (DLS) measures the fluctuations in the intensity of light scattering with time. The decay rate of the time autocorrelation function of these intensity fluctuations is used to directly measure particle translational diffusion coefficients. This can then be related to the particle's hydrodynamic radius. If particles are approximately spherical, then the hydrodynamic radius is equivalent to the geometric radius. Because of its sensitivity and ease of use, PCS is widely used to determine the sizes of small particles (Pecora 2000). As with all scattering methods, distributions of particle size and shape places limits on what can be extracted from PCS data. Currently the analysis method most common is based on an assumption that the time correlation function is essentially a sum of exponential decay times, where each decay rate relates to a different particle diffusion coefficient. The logarithm of such a function can be expanded as a power series in time , with coefficients equal to K_n, known as cumulants. As with our earlier discussion of disorder in EXAFS analysis, K_1 is equal to the average value of the exponent, the reciprocal relaxation time in this case. K_2 is a measure of the dispersion of the reciprocal relaxation time about the average, and so forth. If K_2 and higher cumulants are 0, then there is a single exponential term in the correlation function and a single diffusion rate. Examples of distributions in nanoparticle samples obtained via PCS are given in Kaszuba

(1999). PCS techniques are currently being expanded into the X-ray range (XPCS), e.g., Thurn-Albrecht et al. (1999).

An alternative to PCS is Fabry-Perot interferometry (FPI) which is applicable to particles with considerable geometric anisotropy. FPI uses the frequency broadening of the laser light to determine the rotational diffusion coefficient, and this is related to the particle dimensions. The FPI system is essentially a very high resolution monochromator system that disperses the scattered light into component frequencies. A typical frequency range is 1 MHz to 10 GHz. The incident beam must be efficiently linearly polarized in the plane perpendicular to the scattering plane, and the scattered radiation taken off at right angles to the incident beam to block all light that has not been scattered and depolarized. Hence particle rotation must result in depolarization to be detected, and this is the case only when there is a polarizability change as the particle rotates relative to the incident beam.

"INDIRECT" STRUCTURAL METHODS

Optical spectroscopy (IR, visible, UV, Raman, luminescence, LIBD)

Investigation of the IR-Vis-UV spectrum of nanoparticles can yield considerable information about physical and electronic structure of the materials. IR spectroscopy is mainly useful for examining the bending and stretching vibrational modes of particular molecular units within the particles, e.g., surface dangling bonds or low coordination species, adsorbed or bonded water, and passivating organic coatings. An example of the use of FTIR is the verification of unterminated Se atoms on the surface of CdSe nanoparticles in agreement with EXAFS analysis (Carter et al. 1997). Semiconductor nanoparticles are commonly studied in the UV-Vis range where band gaps and localized electronic states can be probed with such wavelengths. One of the interesting phenomena found in semiconductor materials in this size range is that of "quantum confinement." This means that as particles are reduced in size the delocalization of electrons within energy bands is restricted by the physical size. As delocalization reduces electronic energy, confinement increases electronic state energies and gives rise to an increased HOMO-LUMO (highest unoccupied/lowest unoccupied molecular orbital) gap. The gap energy is roughly equivalent to the energy where absorption begins to increase, presumably via transitions to an exciton state just within the gap. This is well exemplified in CdS nanoparticles (Rockenberger et al. 1997) and CdSe nanocrystals (Carter et al. 1997; Murray et al. 1993), spectra of the latter being shown in Figure 36. Further, as particles decrease in size the absorption spectrum becomes increasingly structured with sharper features, due to a progressive change from a band-like electronic structure to a molecular-like electronic structure. The quantum confinement effect also operates in insulator nanocrystals such as the anatase form of TiO_2 (Reddy et al. 2001). See the chapter by Jacobs and Alivisatos for more details concerning the quantum confinement phenomenon. The nature of particular transition metal sites can often be characterized by interpretation of the IR-Vis electronic absorption spectrum, e.g., determining the distortion of a site due to strain or unusual coordination at a surface that shifts the 3d of other absorption band positions. Interactions between metal ions, and ligand-metal charge transfer can also be evaluated with electronic absorption spectroscopy. However this type of study is possible mainly for transition metal doped semiconductors or insulators, such as TiO_2 and Al_2O_3, and does not appear to have been explored with such types of nanocrystals to date.

A consequence of absorption in the ultraviolet for many nanocrystals is the emission of the absorbed light as luminescence. In these cases, photoluminescence spectroscopy can yield both electronic structure information as well as details of excited electronic

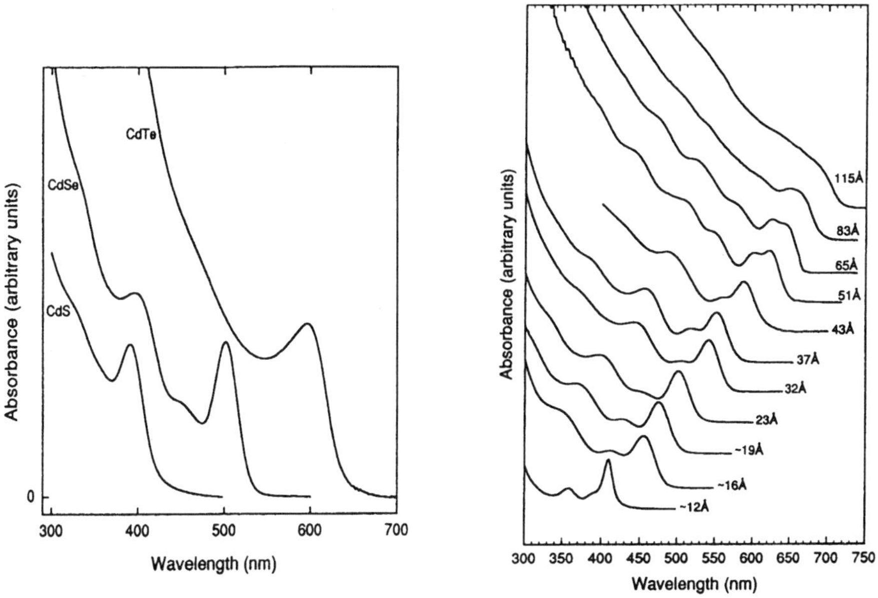

Figure 36. Optical spectra of Cd-X nanoparticles. Left: Effect of composition. Right: Effect of particle size. Quantum confinement effects create the shift in absorption edge. After Murray et al. (1993).

state lifetimes and modes of energy transfer within the structure. Examples of this are photoluminescence of ZnS nanoparticles studied by Wu et al. (1994), and Mn^{2+} doped ZnS nanoparticles by Bhargava et al. (1994). In the latter study, the doped nanocrystals were found to have higher quantum efficiency for fluorescence emission than bulk material, and a substantially smaller excited state lifetime. In the case of environmental nanoparticles of iron and manganese oxides, photoluminescence due to any activator dopant would be quenched by magnetic coupling and lattice vibrations. This reduces the utility of photoluminescence studies to excited state lifetimes due to particle-dopant coupling of various types. The fluorescence of uranyl ion sorbed onto iron oxides has been studied in this way, but not as a function of particle size.

Raman spectroscopy is an optical technique that can give information both on local vibrational states of particular molecular groupings, and on resonances of the entire particle at low frequencies that can be used to estimate particle size (Turkovic et al. 1997). Raman studies have been used to characterize capping materials on nanoparticles, but use of Raman to verify nanoparticle structure, and especially surface structure, has not been thoroughly developed.

Laser induced breakdown detection (LIBD) is a relatively new tool that allows the size distribution of nanoparticles to be determined under dilute conditions where standard optical techniques do not have enough sensitivity. The method is straightforward, and involves setting up a highly focussed laser beam to just below the threshold for plasma formation in the particle medium (generally water). As the threshold energy for breakdown for solids is lower than for liquids, the passage of a colloidal particle into the beam will cause breakdown and the emission of an acoustic wave which can be detected for each breakdown event. The acoustic wave energy is related to particle size, and thus by thorough sampling one can build up a knowledge of the distribution of sizes in a colloid or nanoparticle solution (Bundschuh et al. 2001).

Soft X-ray spectroscopy

At wavelengths shorter than UV, photons are identified as having "vacuum ultraviolet" or VUV energies (>10 eV). Still shorter wavelengths are called "soft" X-rays up to about a few KeV. VUV-soft X-ray spectroscopy is ideal for probing the electronic structure of materials by a wide variety of methods due to the inherent fine structure for absorption edges and valence band spectra at these energies, and the high energy resolution available with grating spectrometers. In particular, density of state calculations for nanoparticles and for nanoparticle surfaces can be examined by direct observation and compared to theoretical calculations. This is an enormous field of research encompassing a large number of experimental technologies, and almost an equal number of theoretical and semi-empirical analysis methodologies. The greatest amount of work in this field has been done on nanoparticulate thin-film semiconductor materials, although dispersed nanoparticulates have also been examined. Examples of the type of work possible is shown in Figures 37 and 38. One question about nanoparticles has been how they react structurally to the effects of large surface forces, such as high effective confining pressures. Hamad et al. (1999) collected soft X-ray Cd and In spectra at the $M_{4,5}$ edges that showed a continuous broadening of features as particle size decreased (Fig. 37). This seemingly goes against the expected trend of spectral structure becoming more clarified and sharp as molecular dimensions are approached. From comparison of the XPS and edge spectra, where the former shows much less broadening, it was deduced that the broadening is due to the final electronic state rather than the core state (initial state). This implies that the broadening is due to aspects of the crystal structure which affect bonding

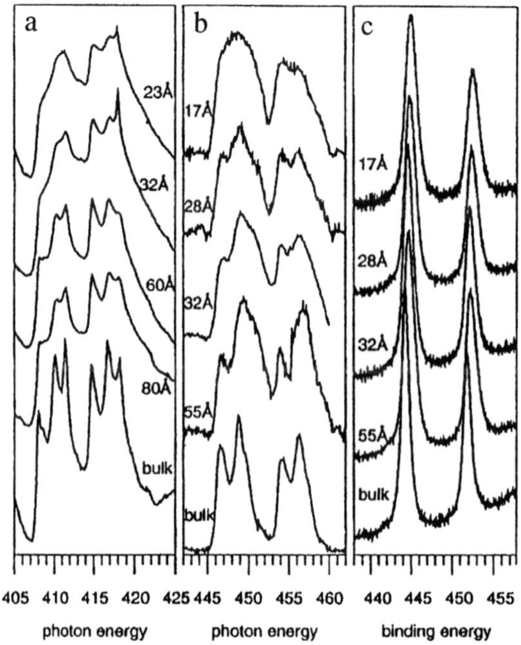

Figure 37. Soft X-ray spectra of (a) Cd and (b) In $M_{4,5}$ edges compared to (c) In 3d XPS spectra. Strong effects of particle size are seen in the soft X-ray edge structure. After Hamad et al. (1999).

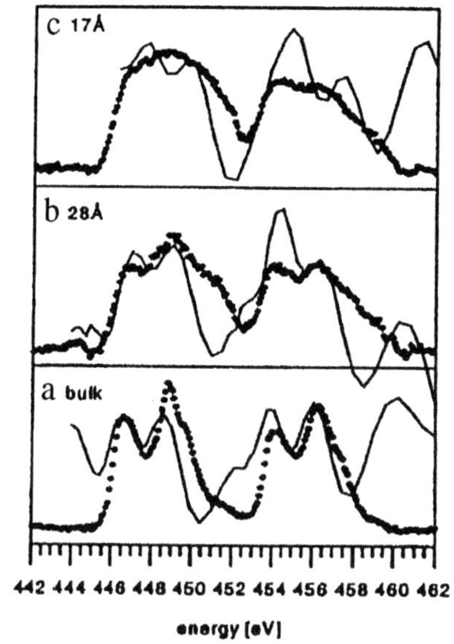

Figure 38. Multiple scattering calculation of the soft X-ray spectrum of different sizes of pherical InAs nanoparticles and InAs bulk (solid lines). The Feff 7 calculations reproduce major features and broadening, but do not model the details well. After Hamad et al. (1999).

electrons. In order to model this effect, Hamad et al (1999) used a multiple scattering calculation based on a nanocrystallite model with relaxed surface bond angles. Such relaxation leads to breakdown of symmetry and an increase in the types of multiple-scattering paths (see XANES discussion earlier) and ultimately to a blurring due to the diversity of unresolvable MS states (Fig. 38). Much less work has been done in soft X-ray spectroscopy of insulators, and essentially none for mineralogical nanomaterials. However, the soft X-ray spectra of anatase (TiO_2) was studied both as bulk and 20-nm nanoparticles by Hwu et al. (1997). These authors found energy shifts and intensity changes on the Ti $L_{2,3}$ edge spectra compared to a rutile (TiO_2) spectrum, but little broadening relative to bulk material on either the Ti $L_{2,3}$ or O K edges. In their case it seems as though the anatase nanocrystals are very little modified from the bulk phase.

Although VUV-soft X-ray techniques yield considerable information on valence and overall electronic structure, they have traditionally been useable only under vacuum conditions (and generally UHV conditions). This means that nanomaterials prepared by aqueous methods must be dried before spectroscopy can be performed, which can seriously affect the nature of the surface chemistry, surface structure and aggregation state. For this reason attempts are being made to develop facilities capable of performing these spectroscopic techniques under environmental conditions, e.g., 20 torr H_2O partial pressure. This is a critical need for studies of geological and soil-related nanomaterials. Such "high pressure" studies can be accomplished in two ways, by use of differential pumping techniques to allow the higher pressure at the sample, but reduced pressure over the rest of the spectrometer, and by using photon-only methodology. The lower pressure is needed to allow electrons ejected from the sample to be measured (e.g., in photoemission studies) or allow passage of very highly absorbed photons. A promising technology is the use of emission spectroscopy to obtain valence structure. A recent general review of much soft X-ray spectroscopy and emission spectroscopy covers the issues in much detail (de Groot 2001).

Mössbauer/NMR spectroscopy

Resonance gamma ray spectroscopy is a powerful technique for the determination of perturbations to the nuclear spin state levels of certain nuclei. It yields similar information to NMR, but is applicable to magnetic systems, and does not generally require the use of strong external magnetic fields. Mössbauer spectroscopy is especially useful for studies of the state of Fe in materials, and has been used extensively in the study of fine particulate magnetic behavior. This technology is explained in detail in the chapter by Rancourt. NMR is a powerful tool for determination of coordination numbers for ^{29}Si species in glasses and fine particles. Relatively new is the use of ^{17}O NMR spectra to define structural linkages in nanoparticles and small titanium oxo clusters, and yield insight into surface Ti-oxygen species (Scolan et al. 1999).

SOME OUTSTANDING ISSUES

Can crystal chemistry systematics be applied to nanoparticles?

Crystal chemical rules have been applied successfully to glass structure and to small units of structure used as models of larger structures. In both cases the rules seem to apply about as well as with extended bulk solids. Burdett et al. (1981) and Burdett and McLarnan (1984) have made an effort to explain some crystal chemical concepts (including the Pauling rules) in terms of modern bonding and quantum theory, with good success. Hence there ought not to be anything inherently difficult in treating nanoparticles much like other crystalline solids. What has yet to be tested is the use of crystal chemical rules to describe surface structures, an aspect crucial to nanocrystal structure, but this is limited by our knowledge of surfaces in the absolute sense. For example, bond valence analysis was used earlier to discuss how a surface might relax with oxygen termination and protonation. At present we have few reliable studies that show if such changes occur and the magnitude of them. Hence we cannot yet test our systematics for accurate predictions.

Can we meaningfully compare nanoparticle surfaces with bulk crystal surfaces (i.e., are nanoparticles just like small units of a bulk crystal surface)?

We know already that nanocrystals do not behave like large surfaces for several reasons: The bulk electrostatics for a small crystallite differs from that of a large flat plate, and the surface energy is not independent of crystallite size. These issues may mean that considerable rethinking of our electrical double layer modeling is required for nanocrystallite surfaces. Also the structure of a small crystallite will be affected by processes that are much slower or nonexistent in large crystal surfaces, such as Ostwald ripening, atomic rearrangements, strain effects and quantum effects. This means we must be cautious in comparing large crystal surfaces with nanocrystal surfaces of the same nominal composition. However, it is important to understand how mineral surface structure and reactivity are altered by size reduction, and this can be approached from both limiting cases: baseline structure and reactivity from the large single crystal surface, and molecular reassembly and dynamics from small molecular clusters.

Is the local structure of water different near nanoparticles? Is this significant in defining nanoparticle surface/bulk structure and perhaps stability?

Water structure differs markedly from the bulk at surfaces, and for crystallites on the order of 1 nm there could be significant structure enforced by strong hydrophilic or hydrophobic interactions. On large surfaces there is evidence for water restructuring and possible densification, both of which damp out with a few nm distance from the surface. Models have also predicted high diffusivity along the surface that is significantly smaller than diffusive mixing in the “layers” or partially ordered water normal to the interface. For small nanoparticles it is possible that solvation structures (see e.g., Israelachvili

1992), more common about highly charged molecules in aqueous solution, might modify the water structure. Water can form certain arrangements of cage-like solvation environments, and these might be favored by particular surface charges, particle diameters and surface terminations. Is it also important to consider the action of hydrogen bonding at nanocrystal surfaces in stabilizing surface structure.

Just as certain molecules are able to passivate the surfaces of semiconducting nanocrystals, water may act as an effective stabilizing agent with certain size nanocrystals under the right conditions.

Do highly defective structures have aspects in common with nanoparticles? For example: Do "nanoporous materials" have similarities to clustered nanoparticles either structurally or energetically?

It is possible that many of the poorly crystallized structures found in natural environments are better defined as aggregates of nanoparticles, than as defective crystal structures. In fact, it is the general desire to define structures within the context of the latter case that limits our openness to consider aggregated and perhaps somewhat recrystallized materials as the stable (or long term metastable) form of minerals. Very critical to dealing with these issues are the thermodynamic properties and reaction kinetics on the molecular scale at the water-mineral interface, most of which are still unavailable and daunting to obtain. An interesting case is ferrihydrite which has seen detailed study to understand its not only defective, but also highly variable microstructure and crystal structure (Drits et al. 1993; Manceau and Drits 1993; Drits et al. 1993).

Outstanding problems that need to be addressed

This chapter has touched on a few of the problems associated with nanoparticle and nanocrystal structure, growth and stability. There is a wide gap in our knowledge of how nanoparticles in the environment form, aggregate and otherwise react. The molecular-level aspects of surface structures, non-stoichiometry, surface relaxation, passivation and surface water structure and dynamics are still being initially explored in simple idealized high purity crystal surfaces. Hence characterization of natural nanoparticle materials will be a consummate challenge for years to come. However we can readily identify areas where study is likely to have important connections for environmental geochemistry. These include sorption and complexation reactions of small Fe- and Mn-oxyhydroxides, surface passivation of these and other nanominerals via natural organic phases, oriented aggregation processes as a new pathway for nanocrystallite growth, biomimetic pathways for the organization of nanocrystals and tissues or organic binding agents, and enzymatic alteration of crystal growth, aggregation and crystal structure stability. Fortunately our tools for these investigations have never been more powerful, and the near future holds much promise for excellent advances in our simulation capacity for surface and crystal structure, bonding, growth and electronic structure.

ACKNOWLEDGMENTS

Jill Banfield contributed mightily to the content of this chapter, and I thank her for many discussions, suggestions, and for initial review. Discussions with Jim Rustad, Hoi-Ying Holman and David Shuh were also most helpful. Carol Taliaferro helped with finalizing corrections to figures, figure captions, references and text. This project was supported by the Office of Science, Office of Basic Energy Sciences, Division of Chemical Sciences, Geosciences Research Program of the U.S. Department of Energy under contract DE-AC03-76-SF00098, and by LDRD funding from Lawrence Berkeley National Laboratory.

REFERENCES

Abruna HD (1991) X-ray Absorption Spectroscopy in the Study of Electrochemical Systems. *In* Electrochemical Interfaces. Abruna HD (ed) VCH Publishers, New York

Albertini G, Carsughi F, Casale C, Fiori F, La Monaca A, Musci M (1993) X-ray and neutron small-angle scattering Investigation of nanophase vanadium-titanium oxide particles. Phil Mag B 68:949-955

Alexandrowicz Z (1993) How to reconcile classical nucleation theory with cluster-cluster aggregation. Physica A 200:250-257

Alivisatos AP (1997) Scaling law for structural metastability in semiconductor nanocrystals. Ber Bunsenges Phys Chem 101:1573-1577

Auer S, Frenkel D (2001) Prediction of absolute crystal-nucleation rate in hard-sphere colloids. Nature 409:1020-1023

Azaroff LV (1968) Elements of X-ray Crystallography. McGraw-Hill, New York

Banfield JF, Welch SA, Zhang H, Ebert TT, Penn RL (2000) Aggregation-based crystal growth and microstructure development in natural iron oxyhydroxide biomineralization products. Science 289:751-754

Benaissa M, Jose-Yacaman M, Xiao TD, Strutt PR (1997) Microstructural study of hollandite-type MnO_2 nano-fibers. Appl Phys Letters 70:2120-2122

Bhargava RN, Gallagher D, Hong X, Nurmikko A (1994) Optical properties of manganese-Doped nanocrystals of ZnS. Phys Rev Letters 72:416-419

Bish DL, Post JE (eds) (1989) Modern Powder Diffraction. Reviews in Mineralogy vol 20, 369 p. Mineralogical Society of America, Washington, DC

Becker R, Döring W (1935) Kinetische behandlung der Keimbildung in übersättigten Dämpfen. Ann Phys 24:719-752

Brasher LL, Kaler EW (1996) A small-angle neutron scattering (SANS) contrast variation investigation of aggregate composition in catanionic surfactant mixtures. Langmuir 12:6270-6276

Brown GE Jr, Heinrich VE, Casey WH, Clark DL, Eggleston C, Felmy A, Goodman DW, Grätzel M, Maciel G, McCarthy MI, Nealson KH, Sverjensky DA, Toney MF, Zachara JM (1999) Metal oxide surfaces and their interactions with aqueous solutions and microbial organisms. Chem Rev 99:77-174

Brown GE Jr , Calas G, Waychunas GA, Petiau J (1988) X-ray absorption spectroscopy: Applications in mineralogy and geochemistry. Rev Mineral 18:431-512

Brown ID, Shannon RD (1973) Empirical bond-strength-bond-length curves for oxides. Acta Crystallogr A29:266-282

Bundschuh T, Knopp R, Kim JI (2001) Laser-induced breakdown detection (LIBD) of aquatic colloids with different laser systems. Colloids Surfaces A 177:47-55

Burdett JK, McLarnan TJ (1984) An orbital interpretation of Pauling's rules. Am Mineral 69:601-621

Burdett JK, Price GD, Price SL (1981) The factors influencing solid state structure. An interpretation using pseudopotential radii maps. Phys Rev B 24:2903-2912

Calas G, Brown GE Jr, Waychunas GA, Petiau J (1987) X-ray absorption spectroscopic studies of silicate glasses and minerals. Phys Chem Minerals 15:19-29

Carter A, Bouldin CE, Kemner KM, Bell MI, Woicik JC, Majetich SA (1997) Surface structure of cadmium selenide nanocrystallites. Phys Rev B 55:13822-13828

Caruso F (2001) Nanoengineering of Particle Surfaces. Adv Mater 13:11-22

Casanove M-J, Lecante P, Snoeck E, Mosset A, Roucau C (1997) HREM and WAXS study of the structure of metallic nanoparticles. J Physique III 7:505-515

Chen LX, Rajh T, Jäger W, Nedeljkovic J, Thurnauer MC (1999) X-ray absorption reveals surface structure of titanium dioxide nanoparticles. J Synch Rad 6:445-447

Cheng B, Kong J, Luo J, Dong Y (1993) Materials Science Progress 7:240-243 (in Chinese)

Chu B, Liu T (2000) Characterization of nanoparticles by scattering techniques. J Nanoparticle Res 2:29-41

Clem PJ, Fisher JC (1958) Acta Metall 3:347

Connolly S, Fullam S, Korgel B, Fitzmaurice D (1998) Time-resolved small-angle X-ray scattering studies of nanocrystal superlattice self-assembly. J Am Chem Soc 120:2969-2970

deGroot F (2001) High-resolution X-ray emission and X-ray absorption spectroscopy. Chem Rev 101:1779-1808

Davis JA, Kent DB (1990) Surface complexation modeling in aqueous geochemistry. Rev Mineral 23: 177-260

Derjaguin BV (1934) Untersuchungen über die Reibung und Adhäsion. Kolloid Zeits 69:155-164

Drits VA, Sakharov BA, Manceau A (1993) Structure of feroxyhite as determined by simulation of X-ray diffraction. Clay Mineral 28:209-222

Drits VA, Sakharov BA, Salyn, AL and Manceau A (1993) Structural model for ferrihydrite. Clay Mineral 28:185-207

Edelstein AS, Cammarata RC (1996) Nanomaterials: Synthesis, Properties and Applications. Institute of Physics, Bristol, UK

Eden M (1961) A two-dimensional growth process. 4th Berkeley Symp Math, Stat, Prob 4:223-239

Eng PJ, Trainor TP, Brown GE, Waychunas GA, Newville M, Sutton SR, Rivers ML (2000) Structure of the hydrated α-Al_2O_3 (0001) surface. Science 288:1029-1033

Frenkel AI (1999) Solving the structure of nanoparticles by multiple-scattering EXAFS analysis. J Synch Rad 6:293-295

Fritsche H-G, Buttner T (1999) Modification of the lattice contraction of small metallic particles by chemical and/or geometrical stabilization. Z Phys Chem 209:93-101

Gao P, Gleiter H (1987) High resolution electron microscope observation of small gold particles. Acta Metall 35:1571-1575

Glusker JP, Trueblood KN (1985) Crystal Structure Analysis. 2nd edition. Oxford Univ Press, Oxford, UK

Gnutzmann V, Vogel W (1990) Surface oxidation and reduction of small platinum particles observed by *in situ* X-ray diffraction. Z Physik D (Atoms, Molecules, Clusters) 12:597-600

Greenwood NN (1970) Ionic Crystals Lattice Defects and Nonstoichiometry. Butterworths, London

Guinier A (1963) X-ray Diffraction in Crystals, Imperfect Crystals, and Amorphous Bodies. W H Freeman, San Francisco

Hadjipanayis GC, Siegel RW (1994) Nanophase Materials: Synthesis-Properties-Applications. NATO ASI Series 260. Kluwer Academic Publ, Dordrecht, The Netherlands

Halsey TC (2000) Diffusion-limited aggregation: A model for pattern formation. Phys Today, Nov 2000, p 36-41

Hamad KS, Roth R, Rockenberger J, van Buuren T, Alivisatos AP (1999) Structural disorder in colloidal InAs and CdSe nanocrystals observed by X-ray absorption near-edge spectroscopy. Phys Rev Letters 83:3474-3477

Harfenist SA, Wang ZL, Alvarez MM, Vezmar I, Whetten RL (1997) Three-dimensional superlattice packing of faceted silver nanocrystals. Nanophase and nanocomposite materials II. Mater Res Soc Symp Proc 457:137-142

Hartman P (1973) Crystal Growth: An Introduction. North-Holland Publishing Co, Amsterdam

Herring C (1951) Some theorems on the free energies of crystal surfaces. Phys Rev 82:87-93

Hettema H, McFeaters JS (1996) The direct Monte Carlo method applied to the homogeneous nucleation problem. J Chem Phys 105:2816-2827

Hiemenz PC, Rajagopalan R (1997) Principles of Colloid and Surface Chemistry, 3rd edition. Marcel Dekker, New York

Hofmeister H, Huisken F, Kohn B (1999) Lattice contraction in nanosized silicon particles produced by laser pyrolysis of silane. Eur Phys J D 9:137-140

Hwu Y, Yao YD, Cheng NF, Tung CY, Lin HM (1997) X-ray absorption of nanocrystal TiO_2. Nanostruc Mater 9:355-358

Hyeon-Lee J, Beaucage G, Pratsinis SE, Vemury S (1998) Fractal analysis of flame-synthesized nanostructured silica and titania powders using small-angle X-ray scattering. Langmuir 14:5751-5756

Israelachvili JN (1991) Intermolecular and Surface Forces, 2nd edition. Academic Press, New York

Israelachvili JN, Pashley RM (1982) The hydrophobic interaction is long range decaying exponentially with distance. Nature 300:341-342

Kaszuba M (1999) The measurement of nanoparticles using photon correlation spectroscopy and avalanche photodiodes. J Nanoparticle Res 1:405-409

Khanna SN, Bucher JP, Buttet J, Cyrot-Lackmann F (1983) Stability and lattice contraction of small platinum particles. Surface Sci 127:165-174

Kirkpatrick RJ (1981) Kinetics of crystallization of igneous rocks. Rev Mineral 8:321-398

Klug HP, Alexander LE (1974) X-ray Diffraction Procedures, 2nd edition. Wiley Interscience, New York

Kortright J, Warburton W, Bienenstock A (1983) Anomalous X-ray scattering and its relationship to EXAFS. Springer Ser Chem Phys 27:362-372

Krill CE, Ehrhardt H, Birringer R (1997) Grain-size stabilization in nanocrystalline materials. *In* Chemistry and Physics of Nanostructures and Related Non-Equilibrium Materials. Ma E, Fultz B, Shull R, Morral J, Nash P (eds) Minerals, Metals and Materials Society, Warrendale, Pennsylvania

Lee PA, Citrin PH, Eisenberger P (1981) Extended X-ray absorption fine structure—its strengths and limitations as a structural tool. Rev Mod Phys 53:769-806

Li D-H, Ganczarczyk J (1989) Fractal geometry of particle aggregates generated in water and wastewater treatment processes. Environ Sci Technol 23:1385-1389

Li JC, Ross DK, Howe LD, Stefanopoulos KL, Fairclough JPA, Heenan R, Ibel K (1994) Small-angle neutron-scattering studies of the fractal-like network formed during desorption and adsorption of water in porous materials. Phys Rev B 49:5911-5917

Liu J (2000) Scanning transmission electron microscopy of nanoparticles. *In* Characterization of Nanophase Materials. Wang ZL (ed) p 81-132. Wiley-VCH, Weinheim, FRG
Linford RG (1973) Surface thermodynamics of solids. *In* Green M (ed) Solid State Surface Science 2: p 1-152. Dekker, New York
Mahanty J, Ninham BW (1976) Dispersion Forces. Academic Press, New York
Manceau A, Drits VA (1993) Local structure of ferrihydrite and ferroxyhite by EXAFS spectroscopy. Clay Mineral 28:165-184
Marasll N, Hunt JD (1998) The use of measured values of surface energies to test heterogeneous nucleation theory. J Crystal Growth 191:558-562
Martin JE, Wilcoxon JP, Odinek J, Provencio P (2000) Control of interparticle spacing in gold nanoparticle superlattices. J Phys Chem B 104:9475-9486
McGrath KM (2001) Probing material formation in the presence of organic and biological molecules. Adv Mater 13:989-992
Meakin P (1992) Aggregation kinetics. Physica Scripta 46:295-331
Milligan WW, Hackney SA, Ke M, Aifantis EC (1993) *In situ* studies of deformation and fracture in nanophase materials. Nanostruct Mater 2:267-276
Miranda Salvado IM, Margaca FMA, Teixeira J (1996) Structure of mineral gels. J Molec Struct 383: 271-276
Montano P, Shenoy GK, Alp EE, Schulze W, Urban J (1986) Structure of copper microclusters isolated in solid argon. Phys Rev Letters 56:2076-2079
Murray CB, Kagan CR, Bawendi MG (1995) Self organization of CdSe Nanocrystallites into three-dimensional quantum dot superlattices. Science 270:1335-1338
Murray CB, Norris DJ, Bawendi MG (1993) Synthesis and characterization of nearly monodisperse CdE (E=S, Se, Te) *S*emiconductor nanocrystallites. J Am Chem Soc 115:8706-8715
Nanda KK, Behera SN, Sahu SN (2001) The lattice contraction of nanometre-sized Sn and Bi particles produced by an electrochemcial technique. J Phys Condens Matter 13:2861-2864
Nikolakis V, Kokkoli E, Tirrell M, Tsapatsis M, Vlachos DG (2000) Zeolite growth by addition of subcolloidal particles: modeling and experimental validation. Chem Mater 12:845-853
Onodera SJ (1992) Lattice parameters of fine copper and silver particles. J Phys Soc Japan 61:2191-2193
Ostwald W (1897) Studien über die Bildung und Umwandlung fester Körper. Z Phys Chemie 22:289-330
Pecora R (2000) Dynamic light scattering measurements of nanometer particles in liquids. J Nanoparticle Res 2:123-131
Penn RL, Banfield JF (1999) Morphology development and crystal growth in nanocrystalline aggregates under hydrothermal conditions: Insights from titania. Geochim Cosmochim Acta 63:1549-1557
Penn RL, Banfield JF (1998a) Imperfect oriented attachment: dislocation generation in defect-free nanocrystals. Science 281:969-971
Penn RL, Banfield JF (1998b) Oriented attachment and growth, twinning, polytypism, and formation of metastable phases: Insights from nanocrystalline TiO_2. Am Mineral 83:1077-1082
Post JE, Bish DL (1989) Rietveld refinement of crystal structures using powder X-ray diffraction data. Rev Mineral 20:277-308
Raghaven V, Cohen M (1975) Solid-state phase transformations. *In* Hannay NB (ed) Treatise on Solid State Chemistry 5:67-128
Reinhard D, Hall BD, Berthoud P, Valkealahti S, Monot R (1998) Unsupported nanometer-sized copper clusters studied by electron diffraction and molecular dynamics. Phys Rev B 58:4917-4926
Reinhard D, Hall BD, Berthoud P, Valkealahti S, Monot R (1997) Size-dependent icosahedral-to-fcc structure change confirmed in unsupported nanometer-sized copper clusters. Phys Rev Letters 79:1459-1462
Reddy M, Reddy G, Manorama SV (2001) Preparation, characterization, and spectral studies on nanocrystalline anatase TiO_2. J Solid State Chem 158:180-186
Rieker T, Hanprasopwattana A, Datye A, Hubbard P (1999) Particle size distribution inferred from small-angle X-ray scattering and transmission electron microscopy. Langmuir 15:638-641
Rietveld HM (1969) A profile refinement method for nuclear and magnetic structures. J Appl Crystallogr 2:65-71
Rockenberger J, Tröger L, Rogach AL, Tischer M, Grundmann M, Eychmüller A, Weller H (1998) The contribution of particle core and surface to strain, disorder and vibrations in thiolcapped CdTe nanocrystals. J Chem Phys 108:7807-7815
Rockenberger J, Tröger L, Kornowski A, Vossmeyer T, Eychmüller A, Feldhaus J, Weller H (1997) EXAFS Studies on the size-dependence of structural and dynamic properties of CdS nanoparticles. J Phys Chem B 101:2691-2701
Sayers DE, Lytle FW, Stern EA (1970) Point scattering theory of X-ray K absorption fine structure. Adv X-ray Analysis 13:248-271

Sayers DE, Stern EA, Lytle FW (1971) New technique for investigating noncrystalline structures: Fourier analysis of the extended X-ray absorption structure. Phys Rev Letters 27:1204-1207
Sayle DC, Watson GW (2000) Simulated amorphisation and recrystallization: Nanocrystallites within meso-scale supported oxides. J Mater Chem 10:2241-2243
Scherrer P (1918) Nachr Göttinger Gesell 98
Schreyer D, Chatelain A (1985) Lattice contraction in small particles of $SrCl_2$:Gd^{3+}: explicit size effect in EPR and Raman spectroscopy. Surface Sci 156:712-719
Schultz, JM (1982) Diffraction for Materials Scientists. Prentice-Hall, Englewood Cliffs, NJ
Schwartz LH, Cohen JB (1987) Diffraction from Materials. Springer-Verlag, Berlin
Scolan E, Magnenet C, Massiot D, Sanchez C (1999) Surface and bulk characterization of titanium-oxo clusters and nanosized titania particles through ^{17}O solid state NMR. J Mater Chem 9:2467-2474
Sekerka RF (1993) Role of instabilities in determination of the shapes of growing crystals. J Crystal Growth 128:1-12
Sekerka RF (1973) Morphological stability. *In* Crystal Growth: An Introduction. Hartman P (ed) p 403-443. North-Holland Publishing Co, Amsterdam
Singer PC, Stumm W (1970) Acidic mine drainage. Science 167:1121-1123
Smoluchowski MV (1918) Versuch einer mathematischen Theorie der koagulationskinetik kolloider Lösungen. Z Phys Chemie 92:129-168
Sposito G (1984) The Surface Chemistry of Soils. Oxford :University Press, New York
Stanley HE, Ostrowsky N (1986) On Growth and Form. NATO ASI Ser E: Applied Sciences, vol 100. Kluwer Academic Publ, Dordrecht, The Netherlands
Stumm W (1987) Aquatic Surface Chemistry. John Wiley, New York
Sun S, Murray CB (1999) Synthesis of monodisperse cobalt monocrystals and their assembly into magnetic superlattices. J Appl Phys 85:4325-4330
Sutherland DN (1970) Chain formation of fine particle aggregates. Nature 226:1241-1242
Suzuki Y, Banfield JF (1999) Geomicrobiology of uranium. Rev Mineral 38:393-432
Talanquer V, Oxtoby DW (1994) Dynamical density functional theory of gas-liquid nucleation. J Chem Phys 100:5190-5200
Thomas DN, Judd SJ, Fawcett N (1999) Flocculation modelling: A review. Water Res 33:1579-1592
Thomas JG, Siegel RW, Eastman JA (1990) Grain boundaries in nanophase palladium: high resolution electron microscopy and image simulation. Scripta Metall Mater 24:201-206
Thurn-Albrecht T, Meier G, Muller-Buschbaum P, Patkowski A, Steffen W, Grubel G, Abernathy DL, Diat O, Winter M, Koch MG, Reetz MT (1999) Structure and dynamics of surfactant-stabilized aggregates of palladium nanoparticles under dilute and semidilute conditions: Static and dynamic X-ray scattering. Phys Rev E 59:642-649
Tranquanda JM, Ingalls R (1983) Extended X-ray-absorption fine-structure study of anharmonicity in CuBr. Phys Rev B 28:3520-3528
Tsunekawa S, Ito S, More T, Ishikawa K, Li Z-Q, Kawazoe Y (2000a) Critical size and anomalous lattice expansion in nanocrystalline $BaTiO_3$ particles. Phys Rev B 62:3065-3070
Tsunekawa S, Ishikawa K, Li Z-Q, Kawazoe Y, Kasuya A (2000b) Origin of anomalous lattice expansion in oxide nanoparticles. Phys Rev Letters 85:3440-3443
Turkovic A, Ivanda M, Popovic S, Tonejc A, Gotic M, Dubcek P, Music S (1997) Comparative Raman, XRD, HREM and SAXS studies of grain sizes in nanophase TiO_2. J Molec Struct 410/411:271-273
ten Wolde PR, Frenkel D (1997) Enhancement of protein crystal nucleation by critical density fluctuations. Science 277:1975-1978
Verwey EJW, Overbeek JTG (1948) Theory of the Stability of Lyophobic Colloids. Elsevier, Amsterdam
Volmer M, Weber A (1925) Klimbildung in übersättigten gebilden. Z Phys Chem 119:277-301
Wadsley AD (1963) Inorganic nonstoichiometric compounds. *In* Mandelcorn L (ed) Nonstoichiometric Compounds. Academic Press, New York
Wang ZL (2000) Characterization of Nanophase Materials. Wiley-VCH, Weinheim, FRG
Warren BE (1969) X-ray Diffraction. Addison-Wesley, Reading, Massachusetts
Warren BE, Averbach BL (1950) The effect of cold-work distortion on X-ray patterns. J Appl Phys 21: 595-599
Waychunas GA (1987) Synchrotron radiation XANES spectroscopoy of Ti in minerals: Effects of Ti bonding distances, Ti valence, and site geometry on absorption edge structure. Am Mineral 72:89-101
Waychunas GA, Fuller CC, Davis JA, Rehr JJ (2001) Surface complexation and precipitate geometry for aqueous Zn(II) sorption on ferrihydrite: II. XANES analysis. Geochim Cosmochim Acta (in press)
Waychunas G, Davis J, Reitmeyer R (1999) GIXAFS study of Fe^{3+} sorption and precipitation on natural quartz surfaces. J Synchrotron Rad 6:615-617

Waychunas GA, Fuller CC, Rea BA, Davis JA (1996) Wide angle X-ray scattering (WAXS) study of "two-line" ferrihydrite structure: Effect of arsenate sorption and counterion variation and comparison with EXAFS results. Geochim Cosmochim Acta 60:1765-1781

Weissmuller J (1996) Characterization by scattering techniques and EXAFS. *In* Nanomaterials: Synthesis, Properties and Applications. Edelstein AS, Cammarata RC (eds) p 219-276. Institute of Physics Bristol, UK

Wilson AJC (1963) Mathematical Theory of X-ray Powder Diffraction. Philips Technical Library, Eindhoven, The Netherlands

Wu M, Gu W, Li W, Zhu X, Wang F, Zhao S (1994) Preparation and characterization of ultrafine zinc sulfide particles of quantum confinement. Chem Phys Letters 224:557-562

Wulff G (1901) Zur Frage der Geschwindigkeit des Wachsthums und der Auflösung der Krystallflächen. Z Kristallogr 34:449-530

Witten TA, Sander LM (1981) Diffusion-limited aggregation, a kinetic critical phenomenon. Phys Rev Letters 47:1400-1403

Xiao TD, Strutt PR, Benaissa M, Chen H, Kear BH (1998) Synthesis of high active-site density nanofibrous MnO_2-base materials with enhanced permeabilities. Nanostruc Mater 10:1051-1061

Yeadon M, Ghaly M, Yang JC, Averback RS, Gibson JM (1998) "Contact epitaxy" observed in supported nanoparticles. Appl Phys Letters 73:3208-3210

Yin JS, Wang ZL (1997) Ordered self-assembling of tetrahedral oxide nanocrystals. Phys Rev Letters 79:2570-2573

Young RA (1993) The Rietveld Method. Oxford University Press, Oxford, UK

Zanchet D, Hall BD, Ugarte D (2000) X-ray Characterization of Nanoparticles. *In* Characterization of Nanophase Materials. Wang ZL (ed) p 13-36. Wiley-VCH, Weinheim, FRG

Zangwill A (1988) Physics at Surfaces. Cambridge University Press, Cambridge, UK

Zhang H, Banfield JF (2000) Understanding polymorphic phase transformation behavior during growth of nanocrystalline aggregates: insights from TiO_2. J Phys Chem B 104:3481-3487

Zhao J, Huggins FE, Feng Z, Huffman GP (1994) Ferrihydrite: Surface structure and its effects on phase transformation. Clays Clay Minerals 42:737-746

Zoltai T, Stout JH (1984) Mineralogy: Concepts and Principles. Burgess Publishing, Minneapolis, Minnesota

5 Aqueous Aluminum Polynuclear Complexes and Nanoclusters: A Review

William H. Casey

Department of Land, Air and Water Resources
and *Department of Geology*
University of California–Davis
Davis, California 95616

Brian L. Phillips

Department of Geosciences
State University of New York at Stony Brook
Stony Brook, New York 11794

Gerhard Furrer

Institute of Terrestrial Ecology
ETH Zürich, Grabenstrasse 3
CH-8952, Schlieren, Switzerland

INTRODUCTION

Most undergraduate students of aqueous geochemistry are told that polynuclear aqueous complexes can largely be ignored because they form only from concentrated metal solutions that are rare at the Earth's surface. However, these polynuclear complexes can serve as models for more-complicated surface structures and are the precursors to nanometric and colloidal solids and solutes. There are many reasons why polynuclear complexes should be foremost in the minds of geochemists, and particularly those geochemists who are interested in molecular information and reaction pathways:

(1) Polynuclear complexes contain many of the structural features that are present at mineral surfaces, including a shell of structured water molecules. Because aqueous nanoclusters tumble rapidly in an aqueous solution, one can use solution NMR spectroscopy to determine the structure and the atomic dynamics in these clusters in ways that are impossible for mineral surfaces.

(2) Some polynuclear complexes are metastable for long periods of time and may represent an important vector for the dispersal of metal contaminants from hazardous waste. The chemical conditions found in many polluted soils: high metal concentrations, elevated temperatures, and either highly acidic or highly alkaline solutions with a large pH-gradient, are needed to synthesize many polynuclear complexes. It is easy to make a solution that is 5 M in dissolved aluminum at $4 < pH < 6$, composed of nanometer-sized clusters that are stable for months or years.

(3) Polynuclear complexes lie at the core of many biomolecules, including metalloproteins such as ferritin and enzymes such as nitrogenase. Recent work has suggested that they are present in natural waters (e.g., Rozan et al. 2000) and serve as nuclei for crystal growth.

(4) Aqueous clusters are sufficiently small that they can serve as experimental models for *ab initio* computer simulations that relate bonding to reactivity.

In this chapter we discuss polynuclear complexes of aluminum. There is a wide variety of multimeric complexes of aluminum, incorporating a number of geochemically

1529-6466/00/0044-0005$05.00

important moieties, that have been well-studied and characterized in both structure and in their interactions with aqueous fluids. Readers interested in a wide range of examples are referred to a recent Ph.D. dissertation by Magnus Karlsson (1998) at the Chemistry Department of the University of Umeå, which contains a comprehensive review of aluminum structures that can be crystallized from solution. We limit the present discussion to complexes for which the structure has been confirmed by NMR spectroscopy or single-crystal X-ray diffraction methods. Likewise, we exclude the large number of multimeric complexes of aluminum that have been synthesized in nonaqueous media.

The logical starting place to discuss the chemistry of aluminum polynuclear complexes is with the simplest molecules. Here we start by looking at the structure of aluminum dimers. Throughout, we employ the formalism that η indicates nonbridging sites and μ_i- sites are ligand atoms that bridge '*i*' metals. We also denote the coordination number of aluminum as 'Al(*i*)' or '$Al(O)_i$' where *i* indicates the number of atoms bonded to aluminum.

METHODS FOR DETERMINING STRUCTURES

X-Ray diffraction

In many cases, a salt of the polynuclear complex can be crystallized from solution, allowing use of traditional X-ray diffraction (XRD) techniques to determine the structure of the complex itself. Furthermore, these salts can sometimes be redissolved to release the multimeric complex into solution intact. The resulting monospecific solutions are particularly useful because the multimer can be probed using spectroscopic methods or even methods of bulk aqueous chemistry.

Nuclear magnetic resonance spectroscopy

For polynuclear complexes of aluminum, ^{27}Al-NMR spectroscopy has been used extensively to characterize the structure of the complexes as well as the speciation of the aqueous fluids. The characteristics of NMR spectroscopy—nucleus specific, quantitative intensities, and sensitivity to only short-range structure—combined with the high natural abundance of ^{27}Al make this a powerful technique at millimolar concentrations. However, the quadrupolar nature of the ^{27}Al nucleus (spin number I = 5/2) introduces some complications in spectral interpretation that are worth mentioning here.

In principal, a peak will appear in the NMR spectrum for each structurally and/or chemically distinct site for each type of molecule present in solution. However, the number of distinct resonances observed is usually less than this for two reasons. First, resolution is usually limited by the large peak widths. The ^{27}Al NMR peak widths in solution are usually determined by the strength of the quadrupole coupling and are proportional to the product of the quadrupole coupling constant squared, $(C_q)^2$, and the rotational correlation time for the complex, τ_c. The C_q is proportional to the electric field gradient at the nucleus. Hence, sites in distorted environments will display large C_q, yielding broad peaks in solution. Highly symmetric sites, such as the $Al(O)_4$ of Al_{13} or the $Al(H_2O)_6^{3+}$ complex, have small C_q and yield sharp ^{27}Al NMR peaks. Because the rotational correlation times increase approximately in proportion to the volume of the complex, ^{27}Al NMR peaks for large polynuclear complexes and be quite broad - sometimes too broad to detect by standard pulse-Fourier transform NMR techniques. This can lead to "missing" intensity from NMR-"invisible" nuclei in distorted environments of large complexes. In some cases, raising the temperature improves resolution and the observability of broad peaks by reducing τ_c.

Secondly, NMR averages over chemical exchange processes that are faster than the separation of the peaks (in frequency units) arising from the exchanging species in the limit of no exchange. Species undergoing rapid interconversion yield only a single NMR peak which occurs at the weighted average position of the constituent species. For example, the monomeric hexaqua Al^{3+} complex and its monomeric hydrolysis products (principally, $Al(OH)^{2+}$) are related by simple proton exchange between the bound waters and solvent. This proton exchange is very fast (10^9 s^{-1}) so that the rate of interconversion among the hydrolysis products is much larger than the separation of peaks of the corresponding species ($\Delta\nu < 10^3$ s^{-1}). Thus, ^{27}Al NMR spectra contain only a single peak for the monomeric aqua complexes, the position of which varies with pH, corresponding to the weighted average of the chemical shifts of the constituent complexes.

Potentiometry

In all cases, potentiometric titrations provide important constraints on the metal speciation in aqueous fluids. Although it cannot provide the structure of the complexes, potentiometry constrains the stoichiometries of the possible complexes, and changes with metal concentration, ligand concentration, and pH.

STRUCTURES

Aluminum monomers

Aluminum is a hard Lewis acid and coordinates strongly with hard Lewis bases such as oxide and fluoride. Therefore the aqueous complexes of aluminum are dominated by hydrolysis species (e.g., $Al(H_2O)_6^{3+}$, $AlOH(H_2O)_5^{2+}$, ...) and complexes to organic ligands like carboxylates. The points that are important for this chapter is that Al(III) can change coordination number easily in the range 6 to 4 and it is highly charged. While Al(III) in $Al(H_2O)_6^{3+}$ is 6-coordinated to oxygens, it becomes coordinated to 4 oxygens in the $Al(OH)_4^-$ complex that becomes appreciable at near-neutral pH and above. This point is important because some nanoclusters often require mixed coordination states. Aluminum forms multimers easily at concentrations of 10^{-5} M and above.

Aluminum dimers

Several solids containing isolated dihydroxy-bridged $Al(O)_6$ dimers are known and can be crystallized by slow growth from solution. There is strong evidence that dissolution of some of these solids releases dimers intact, which then dissociate slowly in the aqueous phase. The earliest work was provided by Johansson (1962a), who reported a method of crystallizing $Al_2(OH)_2(SeO_4)_2(H_2O)_{10}(s)$, which consists of layers of isolated $(H_2O)_4Al(\mu_2\text{-}OH)_2Al(H_2O)_4$ dimers. Oxygens in the selenate counterions do not bond directly to aluminums. Aluminum dimers of stoichiometry $Al_2(\mu_2\text{-}OH)_2(H_2O)_8^{4+}(aq)$ were originally suspected in hydrolyzed $AlCl_3$-NaOH solutions (e.g., Baes and Mesmer 1976; Akitt et al. 1972), but the dimers are difficult to identify by ^{27}Al-NMR and their presence was later questioned from potentiometric data (Öhman and Forsling 1981). Akitt and Elders (1988) later crystallized the dimer sulfate salts and redissolved them in dilute solution, which yields a peak near +4 ppm in ^{27}Al-NMR spectra. They concluded, however, that the dimer decomposes in solution and hydrolyzes to form a higher order complex, such as a trimer, and monomeric species (see also Schönherr et al. 1983). In summary, the existence of the dimer in solution is highly suspected but unproven.

Like most polynuclear *oxy* complexes and the surfaces of oxide minerals, these dimers are amphoteric in water and can dissociate by pathways that involve either proton uptake at a hydroxyl bridge or proton release from a bound water molecule. Therefore the stability of a dimer relates closely to its Brønsted acid-base chemistry. The decomposition pathways for hydroxy-bridged dimers are well understood from studies of inert-metal compounds (Springborg 1988). One pathway for dissociation is independent of pH and probably involves a nucleophilic attack on the metal by an incoming water molecule, which is probably assisted by partial dissociation of $Al(O)_6$ because aluminum is coordinatively saturated. This pH independent reaction may be assisted by an internal proton transfer. A second pathway involves direct protonation of a μ_2-OH bridge to form bridging water molecule. The protonation step is much faster than the subsequent dissociation of the bridge, so that protonation usually proceeds to equilibrium such that the fraction of bridges protonated varies with solution composition and introduces a distinct pH dependence to the reaction. At some point in the dissociation of the Al-O bond to the μ_2-OH_2^+ bridge one of the aluminum atoms becomes underbonded and coordinates to an incoming water molecule. Bound water molecules can affect the stability of the complex by either stabilizing the hydroxyl bridge by hydrogen bonding, or by deprotonating and forming a bound hydroxyl that donates negative charge to the metal center, thereby weakening other bonds to oxygens.

An interesting series of aluminum dimers can be crystallized with carboxylate ligands (Fig. 1) that replace bound water molecules or a μ_2-OH bridge. The carboxylate ligands are homologous and differ from each other by simple changes in one or two functional groups that are coordinated to a central quaternary amine. Nitrilotriacetic acid (NTA) consists of three ethylcarboxylate ligands coordinated to the quaternary amine whereas iminodiacetic acid (IDA) has only two ethylcarboxylates. Hydroxyethyliminodiacetic acid (HEIDI) has an ethyl alcohol replacing one of the ethylcarboxylate ligands. The homologous series of three aluminum dimers are:

(i) $Al_2(\mu_2\text{-}OH)_2(\eta\text{-}OH_2)_2(IDA)_2^0(aq)$ (Petrosyants et al. 1995; Yokoyama et al. 1999)
(ii) $Al_2(\mu_2\text{-}OH)_2(NTA)_2^{2-}(aq)$ (Valle et al. 1989)
(iii) $[Al(HEIDI)(H_2O)]_2(aq)$ (Heath et al. 1995; Jordan et al. 1996)

In this series of aluminum dimers the structure is modified systematically by slight changes in functional groups on the coordinating ligand. The structures of the three dimers are shown in Figure 1 and differ in the number of bonded water molecules and bridging hydroxyls. By analogy with the stability of inert-metal dimers, the presence of a water molecule can control the rates of dissociation of the dimers. These three similar dimers probably exhibit dramatically different stabilities and rates of oxygen exchanges. These dimers can dissociate in an aqueous solution either by protonation of the μ_2-OH sites or by deprotonation of a bonded water, which generally increases the rates of metal-ligand dissociation as a result of the increased charge donation from the η-OH to the metal center.

Heath et al. (1995) and Jordan et al. (1996) report that the $[Al(HEIDI)(H_2O)]_2(aq)$ dimer is stable for days to weeks in an aqueous solution after redissolution of the crystals and yields a distinct and relatively narrow peak at 27 ppm in ^{27}Al-NMR spectra. There is less information about the products of dissolution of the $Na_2Al_2(OH)_2(NTA)_2$ and $Al_2(OH)_2(OH_2)_2(IDA)_2$ crystals. Petrosyants et al. (1995) report that a fraction of the $Al_2(OH)_2(OH_2)_2(IDA)_2^o(aq)$ dimer dissociates as the crystals dissolve, and that this fraction varies with pH, but their study is cursory. Although there are several ^{27}Al-NMR studies of titrated Al-NTA solutions, there are no studies of the kinetic stability of aluminum dimers released by dissolution of $Na_2Al_2(OH)_2(NTA)_2$ crystals.

Existence of carboxylate-Al(III)-OH dimers has been inferred from a large number of very careful potentiometric studies of aluminum complexation by organic acids (e.g., oxalic acid—Sjöberg and Öhman 1985; lactic acid—Marklund and Öhman 1990; phthalic acid—Hedlund et al. 1987a; carbonic acid—Hedlund et al. 1987b). None of these postulated dimeric structures have been yet confirmed by X-ray structural study of

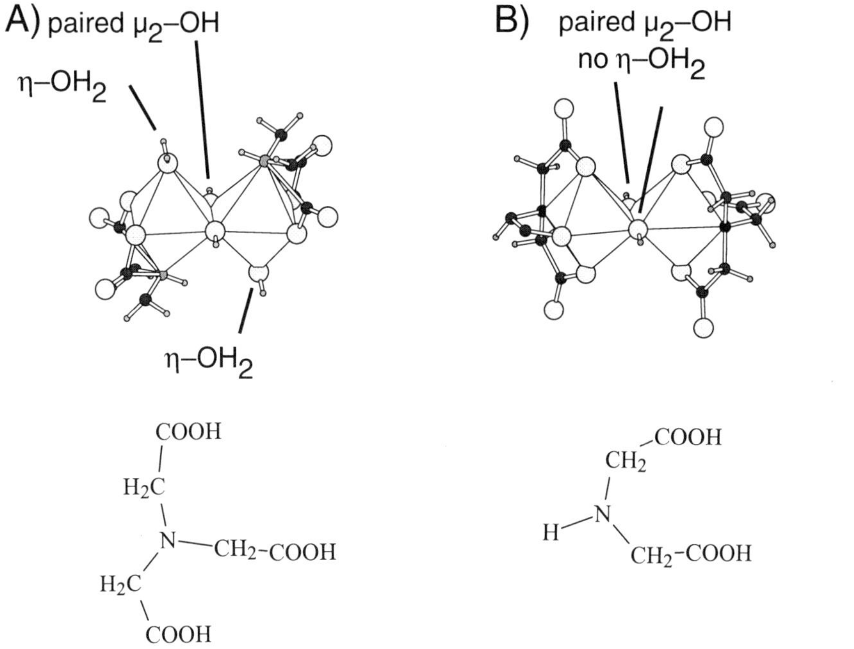

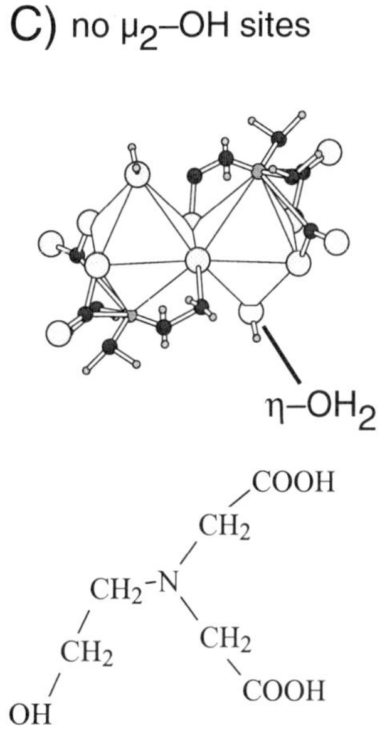

Figure 1. Aluminum aminocarboxylate dimers (ball and spoke drawings) with the ligands shown below.

(A) The aluminum-IDA dimer with stoichiometry $Al_2(\mu_2\text{-}OH)_2(\eta\text{-}OH_2)_2(IDA)_2^{\circ}$.

(B) The $Al_2(\mu_2\text{-}OH)_2(NTA)_2$ dimeric complex that exists within $Na_2Al_2(OH)_2(NTA)_2$ crystals.

(C) The aluminum dimer with HEIDI ligand has bridging *oxo* groups and no μ_2-OH bridges.

a crystal grown from these solutions. Growing aluminum-carboxylate crystals is notoriously difficult. Upon evaporation, the solutions tend to form viscous gels rather than crystals suitable for X-ray structure analysis, although some aluminum monomers have been crystallized (e.g., Bombi et al. 1990; Matzapetakis et al. 1999; Tapporo et al. 1996; Golic et al. 1989).

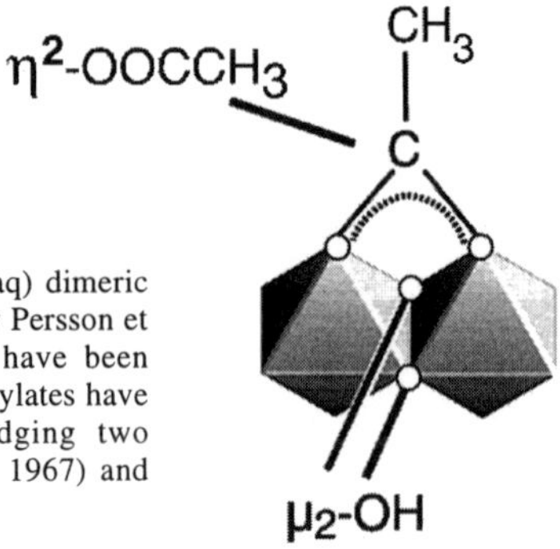

Figure 2. The structure of the $[Al_2(\mu_2\text{-}OH)_2OAc]^{3+}(aq)$ dimeric complex suggested by Öhman (1991) and discussed by Persson et al. (1998). Although no aluminum-acetate solids have been crystallized with this structure, other aluminum carboxylates have been found to have the carboxylate moiety bridging two aluminums, such as dawsonite (Frueh and Golightly 1967) and alumoxanes (Bethley et al. 1997).

Akitt et al. (1989b) examined ^1H-, ^{13}C- and ^{27}Al-NMR spectra of aluminum-acetate solutions and concluded that a dimeric complex with a ratio of OH_2:OH of 1:3 is dominant, which is consistent with the results of potentiometry (Öhman 1991). Persson et al. (1998) provide convincing evidence that the dimer has the stoichiometry: $[Al_2(\mu_2\text{-}OH)_2OAc]^{3+}$ with dihydroxy bridges and the acetate ion bridging two aluminums (Fig. 2). Öhman (1991) originally suggested the structure by analogy with the mineral dawsonite, which contains a carbonate group that bridges two Al(6) (Frueh and Golightly 1967), and this structure is consistent with the stoichiometry proposed by Akitt et al. (1989b) based on ^1H-NMR results. The bridging character of the carboxylate is consistent with other results on noncrystalline, aluminum solids (e.g., Callender et al. 1997) and other alkyl-substituted alumoxanes that can be crystallized (e.g., Koide and Barron 1995). Attempts to crystallize the acetate dimer have so far failed (Karlsson et al. 1998).

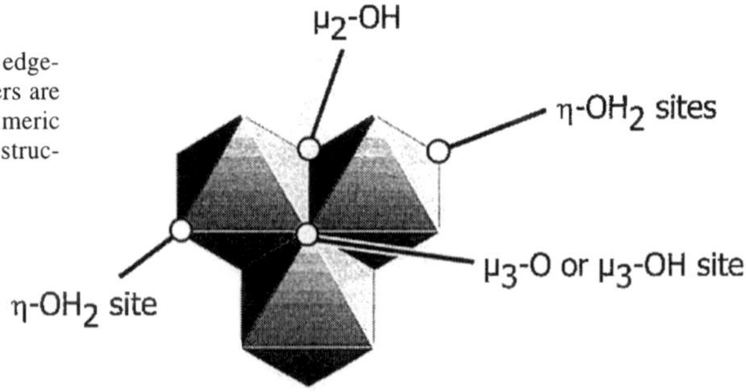

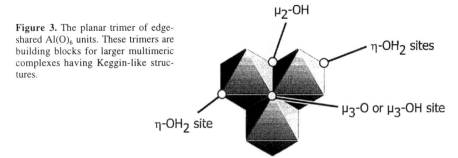

Figure 3. The planar trimer of edge-shared $Al(O)_6$ units. These trimers are building blocks for larger multimeric complexes having Keggin-like structures.

Aluminum trimers and tetramers*

Planar $^{VI}Al(III)$ trimers are likely key building blocks for the larger nanoclusters but have not been crystallized. Potentiometric studies suggest that hydrolysis of a concentrated aluminum solution creates a trimer complex (Fig. 3) having the stoichiometry $Al_3(OH)_4^{5+}$. However, crystals of a trimeric salt have never been isolated from these solutions. A ligated version of a planar trimer has been isolated and crystallized by Feng et al. (1990) using citrate as ligand (Fig. 4). Evidence from potentiometric titrations (e.g., Öhman and Forsling 1981; Hedlund et al. 1987b) indicate that the trimer complex exists in solutions at $4 < pH < 5$ but never reaches a large concentration, even in concentrated aluminum solutions, probably because they condense

* *Note added in proof:* A new multimeric Al(III) complex was synthesized by the Powell group using an organic reagent (hpdta). See Schmitt et al. (2001) *Agnew Chem In'l Ed* 40:3578-3581.

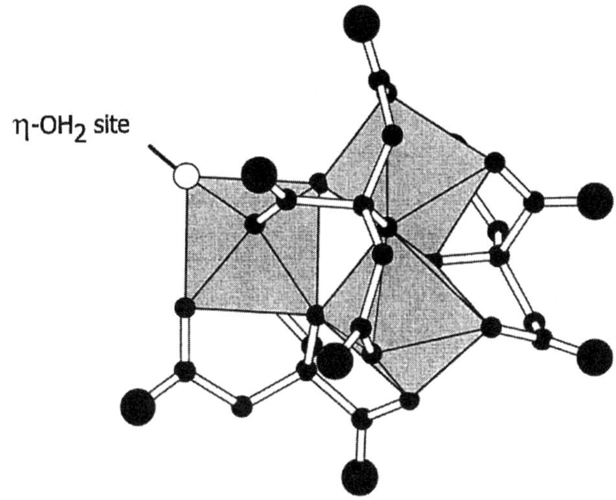

Figure 4. The structure of the aluminum citrate trimeric complex (Feng et al. 1990). There are no bridging hydroxyls and a single bound water molecule.

to form larger clusters such as the Al_{13} (see below). There is some spectroscopic hint of the presence of these trimers in concentrated Al(III) solutions as there is sometimes a relatively narrow peak in these solutions that is evident at +11 ppm, particularly for solutions in which Al_{13} is being formed or decomposed (see Figs. 8 and 10 below). Assignment of this peak to the planar trimer is suggested by the similarity of the chemical shift to that of the $Al(O)_6$ of the Al_{13} complex, which is composed of four of these trimers linked together, and the small peak width, consistent with a relatively small complex. At pH > 5, the $AlO_4Al_{12}(OH)_{24}(H_2O)_{12}{}^{7+}(aq)$ tridecamer complex (Al_{13}) dominates (see below), and at pH < 4 monomers predominate.

Baker-Figgis Keggin-like structures

The dominant multimeric aluminum complex in fresh aqueous solutions is the Al_{13} complex, which is synthesized as a relatively concentrated (≈0.1 M) aluminum solution is titrated with a base to $2.1 \leq OH/Al \leq 2.5$. This molecule has a ε-Keggin-like structure (Fig. 5) and contains a central tetrahedral $Al(O)_4$ unit surrounded by twelve $Al(O)_6$ octahedra. There are twelve η-OH_2 sites and two distinct sets of 12 μ_2-OH at the shared edges of $Al(O)_6$ octahedra. These two sets differ in their positions relative to the μ_4-O groups. One site, labeled μ_2-OH^a, lies *cis* to two μ_4-O groups. The other site, labeled μ_2-OH^b, lies *cis* to one μ_4-O site. The structure consists of four planar trimeric groups (cf. Fig. 3) linked together by the central $Al(O)_4$ site. Presence of the Al_{13} complex is easily established because it gives a distinct narrow peak at 62.5 ppm in the ^{27}Al-NMR spectra that arises from the highly symmetric $Al(O)_4$ site in the center of the complex. In solution, the $Al(O)_6$ sites give rise to a peak near +10 ppm that is very broad because of the large quadrupolar coupling constant (ca 10 MHz). Nonetheless, this resonance can be easily observed at elevated temperature, which reduces the peak width due to decreases in the rotational correlation time.

This Al_{13} molecule has been found in soils (Hunter and Ross 1991), cultivated plants (Rao and Rao 1992), in deodorants (see below) and is the essential ingredient of the aluminum reagents that are used to clarify water. The pioneering work on this molecule,

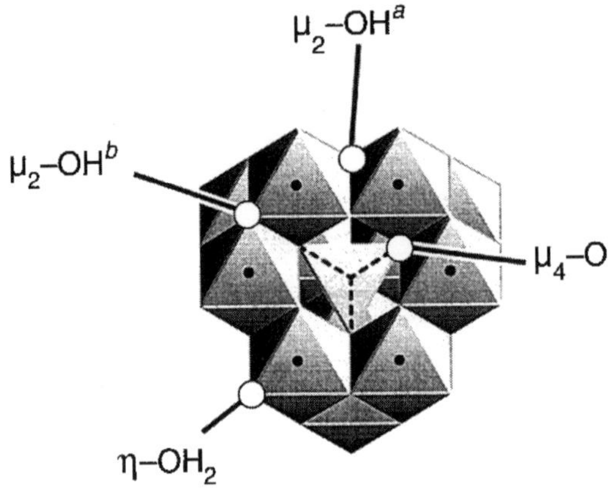

Figure 5. Polyhedral representation of the ε-Keggin-like structures of the Al_{13}, $GaAl_{12}$, or $GeAl_{12}$ molecules. The structure has a tetrahedral $M(O)_4$ unit [M = Al(III), Ga(III) or Ge(IV)] surrounded by twelve AlO_6 octahedra. There are twelve equivalent η-OH_2 sites and two distinct sets of 12 μ_2-OH at the shared edges of AlO_6 octahedra. These two sets differ in their positions relative to the μ_4-O groups. One site, labeled μ_2-OH^a, lies *cis* to two μ_4-O groups. The other site, labeled μ_2-OH^b, lies *cis* to one μ_4-O site. Dots are added to the six-membered ring of $Al(O)_6$ octahedra to make them easier to identify.

including methods of synthesis and structure, was by Johanssen (1960, 1962b,c), but the work was considerably extended by the research teams of Schönherr and Bradley in a series of remarkable papers that examine metal substitutions and polymerization (e.g., Schönherr and Görz 1983; Thomas et al. 1987; Bradley et al. 1990a,b, 1992, 1993).

A wide range of ε-Keggin molecules have been postulated to form in aqueous solutions that cannot be crystallized for X-ray structure analysis, including the corresponding complexes of Ga(III) (Bradley et al. 1990; Michot et al. 2000) and Fe(III) (Bradley and Kydd 1993). Analogues of the Al_{13} molecule with various metals substituted into the central tetrahedrally coordinated site of the Al_{13} (i.e., MAl_{12}) have also been reported, including Ge(IV) (Schönherr and Görz 1983), Mn(II) (Kudynska et al. 1993), Ga(III) (Thomas et al. 1987; Bradley et al. 1990a); and Fe(III) (Oszkó et al. 1999). Parker et al. (1997) attempted to synthesize various MAl_{12} molecules and concluded that only the substitution of Ga(III) for Al(III) in the tetrahedral site to form $GaAl_{12}$ was unequivocal (see also Nagy et al. 1995). Crystals of $Na[GaO_4Al_{12}(OH)_{24}(H_2O)_{12}(SeO_4)_4]\cdot x(H_2O)$ and $Na[AlO_4Al_{12}(OH)_{24}(H_2O)_{12}(SeO_4)_4]\cdot x(H_2O)$ can be grown by hydrolysis of $AlCl_3$+$GaCl_3$ solutions followed by addition of selenate to induce crystallization. Building on previous work by Schönherr and Görz (1983), Lee et al. (2001) synthesized a selenate salt of the Ge(IV)-substituted Keggin ($GeAl_{12}$) that has a slight structural distortion from the cubic symmetry exhibited by Al_{13} and $GaAl_{12}$. The salts of the Al_{13}, $GaAl_{12}$ and $GeAl_{12}$ Keggins dissolve to release the complexes intact, which allowed Phillips et al. (2000) to show that the kinetics of oxygen exchange between these molecules and water can be determined via ^{17}O-NMR. These exchange rates show remarkable differences, both between distinct sites on a molecule and between different molecules, which are discussed in a separate section, below.

Aging of aqueous solutions of Al_{13} molecules yields larger molecules (see the next section) that are made up of dimers of δ-Keggin structure, which have one trimeric group rotated so that its $Al(O)_6$ polyhedra share corners with the rest of the molecule, not edges as in the ε-Keggin structure. The δ-Keggin isomer of the Al_{13} molecule is a key intermediate in the formation of larger polyoxocations of aluminum (see below).

The flat aluminum tridecamers

Soil scientists have long suspected that flat aluminum polymers form on clay surfaces and interlayers (see review by Huang 1988). These nanoparticles have not been detected directly because they give no NMR signal and their existence is inferred from

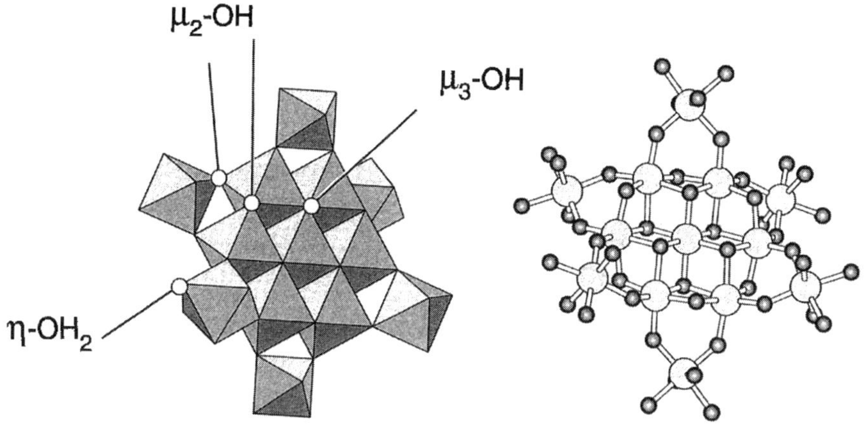

Figure 6. Polyhedral representation (left) and a ball-and-stick model (right) of the flat $Al_{13}(OH)_{24}(H_2O)_{24}$ moiety very recently synthesized by Seichter et al. (1998), Heath et al. (1995) and Karlsson (1998) independently. There are hydroxyls that bridge edge-shared $Al(O)_6$ groups [identified as μ_2-OH sites] and bridges at corner-shared $Al(O)_6$ groups [μ_2-OH' sites]. In the synthesis method of Heath et al. (1985) eighteen of the 24 water molecules are replaced by functional groups in the HEIDI carboxylate complex, leaving one set of six identical bound water molecules, two of which are identified in the figure as η-OH_2 sites.

the rate of reaction of extractable aluminum with the Ferron indicator (e.g., Wang and Hsu 1994).

Several research groups have independently synthesized a flat tridecamer complex, which consists of a central sheet of seven edge-shared $Al(O)_6$ surrounded by six corner-shared $Al(O)_6$ moieties (Fig. 6). Although this tridecamer can be crystallized as a chloride salt (Karlsson 1998; Seichter et al. 1998), it is much easier to form as a carboxylate solid where HEIDI (hydroxyethyliminodiacetic acid) ligands replace 18 of the water molecules (Heath et al. 1995; Jordan et al. 1996). Although it is easier to synthesize the tridecamer as the carboxylate salt, it is extraordinarily difficult to confirm the resulting structure because the crystals are twinned to an extremely fine scale. Active research is underway to synthesize this molecule using ligands different than the HEIDI that might eliminate the twinning problem and via hydrothermal methods (Allouche et al. 2001). A trivalent carboxylate complex of this molecule with the stoichiometry: $Al_{13}(\mu_3\text{-}OH)_6(\mu_2\text{-}OH)_{12}(H_2O)_6{\cdot}(HEIDI)_6^{3+}(aq)$ is apparently stable in an aqueous solution for days at room temperature and over a wide range in pH (Jordan et al. 1996).

Al_{30} polyoxocations

As discussed above, the presence of Al_{13} is usually identified via a +62.5 ppm peak in the ^{27}Al-NMR spectra that arises from the central $Al(O)_4$ site in the ε-Keggin-like Al_{13} structure. Fu et al. (1991) and Nazar et al. (1992) noticed that additional peaks for the $Al(O)_4$ occur in the ^{27}Al-NMR spectra when Al_{13} solutions were reacted for extended periods of time at temperatures of 85-90°C and polymerized by titration with base. Peaks in the ^{27}Al-NMR spectra appear at +64.5, +70.2, and +75.6 ppm as a peak near 0 ppm increases in intensity. The 0 ppm peak arises from the monomeric aluminum hydrolysis complexes, mostly Al^{3+} + $Al(OH)^{2+}$ at $4 < pH < 5$. The peak at +64.5 ppm is weak and transient and that at 75.6 ppm only appears after extended periods of reaction. The dominant peak is at 70.2 ppm.

The authors interpreted their results to indicate release of aluminum monomers as the Al_{13} complexes polymerize to form larger complexes. Because the stoichiometries and structures were not then known, these molecules are referred to as AlP_1 (+64.5 ppm), AlP_2 (+70.2 ppm) and AlP_3 (+76.5 ppm), based on their observable Al(4) NMR signals. Because the resonance at +64.5 ppm appears, then disappears as those at +70.2 ppm and 0 ppm grow, the authors concluded that the AlP_1 complex is a transient intermediate and that the AlP_2 complex was more stable. The AlP_1, AlP_2 and AlP_3 complexes could be isolated using gel permeation.

Two research groups (Allouche et al. 2000, 2001; Rowsell and Nazar 2000) recently crystallized a large polyoxocation that, when dissolved, gives ^{27}Al-NMR peaks identical to those of the AlP_2 complex and similar gel permeation elution. The structure comprises a dimer of two Al_{13} complexes that are linked together (Fig. 7), but one trimeric group in each Al_{13} is rotated 60° (to form the δ-Keggin of Al_{13} isomer) so that they no longer

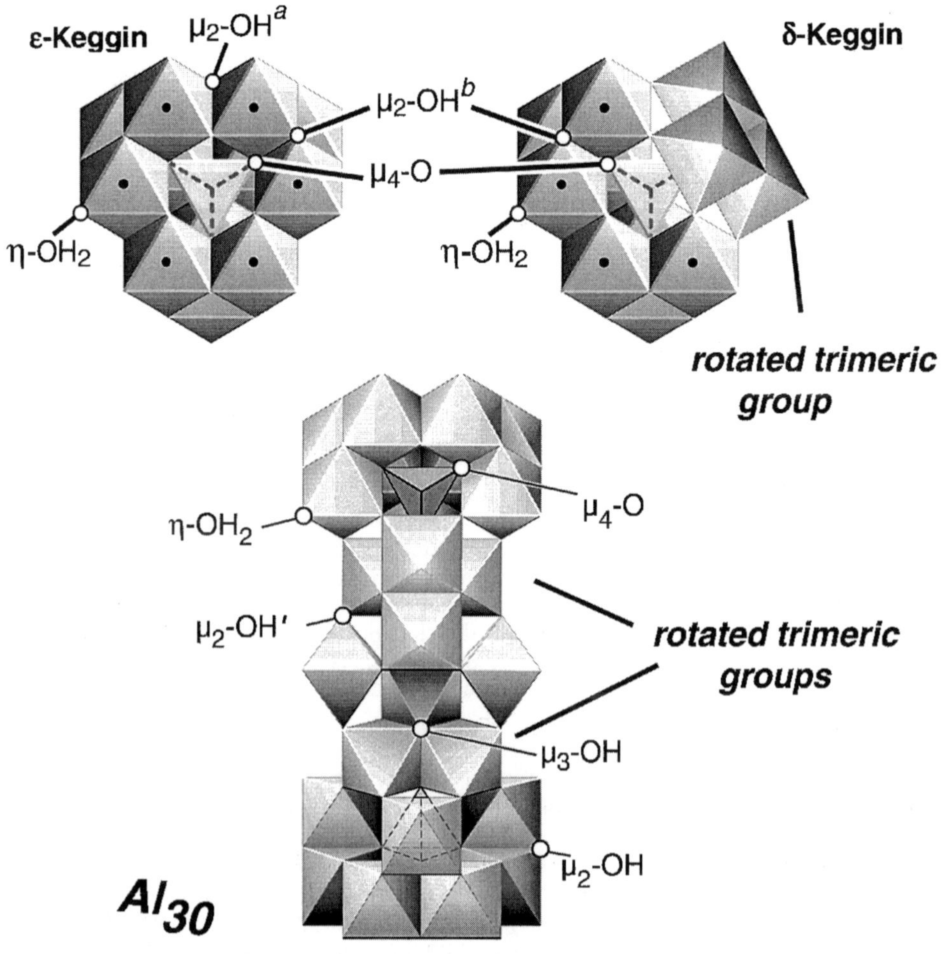

Figure 7. The ε-Keggin (top, left) and the δ-Keggin isomers (top, right) of the Al_{13} complex in polyhedral representation. The ε- and δ-Keggin isomers differ by rotation of a single trimeric group of $Al(O)_6$ octahedra by 60°. These combine to form the Al_{30} structure (bottom), which consists of two δ- isomers linked by four AlO_6 octahedra.

share edges with the other trimeric groups. The stoichiometry of the complex is: $Al_{30}O_8(OH)_{56}(H_2O)_{24}^{18+}$(aq) and, although it is yet unappreciated by geochemists, it may be an enormously useful model for the surface sites on minerals. The molecule is approximately 2 nm in length, which places it close to the size of some oxide colloids.

This Al_{30} complex can be easily formed by heating a solution of Al_{13} for a few days at 85°C or by storing a stock Al_{13} solution for a decade or so. In Figure 8 the ^{27}Al-NMR spectra of 11-year-old solutions of Al_{13} that had aged at room temperature to form minor (bottom) or major (top) amounts of gibbsite, which was periodically removed by filtration. There are two important points to note in this figure. First, the Al_{13} had reacted to form the larger Al_{30} complex, which has a small broad peak near +71 ppm, and secondly, that the $Al(OH)_3$ solid formed primarily at the expense of the Al_{13} and the Al^{3+} monomers. When interpreting these ^{27}Al-NMR spectra it is important to remember that only the $Al(O)_4$ sites are easily resolved, whereas the $Al(O)_6$ sites in the large complexes give broad overlapping peaks. In the case of the Al_{13}, the peak at +62.5 ppm corresponds to only 1/13 of the total aluminum and for Al_{30} the peak at +71 ppm corresponds to 1/15 of the total aluminum. The +71 ppm signal is broad because the symmetry of the coordination sphere of the $Al(O)_4$ in the Al_{30} is lower than in the Al_{13} polymer.

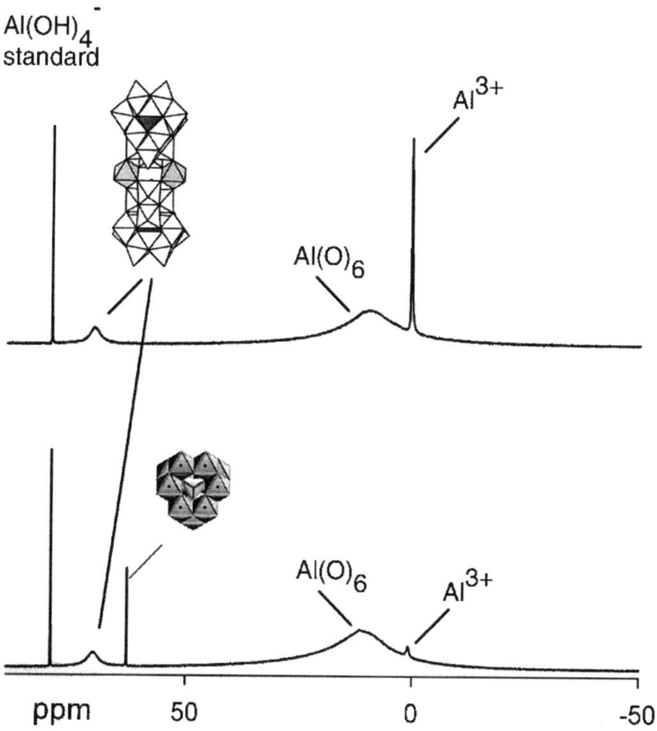

Figure 8. ^{27}Al-NMR spectra of two 11-year-old stock solutions originally synthesized with 7.8 mM Al_{13} (ΣAl = 0.1 M) and stored at room temperature. The solutions had a final pH value of 4.1 (top spectrum) and 4.7 (bottom spectrum). Aluminum monomers have a sharp peak near 0 ppm, the Al_{13} exhibits a sharp peak near 63 ppm, the Al_{30} ($Al_{30}O_8(OH)_{56}(H_2O)_{24}^{18+}$) has a relatively broad peak near 70 ppm. The broad peak denoted $Al(O)_6$ corresponds to octahedrally coordinated Al(III) in the multimers. The top spectrum corresponds to a sample that precipitated large amounts of solid which was predominately gibbsite. The spectra were collected at 70°C. The peak at 80 ppm corresponds to $Al(OH)_4^-$ in an external standard.

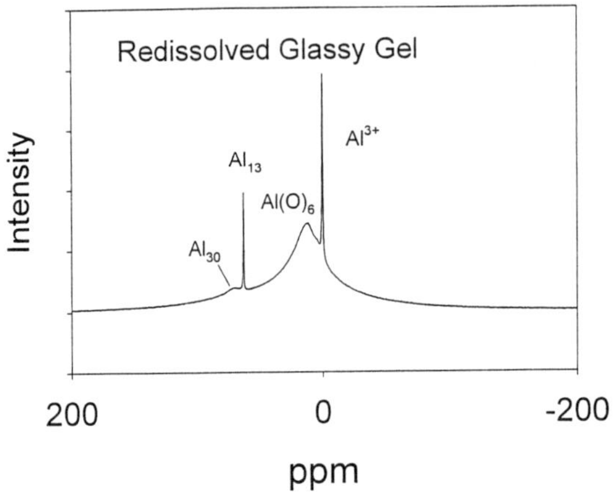

Figure 9. ^{27}Al-NMR spectra of a concentrated solution of aluminum chlorhydrate ($Al_2(OH)_5Cl$). Note the similarity of this spectrum with those in Figure 8. Aluminum monomers have a sharp peak near 0 ppm, the Al_{13} exhibits a sharp peak near 63 ppm, the Al_{30} ($Al_{30}O_8(OH)_{56}(H_2O)_{24}{}^{18+}$) has a relatively broad peak near 70 ppm. The broad peak denoted $Al(O)_6$ corresponds to octahedrally coordinated Al(III) in these multimers.

These large nanoclusters of aluminum form readily in an aqueous solution, even at room temperature, and provide useful models for the surface sites on aluminum (hydr)oxide minerals, as discussed below. They also are probably much more common outside the laboratory, as suggested by the ^{27}Al-NMR spectrum shown in Figure 9, which corresponds to a solution made with commercial aluminum chlorhydrate [$Al_2(OH)_5Cl$], which is a principal ingredient in antiperspirant. Comparison with the two spectra shown indicates that the commercial aluminum chlorhydrate contains a high concentration of both the Al_{13} and Al_{30} species. These molecules prevent perspiration by hydrolyzing in skin pores to form aluminum hydroxide solids (Fitzgerald and Rosenberg 1999). Other types of polymers are created by adding sulfate or silicate to a concentrated aqueous solution, but these are not easily characterized via either NMR or single-crystal XRD methods.

REACTIVITIES OF KEGGIN-LIKE ALUMINUM POLYOXOCATIONS

Most of the reactions that concern Earth scientists are ligand-exchange reactions, where one ligand in the inner-coordination sphere of a metal is replaced with another. This applies also for mineral dissolution and adsorption reactions at mineral surfaces. The Al_{13} molecule has been used by several research groups as a model for the surfaces of aluminum (hydr)oxide phases because it contains structural environments that resemble those found at common mineral surfaces, including the surface of γ-Al_2O_3 or the rings of $Al(O)_6$ octahedra on the basal surfaces of $Al(OH)_3$ solids or dioctahedral clays. This point was made explicitly by Wehrli et al. (1990), Bradley et al. (1993) and Casey et al. (2000). Furthermore, the surface chemistry of the Al_{13} complex itself may be important to environmental sciences. Furrer (1993) and Lothenbach et al. (1997) have studied the adsorptive properties of the Al_{13} molecule and show that it forms complexes with several heavy metals, including Pb(II). These complexes may be a key vector for dispersing heavy metals into the environmental at metal-polluted sites. Such ternary complexes may condense into transportable colloids as the acidic- and aluminum-rich solutions are diluted with unpolluted waters.

Rates of oxygen-isotopic exchange in ε-Keggin molecules

There are four types of oxygens in the topology of the Al_{13}, $GaAl_{12}$, and $GeAl_{12}$ molecules (Fig. 5). Phillips et al. (2000), building on earlier work by Thompson et al. (1987), showed that the rate of exchange of three of these sets of oxygens with solvent waters could be determined by ^{17}O-NMR on the Al_{13} complex, leading to a series of papers on this molecule and the MAl_{12} substituted versions. These oxygen exchange reactions are the elementary, or nearly so, steps that underlie geochemically important processes such as adsorption, polymerization, and dissolution. We summarize the key results in this section.

Bound water molecules (η-OH$_2$). The rates of exchange of oxygen between η-OH_2 sites and bulk solution can be measured using the ^{17}O-NMR line-broadening method (Swift and Connick 1962) that has also been used to determine the rates of solvolysis of aluminum monomer complexes. All of the direct measurements of rates of dissociation of bound waters for complexes of aluminum are compiled in Table 1. The data for exchange of bound waters from the $GeAl_{12}$ molecule (from Lee et al. 2001b) is only approximate because the reaction rate contains a small pH-dependence that has not yet be resolved and probably results from the presence of a partly deprotonated species. The $GeAl_{12}$ molecule is a stronger Brønsted acid than either the Al_{13} or $GaAl_{12}$. In similar chloride solutions, the conditional pK_a for dissociating protons from η-OH_2 sites on the $GeAl_{12}$ is almost 0.5 units smaller than for the Al_{13} or $GaAl_{12}$ (Lee et al., in preparation).

Several important features are evident in the rates of these reactions. Most importantly, rates of exchange of η-OH_2 sites from the ε-Keggin molecules fall in the

Table 1. A compilation of rate coefficients and activation parameters for exchange of water molecules from the inner-coordination sphere of Al(III) complexes to the bulk solution, as determined from ^{17}O-NMR. Included in these data are estimates for the substituted ε-Keggin molecules.

Species	k_{ex}^{298} (s^{-1}) (±1σ)	$\Delta H^{\ddagger}$ ($kJ \cdot mol^{-1}$)	$\Delta S^{\ddagger}$ ($J \cdot K^{-1} \cdot mol^{-1}$)	*Source*
		Monomeric complexes		
$Al(H_2O)_6^{+3}$	1.29 (±.03)	85 (±3)	42 (±9)	Hugi-Cleary et al. (1985)
$Al(H_2O)_5OH^{2+}$	31000 (±7750)	36 (±5)	-36 (±15)	Nordin et al. (1999)
$AlF(H_2O)_5^{2+}$	240 (±34)	79 (±3)	17 (±10)	Yu et al. (2001)
$AlF_2(H_2O)_4^{+}$	16500 (±980)	65 (±2)	53 (±6)	Yu et al. (2001)
$Al(ssal)^{+}$	3000 (±240)	37 (±3)	-54 (±9)	Sullivan et al. (1999)
$Al(sal)^{+}$	4900 (±340)	35 (±3)	-57 (±11)	Sullivan et al. (1999)
$Al(mMal)^{+}$	660 (±120)	66 (±1)	31 (±2)	Casey et al. (1998)
$Al(mMal)_2^{-}$	6900 (±140)	55 (±3)	13 (±11)	Casey et al. (1998)
$Al(ox)^{+}$	109 (±14)	69 (±2)	25 (±7)	Phillips et al. (1997b)
		Multimeric complexes		
Al_{13}	1100 (±100)	53 (±12)	-7 (±25)	Phillips et al. (2000); Casey et al. (2001)
$GaAl_{12}$	227 (±43)	63 (±7)	29 (±21)	Casey & Phillips (2001)
$GeAl_{12}$	190 (±43)	56 (±7)	20 (±21)	Lee et al. (2001b)

Abbreviations: ox = oxalate; ssal = sulfosalicylate; sal = salicylate; mMal = methylmalonate; $Al_{13} = AlO_4Al_{12}(OH)_{24}(H_2O)_{12}^{7+}(aq)$; $GaAl_{12} = GaO_4Al_{12}(OH)_{24}(H_2O)_{12}^{7+}(aq)$; $GeAl_{12} = GeO_4Al_{12}(OH)_{24}(H_2O)_{12}^{8+}(aq)$

same general range as those of hydration waters from the dissolved monomers, which range from 1 to 30,000 s^{-1}. This point is important if the large polyoxocations are regarded as models for mineral surfaces because it means that rates of dissociation of water molecules from fully protonated aluminum (hydr)oxide minerals will probably also fall into this range. Although the rates still span a range of 10^4, one needs to remember that the entire range of dissociation rates for different metals spans a range larger than 10^{15} (Richens 1997). Furthermore, the important range of reaction rates is probably closer to 10^2 (100-30,000 s^{-1}) if one excludes the highly charged Al^{3+} complex.

The similarity in reaction rates for waters bound at monomers and nanoclusters is important because many ligand-exchange reactions involving Al(III) in neutral to acidic solutions are dissociative; that is, the incoming new ligand has little influence over the reaction rate and the rate of the overall process is controlled by dissociation of a Al-O bond. In acidic solutions the Al(III) is coordinatively saturated. Thus one expects the rates of many ligand-exchange processes at aluminum nanoclusters to fall in the same overall range as rates of loss of hydration waters from monomeric species. The rates can be greatly enhanced by deprotonation reactions (cf. the rates of exchange for the $Al(OH)^{2+}$ complex with the isoelectronic AlF^{2+}) but the same processes that labilize the bound waters in monomeric complexes also affect larger molecules and perhaps even mineral surfaces. In basic solutions, associative reactions may become important because of the reduced coordination number of the Al(III) center. This may explain the apparently very rapid rate of OH^- exchange from $Al(OH)_4^-(aq)$, which is much too fast to measure via NMR.

Bridging hydroxyls (μ_2-OH). The rates of oxygen exchange between the μ_2-OH bridges and solvent waters has also been determined for the ε-Keggin molecules using an isotopic equilibration technique and ^{17}O-NMR for detection. In these experiments, crystalline selenate salts of the MAl_{12} (M=Al(III), Ga(III), Ge(IV)) Keggin were redissolved in ^{17}O-enriched water. The crystals were synthesized from solutions that had a natural abundance of $H_2{}^{17}O$. The ^{17}O-NMR signal from the freshly dissolved molecules is too low to observe until they exchanged oxygen with $H_2{}^{17}O$ in the bulk solution. These experiments demonstrate that the average lifetime for an μ_2-OH bridge at 298 K varies from less than a minute to several weeks and is much shorter than the lifetime of the complex in solution. The exchange rates are very strongly affected by metal substitutions in the center of the nanoclusters (Table 2). These metal substitutions are three bonds away from the oxygens that exchange with the bulk solution and do not significantly change the Al-O bond lengths for the μ_2-OH bridges. The reactivity of these μ_2-OH bridges do appear to reflect changes in bonding caused by mismatch of the central tetrahedral metal sites with the rest of the molecule (compare bond lengths in Tables 3 and 4).

There are several important features to note about the kinetics of isotopic exchange of the μ_2-OH bridges in the Al_{13}, $GaAl_{12}$ and $GeAl_{12}$ molecules.

(1) Within a given molecule, the two topologically distinct μ_2-OH bridges (labeled μ_2-OH^a and μ_2-OH^b in Fig. 5) react at rates that differ considerably. Characteristic times for exchange of the two μ_2-OH sites in the $GaAl_{12}$ molecule at 298K are $\tau_{298} \approx 15.5$ and $\tau_{298} \approx 680$ h, respectively, whereas the corresponding times are $\tau_{298} \approx 1$ minute and $\tau_{298} \approx 17$ h for the Al_{13} molecule (Phillips et al. 2000; Casey et al. 2000). One of the μ_2-OH bridges in the $GeAl_{12}$ molecule exchanges at rates too fast to measure (Lee et al. 2001b) but the other exhibits $\tau_{298} \approx 25$ min. Extrapolated to 298K, the observed rates of exchange of μ_2-OH sites in these molecules spans a factor of about 10^5 and is probably much larger if the more reactive set of μ_2-OH sites in the $GeAl_{12}$ molecule are included.

Table 2. Rates of exchange of oxygens between bulk solution and μ_2-OH bridges in the Al_{13}, $GaAl_{12}$ and $GeAl_{12}$ molecules. The two sets of μ_2-OH sites are identified as μ_2-OH^{fast} and μ_2-OH^{slow} because reactivities cannot yet be assigned to the structural positions μ_2-OH^a and μ_2-OH^b indicated in Figure 5.

Molecule		$\Delta H^{\ddagger}$	$\Delta S^{\ddagger}$	*Sources*
	(s^{-1})	($kJ \cdot mol^{-1}$)	($kJ \cdot mol^{-1} \cdot K^{-1}$)	
		Al_{13}		(1),(2)
μ_2-OH^{fast}	$1.6(\pm 0.4) \cdot 10^{-2}$	204(±12)	403(±43)	
μ_2-OH^{slow}	$1.6(\pm 0.1) \cdot 10^{-5}$	104(±20)	5(±4)	
		$GaAl_{12}$		(3)
μ_2-OH^{fast}	$1.8(\pm 0.08) \cdot 10^{-5}$	98(±2.6)	-8(±8.5)	
μ_2-OH^{slow}	$4.1(\pm 0.2) \cdot 10^{-7}$	125(±3.7)	54(±12)	
		$GeAl_{12}$		(4)
μ_2-OH^{fast}		(complete in minutes or less)		
μ_2-OH^{slow}	$6.6(\pm 0.2) \cdot 10^{-4}$	82(±2)	-29(±7)	

(1) Phillips et al. 2000 (2) Casey et al. 2000 (3) Casey & Phillips 2001 (4) Lee et al. 2001

Table 3. Structural data for Al_{13}, $GaAl_{12}$ and $GeAl_{12}$ selenate or sulfate crystals. The subscript abbreviations are T = tetrahedral; O = octahedral. The standard deviation for each bond parameter, when available, is given in parentheses and corresponds to the last place in the value. Data for Al_{13} and $GaAl_{12}$ are from Parker et al. (1997) and data for the $GeAl_{12}$ are from Lee et al. (2001). A range of values is given for the μ_2-OH^a and μ_2-OH^b sites in the $GeAl_{12}$ molecule because two crystallographically distinct octahedral aluminum sites appear in the structure of this crystal, but not in the Al_{13} or $GaAl_{12}$ salts.

Moiety	*Bond length* (Å)			*Structural site*
	M=Al(III)	*M=Ga(III)*	*M=Ge(IV)*	
M_T-O	1.831(4)	1.879(5)	1.809(8)	μ_4-O
Al_o-O	2.026(4)	2.009(6)	2.103	μ_4-O
Al_o-O	1.857(6)	1.852(6)	1.843-1.854	μ_2-OH^a
Al_o-O	1.857(6)	1.869(7)	1.869-1.841	μ_2-OH^b
Al_o-O	1.961(4)	1.962(6)	1.928	η-OH_2 (bound water)

(2) The rates of exchange of the two μ_2-OH sites in the $GaAl_{12}$ and Al_{13} molecules do not depend on solution pH over the relatively narrow experimental range ($4.1 < pH < 5.4$). The absence of a first-order dependence on proton concentration for oxygen exchange of μ_2-OH sites in the $GaAl_{12}$ and Al_{13} molecules helps to identify the mechanism of dissolution of these molecules (see below). The rates of exchange of the one measurable μ_2-OH site in the $GeAl_{12}$ molecule exhibits a near-first-order dependence on solution pH.

Table 4. Bond lengths for $Al(O)_4$, $Ga(O)_4$, $Ge(O)_4$ in oxides and in the MAl_{12} Keggins.

Site	*d(M-O)* Å	*Phase*	*Source*
Al	1.831(4)	Al_{13}	Parker et al. (1997)
	1.77	α-Al_2O_3	Ollivier et al. (1997)
	1.777	γ-Al_2O_3	Zhou & Snyder (1991)
	1.761	$Y_3Al_5O_{12}$ garnet	Euler and Brace (1965)
Ga	1.879(5)	$GaAl_{12}$	Parker et al. (1997)
	1.849	$Y_3Ga_5O_{12}$ garnet	Euler & Brace (1965)
	1.83-1.863	β-Ga_2O_3	Aahman et al. (1996)
Ge	1.809(8)	$GeAl_{12}$	Lee et al. (2001a)
	1.751(4)	GeO_2 quartz structure	Yoshiasa et al. (1999)

(3) The activation energies for oxygen exchange are similar to those determined for exchange of μ_2-OH sites in other inert-metal complexes (e.g., Springborg 1988) and are much higher than the activation energies for dissolution of the molecule (see below).

(4) The reactive and less-reactive μ_2-OH sites cannot yet be assigned to the structurally distinct μ_2-OH^a and μ_2-OH^b sites in any of the ε-Keggin structures.

(5) The two μ_2-OH sites are *generally* less labile in the $GaAl_{12}$ complex than in Al_{13}. Furthermore, the decrease in reactivity of the two μ_2-OH sites is not uniform as Ga(III) substitutes for Al(III) in the central tetrahedral site. The labilities of the two μ_2-OH sites within each molecule is reduced considerably when Ga(III) is substituted for Al(III) in the structure. For the Al_{13} complex: $\frac{\tau_{298}^{fast}}{\tau_{298}^{slow}} \approx 10^3$ but for the $GaAl_{12}$ molecule: $\frac{\tau_{298}^{fast}}{\tau_{298}^{slow}} \approx 44$. (The ratio cannot be determined for the $GeAl_{12}$ molecule since τ_{298}^{fast} is unknown.)

(6) Bonding between the central tetrahedral metal (Al(III), Ga(III), or Ge(IV)) controls the reactivity of these bridging hydroxyls. Lee et al. (2001b) pointed out that the <Ga-O> bond lengths in the least reactive molecule ($GaAl_{12}$) are closest to those observed for $Ga(O)_4$ in other oxides. The Al-O and Ge-O bond lengths in the $Al(O)_4$ and $Ge(O)_4$ sites of the Al_{13} and $GeAl_{12}$ molecules are larger than in crystalline solids that contain these metals in tetrahedral coordination to oxygens (Table 3).

(7) The mechanism of isotopic oxygen between the μ_2-OH sites and solution is not known with certainty, but probably involves protonation, exchange of the bridging water molecule with a bulk water, followed by deprotonation. For pH-independent pathways (Al_{13}, $GaAl_{12}$) the proton transfer may be internal to the molecule but for the $GeAl_{12}$, which exhibits a pH dependence to oxygen exchange, proton transfer is from the bulk solution.

The μ_4-O sites. For the Al_{13}, $GaAl_{12}$ and $GeAl_{12}$ ε-Keggin molecules there is no evidence for exchange of oxygen between the μ_4-O sites and solution unless the molecule is dissolved and then reassembled. The μ_4-O gives a distinct ^{17}O-NMR signal that is not observed if crystals of isotopically normal MAl_{12} Keggins are dissolved in ^{17}O-enriched water, unless the pH is lowered by acid addition so that some of the molecules dissolve and reform. This point is key for understanding the mechanisms of dissolution since all oxygens on the molecule except those in the μ_4-O sites exchange more rapidly than the molecule dissolves.

Decomposition of the ε-Keggin Molecules

The rate law for proton-promoted dissolution of the Al_{13} molecule has been determined through experiments by Furrer and colleagues (Wehrli et al. 1990; Furrer et al. 1999; Amirbahman et al. 2000), but no data yet exist for the $GaAl_{12}$ or $GeAl_{12}$ molecules. These data allow geochemists to compare dissolution of the molecule with the rates of dissociation of some of the Al-O bonds, which cannot be done with any other system.

There are several important differences between the rates of dissolution of the Al_{13} molecule and the rates of Al-O bond dissociation inferred from oxygen-exchange rates discussed above.

(1) The dissolution rate of the Al_{13} molecule exhibits pH regions of first order and, at pH < 2.5, second order, dependencies on dissolved proton concentration. In the pH range where ^{17}O exchange between bulk water and Al_{13} can be measured ($4.6 < pH < 5.4$), the rate of proton-promoted Al_{13} decomposition is first order in proton concentration, while the rates of exchange of the μ_2-OH sites are independent of pH.

(2) The activation energies for dissolution of Al_{13} depend on the solution pH and are much smaller than the activation energies determined for oxygen exchange at its μ_2-OH sites. For dissolution the experimental activation energies vary from 31.4 $kJ{\cdot}mol^{-1}$ at pH = 2.4 to 16 $kJ{\cdot}mol^{-1}$ at pH = 3.4 (Furrer et al. 1999), whereas the activation enthalpy for exchange of the more-labile μ_2-OH site in the Al_{13} molecule is: 204(±12) $kJ{\cdot}mol^{-1}$ and that for the less-labile site: 104(±20) $kJ{\cdot}mol^{-1}$ (Phillips et al. 2000; Casey et al. 2000). Activation energies of ca. 100 $kJ{\cdot}mol^{-1}$ are typical for hydroxo bridge dissociation in other metal complexes (Springborg 1988), which suggests that equilibrium protonation is contributing much enthalpy to the dissolution reaction.

(3) Cleavage of the μ_2-OH sites cannot control the overall reaction rate in dissolving the molecule because the μ_2-OH sites exchange many times before the Al_{13} molecule dissolves. Near pH 5 and 298 K, the average lifetime of the complex is estimated at ~600 h by extending the dissolution data of Furrer et al. (1999), whereas the lifetime for the less-labile μ_2-OH is about 17 h at the same conditions.

A more likely possibility is that protonation of the μ_4-O atoms is the rate-limiting step in Al_{13} decomposition. Since no intermediate products have been identified so far, we can only speculate about the actual decomposition mechanism. The proton-promoted decomposition might involve proton diffusion via the outer oxygens in the molecule, such as the μ_2-OH bridges, to one of the central μ_4-O atoms. Such a mechanism is supported by studies of proton mobility in γ-Al_2O_3 solids (Sohlberg et al. 1999). A mechanism involving proton transfer from the μ_2-OH bridges to the μ_4-O atoms is consistent with the observation that the μ_4-O are not exchanged in the intact molecule.

One suspects that the mechanisms of dissolution may vary across the series of ε-Keggin molecules. The oxygens on the $GaAl_{12}$ molecule are generally much less labile than those of the Al_{13}, and the Al_{30} polyoxocation appears to be more stable thermodynamically than the Al_{13}. One wonders whether their dissolution rates are also slower. The most interesting of these ε-Keggin molecules may be the $GeAl_{12}$, the oxygens of which are much more labile than those of either the Al_{13} or $GaAl_{12}$. The $GeAl_{12}$ molecule also exhibits a pH dependence to the rates of exchange of the μ_2-OH sites, a pH dependence to the rates of exchange of η-OH_2 sites, and a μ_4-O site that is even more highly overbonded (formal charge of +0.5 when unprotonated). A comparison of dissolution of the $GeAl_{12}$ molecule to the rates of oxygen exchange may be very fruitful for understanding mineral dissolution.

Mechanisms of formation of the ε-Keggin molecules

The ε-Keggin molecules can be viewed as a set of four $[^{VI}Al_3(OH)_4]^{5+}$ planar trimers that are bonded to a common $M(O)_4$ (Fig. 5). The fact that the Al_{13} complex forms so readily during a titration suggests that it assembles from molecules that already exist in solution and that contain relatively highly coordinated oxygens. Thus, most authors conclude that the planar trimers are the fundamental building blocks for the ε-Keggin molecules. However, one of the pressing problems in aqueous aluminum chemistry is the unequivocal detection of these trimers in solution since the ^{27}Al-NMR data are not clear.

Nevertheless, there are two schools of thought about the pathways to assemble these trimers into the intact molecule:

(1) In one model, the trimers condense around an Al(III) monomer that is initially hexacoordinated, but that loses coordinated oxygens as the molecule forms (see Jolivet 1994; Jolivet et al. 2000). Michot et al. (2000) used a variation of this pathway to interpret spectroscopic data on the growth of the $GaO_4Ga_{12}(OH)_{24}(H_2O)_{12}^{7+}(aq)$ (Ga_{13}) complex. They concluded that the molecule forms from condensation of several planar trimeric $Ga_3(OH)_4^{5+}(aq)$ groups around a metastable tetrameter (Fig. 10). The exact stoichiometry of the tetramer is unclear but there are two candidates: (1) a molecule containing a planar trimer with a single edge-sharing $M(O)_6$ group that is offset from the plane of the trimer; or (2) a tetramer with a single corner-shared $M(O)_6$ group. The latter tetramer has been detected in solution but the former tetramer already contains a μ_4-O site. In this tetramer the offset $Al(O)_6$ can easily form the core of the ε-Keggin molecule by loss of two bound waters and by bonding to the incoming trimers (Fig. 10).

(2) A second model invokes the presence of excess $Al(OH)_4^-$ ion in solution during synthesis (e.g., Akitt and Farthing 1981; Bertsch 1987; Kloprogge et al. 1992) as a source of $Al(O)_4$. In this model, the trimers condense around the $Al(OH)_4^-$ monomer created locally at the site of poor titrant mixing. Both of these models are reasonable and are concerned with a detail in the pathway. The most important implication of Michot et al. (2000) is that multimers form quickly in an aqueous solution. Molecules such as the Al_{13}, $GaAl_{12}$ and $GeAl_{12}$ probably form within seconds to minutes at the appropriate conditions of pH and total metal concentration.

Mechanisms of formation of the Al_{30} molecules

The pathways for forming the Al_{30} molecule are less-well known than for the ε-Keggin molecules, but it is clear that the first step is formation of a small metastable concentrations of δ-Keggin molecule. It is likely that these molecules form with increased temperature (or long periods of time) and link together via four monomeric $Al(H_2O)_6^{3+}$ groups (or hydrolyzed species) to form the large $Al_{30}O_8(OH)_{56}(H_2O)_{26}^{18+}(aq)$ complex, which has enhanced thermal stability.

CATALYSIS BY OXIDE NANOCLUSTERS

By definition a *catalyst* accelerates the rate of a reaction but is not consumed by it. Many aqueous oxide surfaces have catalytic properties and small nanoclusters are particularly effective because they have so many of their atoms very near the interface with water. Therefore a gram of nanocluster may be many times more effective than a

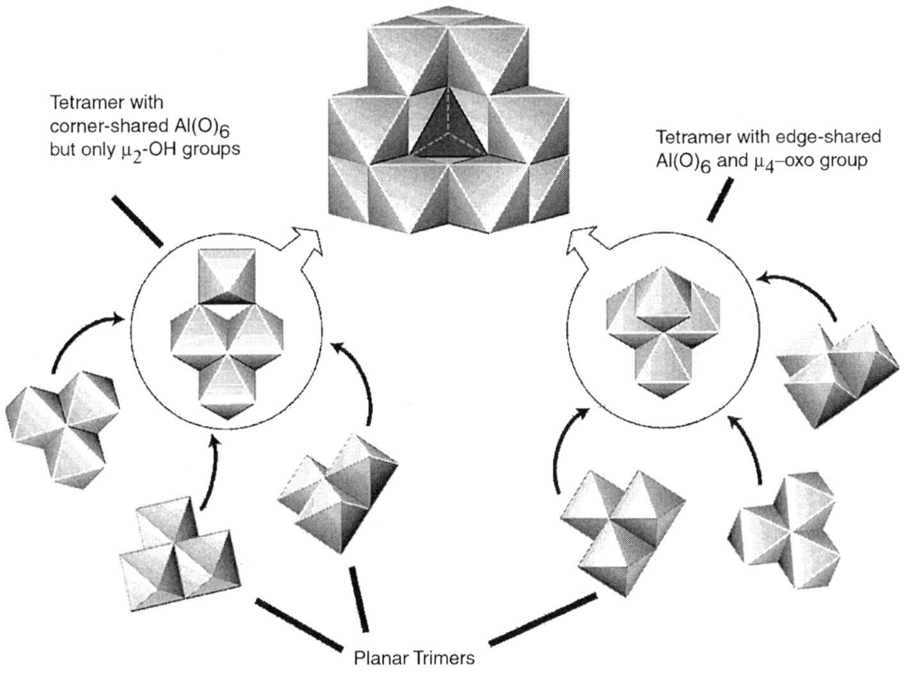

Figure 10. Two pathways proposed to form the Keggin-like molecules (e.g. Michot et al. 2001). The key reactants are sets of planar trimers that react with a yet-unidentified tetramer to form the ε-Keggin structure. One tetramer consists of a corner-shared $M(O)_6$ octahedra bonded to a planar trimer; this tetramer is detected in EXAFS spectra (from Michot et al. 2001). The other tetramer has a $M(O)_6$ group that shares edges with a planar trimer. This tetramer is yet undetected spectroscopically, but contains a μ_4-O group that may facilitate the formation of the ε-Keggin molecule since the $M(O)_6$ group can become the central $M(O)_4$ core of the ε-Keggin molecule by loss of two bound waters and rotation.

gram of coarser material. The subject of catalysis is immense and will be covered in detail by Anastasio and Martin later in this volume (Chapter 8). Some of the catalytic features of oxides are enumerated below.

(1) Nanoclusters provide sites of Brønsted acidity that accelerate rates of proton-catalyzed reactions. Clays have environments of very strong acidity in the interlayers while the surfaces of most oxide minerals are weak acids. An open question is the extent to which nanoparticles exhibit different Brønsted acidities than the corresponding bulk phases. On one hand, nanoclusters have very high surface areas so that the near-surface metal-oxygen bonds are probably strained, which will be manifested in the proton affinities of the oxygens. One the other hand, the ε-Keggin molecules exhibit Brønsted acidities that are similar to aluminum (hydr)oxide colloids, suggesting that acidity does not change with size.

(2) Metals at nanocluster surfaces can coordinate to functional groups in an adsorbate molecule and labilize them. By donating charge, surface metals on a nanocluster modify the susceptibility (or Brønsted acidity) of key atoms in the functional group. For the reaction to be catalytic, the hydrolyzed adsorbate must then detach from the nanoparticle surface.

(3) Metals at the nanoparticle surface can accelerate the exchange of ligands from an adsorbate, so that the adsorbate then has a temporarily unsaturated metal center. These unsaturated metal centers are very reactive and can bond to other ligands. Again, the adsorbate must detach to recycle the catalytic site on the nanocluster.

(4) For clusters containing transition metals, there are closely spaced electron orbitals that can assist in electron-redistribution or electron-exchange reactions that involve excited states or slightly varying electron configurations. Stated differently, the LUMO orbitals in transition metal complexes are sufficiently close to HOMO orbitals that they can store electrons at low energy cost. Thus, metal centers on nanoclusters can accelerate reactions that involve electron redistribution.

(5) Some nanoclusters, particularly porous nanoclusters, confine reactants into close proximity and accelerate reaction rates via mass action.

For geochemists, a distinction must be made between proton *catalysis* and proton *induction*. Many reactions of geochemical interest involve the cleavage of a metal-oxide bond and these are commonly enhanced by protons. An example is provided by the proton-enhanced cleavage of a μ_2-OH bridge between two hydrated metals, making up a dimer. The first step is equilibrium protonation of the μ_2-OH bridge to form a bridging water molecule, which is a weak bridge:

$$[(H_2O)_5M_1(\mu_2-OH)M_2(OH_2)_5]^{(2z-1)+} + H^+(aq) \xleftrightarrow{\text{equilibrium}}$$
$$[(H_2O)_5M_1(\mu_2-OH_2)M_2(OH_2)_5]^{2z+}$$

An incoming water molecule coordinates to one of the metals, causing the dimer to dissociate:

$$[(H_2O)_5M_1(\mu_2-OH_2)\,M_2(OH_2)_5]^{2z+} + H_2O \longleftrightarrow$$
$$[(H_2O)_5M_1(\mu_2-OH_2)\,(H_2O)M_2(OH_2)_5]^{2z+}$$
$$[(H_2O)_5M_1(\mu_2-OH_2)\,(H_2O)M_2(OH_2)_5]^{2z+} \xleftrightarrow{\text{slow}} M_1(H_2O)_6^{z+} + M_2(H_2O)_6^{z+}$$

leading to release of hydrated metals. Note that the role of the proton is *inductive*, not *catalytic,* because it is not recycled. It would be recycled only if one of the hydrated metals dissociates a proton after the reaction, which is unlikely in the case of oxide bridge dissociation because the monomers released by the reaction are almost always weaker Brønsted acids than the oxide bridge in the dimer. Because protonation/deprotonation reactions are so much more rapid than most bond-breaking steps, the actual protonation state of the immediate reaction products is commonly impossible to ascertain, but the reactions are rarely truly catalytic.

CONCLUSIONS

The polynuclear complexes of aluminum can help answer exactly the questions many geochemists pose about reactions at mineral surfaces, including: (1) How does the reactivity of a metal-oxygen bond relate to covalency and coordination geometry? (2) How rapidly do different metal-oxygen bonds dissociate at a mineral surface? (3) What are the molecular pathways for metal or ligand adsorption? (4) How can we infer molecular information from dissolution experiments? (5) Do the macroscopic properties of a surface in water, such as Brønsted acidity, change with size of the molecule? (6) How do the microscopic properties of a molecule, such as the rates of bridge dissociation, relate to calculable bond properties, such as electronic charge densities? (7) Over what time scales are different types of bond dissociations complete? (8) Can geochemists predict rates of multi-step reactions?

These questions can be answered directly with nanocluster solutes. Some can be easily synthesized and are metastable in solution for experimentally long periods of time. They have structural features in common with aluminum (hydr)oxide minerals so that ideas about reaction pathways at mineral surfaces can be examined in detail. Most importantly, they are sufficiently small that computer-based models can be used to understand the reactions.

ACKNOWLEDGMENTS

The article benefited considerably from a review by Francis Taulelle, and from the editors of the volume. The authors also thank Magnus Karlsson who organized some of these structures as part of his Ph.D. thesis as the University of Umeå and Professors A. Powell, S. Heath and L. Nazar, who provided structures from their unpublished work. This work was supported by the U.S. National Science Foundation via grant EAR 98-14152 and by the U.S. Department of Energy via grant DE-FG03-96ER14629.

REFERENCES

Aahman J, Svensson G, Albertsson J (1996) A reinvestigation of β-gallium oxide. Acta Crystallogr, Cryst Struct Comm C52:1336-1338

Akitt JW (1989) Multinuclear studies of aluminum compounds. Progr NMR Spectros 21:1-149

Akitt JW, Elders JM, Fontaine XLR, Kundu AK (1989b) Multinuclear magnetic resonance studies of the hydrolysis of aluminum(III). Part 10: Proton, carbon, and aluminum-27 spectra of aluminum acetate at very high magnetic field. J Chem Soc Dalton Trans 1989:1897-1901

Akitt JW, Elders JM (1988) Multinuclear magnetic resonance studies of the hydrolysis of aluminium(III). Part 8: Base hydrolysis monitored at very high magnetic field. J Chem Soc Dalton Trans 1988: 1347-1355

Akitt JW, Farthing A (1981) Aluminum-27 nuclear magnetic resonance studies of the hydrolysis of aluminum(III). Part 4: Hydrolysis using sodium carbonate. J Chem Soc Dalton Trans 1981:1617-1623

Akitt JW, Greenwood NN, Khandelwal BL, Lester GD (1972) ^{27}Al nuclear magnetic resonance studies of the hydrolysis and polymerization of the hexa-aquo-aluminum(III) cation. J Chem Soc Dalton Trans 1972:604-610

Allouche L, Gérardin C, Loiseau T, Férey G, Taulelle F (2000) Al_{30}: A giant aluminum polycation. Angew Chem Int'l Edition 39:511-514

Allouche L, Huguenard C, Taulelle F (2001) 3QMAS of three aluminum polycations: space group consistency between NMR and XRD. J Phys Chem Solids 62:1525-1531

Amirbahman A, Gfeller M, Furrer G (2000) Kinetics and mechanism of ligand-promoted decomposition of the Keggin Al_{13} polymer. Geochim Cosmochim Acta 64:911-919

Baes CF, Mesmer RE (1976) The Hydrolysis of Cations. John-Wiley, New York

Banfield JF, Welch SA, Zhang H, Ebert TT, Penn RL (2000) Aggregation-based crystal growth and microstructure development in natural iron oxyhydroxide biomineralization products. Science 289:751-754

Bertsch PM (1987) Conditions for Al_{13} polymer formation in partially neutralized aluminum solutions. Soil Sci Soc Am J 51:825-828

Bertsch PM, Parker DR (1995) Aqueous polynuclear aluminum species. *In* The Environmental Chemistry of Aluminum. Sposito G (ed) CRC Press, Boca Raton, Florida, p 117-168

Bethley CE, Aitken CL, Harlan CJ, Koide Y, Bott SG, Barron AR (1997) Structural characterization of dialkylaluminum carboxylates: models for carboxylate alumoxanes. Organometallics 16:329-341

Bombi, GG, Corain, B, Abdiqafar, AS.-O, Valle, GC (1990) The speciation of aluminum in aqueous solutions of aluminum carboxylates. Part 1: X-ray molecular structure of $Al[OC(O)CH(OH)CH_3]_3$. Inorganica Chimica Acta 171:79-83

Bradley SM, Kydd RA, Yamdagni R (1990a) Study of the hydrolysis of combined Al^{3+} and Ga^{3+} aqueous solutions: formation of an extremely stable $GaO_4Al_{12}(OH)_{24}(H_2O)_{12}{}^{7+}$ polyoxocation. Magn Res Chem 28:746-750

Bradley SM, Kydd RA, Yamdagni R (1990b) Detection of a new polymeric species formed through the hydrolysis of gallium(III) salt solutions. J Chem Soc Dalton Trans 1990:413-417

Bradley SM, Kydd RA, Fyfe CA (1992) Characterization of the $GaO_4Al_{12}(OH)_{24}(H_2O)_{12}{}^{7+}$ polyoxocation by MAS NMR and infrared spectroscopies and powder X-ray diffraction. Inorg Chem 31:1181-1185

Bradley SM, Kydd RA, Howe RF (1993) The structure of Al-gels formed through base hydrolysis of Al^{3+} aqueous solutions. J Coll Interf Sci 159:405-422

Bradley SM, Kydd R (1993) Comparison of the species formed upon base hydrolysis of gallium(III) and iron(III) aqueous solutions: the possibility of existence of an $[FeO_4Fe_{12}(OH)_{24}(H_2O)_{12}]^{7+}$ polyoxocation. J Chem Soc Dalton Trans 1993:2407-2413

Callender RL, Harlan CJ, Shapiro NM, Jones CD, Callahan DL, Wiesner MR, MacQueen DB, Cook R, Barron AR (1997) Aqueous synthesis of water-soluble alumoxanes: environmentally benign precursors to alumina and aluminum-based ceramics. Chem Materials 9:2418-2433

Casey WH, Phillips BL, Karlsson M, Nordin S, Nordin JP, Sullivan DJ, Neugebauer-Crawford S (2000) Rates and mechanisms of oxygen exchanges between sites in the $AlO_4Al_{12}(OH)_{24}(H_2O)_{12}{}^{7+}(aq)$ complex and water: Implications for mineral surface chemistry. Geochim Cosmochim Acta 64: 2951-2964

Casey WH, Phillips BL (2000) The kinetics of oxygen exchange between sites in the $GaO_4Al_{12}(OH)_{24}(H_2O)_{12}{}^{7+}(aq)$ molecule and aqueous solution. Geochim Cosmochim Acta 65:705-714

Casey WH, Phillips BL, Nordin JP, Sullivan DJ (1998) The rates of exchange of water molecules from Al(III)-methylmalonate complexes: The effect of chelate ring size. Geochim Cosmochim Acta 62:2789-2797

Euler F, Bruce JA (1965) Oxygen coordinates of compounds with garnet structure. Acta Crystallogr 19:971-978

Feng TL, Gurian PL, Healy MD, Barron AR (1990) Aluminum citrate: isolation and structural characterization of a stable trinuclear complex. Inorg Chem 29:408-411

Fitzgerald JJ, Rosenberg AH (1999) Chemistry of aluminum chlorohydrate and activated aluminum chlorohydrate. *In* Antiperspirants and Deodorants, 2nd edition. Laden K (ed) Marcel Dekker, New York, p 83-136

Frueh AJ, Golightly JP (1967) The crystal structure of dawsonite $NaAl(CO_3)(OH)_2$. Can Mineral 9:51-56

Fu G, Nazar LF, Bain AD (1991) Aging processes of alumina sol gels-characterization of new aluminum polyoxycations by Al-27 NMR spectroscopy. Chem Mater 3:602-610

Furrer G (1993) New aspects on the chemistry of aluminum in soils. Aquatic Sci 55:281-290

Furrer G, Gfeller M, Wehrli B (1999) On the chemistry of the Keggin Al_{13} polymer-kinetics of proton-promoted decomposition. Geochim Cosmochim Acta 63:3069-3076

Furrer G, Trusch B, Müller C (1992a) The formation of polynuclear Al_{13} under simulated natural conditions. Geochim Cosmochim Acta 56:3831-3838

Furrer G, Ludwig P, Schindler W (1992b) On the chemistry of the Keggin Al_{13} polymer: I. Acid-base properties. J Colloid Interface Sci 149:56-67

Golic L, Leban I, Bulc N (1989) The structure of sodium bis(tetraethylammonium) tris(oxalato)aluminate(III) monohydrate. Acta Crystallogr C45:44-46

Heath SL, Jordan PA, Johnson ID, Moore JR, Powell AK, Helliwell M (1995) Comparative X-ray and ^{27}Al NMR spectroscopic studies of the speciation of aluminum in aqueous systems: Al(III) complexes of $N(CH_2CO_2H)_2(CH_2CH_2OH)$. J Inorg Biochem 59:785-594

Hedlund T, Bilinski H, Horvath L, Ingri N, Sjöberg S (1987a) Equilibrium and structural studies of silicon(IV) and aluminum(III) in aqueous solution. 16: Complexation and precipitation reactions in the H^+-Al^{3+}-phthalate system. Inorg Chem 27:1370-1374

Hedlund T, Sjöberg S, Öhman L-O (1987b) Equilibrium and structural studies of silicon(IV) and aluminum(III) in aqueous solution. 15: A potentiometric study of speciation and equilibria in the Al^{3+}-$CO_2(g)$-OH^- system. Acta Chem Scand 41:197-207

Huang PM (1988) Ionic factors affecting aluminum transformations and the impact on soil and environmental sciences. Adv Soil Science 8:1-78

Hugi-Cleary D, Helm L, Merbach AE (1985) Variable temperature and variable pressure ^{17}O NMR study of water exchange of hexaquaaluminum(III). Helv Chim Acta 68:545-554

Hunter D, Ross DS (1991) Evidence for a phytotoxic hydroxy-aluminum polymer in organic soil horizons. Science 251:1056-1058

Johansson G (1960) On the crystal structures of some basic aluminum salts. Acta Chem Scand 14:771-773

Johansson G (1962a) The crystal structures of $[Al_2(OH)_2(H_2O)_8](SO_4)_2 \cdot 2H_2O$ and $[Al_2(OH)_2(H_2O)_8](SeO_4)_2 \cdot 2H_2O$. Acta Chem Scand 16:403-420

Johansson G (1962b) On the crystal structure of the basic aluminum sulfate $13 \cdot Al_2O_3 \cdot 6SO_3 \cdot xH_2O$. Arkiv. Kemi 20:321-342

Johansson G (1962c) On the crystal structure of the basic aluminum selenate. Arkiv. Kemi 20:305-319

Johansson G, Lundgren G, Sillén LG, Söderquist R (1960) On the crystal structure of a basic aluminum sulfate and the corresponding selenate. Acta Chem Scand 14:769-771

Jolivet J-P (1994) De la Solution à l'Oxyde, 3rd ed, Masson, Paris

Jolivet J-P, Henry M, Livage J (2000) Metal-oxide Chemistry and Synthesis. John-Wiley, New York (translation by E Bescher)

Jordan PA, Clayden NJ, Heath SL, Moore GR, Powell AK, Tapparo A (1996) Defining speciation profiles of Al^{3+} complexed with small organic ligands: the Al^{3+}-heidi system. Coord Chem Rev 149:281-309

Karlsson M (1998) Structure studies of aluminum(III) complexes in solids, in solutions, and at the solid/water interface. PhD thesis, University of Umeå

Karlsson M, Bostrom D, Clausen M, Öhman LO (1998) Equilibrium and structural studies of silicon(IV) and aluminium(III) in aqueous solution. 34. A crystal structure determination of the Al(methylmalonate)(2)(CH3OH)(2)(-) complex with Na^+ as counter-ion. Acta Chem Scand 52: 1116-1121

Kloprogge JT, Seykens D, Jansen JBH, Geus JW (1992) A ^{27}Al nuclear magnetic resonance study on the optimalization of the development of the Al13 polymer. J Non-crystalline Solids 142:94-102

Koide Y, Barron AR (1995) $[Al\text{-}5(^tBu)_5(\mu_3\text{-}OH)_2(\mu\text{-}OH)_2(\mu\text{-}O_2CPh)_2]$: A model for the interaction of carboxylic acids with boehmite. Organometallics 14:4126-4029

Kudynska J, Buckmaster HA, Kawano K, Bradley SM, Kydd RA (1993) A 9 GHz cw-electron paramagnetic resonance study of the sulphate salts of tridecameric $[Mn_xAl_{13-x}O_4(OH)_{24}(H_2O)_{12}]^{(7-x)+}$. J Chem Phys 99:3329-3334

Landry CC, Pappé N, Mason M, Apblett AW, Tyler AN, MacInnes AnN, Barron AR (1995) From minerals to materials-synthesis of alumoxanes from the reaction of boehmite with carboxylic acids. J Mater Chem 5:331-341

Lee AP, Phillips BL, Olmstead MM, Casey WH (2001a) Synthesis and characterization of the $GeO_4Al_{12}(OH)_{24}(O\ H_2)_{12}{}^{8+}$ polyoxocation. Inorg Chem. 40:4485-4487

Lee AP, Phillips BL, Casey WH (2001b) The kinetics of oxygen exchange between sites in the $GeO_4Al_{12}(OH)_{24}(OH_2)_{12}{}^{8+}$(aq) molecule and aqueous solutions. Geochim Cosmochim Acta (submitted)

Lothenbach B, Furrer G, Schulin R (1997) Immobilization of heavy metals by polynuclear aluminum and montmorillonite compounds. Environ Sci Technol 31:1452-1462

Marklund E, Öhman L-O (1990) Equilibrium and structural studies of silicon(IV) and aluminum(III) in aqueous solution. 24 A potentiometric and ^{27}Al NMR study of polynuclear aluminum(III) hydroxo complexes with lactic acid. Acta Chem Scand 44:228-234

Matzapetakis M, Raptopoulou CP, Terzis A, Lakatos A, Kiss T, Salifoglou A (1999) Synthesis, structural characterization, and solution behavior of the first mononuclear, aqueous aluminum citrate complex. Inorg Chem 38:618-619

Michot LJ, Montargès-Pelletier E, Lartiges BS, d'Espinose de la Caillerie B, Briois V (2000) Formation mechanism of the Ga_{13} Keggin ion: A combined EXAFS and NMR study. J Am Chem Soc 122: 6048-6056

Montarges E, Michot LJ, Lhote F, Fabien T, Villieras F (1995) Intercalation of Al_{13}-polyethylene oxide complexes into montmorillonite clay. Clays Clay Minerals 43:417-426

Nagy JB, Bertrand J-C, Palinko I, Kiricsi I (1995) On the feasibility of iron or chromium substitution for aluminum in the Al_{13}-Keggin ion. J Chem Soc Chem Comm 1995:2269-2270

Nordin JP, Sullivan DJ, Phillips BL, Casey,WH (1998) An ^{17}O-NMR study of the exchange of water on $AlOH(H_2O)_5{}^{2+}$(aq). Inorg Chem 37:4760-4763

Nordin JP, Sullivan DJ, Phillips BL, Casey WH (1999) Mechanisms for fluoride-promoted dissolution of bayerite [β-Al(OH)$_3$(s)] and boehmite [γ-AlOOH(s)]-^{19}F-NMR spectroscopy and aqueous surface chemistry. Geochim Cosmochim Acta 63:3513-3524

Öhman L-O, Forsling W (1981) Equilibrium and structural studies of silicon(IV) and aluminium(III) in aqueous solutions. 3. A potentiometric study of aluminum(III) hydrolysis and aluminum(III) hydroxo carbonates in 0.6 M Na(Cl). Acta Chem Scand A 35:795-802

Öhman L-O (1991) Equilibrium and structural studies of silicon(IV) and aluminum(III) in aqueous solutions. 27. Al^{3+} complexation to monocarboxylic acids. Acta Chem Scand 45:258-264

Ollivier B, Retoux R, Lacorre P, Massiot D, Ferey G (1997) Crystal structure of kappa-alumina: an X-ray powder diffraction, TEM and NMR study. J Mater Chem 7:1049-1056

Oszkó A, Kiss J, Kiricsi I (1999) XPS investigation on the feasibility of isomorphous substitution of octahedral Al^{3+} for Fe^{3+} in Keggin ion salts. Phys Chem Chem Phys 1:2565-2568

Parker WO Jr, Millini R, Kiricsi I (1997) Metal substitution in Keggin-type tridecameric aluminum-oxo-hydroxy clusters. Inorg Chem 36:571-576

Parker WO Jr, Millini R, Kiricsi I (1995) Aluminum complexes in partially hydrolyzed aqueous $AlCl_3$ solutions used to prepare pillared clay catalysts. Appl Catalysis A: General 121:L7-L11

Persson P, Karlsson M., Öhman L-O. (1998) Coordination of acetate to Al(III) in aqueous solution and at the water-aluminum hydroxide interface: a potentiometric and attenuated total reflectance FTIR study. Geochim Cosmochim Acta 62:3657-3668

Petrosyants SP, Malyarik MA, Ilyukhin AB (1995) Complexation of aluminum and gallium with aminoacetic acid: crystal structure of diaquadi-μ-hydroxobis-(iminodiacetato)dialuminum(III) and potassium bis(iminodiaceto)gallate(III). Russ J Inorg Chem 40:769-775

Phillips BL, Casey WH, Neugebauer-Crawford S (1997a) Solvent exchange in $AlF_x(H_2O)_{6-x}^{3-x}$(aq) complexes: ligand-directed labilization of water as an analogue for ligand-induced dissolution of oxide minerals. Geochim Cosmochim Acta 61:3041-3049

Phillips BL, Neugebauer-Crawford S, Casey WH (1997b) Rate of water exchange between $Al(C_2O_4)(H_2O)_4^+$(aq) complexes and aqueous solution determined by ^{17}O-NMR spectroscopy. Geochim Cosmochim Acta 61:4965-4973

Phillips BL, Casey WH, Karlsson M (2000) Bonding and reactivity at oxide mineral surfaces from model aqueous complexes. Nature 404:379-382

Rao GV, Rao KSJ (1992) Evidence for a hydroxy-aluminum polymer (Al_{13}) in synaptosomes. FEBS 311:49-50

Richens DT (1997) The Chemistry of Aqua Ions. Wiley, New York, 592 p

Rozan TF, Lassman ME, Ridge DP, Luther GW (2000) Evidence for iron, copper and zinc complexation as multinuclear sulphide clusters in oxic rivers. Nature 406:879-882

Rowsell J, Nazar LF (2000) Speciation and thermal transformation in alumina sols: structures of the polyhydroxyoxoaluminum cluster $[Al_{30}O_8(OH)_{56}(H_2O)_{26}]^{18+}$ and its δ-Keggin moieté. J Am Chem Soc 122:3777-3778

Schönherr S, Görz H, Gessner W, Bertram R (1983a) Protoyzevorgänge in wässrigen aluminiumchlorid-lösungen. Z Chemie 23:429-434

Schönherr SV, Görz H (1983b) Über das Dodekaaluminogermaniumsulfate $[GeO_4Al_{12}(OH)_{24}(H_2O)_4]$-$(SO_4)_4 \cdot xH_2O$ (1983). Z anorg allg Chem 503:37-42

Seichter W, Mögel H-J, Brand P, Salah D (1998) Crystal structure and formation of the aluminum hydroxide chloride $[Al_{13}(OH)_{24}]Cl_{15} \cdot 13H_2O$. Eur J Inorg Chem 1998:795-797

Sjöberg S, Öhman L-O (1985) Equilibrium and structural studies of silicon(IV) and aluminum(III) in aqueous solution. Part 13: A potentiometric and ^{27}Al nuclear magnetic resonance study of speciation and equilibrium in the aluminium(III)-oxalic acid-hydroxide system. J Chem Soc Dalton Trans 1985: 2665-2669

Sohlberg K, Pennycook SJ, Pantelides ST (1999) Hydrogen and the structure of the transition aluminas. J Am Chem Soc 25:7493-7499

Smyth JR, Bish DL (1988) Crystal structures and cation sites of the rock-forming minerals. Allen and Unwin, London, 332 p

Springborg J (1988) Hydroxo-bridged complexes of chromium(III), cobalt(III), rhodium(III) and iridium(III). Adv Inorg Chem 32:55-169

Sullivan DJ, Nordin JP, Phillips BL, Casey WH (1999) The rates of water exchange in Al(III)-salicylate and Al(III)-sulfosalicylate complexes. Geochim Cosmochim Acta 63:1471-1480

Thomas B, Görz H, Schönherr, S (1987) Zum NMR-spektroscopishen nachweis von dodecaalumino-gallium-ionen. Z Chemie 27:183

Thompson AR, Kunwar AC, Gutowsky HS, Oldfield E (1987) Oxygen-17 and aluminum-27 nuclear magnetic resonance spectroscopic investigations of aluminum(III) hydrolysis products. J Chem Soc Dalton Trans 1987:2317-2322

Tapparo A, Heath SL, Jordan PA, Moore GR, Powell AK (1996) Crystal structure and solution-state study of $K[Al(mal)_2(H_2O)_2] \cdot 2H_2O$ (H_2mal = malonic acid). J Chem Soc Dalton Trans 1996:1601-1606

Valle GC, Bombi GG, Corain B, Favarato M, Zatta P (1989) Crystal and molecular structures of diaqua(nitrilotriacetato)aluminum(III) and di-μ-hydroxo-bis(nitrilotriacetato)dialuminate(III) dianion. J Chem Soc Dalton Trans 1989:1513-1517

Vicente MA, Lambert J-F (1999) Al-pillared saponites. Part 4: Pillaring with a new Al_{13} oligomer containing organic ligands. Phys Chem Chem Phys 1:1633-1639

Wang W-Z, Hsu PH (1994) The nature of polynuclear OH-Al complexes in laboratory hydrolyzed and commercial hydroxyaluminum solutions. Clays Clay Minerals 42:356-368

Wehrli BE, Wieland E, Furrer G (1990) Chemical mechanisms in the dissolution kinetics of minerals: The aspect of active sites. Aquatic Sci 52:3-31

Yokoyama T, Abe H, Kurisaki T, Wakita H (1999) ^{13}C and ^{27}Al NMR study on the interaction between aluminum ion and iminodiacetic acid in acidic aqueous solutions. Analyt Sci 15:393-395

Yoshiasa A, Tamura T, Kamishima O, Murai K-I, Ogata K, Mori H (1999) Local structure and mean-square relative displacement in SiO_2 and GeO_2 polymorphs. J Synchrotron Rad 6:1051-1058

Yu P, Phillips BL, Casey WH (2001) Water exchange on fluoroaluminate complexes in aqueous solution: A variable temperature multinuclear NMR study. Inorg Chem 40:4750-4754

Zhou R-S, Snyder RL (1991) Structures and transformation mechanisms of the η, γ and θ transition aluminas. Acta Crystallogr B47:617-630

6 Computational Approaches to Nanomineralogy

James R. Rustad
William R. Wiley Environmental Molecular Sciences Laboratory
Richland, Washington 99352

Witold Dzwinel
AGH University of Mining and Metallurgy
Institute of Computer Science
Krakow, Poland

David A. Yuen
Department of Geology and Geophysics
and *Minnesota Supercomputing Institute*
University of Minnesota
Minneapolis, Minnesota 55415

INTRODUCTION

Nanomineralogy is concerned with the behavior of minerals on length scales between 10 Å and 1 micron. Within the realm of computational science, molecular modeling methods have been working at the lower end of this scale for more than 50 years and comprise a relatively mature field even within the geosciences community (see Cygan and Kubicki 2001). Somewhere near the upper end of this scale, continuum approaches using bulk thermodyamics and homogeneous transport properties (diffusion, viscous flow and heat flow, and elastic moduli) can be used effectively (Turcotte and Schubert 1982). From one perspective, the nanoscale regime is the theoretical/ computational no-man's land between atomistic and continuum scales, in which atoms cannot quite be ignored and continuum models cannot quite be applied. More generally, it is the simultaneous consideration of multiple scales, each requiring different methods, which, from the computational point of view, provides the driving force and makes this field exciting. The focus of this volume is on nanoscale phenomena in mineralogy and geochemistry in low-temperature, near-surface environments. This is indeed a rich area for multiscale investigations, as shown in Figure 1. At the finest scales, one is interested in the spatio-temporal variability of the collective wave functions (or density) of electrons. At a simple level, this results in the formation of chemical bonds. Even at this scale, consideration of the detailed aspects of electron density topology is an emerging area (Bader 1990; Gibbs et al. 1998; Rescigno et al. 1999; Blanco et al. 2000; Espinosa and Molins 2000). As more than a few atoms begin to interact, complexity begins to be revealed in molecular arrangements as well as the electronic structure. In mineralogy, we have polynuclear clusters (such as discussed in detail by Casey and Furrer in this volume) and nanoporous minerals, such as cacoxenite and zeolites (Gier and Stucky 1991; Patarin and Kessler 2000). In materials science and chemistry, we have the familiar nanotubes (Mintmire and White 1998) and buckyballs (Lof et al. 1992). The field of supramolecular chemistry also operates in this regime and is concerned with controlled assembly of these building blocks into chemical machines (Varnek et al. 2000). These polynuclear molecules also form building blocks for familiar oxide minerals (Schwertmann et al. 1999; Cannas et al. 2001; Casey and Furrer, this volume). The process of assembly of

1529-6466/00/0044-0006$05.00

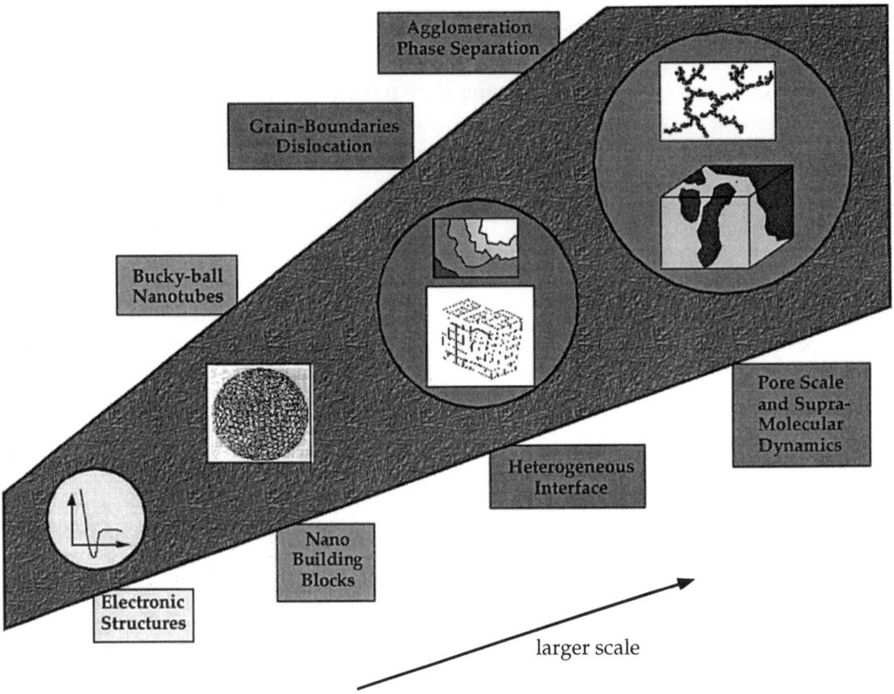

Figure 1. Cross-scaling processes in low-temperature geochemisty and mineralogy.

these molecules leads to the elaborate crystal shapes and distributions familiar in nonequilibrium low-temperature aqueous environments (Dixon and Weed 1989). Familiar heterogeneities at this scale include surfaces with terraces and kinks, dislocations, and grain boundaries, including aqueous mineral interfaces. At still larger scales, we have heterogeneities associated with agglomerating mineral assemblages and other forms of phase separation, such as exsolution phenomena. Agglomeration in itself is a multiscale process that, at very small scales, plays an important role in the assembly of single oxide crystals (Banfield et al. 2000; Penn et al. 2001a). At larger scales, this leads us to the pore-scale regime of concern in reactive transport investigations (Yabusaki et al. 2001).

Computational methods capable of spanning these scales take two types of approaches. The more familiar is the serial approach. For example, electronic structural calculations on small systems are used to parameterize tight binding or classical potential functions. Molecular dynamics (MD) or Monte Carlo methods are applied to the potential functions to obtain transport phenomena or thermodynamic quantities. In some cases, transport coefficients derived in this way are used in continuum models at the next level. Systems with moving inhomogeneities, such as crack propagation, or highly collective problems having complex structural characteristics often require a multiresolution approach that combines *simultaneously* continuum, atomistic, and electronic structural components (Broughton 1999). In such systems, it is often not possible to isolate scales and eliminate them through averaging or through parameterization. Even before doing any physicochemical modeling, just specifying the initial conditions in an inherently multiscale system is a challenging task. The multiresolution aspects are as important in time as in space (if not more so).

What motivation exists for building a multiscale understanding of mineralogical systems? For the geoscientist interested in near-surface environments, probably the most compelling reason to study these multiscale phenomena from a simulation perspective is to sort out the various factors contributing to chemical reactivity in terms of both energetics and rates. The sorption of chromate, or even the uptake of protons in these morphologically intricate minerals, involves highly coupled contributions from electronic structure, surface topography, and long-range viscoelastic and solvent effects. For example, even something as fundamental as the point of zero charge of goethite varies over 2 orders of magnitude as a function of the mesoscopic hetereogeneity of the phases involved (D. Sverjensky, personal communication). It is very difficult, if not impossible, to isolate and uncouple these contributions experimentally; understanding these processes almost always requires the use of model systems, where the contributions can be theoretically decoupled. At present, the overall mechanisms and magnitudes of nanoscale influences on chemical reactivity are not understood even qualitatively.

We take a decidedly multiscale perspective in this review and de-emphasize some natural links between molecular modeling and nanoscience such as the evaluation of surface energies (Wasserman et al. 1997) and polynuclear ion structure and reactivity (Rustad et al. 2000). These are important connections, but, from the point of view of computation, they are covered as completely as presently possible in *Reviews in Mineralogy and Geochemistry*, Volume 42, on molecular modeling theory (Cygan and Kubicki 2001). The downside of this emphasis is that, at least in mineralogy, the field is wide open. There are no papers focused on multiscale approaches to computational modeling in the mineralogical sciences, and far too few papers are focused on multiscale description of mineralogical systems. This more or less remains true even if we open the scope to all of chemical physics. At least as far as modeling is concerned only a few research groups are taking on these types of problems. On the other hand, the multiscale approach is much better developed in the geophysical literature (Yuen et al. 2000).

This chapter is therefore aimed at giving the Earth scientist or mineralogist a simple introduction to the techniques and concepts necessary for approaching nanoscale phenomena such as are associated, for example, with the largest three scales in Figure 1. We begin with an elementary introduction to scaling concepts and the computational description of complex patterns, for example, using wavelets. We then review multiscale computational physicochemical modeling techniques for solids and fluids with applications in interfacial reactivity, phase separation, and particle agglomeration. Examples are drawn from the chemical physics literature that should resonate with the geochemist or mineralogist.

MULTISCALE DESCRIPTION OF COMPLEX SURFACES

For the kinds of problems characterizing the middle regions of Figure 1, even just model construction presents a serious difficulty. As we pass through the nanoscale regime, we move from systems having simple, straightforward initial conditions, such as the ideally terminated slab shown in Figure 2a, to those that do not, such as the partially oxidized iron nanoparticle shown in Figure 2b. The easy "virtual sample preparation" tasks needed for Figure 2a become considerably more challenging in Figure 2b. The essential problem is one of timescales: even if the molecular-scale phenomena one might wish to simulate occur, or can be made to occur, on nanosecond MD time scales, the processes creating the surface shown in Figure 2b occur on much longer time scales, far outside those in which there is any hope of simulating the evolution of the surface structure directly. Setting up these problems will require close integration of different microscopies, techniques of imaging science, and simulation.

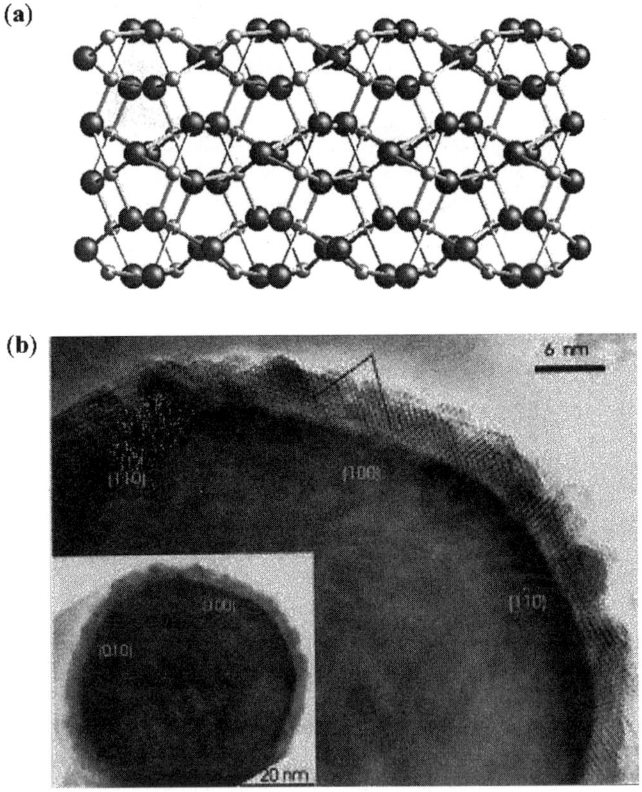

Figure 2. (a) Idealized surface of hematite (001); (b) oxidized nanoparticle of zero-valent (metallic) iron with oxide rind [used by permission of the editor of *Applied Physics Letters*, from Kwok et al. (2000), Fig. 2, p. 3972]. While the initial conditions for (a) are easily produced, those for (b) are extremely difficult to produce, even if techniques are available that can model the system shown in (b).

Indeed, while glancing through atlases of mineral surface micrographs such as are found in Schwertmann and Cornell (1991), Cornell and Schwertmann (1992), or Dixon and Weed (1989), and through more recent work on biogeochemically produced minerals (Frankel and Blakemore 1991; Orme et al. 2001; Banfield et al. 2000; Zachara et al. 1998; Maurice et al. 2000) and minerals formed from aggregation of nanocrystallites (Penn et al. 2001b), one cannot help being struck by the morphological complexity of so-called "kinetically roughened" mineral surfaces formed in non-equilibrium low-temperature environments. One clearly needs new tools in addition to x-ray crystallography to characterize these systems. Although the terms "unit cell," "Bravais lattice," "space group," and "Miller index" are familiar to any geology student, the terms, "self-affine," "mother wavelet," and "dynamic roughness exponent" are not. But such concepts are needed to approach the types of problems associated with Figure 1.

Scaling concepts

All of us have probably seen impressive atomic scale micrographs of nearly perfect silicon or gallium arsenide crystals. These pictures convey the concept of a perfect surface perturbed by a set of relatively well-defined defects. Kinetically roughened natural mineral surfaces in low-temperature environments are far more complex, with "defects" occurring on all scales. What is a kinetically roughened surface? A good pedagogical analogy is the game of *Tetris*, which many people have played and which has been used as a model for the compaction of dry granular media (Caglioti et al. 1997). One begins with an idealized flat surface on a square lattice. Growth units (possibly of several different shapes) are introduced at a certain rate from above, and the player manipulates the x positions and orientations of these units to obtain a surface that is as uniform as possible. When any value of the height $h(x)$ exceeds a maximum value, the game is over. As a reward for achieving an efficient packing density, if a fully occupied layer forms in the system, the maximum height is increased one unit. The score is based on the number of units deposited before the end of the game. At a very slow rate of growth, optimal fits are found easily and the surface remains relatively smooth as a function of time. At very fast rates of growth (or for complex distributions of growth units), the game becomes more difficult, the growth surface becomes quite rough, and the game ends quickly. The amount of roughening clearly scales with the rate of introduction of the growth units; hence, the term kinetic roughening.

Two fundamental scaling parameters characterize the morphology of rough surfaces as a function of time t and particle size L. First, the width of the interface ($w = h_{max} - h_{min}$) is generally found, empirically, to scale with time according to a power law:

$$w(L,t) \propto t^{\beta} \tag{1}$$

β is known as the growth exponent. As time goes on, the surface gets rougher in the sense that its width increases. However, it is also generally observed that this roughness increases only up to a certain saturation point, after which the surface roughness is constant at w_s.

Second, it is found that the larger the particle, the greater the roughness at the saturation point. This is also found (again empirically) to follow a power law:

$$w_s(L) \propto L^{\alpha} \tag{2}$$

α is known as the roughness exponent.

What is the fundamental machinery behind the power-law scaling? As a very simple illustration, imagine taking the trace of a surface of a small particle and magnifying it uniformly in all directions. Clearly in this case, w_s scales as the first power of L and α=1. The uniform scaling indicates a self-similar surface. The scaling need not be uniform. For example, consider Figure 3, consisting of a nested arrangement of lines recursively copied onto itself over four decades in scale. Moving upward, at each iteration, the length scale expands by a factor of 4 and the height by a factor of 2. Hence, the scaling still follows a power law, but now the roughness exponent is one-half. In this case, the surface is said to be self-affine, a less restrictive sort of self-similarity in which the scaling relation is not uniform.

Self-similarity is a key concept in understanding scaling (Falconer 1990; Barenblatt 1996; Turcotte 1997; Meakin 1998). The origin of self-similarity or self-affinity can be roughly understood in terms of fundamental growth units assembling themselves into larger units. These larger units then constitute a new set of fundamental particles that assemble themselves into still larger units, and so on. The situation is not always so transparent, but somewhere backstage this type of process goes on to generate the power-law scaling.

What about the existence of w_s? What determines the saturation? Surprisingly, this

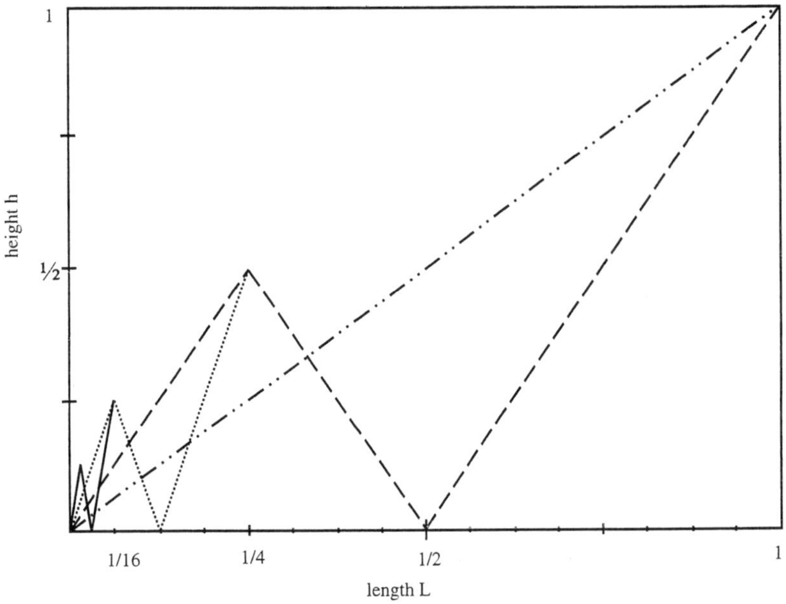

Figure 3. Recursive replacement of the zigzag motif over four decades in length scale (dash-dot-dot, coarse dash, fine dash, solid; *largest scale shown only partially*). This generates a self-affined surface with a growth exponent equal to one-half. Modified from Barabasi and Stanley (1995).

question is not easily answered. The origin lies in finite size effects and correlations in surface topography due to frustration of atomic surface rearrangements. When the correlation length, the length scale over which height information is communicated, approaches the system size, the width saturates. Clearly, as L approaches infinity, so does that saturation width. As L shrinks, the width becomes vanishingly small. This is perhaps evident in the regular surfaces of very small nanoparticles (Penn and Banfield 1998). These are only the most elementary concepts in an immense literature with several excellent textbooks, including Barabasi and Stanley (1995) and Meakin (1998).

In mineralogy, very little work quantifying these scaling laws has been done, and no studies to date have considered the dynamic scaling problem as applied to surfaces. In part, this is due to the difficulty of obtaining contiguous data over sufficiently long length scales at sufficiently high resolution to determine accurate scaling exponents. Pioneering examples are work done on goethite by Weidler et al. (1998) and Liu and Huang (1999). Titanium dioxide has received attention in the photocatalysis community (Xagas et al. 1999; Lee 2001). Cardone et al. (1999) have studied MnO_2 and chalcopyrite from the standpoint of fractal geometry. Much more experimental work on multiscale aspects of mineral surfaces and aggregates will doubtless be forthcoming.

Wavelets and multiscale description of surfaces and interfaces

One of the most crucial issues facing scientists today is the flood of data generated by more accurate laboratory measurements and higher-resolution numerical simulations. If we are to perceive some succinct patterns buried in the large arrays of numbers or pixels, fast and efficient techniques are required. We cannot afford to be looking at the data at full resolution all the time because of the time-consuming process in visualizing gigabytes to terabytes of data. Feature extraction is a technique whereby we can distill the most essential aspects of the data, such as the outline of a skeleton in a biological organism or the peaks and valleys of a complicated terrain. Wavelets (e.g., Holschneider 1995; Bowman and Newell 1998) mostly developed over the past 17 years present an ideal and relatively easy-to-master tool for extracting certain outstanding scales of interest. They are numerical filters able to zoom in and out in both a given location in physical space and magnification and scale in wavelet space. On the other hand, the traditional method of Fourier analysis yields only a global type of information and loses local knowledge, such as the place or time of the particular phenomenon. Wavelets are linear mathematical transformations (e.g., Resnikoff and Welss 1998) that can analyze both temporal signals and spatial images at different scales. The wavelet transform is sometimes called a mathematical microscope. Large wavelets give an approximate image of the signal, while smaller and smaller wavelets zoom in on small details.

Until recently, most of the applications of wavelets in geoscience have been focused on geophysical applications—for example, the use of one-dimensional wavelets to analyze time-series of the Chandler wobble or one-dimensional spatial tracks such as topography and gravity anomalies. Recently, fast multidimensional wavelet transforms (Bergeron et al. 1999, 2000a,b; Yuen et al. 2000), based on second-derivatives of the Gaussian function, have been developed, allowing us to construct rapidly two- and three-dimensional wavelet-transforms of geophysically relevant fields, such as geoid anomalies, temperature-fields in high-Rayleigh-number convection, and mixing of passive heterogeneities. These same techniques will be useful also for characterization of mineralogical/geochemical systems. They have already been put to use in materials science investigations involving high-resolution transmission electron microscopy (HRTEM) (Jose-Yacaman et al. 1995) and in atomic force microscopy (AFM) (Duparre et al. 1999; Moktadir and Sato 2000).

We can define the wavelet transform *W(a, b)* as the transformation of a signal *f* in Cartesian space by the three-dimensional integral

$$W(a,\vec{b}) = \frac{1}{a^{3/2}} \int_0^{Lx} \int_0^{Ly} \int_0^{Lz} f(x)\psi\left(\frac{(x-\vec{b})}{a}\right) d^3x \tag{3}$$

where L_x, L_y, and L_z are the lengths of the periodic box in Cartesian space, a is the scale or the magnification and b is the position vector. The mother wavelet, also known as the convolution kernel or filter (Yuen et al. 2001), is given by $\psi(a,b)$. We have used the higher-order derivatives of the Gaussian function for computational purposes because of its analytical advantage in the transformed domain during the convolution process (Yuen et al. 2001). Instead of having a single parameter, the wave number k in Fourier analysis, there are now two parameters in isotropic wavelets, namely, a and b. This will increase the dimensionality of wavelet transform by one. Thus, a one-dimensional wavelet transform needs a two-dimensional plane (a, b) for describing its distribution, while a two-dimensional wavelet transform would need a three-dimensional volume (a, b_x, b_y) for portraying its multiscale distribution on a plane. For one-dimensional data, these plots are commonly shown for time series (Gibert et al. 1998; Vecsey and Matyska 2000) or in the spatial domain over the Laurentide ice sheet (Simons and Hager 1997).

Wavelet analysis has great potential in image processing applications of interest in mineralogy (see Moktadir and Sato (2000) for an illustrative example for silicon). As an illustration, in Figure 4 we show a version of Equation (3) over a one-dimensional trace across a two-dimensional AFM image of a hematite surface where there are some traces of bacterially mediated reduction reactions. One-dimensional wavelets with the second-derivative of the Gaussian function, also known as Mexican-hat wavelets because of their

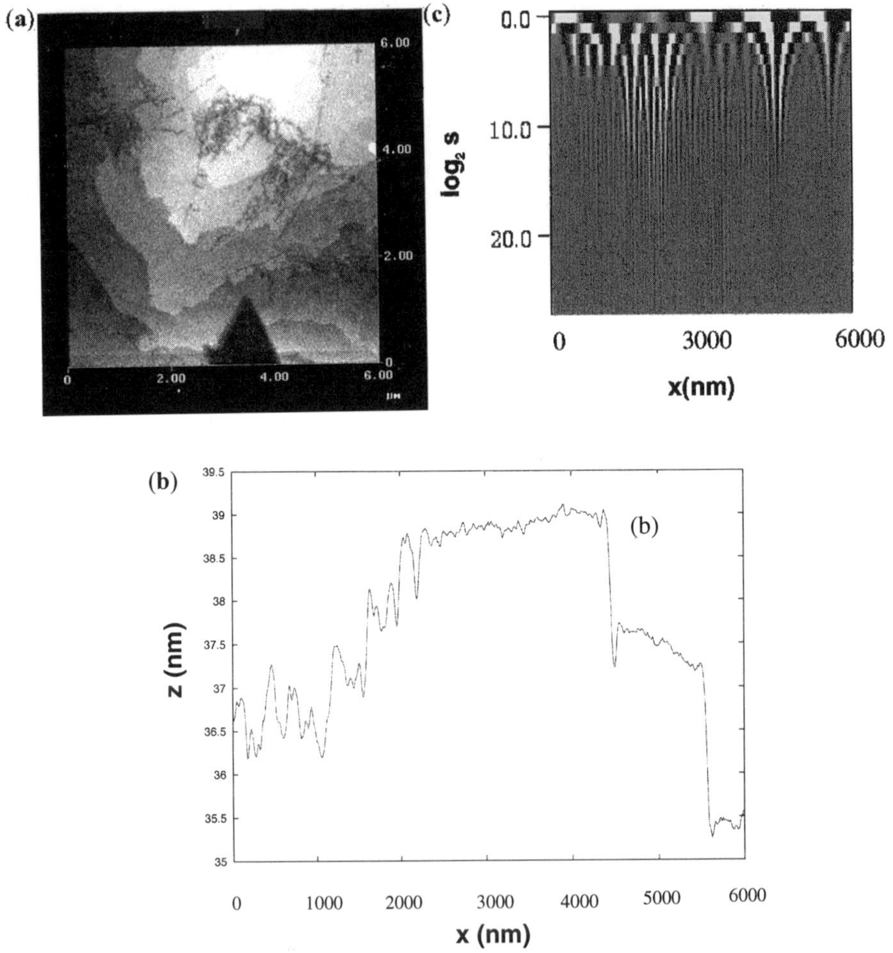

Figure 4. Multiscale aspects of oxide surfaces in aqueous environments. (b) represents a single trace across a hematite (001) surface undergoing biotically mediated reductive dissolution (a). Relatively clean steps 2.6 nm in size on the left of the figure give way to a complex surface morphology on the right side of the image where biotic dissolution was pervasive. The wavelet transform, to $1/2^7$ levels, is shown in (c). (b) provided by Kevin Rosso, Pacific Northwest National Laboratory.

shape, are used over this path. The inset shows the contours of the function $W(a,b)$ plotted over the (a,b) plane, with the ordinate being the scale and the abscissa the position b along the path. Long-length scale features are shown for a near zero, with smaller-scale features being displayed with increasing value of the scale along the descending direction of the ordinate. One can discern clearly the two sharp peaks of high strength in $W(a,b)$ and how they correlate directly with the two precipitous drops in the path topography. Also note the loss of the long length-scale features in the biologically dissolved region.

One can improve on the resolvability of the fine structures in Figure 4 by using a mapping involving the display of the two proxy quantities $E_{max}(x)$ and $k_{max}(x)$ as a function of the horizontal axis x, where E_{max} is the maximum of the L2-norm of the

W(a,b) over all scales *a*, and k_{max} is the local wave number associated with the scale a_{max}, where E_{max} takes place. This is a form of data-compression, in which one focuses only on the maximum strength of the wavelet signal. It has been used successfully to look at three-dimensional data sets such as three-dimensional tomographical models of the mantle (Bergeron et al. 1999) and gravity signals on the Earth's surface (Yuen et al. 2001). These same techniques, which have proved to be successful in global geophysics, can be brought to bear also on nanoscale phenomena in mineral physics and be equally promising, because we are dealing with Cartesian geometry. Wavelets have been applied in similar contexts in the analysis of shell growth in biomineralization (Toubin et al. 1999).

In the future, high-performance computing will play an increasingly important role in image analysis. One can imagine eventually taking Figure 2b and "photocopying" an atomically resolved representation suitable for use, for example, as initial conditions in a molecular simulation. Such applications would require integration of multiple imaging methods (AFM, scanning tunneling microscopy [STM], other scanning probe microscopies, nuclear magnetic resonance [NMR], HRTEM) with multiscale extrapolations to selective atomic-scale resolution.

MULTISCALE SIMULATION METHODS FOR SOLIDS

Molecular simulation methods at various scales are shown in Figure 5. Many of these methods are undoubtedly familiar to the reader. At the smallest scale are the electronic structure methods used to describe the formation of chemical bonds. These include Molecular Orbital (MO), density functional (DFT), and the so-called "tight binding" (TB) or semi-empirical methods. In molecular dynamics (MD) methods, the electronic degrees of freedom are parameterized out of the system, and we are left with atoms that interact with each other with pair-wise or higher-order interaction potentials. At the largest scales, the atoms are parameterized out of the system, and we are left with a mesh of finite elements that respond elastically (or viscoelastically) to driving forces at the boundaries of the system. On the left-hand side of Figure 5 lie the so-called kinetic Monte Carlo (KMC) methods, in which the atoms remain but the atomic *degrees of freedom* are projected out of the system, and in which the atoms are assembled not of their own accord but according to assumed probabilities for growth, dissolution, or reorganization events.

When phenomena at different scales are highly coupled, multiscale modeling methods, which simultaneously treat processes at disparate scales, are required. Many such examples are concerned with dynamically evolving chemical bond formation/dissociation in the presence of a continuum field. Crack propagation is a classic example of this type of inherently multiscale problem, involving close coupling between bond-breaking processes within the crack, elastically nonlinear response near the breaking bonds, and elastic or viscoelastic response at still longer scales. As the crack propagates, regions once allowing a simple continuum description may require atomistic treatment at later times, depending on where the crack decides to propagate. Here, as in other problems, much is gained by allowing for dynamic or adaptive multiresolution in which the relevant scale changes with time in a given region.

Another more mineralogically relevant example is crystal growth by assembly of polynuclear clusters or nanoparticles in aqueous solution (Penn and Banfield 1998). As the particles or clusters approach each other, hydrolysis of outer hydroxide functional groups yields water molecules that are eliminated progressively from the interfacial region. At some later time, the particles are attached and no longer require an atomistic description at the shared interface. The oxidation of a metallic nanoparticle such as

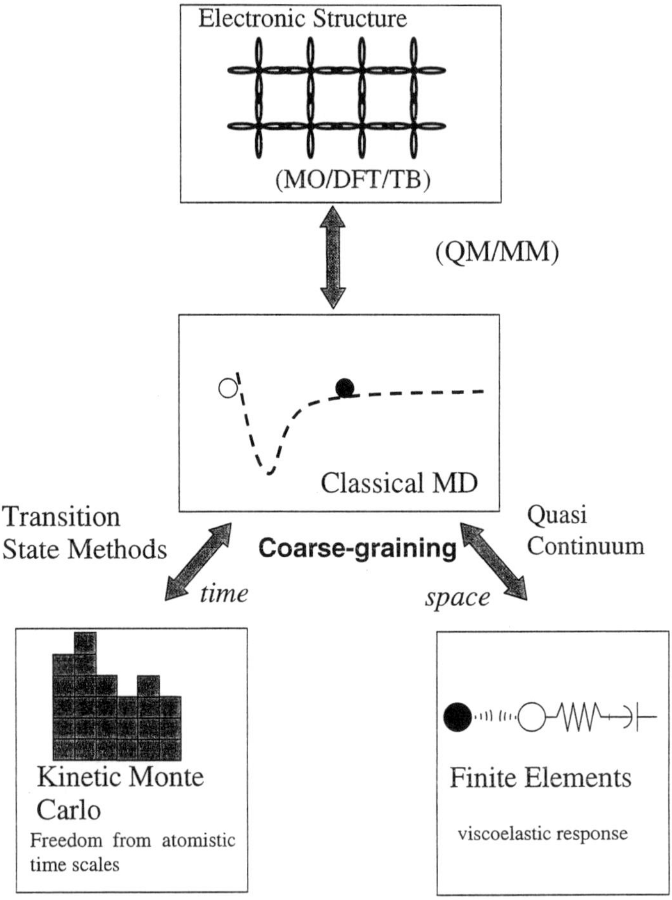

Figure 5. Molecular modeling methods at various scales. MO—molecular orbital; DFT—density functional theory; TB—tight binding; QM/MM—hybrid quantum mechanics/molecular mechanics; MD—molecular dynamics.

shown in Figure 2b would be another example of an inherently multiscale problem.

Large-scale molecular dynamics methods

The simplest approach to multiscale modeling is simply to enlarge the size of an MD calculation so that it spans many scales. Beginning in approximately 1990, million-atom MD simulations were being carried out on parallel computers (e.g., Swope and Anderson 1990). Parallelization is relatively simple for particles subject to short-range interactions such as exhibited by Lennard-Jones atoms (Smith 1991; Plimpton 1995). One simply sorts the atoms into different regions and carries out the force evaluations for each region in parallel with the other regions. Long-range coulombic interactions essentially require that interactions be computed for all *i, j* pairs in the system, with no cutoffs allowed. Some researchers, however, do cut these interactions at some finite range. Alternatively, one can obtain machine precision for the coulomb sum using the fast multipole method (FMM) developed by Greengard and Rokhlin (1987). In multipole methods, the long-range interactions are expanded in multipoles, eliminating the need to evaluate every *i, j* interaction. The multipole methods were used originally in the context of gravitational simulations (Barnes and Hut 1986). What distinguishes the Greengard-Rokhlin FMM is the introduction of a local expansion in addition to the far-field multipole expansion. When only the far-field expansion is used, one obtains $O(N \log N)$ scaling. When both the far-field and local expansions are used, the method scales as $O(N)$. See Kalia et al. (2000) for discussion of the FMM as well as strategies for parallel implementation. While the FMM is the method of choice for systems with N $O(10^6)$, others such as particle-mesh Ewald summation (Straatsma et al. 2000) may be competitive for smaller sample sizes.

Another promising approach is to use specialized hardware (GRAPE) designed at the chip-level for rapid computations of pair-wise energy and force evaluations (Higo et al. 1994, Hut and Makino 1999; also see http://mdm.atlas.riken.go.jp)

An example of some interest in mineralogy is the calculation of the oxidation of a metallic Al nanoparticle carried out by Campbell et al. (1999). In this study, a spherical crystalline metallic Al particle with a radius of 20 nm (252 158 atoms) was placed in an atmosphere of 265 360 O_2 atoms confined in a hypothetical container of radius 40 nm (at which reflecting boundary conditions are imposed). The equations of motion for the system were integrated for approximately 500 ps. The system was maintained at a temperature of 400 K. The major findings were that large stress gradients on the order of 1 GPa/nm controlled the diffusion of atoms into the nascent oxide layer. The saturated oxide thickness was in good agreement with experimental measurements of oxide layer thickness measurements in real systems. The oxide layer formed was a mixed octahedral-tetrahedral amorphous phase with a density about 75% of corundum.

As impressive as this calculation is, it still leaves us well short of being able to describe the system illustrated in Figure 2b, where the oxide layer is clearly crystalline. The thickness of the oxide layer may not be strongly influenced by the crystallinity, but certainly the chemical reactivity of the oxide layer will be influenced by nanometer-scale heterogeneities. The question of the actual structure of the oxide layer is entirely outside the range of direct MD simulation, even with the best computers and state-of-the-art computational techniques. Thus, if we want to use simulation to help answer geochemical questions—for example, why do iron nanoparticles of different sizes give rise to different reaction products in the reduction of nitrate (Choe et al. 2000) and CCl_4 (Lien and Zhang 1999)—a different approach is required.

Coupling methods

In some cases, it may be desirable to treat different parts of the system under consideration using different methods and use a truly multiresolution technique. This approach is illustrated in Figure 6. The example used here is the attachment of two oxide particles in aqueous solution. During this process, core atoms can probably be treated differently from reactive surface atoms. One might, in fact, abandon an atomistic representation of the core in favor of a continuum model. Solvent far removed from the active regions also could be treated by continuum methods. However, the interfacial water, especially that generated from dewatering reactions 2 SOH $\rightarrow$ SOS + H_2O (where S represents a surface cation) occurring during grain boundary healing, will need to be treated atomistically.

The general approach of simultaneously treating multiple-length scales has been pioneered by several research groups (Tadmor et al. 1993; Bulatov et al. 1998; Broughton et al. 1999; Tadmor et al. 2000; Rafii-Tabar 2000; Rudd and Broughton 2000). The general idea, of course, is to embed one type of calculation into another type of calculation. The concept is familiar in computational chemistry (Cramer and Truhlar

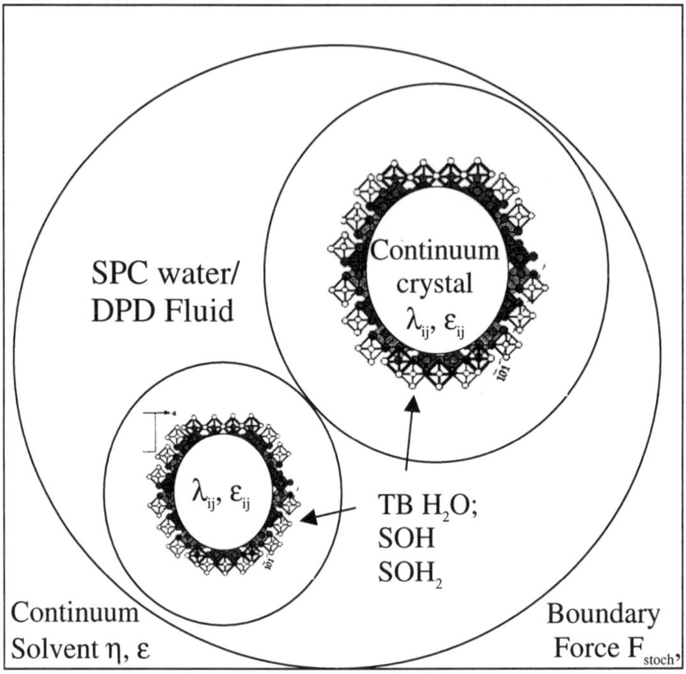

Figure 6. Hypothetical coupling of length scales in the attachment of two hydroxylated oxide nanoparticles in aqueous solution. The crystal cores are represented by continuum finite elements with elastic moduli λ_{ij} and dielectric tensor ε_{ij}. The far-field continuum solvent has viscosity η, dielectric constant ε, and exerts random boundary forces F_{stoch} on the fluid inside the large sphere modeled using particle methods. SPC is a simple point charge model for water. Dissipative particle dynamics or Lattice Boltzman methods also may be used here. Inside the small spheres are reactive water molecules modeled using tight-binding (TB) approaches. TB is also used to treat reactive surface functional groups.

1991; Fisher et al. 1991; Gao and Xia 1992; Tomasi and Persico 1994)). Broughton et al. (1999) provide a very clear and detailed discussion of the various issues involved in coupling the electronic structure, molecular dynamics, and finite element regions; we refer the reader to that paper for further discussion of these issues.

Here it suffices to say that the coupling interface between the finite element mesh and the atomic coordinates in the molecular dynamics is accomplished through resolving the near part of the finite element mesh on the atomic coordinates. On either side of the interface, the atoms and the finite element mesh overlap. Finite element cells that intersect the interface and atoms that interact across the interface each contribute to the Hamiltonian at half strength. See Rudd and Broughton (1998) for further discussion of the relationship between molecular dynamics and finite element methods.

In the studies discussed in Broughton et al. (1999), semi-empirical tight binding methods were used to represent the electronic structure. Tight binding methods are gaining popularity in large-scale simulations requiring a representation of the electronic structure (Schelling and Halley 1998; Frauenheim et al. 2000). In tight binding methods, orbitals ϕ_{ia} are imagined at each atomic center *i*,· for example, one might consider an *s* and p_x, p_y, p_z orbitals on an oxygen atom. The Hamiltonian matrix elements $\langle\phi_{ia}|H|\phi_{jb}\rangle$ are then parameterized based on empirical data or density functional theory (see Slater and Koster 1954; Papaconstantopolous 1986). There are several reasons for using tight binding methods for the electronic structural part of the problem. Foremost, they are the most generally applicable and most commonly used electronic structure methods that can be applied to from hundreds to thousands of atoms in a reasonable amount of computational time. Second, the inherent flexibility of the tight binding approach allows creativity in coupling the tight binding region to the MD region.

In Broughton et al. (1999), the boundary between the tight binding region and the MD region was occupied by fictitious atoms ("silogen" atoms) whose tight binding parameters are modified to give the correct Si-Si bond length, binding energies, and forces. This is similar to the familiar technique of terminating a cluster representation of an extended system using hydrogen atoms (Gibbs 1982) but now with the possibility of modifying the hydrogen atoms to couple with the MD region. These modified atoms are not coupled directly with each other in the tight binding Hamiltonian but they experience forces due to other non-silogen atoms in the tight binding region as well as from the MD region.

Transition state searching and kinetic Monte Carlo techniques

The major limitation of the approaches to multiscale modeling discussed thus far is the timescale. In each of these examples, there are atomic vibrations (on the order of 10^{-14} seconds) that need to be followed. This pins down the total simulation time to $O(10^{-9})$ seconds for reasonable calculations. There are many clever multiple time step methods for improving efficiency (e.g., Nakano 1999) by using a quaternion/normal mode representation for atoms that are simply vibrating or rotating, but this buys only a factor of $O(10)$.

In some kinds of systems—for example the "mineral part" (or even the near interfacial water) of the mineral-water interface—the time evolution of the system is characterized by uneventful periods of atoms vibrating around some average location. Only infrequently do some of them collectively assemble, pass through a transition state, and make an interesting conformational move. One needs to somehow "fast-forward" through the tedious parts of the simulation, find the interesting scenes, and use those to advance the system from state to state. The idea is to search explicitly for transition states instead of randomly stumbling around and taking the first one that comes along.

For small numbers of particles, there are "mode following" methods familiar in quantum chemistry, based on either explicit evaluation of the matrix of second derivatives or guesses at this matrix using first derivatives, and using the normal modes to locate transition states. This explicit method scales poorly with the number of atoms and is not feasible in large-scale simulations. Henkelman and Jónsson (1999) have suggested the so-called "dimer method" for saddle point searches. In this method, two system images slightly displaced from one another are used to define the direction of lowest curvature. Having defined this direction, the dimer is translated based on an effective force in which the force component along the direction of lowest curvature is inverted, driving the system toward a saddle point. Sørenson and Voter (2000) have suggested a method (temperature accelerated dynamics) in which multiple transition paths are located through standard high-temperature MD simulations restricted to a single conformational state. The simulation performs periodic quenching to check whether a new state has been obtained; the criterion is changes in atomic positions in the quenched configurations. If a new configuration is detected, the transition state is located using the nudged elastic band method Henkelman et al. (2000) that solves the problem of locating a transition state between any two local minima. Using these methods, processes on the order of hours can be investigated with molecular modeling approaches.

As yet, these methods have not been applied in very large-scale simulations, but Henkelman and Jónsson (1999) have shown that the method is relatively insensitive to increasing the phase-space dimensionality of the system, so long as the system is large enough to allow all collective relaxations to take place.

The limitations of this theory are: (1) the applicability of harmonic transition state theory (which is rarely an issue for the kind of accuracies typically required in geochemical/mineralogical problems), and (2) the sparsity of transition states: the dimer method, as presently formulated, finds *any* transition state. If many of these are not of interest, as might be the case for diffusion barriers on the "water" side of the mineral-water interface, the method would be impractical. This points out another advantage to the multiresolution approach: keeping the extra degrees of freedom of atoms in a region where one could get by with a continuum approach would, for example, require a reformulation of the dimer method.

In some cases, it may be possible to guess at the complete registry of configuration possibilities and kinetic processes. For example, atoms may be fixed on a specific lattice, and events such as attachment or dissolution may be assumed to occur with a specified probability on this fixed lattice. The system then evolves via Monte Carlo simulation over the degrees of freedom permitted by the events that are assumed to occur (Voter 1986). This type of simulation is referred to as the "kinetic Monte Carlo" (KMC) approach. There have been several applications using this method to simulate crystal growth in mineralogical systems (Blum and Lasaga 1987; Wehrli 1989; McCoy and LaFemina 1997). The obvious drawback to this method is that the elementary processes can come only from the imagination of the investigator. Some surface relaxation processes in simple metal systems are highly collective, and the fundamental events required for the KMC scheme can be very hard to guess in advance (Feibelman 1998). Similarly, collective behavior may be indicated on oxide surfaces in the work of Henderson et al. (1998), in which complete mixing of singly and triply coordinated oxide surface sites was observed to take place in hours to minutes. Because the activation barrier for a single exchange is probably prohibitive, the mechanism by which this mixing takes place is almost certainly highly collective. Despite its shortcomings, the KMC method is really the only approach for large problems on long time scales. Using the transition state searching method in conjunction with the KMC method, to identify collective conformer rearrangements for a traditional KMC approach, may be practical.

MULTISCALE COMPUTATIONAL METHODS FOR FLUIDS

Solvent effects on chemical reactivity are very large. For example, it takes approximately 395 kcal/mol to dissociate a water molecule in the gas phase but approximately 25 kcal/mol to do this in water at 298 K. Solvent obviously exerts a profound influence on the rates of chemical reactions as well. Nanoscale hydrodynamic effects couple strongly with nanoscale surface structure heterogeneities (including nanopores) and give rise to strong intrinsically nanoscale solvent effects. Of course, these effects must be described with multiscale models for fluids, which have evolved along lines somewhat different from those for solids. However, the general trend is similar, in that discrete particle methods are used to simulate fluids in complex, irregular environments. Here, we provide a review of discrete particle methods in simulation of complex fluids. The techniques are summarized in Figure 7.

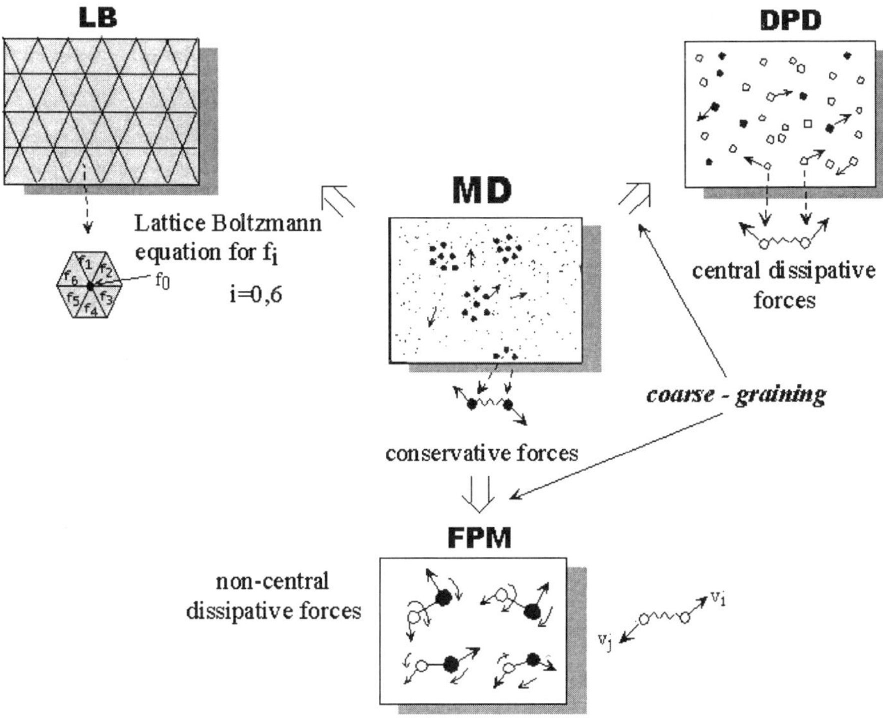

Figure 7. Discrete particle methods for simulation of fluids. LB (lattice Boltzmann), MD (molecular dynamics), DPD (dissipative particle dynamics), FPM (fluid particle model).

Dissipative particle dynamics

Microscopic techniques such as MD and MC are very useful in studying interactions between "primary particles" that form molecules and microstructures. However, MD becomes demanding for simulating larger systems. Most of the computations yield information on microscopic fluctuations, which are inessential in scales of complex fluids ordering. The extension of discrete-particle methods to larger spatial scales can be realized by changing the notion of the inter-particle interaction potential by treating a large-sized particle as a cluster of computational molecules. This idea of upscaling has been followed in the dissipative particle dynamics (DPD) method.

Dissipative particle dynamics (Hoogerbrugge and Koelman 1992) is one of the mesoscopic techniques based on the discrete particles paradigm (SPH, smoothed particle hydrodynamics is another; Monaghan 2000). This off-lattice algorithm (meaning the particles can be anywhere) was inspired by the idea of coupling the advantages of both the molecular dynamics and lattice-gas methods. Unlike atoms in molecular dynamics, the dissipative particles employed in DPD represent mesoscopic portions of a real fluid. Particles can be viewed as "droplets" of liquid molecules with an internal structure and with some internal degrees of freedom. The forces acting between dissipative particles are central and consist of the superposition of conservative $\mathbf{F}_C$, dissipative $\mathbf{F}_D$ and Brownian $\mathbf{F}_B$ components. As in molecular dynamics, the temporal evolution of the particle ensemble obeys the Newtonian equations of motion.

As shown in Hoogerbrugge and Koelman (1992), Marsh et al. (1997), and Español (1998), the interactions among DPD particles are postulated from simplicity and symmetry principles. The DPD forces are mesoscopic because they only resolve the center-of-mass motion of the droplets and do not give any detailed description of their internal degrees of freedom. As was proved in Español (1998), the one-component DPD system obeys the fluctuation dissipation theorem defining the relationship between the dissipative and Brownian forces. Marsh et al. (1997) have given a solid background for DPD as a statistical mechanics model. It provides explicit formulae for the transport coefficients in terms of particle interactions.

These principles ensure correct hydrodynamic behavior of DPD fluid. The advantage of DPD over other methods lies in the possibility of matching the scale of discrete-particle simulation to the dominant spatio-temporal scales of the entire system. For example, in MD simulation the timescales associated with evolution of heavy colloidal particles are many orders of magnitude larger than the temporal evolution of solvent particles. If the solvent molecules are coarse-grained into DPD droplets, they evolve much more slowly and are able to match the time scales close to those associated with the colloidal particles.

Dissipative particle dynamics can be employed also for simulating hydrodynamic instabilities. Dzwinel and Yuen (1999) present the algorithm applied for simulation of thin film falling down the inclined plane. In other studies, such as Clark et al. (2000), Dzwinel and Yuen (2001), and Boryczko et al. (2000), more challenging problems are attacked; for example, droplet breakup and mixing in complex fluids.

In Dzwinel and Yuen (2000a,b; 2001), it was demonstrated that DPD fits very well for simulating multiresolution structures of complex fluids. Typical examples of complex fluids with large molecular structure include microemulsions, micellar solutions, and colloidal suspensions like blood, ink, milk, fog, paints, and partially crystalline magmatic melts (Larson 1999).

For complex fluids, the gap in the spatio-temporal scales between the smallest microstructures and the largest structures is much smaller than for simple fluids. Dzwinel and Yuen (2000b) have shown that by using moderate number of particles, we can simulate in two dimensions multiresolution structures ranging frommicellar arrays to the large colloidal agglomerates.

Agglomeration of particles

Despite the general emphasis on atom-by-atom growth, crystalline materials can grow also by accumulation of units of material larger than a single atom or small atomic cluster. Growth of single crystals by oriented aggregation of clay platy particles is well known (e.g., fundamental particle theory for formation of interlayered layer silicates), as is ordered crystallization of large organic atomic clusters such as proteins and large inorganic polynuclear clusters such as Al_{13} (see review by Casey and Furrer, this volume) and porphyrins (Lauceri et al. 2001). More recently, the importance of growth of nanocrystals of oxides and zeolites by crystallographically specific attachment and interface elimination has been reported (Penn and Banfield 1998; Banfield et al. 2000).

Homogenous nucleation is a fundamental step in the crystallization of many solids. The mechanisms by which nuclei form, as well as the pathways for subsequent growth, are of basic scientific interest. Control over these phenomena is essential for materials design and prediction of how materials properties will evolve over time.

Theoretical treatment of nucleation processes has evolved along two directions. Hettema and McFeaters (1996) refer to these, respectively, as "classic nucleation theory" and "the kinetic approach." The kinetic approach is based on a set of time-dependent coupled ordinary differential equations (the Smoluchowski equation) and chemical rate constants (kernels) that are assumed to have non-integer scaling properties with cluster size. The nucleating systems are described in terms of the distribution of cluster sizes as a function of time.

The Smoluchowski equation is given by

$$\frac{dn_k}{dt} = \frac{1}{2} \sum_{i+j=k} \beta(i,j) n_i n_j - \sum_{i=1}^{\infty} \beta(i,k) n_i n_k \tag{4}$$

where, in the first term in Equation (4), i and j represent discrete particle sizes that combine to form a particle of size k. The second term represents the loss of particles of size k due to aggregation with other particles of size i. $\beta(i,j)$ is the kernel that is the rate of collision between particles i and j.

Most numerical techniques employed for aggregation simulation are based on the equilibrium growth assumption and on the Smoluchowski theory. As shown in Meakin (1988, 1998), analytical solutions for the Smoluchowski equation have been obtained for a variety of different reaction kernels; these kernels represent the rate of aggregation of clusters of sizes x and y. In most cases, these reaction kernels are based on heuristics or semi-empirical rules.

For complex kernels and for collecting information about aggregation kinetics, many simulation techniques were devised (Meakin 1998). They are based mainly on diffusion-limited aggregation (DLA) and ballistic off-lattice and on-lattice methods. These methods still are far from physical realism. They can be useful for investigating static fractal structures of large agglomerate in the absence of solvent. Moreover, a low initial concentration of colloidal particles has to be assumed. The rheological properties of solvent and the mechanisms of aggregation change with increasing concentration of particles. The reaction kernels poorly reflect actual cluster-cluster and cluster-solvent interactions, which vary with time. Especially, the kernels must be different for a perfectly mixed system and for a system with well-established clusters.

In Dzwinel and Yuen (2000b), we presented a numerical model in which both the colloidal particles and solvent are represented by interacting particles. The time scales associated with evolution of heavy colloidal particles are many orders of magnitude larger than the temporal evolution of solvent molecules. Therefore, we employ DPD for simulating solvent to bridge the two disparate scales. The DPD particles are larger and evolve much slower than molecules, making it possible to match the time scales close to those associated with the colloidal particles. The difference between colloidal and DPD particles consists in the different particle-particle interactions employed. To avoid fluidization of the colloidal particle system and facilitate aggregation, the colloid-colloid interactions should be conservative; they should posses a hard-sphere core with a very short-ranged adhesive part. They model the electro-chemical and depletion interactions acting between colloidal particles in the real solid-in-fluid mixtures.

As shown in Figure 8, by introducing conservative Lennard-Jones interactions between the colloid particles, we can simulate the spontaneous creation of micelles and their clustering in two-dimensional crystal arrays (see Fig. 8a) and large-scale fractal agglomerates (Fig. 8b). The growth of the fractal agglomerates is reflected by the scaling properties of mean cluster size $S(t)$—expressed in number of particles—with time. The sophisticated shapes of cluster can be recognized by using a clustering procedure based on the mutual nearest-neighboring distance concept. This procedure is outlined in Dzwinel and Yuen (2000b).

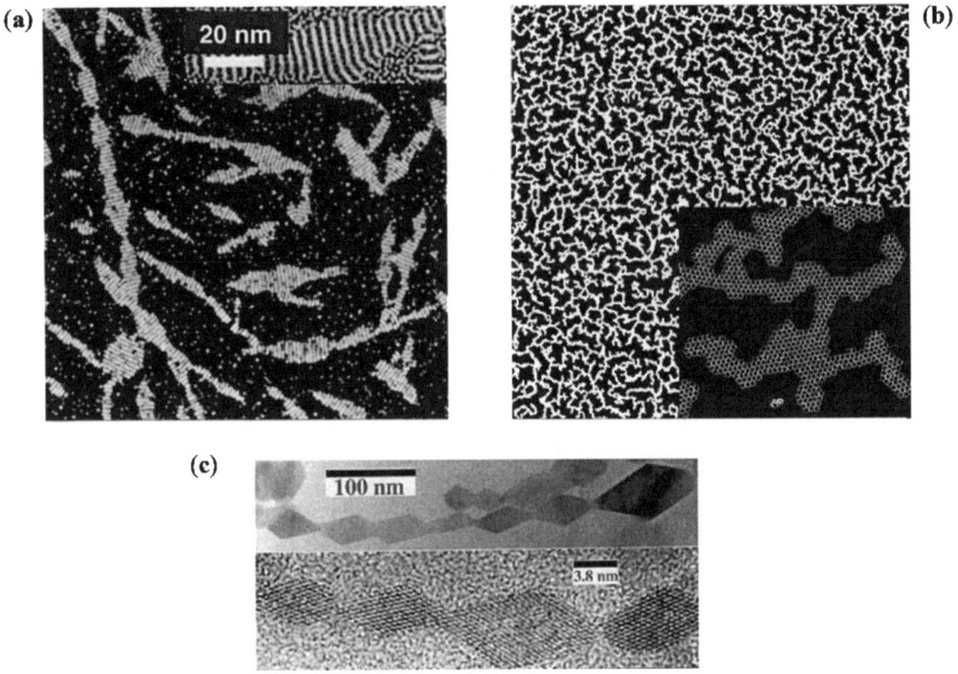

Figure 8. Snapshots from spontaneous agglomeration simulated by using an MD-DPD algorithm. (a) Crystal arrays compared to (c) chain of titania nanocrystals in pH 3 solution. (b) Colloidal aggregates [(c) used with permission of Elsevier, publisher of *Geochimica et Cosmochimica Acta*, from Penn and Banfield (1999), Fig. 6, p. 1552; see Fig. 19, p. 43, this volume, for a better figure.]

In Dzwinel and Yuen (2000c), we show that in DPD fluid with high concentration of colloidal particles, the growth of average size of agglomerate can be described by the power law $S(t) \propto t^{\kappa}$. For $\kappa = 0.5$, the intermediate DLA regime was found, which spans a relatively long time. The length of this relaxation time depends on physical properties of solvent as well as concentration of colloidal particles. The character of cluster growth varies with time, and the value of κ shifts for longer times from 0.5 to ~1. This result agrees well with the theoretical prediction for diffusion-limited cluster-cluster aggregation. It says that for $t \to \infty$, the value of κ approaches 1 for a low colloidal particle concentration. As shown in Dzwinel and Yuen (2000c), this process cannot be asymptotic for a larger concentration of colloidal particles.

Coarse-graining dissipative particle dynamics: fluid particle model

A serious drawback of DPD is the absence of a drag force between the central particle and the second one orbiting about the first particle. The dissipative force $\mathbf{F_D}$ representing the dot product of differential velocities between interacting particles and their relative position vector is then equal to zero. This relative motion may produce a net drag only when many particles are participating at the same time (Español 1998). This cumulative effect requires more particles to be involved and reduces the computational efficiency of the DPD method.

With the aim of coarse-graining DPD, the fluid particle method (FPM), a non-central force has been introduced that is proportional to the difference between the velocities of the particles (Español 1998). This force exerts additional drag, which produces rotational motion. This would allow for the simulation of physical effects associated with rotational diffusion and rotation of the colloidal beds resulting from hydrodynamics or their mutual interactions.

The fluid particles possess several attributes as mass m_i, inertia, position $\mathbf{r}_i$, translational and angular velocities, $\mathbf{v}$, $\boldsymbol{\omega}$, and a force law. The "droplets" interact with each other by forces dependent on the type of particles. We use the two-body, short-ranged force $\mathbf{F}$ as it is postulated in Español(1998). This type of interaction is a sum of conservative force $\mathbf{F}^C$, two dissipative forces with translational and rotational parts $\mathbf{F}^T$ and $\mathbf{F}^R$, and a random Brownian force $\tilde{\mathbf{F}}$; that is,

$$\mathbf{F}_{ij} = \mathbf{F}_{ij}^C + \mathbf{F}_{ij}^T + \mathbf{F}_{ij}^R + \tilde{\mathbf{F}}_{ij} \tag{5}$$

$$\mathbf{F}_{ij}^C = -F\left(r_{ij}\right) \cdot \mathbf{e}_{ij} \tag{6}$$

$$\mathbf{F}_{ij}^T = -\gamma \cdot m\mathbf{T}_{ij} \cdot \mathbf{v}_{ij} \tag{7}$$

$$\mathbf{F}_{ij}^R = -\gamma \cdot m\mathbf{T}_{ij} \cdot \left(\frac{1}{2}\mathbf{r}_{ij} \times \left(\omega_i + \omega_j\right)\right) \tag{8}$$

$$\tilde{\mathbf{F}}_{ij}dt = \left(2k_B T\gamma \cdot m\right)^{1/2} \mathbf{W}_{\mathbf{ij}} \cdot \mathbf{e}_{ij} \tag{9}$$

$$\mathbf{T}_{ij} = A\left(r_{ij}\right)\ 1 + B\left(r_{ij}\right)\ \mathbf{e}_{ij}\mathbf{e}_{ij} \tag{10}$$

where F, A, and B are weighting functions and $\mathbf{W}_{\mathbf{ij}}$ is a stochastic tensor defined in Español (1998).

As shown in Figure 9, the FPM model represents a generalization not only of DPD but also of the MD technique. It can be used as DPD by setting the noncentral forces to zero, or as MD by dropping the dissipative and Brownian components. These three techniques also can be combined into one three-level hybrid model. As shown in Figure 9, the three-level system consists of three different procedures representing each technique invoked in dependence on the type of particle interactions. We define three types of particles:

"Hard" colloidal particles. The interactions between colloidal particles can be simulated by a soft-sphere, energy-conserving force with an attractive tail. The force vanishes for particles separated by the distance greater than 2.5 λ (λ is a characteristic length, equal to the average distance between particles).

"Soft" dissipative particles. They represent "clusters" of molecules located in the closest neighborhood of the colloidal particles with an interaction range $\geq$2.5 λ. The DPD-DPD and DPD-MD interactions represent only the two-body central forces.

Fluid particles (FP) are the "lumps of fluid" particles in the bulk solvent, with interaction range $\leq$1.5 λ. Noncentral forces are included within this framework.

Mesocopic flows are important to understand because they hold the key to the interaction between the macroscopic flow and the microstructural inhomogeneities. This is especially true in colloidal flows, which involve colloidal mixtures, thermal fluctuations and particle-particle interactions. Dynamic processes occurring in the granulation of colloidal agglomerate in solvents are severely influenced by coupling between the dispersed microstructures and the global flow. On the mesoscale, this

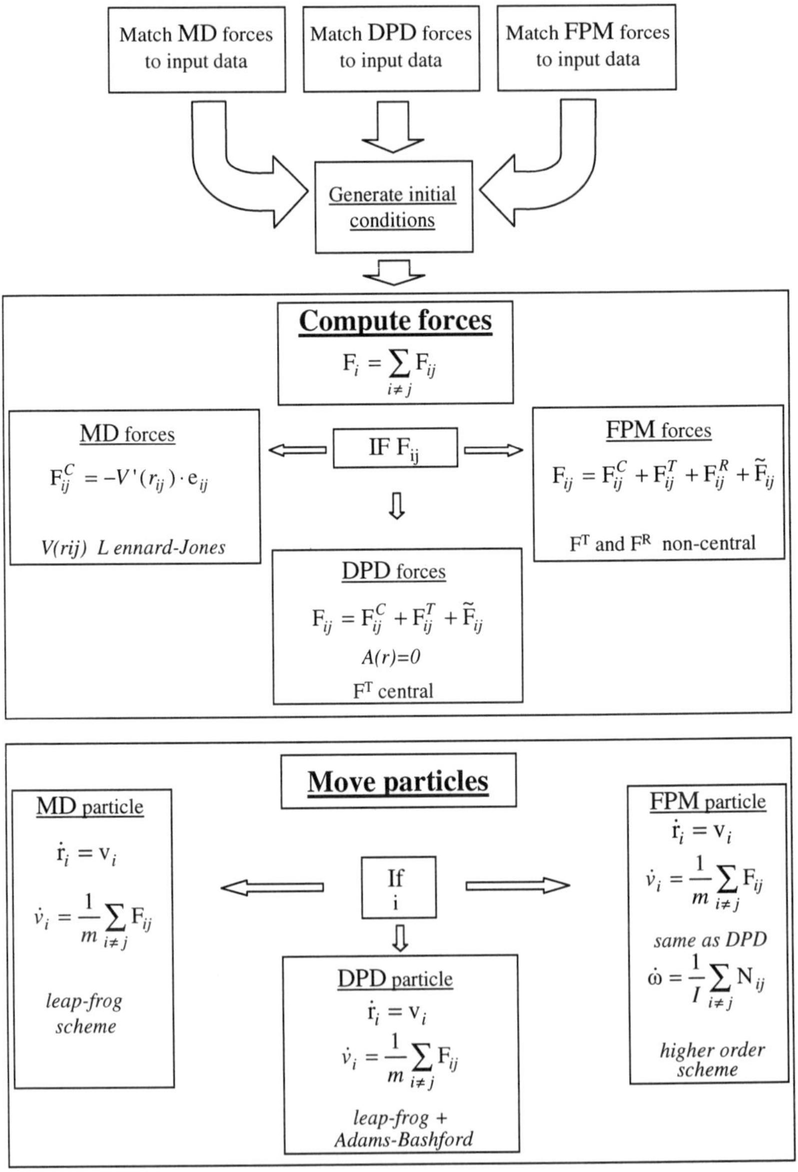

Figure 9. Three-level fluid particle model for coarse-graining DPD simulations.

coupling is further exacerbated by thermal fluctuations, excluded volume effects, cohesive interactions between colloidal beds, and hydrodynamic interactions between colloidal beds and the solvent.

The dispersion of an aggregated composite generally proceeds through three stages, which usually occur with some degree of overlap:

Imbibition – consisting of spreading off the liquid solvent into the colloidal cluster and reducing the cohesive forces between the colloidal beds.

Fragmentation – consisting of

shatter – producing a large number of smaller fragments in a single event,

rupture – breakage of a cluster into several fragments of comparable size,

erosion – gradual shearing off of small fragments of comparable size (Ottino et al. 2000)

Aggregation – the reverse of dispersion. Two traditional mechanisms can be recognized:

nucleation –the gluing together of primary particles due to the attractive forces,

coalescence – the combination of two larger agglomerates to form a granule.

In the framework of the FPM model, we have simulated the dispersion and agglomeration microstructures, which appear during acceleration of a slab in FPM fluid. Figure 10 displays the moment of the slab disintegration and agglomeration of the slab remnants. The fragmentation occurs due to a shatter mechanism, which is generated by accumulation of a large amount of energy during compression and its fast release in the decompression stage.

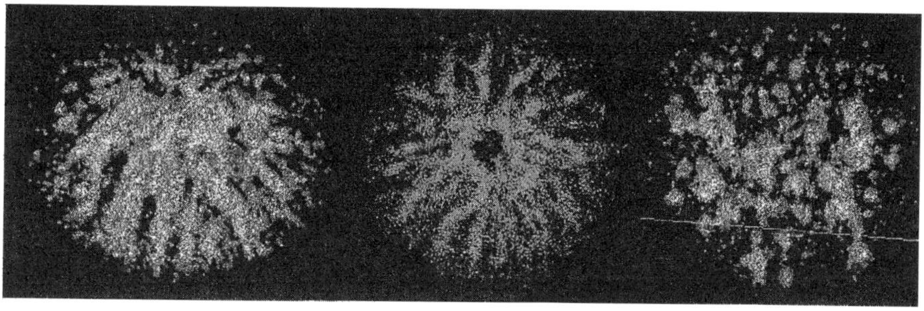

(a) (b) (c)

Figure 10. Dispersion of colloidal slab made of "hard" MD particles in FPM fluid. (a) Dispersion micro-structures (DataExplorer surface). (b) View from below (particles are displayed). (c) Agglomeration microstructures.

In Figure 11, we display the changes of the largest cluster size and thermodynamic pressure with time. Initial compression of the slab and its subsequent decompression causes the largest cluster disintegrating. The following decompression wave is too weak for disrupting cohesion forces, which contributes to the agglomeration of the slab remnants into larger droplets.

Within the FPM, we can extend further the capabilities of the discrete particle method to the mesoscopic regime and show that they are competitive to standard simulation techniques with continuum equations. These methods establish a foundation for cross-scale computations ranging from nanoscales to microns and can provide a framework for studying the interaction of microstructures and large-scale flow, which may be of value in blood flow and other applications in polymeric flows (Banfield et al. 2000; Schwertman et al 1999; Hiemstra and VanReimsdijk 1999).

OUTLOOK

Indeed, the energy and enthusiasm generated by the scientific community's recent focus on nanoscale phenomena presents an opportunity for making a significant leap in

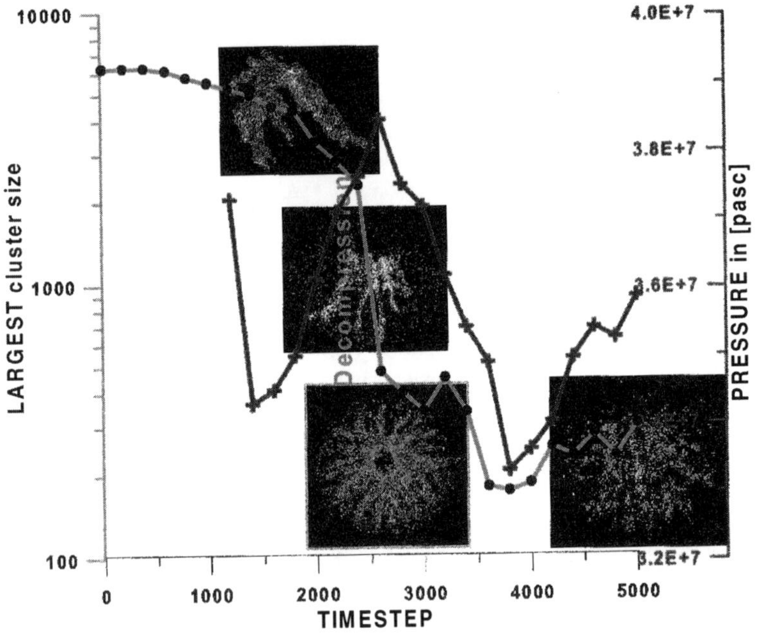

Figure 11. Dispersion of the colloidal slab in the FPM fluid. The plot presents the changes of the largest cluster size and thermodynamic pressure for the particle system with time. Simulation results come from FPM-MD two-level model.

the computational geosciences. In the last few years, revolutionary changes have occurred in the types of problems that can be investigated with computational methods because of the growth in computational power. Higher-resolution calculations and laboratory measurements also produce a data deluge, which we must confront if we are to comprehend fully all of the information. Modeling, feature extraction, and visualization techniques all undoubtedly will play important roles in developing conceptual models of what is really important in determining structure-reactivity relationships at the nanoscale. As is often the case in other contexts, we will need to motivate unique interdisciplinary combinations of research techniques not found in a typical academic department, such as molecular modeling, signal processing, scientific visualizaiton, pattern recognition, and artificial intelligence, as well a techniques in the laboratory, like scanning probe and HRTEM microscopies. Such a mode of scientific operation demands a new way of thinking.

ACKNOWLEDGMENTS

Authors Rustad and Yuen acknowledge support from the Geosciences Division, U.S. Department of Energy Office of Basic Energy Sciences. The authors are grateful to Kevin Rosso, Pacific Northwest National Laboratory, for providing AFM images for example calculations. We thank Andy Felmy, Dave Dixon, Michel Dupuis, and Eric Bylaska of Pacific Northwest National Laboratory and to the editors of this volume, Alex Navrotsky and Jill Banfield, for helpful comments on improving the manuscript. We thank Andrea Currie and Jamie Benward of Pacific Northwest National Laboratory and Heather Crull of the University of Minnesota for help in manuscript preparation.

REFERENCES

Bader RFW (1990) Atoms in Molecules. Oxford: Clarendon Press

Banfield JF, Welch SA, Zhang HZ, Ebert TT, Penn RL (2000) Aggregation-based crystal growth and microstructure development in natural iron oxyhydroxide biomineralization products. Science 289:751-754

Barabasi A-L, Stanley HE (1995) Fractal Concepts in Surface Growth. Cambridge: Cambridge University Press

Barenblatt GI (1996) Scaling, Self Similarity, and Intermediate Asymptotics. Cambridge: Cambridge University Press

Barnes J, Hut PA (1986) Hierarchical O(N-log-N) force-calculation algorithm. Nature 324:446-449

Bergeron SY, Vincent AP, Yuen DA, Tranchant BJS, Tchong C (1999) Viewing seismic velocity anomalies with 3-D continuous Gaussian wavelets. Geophys Res Lett 26:2311-2314

Bergeron SY, Yuen DA, Vincent AP (2000a) Capabilities of 3-D wavelet transforms to detect plume-like structures from seismic tomography. Geophys Res Lett 27:3433-3436

Bergeron SY, Yuen DA, Vincent AP (2000b) Looking at the inside of the Earth with 3-D wavelets: A new pair of glasses for geoscientists. Electronic Geosci 5:3

Blanco MA, Costales A, Pendas AM, Luana V (2000) Ions in crystals: The topology of the electron density in ionic materials. V. The B1-B2 phase transition in alkali halides. Phys Rev B 62:12028-12030

Blum AE, Lasaga AC (1997) Monte Carlo simulation of surface reaction rate laws in aquatic surface chemistry. *In* Chemical Processes and the Particle-Water Interface. Stumm W (ed) p 255-292. New York: John Wiley & Sons

Boryczko K, Dzwinel W, Yuen D (2000) Mixing and droplets coalescence in immiscible fluid: 3-D dissipative particle dynamics model, UMSI 2000/142. Minneapolis: University of Minnesota Supercomputing Institute

Bowman C, Newell AC (1998) Natural patterns and wavelets. Rev Modern Phys 70:289-301

Broughton JQ, Abraham FF, Bernstein N, Kaxiras E (1999) Concurrent coupling of length scales: Methodology and application. Phys Rev B 60:2391-2403

Bulatov V, Abraham FF, Kubin L, Devincre B, Yip S (1998) Connecting atomistic and mesoscale simulations of crystal plasticity. Nature 391:669-672

Caglioti E, Loreto V, Herrmann HJ, Nicodemi M (1997) A "tetris-like" model for the compaction of dry granular media. Phys Rev Lett 79:1575-1578

Campbell T, Kalia RK, Nakano A, Vashishta P, Ogata S, Rodgers S (1999) Dynamics of oxidation of aluminum nanoclusters using variable charge molecular dynamics simulations on parallel computers. Phys Rev Lett 82:4866-4869

Cannas C, Concas G, Falqui A, Musinu A, Spano G, Piccaluga G (2001) Investigation of the precursors of gamma-Fe_2O_3 in Fe_2O_3/SiO_2 nanocomposites obtained through sol-gel. J Non-crystal Solids 286:64-73

Cardone P, Ercole C, Breccia S, Lepidi A (1999) Fractal analysis to discriminate between biotic and abiotic attacks on chalcopyrite and pyrolusite. J Microbiol Methods 36:11-19

Choe S, Chang YY, Hwang KY, Khim J (2000) Kinetics of reductive denitrification by nanoscale zero-valent iron. Chemosphere 41:1307-1311

Clark AT, Lal M, Ruddock JN, Warren PB (2000) Mesoscopic simulation of drops in gravitational and shear fields. Langmuir 16:6342-6350

Cornell RM, Schwertmann U (1996) The Iron Oxides. Weinheim, FRG: Wiley-VCH

Cramer CJ, Truhlar DG (1991) General parameterized SCF model for free-energies of solvation in aqueous-solution. J Am Chem Soc 113:8305-8311

Cygan R, Kubicki JD (eds) (2001) Molecular Modeling Theory and Applications in the Geosciences. Reviews in Mineralogy and Geochemistry, vol 42. Washington, DC: Mineralogical Society of America

Dixon JB, Weed SB (1989) Minerals in Soil Environments. Madison, Wisconsin: Soil Science Society of America

Duparre A, Notni G, Recknagel RJ, Feigl T, Gliech S (1999) High resolution topometry in conjunction with macro structures. Technisches Messen 66:437-446

Dzwinel W, Yuen DA (2000a) A two-level, discrete-particle approach for simulating ordered colloidal structures. J Colloid Interface Sci 225:179-190

Dzwinel W, Yuen DA (2000b) Matching macroscopic properties of binary fluid to the interactions of dissipative particle dynamics. Int'l J Modern Phys C 11:1-25

Dzwinel W, Yuen DA (2000c) A two-level, discrete particle approach for large-scale simulation of colloidal aggregates. Int'l J Modern Phys C 11:1037-1061

Dzwinel W, Yuen DA (1999) Dissipative particle dynamics of the thin-film evoluation in mesoscale. Molecular Simul 22:369-395

Dzwinel W, Yuen DA (2001) Mixing driven by Rayleigh-Taylor instability in the mesoscale modeled with dissipative particle dynamics. Int'l J Modern Phys C 12:91-118.

Frauenheim T, Seifert G, Elstner M, Hajnal Z, Jungnickel G, Porezag D, Suhai S, Scholz R (2000) A self-consistent charge density-functional based tight-binding method for predictive materials simulations in physics, chemistry, and biology (2000) Physica Status Solidi B 217:41-62

Español P (1998) Fluid particle model. Phys Rev E 57:2930-2948

Espinosa E, Molins E (2000) Retrieving interaction potentials from the topology of the electron density distribution: The case of hydrogen bonds. J Chem Phys 113:5686-5694

Falconer K (1990) Fractal Geometry: Mathematical Foundations and Applications. New York: John Wiley and Sons

Feibelman PJ (1998) Interlayer self-diffusion on stepped Pt(111). Phys Rev Lett 81:168-171

Fisher AJ, Harding JH, Harker AH, Stoneham AM (1991) Embedded cluster calculations of defect processes. Rev Solid State Sci 2:133-147

Frankel RB, Blakemore RP (eds) (1999) Iron Biominerals. New York: Plenum Press

Gao JL, Xia XF (1992) A priori evaluation of aqueous polarization effects through Monte-Carlo QM-MM simulations. Science 258:631-635

Gibbs GV (1982) Molecules as models for bonding in silicates. Am Mineral 67:421-450

Gibbs GV, Hill FC, Boisen MB, Downs RT (1998) Power law relationships between bond length, bond strength and electron density distributions. Phys Chem Minerals 25:585-590

Gibert D, Holschneider M, Le Mouel J-L (1998) Wavelet analysis of the Chandler Wobble. J Geophys Res 103:27,069-27,089

Gier TE, Stucky GD (1991) Low-temperature synthesis of hydrated zinco(beryllo)-phosphate and arsenate molecular-sieves. Nature 349:508-510

Greengard L, Rokhlin V (1987) A fast algorithm for particle simulations. J Comput Phys 73:325-348

Henderson MA, Joyce SA, Rustad JR (1998) Interaction of water with the (1×1) and (2×1) surfaces of α-Fe_2O_3(012). Surface Sci 417:66-81

Henkelman G, Jonsson H (1999) A dimer method for finding saddle points on high dimensional potential surfaces using only first derivatives. J Chem Phys 111:7010-7022

Henkelman G, Uberuaga BP, Jonsson H (2000) A climbing image nudged elastic band method for finding saddle points and minimum energy paths J Chem Phys 113:9901-9904

Hettema H, McFeaters JS (1996) The direct Monte Carlo method applied to the homogeneous nucleation problem. J Chem Phys 10:2816-2827

Hiemstra T, Van Riemsdijk WH (1999) Effect of different crystal faces on experimental interaction force and aggregation of hematite. Langmuir 15:8045-8051

Higo J, Endo S, Nagayama K, Ito T, Fukushige T, Ebizuzaki T, Sugimoto D, Miyagawa H, Kitamura K, Makino J (1994) Application of a high-performance, special-purpose computer, GRAPE-2A, to molecular-dynamics. J Comput Chem 15:1372-1376

Holschneider M (1995) Wavelets: An Analysis Tool. Oxford: Oxford University Press

Hoogerbrugge PJ, Koelman JMVA (1992) Simulating microscopic hydrodynamic phenomena with dissipative particle dynamics. Europhys Lett 19:155-160

Hut P, Makino J (1999) Computational physics—Astrophysics on the GRAPE family of special-purpose computers. Science 283:501-505

Jose-Yacaman M, Diaz G, Gomez A (1995) Electron-microscopy of catalysts—The present, the future and the hopes. Catalysis Today 23:161-199

Kalia RK, Campbell TJ, Chatterjee A, Nakano A, Vashishta P, Ogata S (2000) Multiresolution algorithms for massively parallel molecular dynamics simulations of nanostructured materials. Computer Phys Commun 128:245-259

Kwok YS, Zhang XX, Qin B, Fung KK (2000) High-resolution transmission electron microscopy study of epitaxial oxide shell on nanoparticles of iron. Appl Phys Lett 77:3971-3973

Larson RG (1999) The Structure and Rheology of Complex Fluids. New York: Oxford University Press

Lauceri R, Campagna T, Raudino A, Purrello R (2001) Porphyrin binding and self-aggregation onto polymeric matrix: A combined spectroscopic and modelling approach. Inorganica Chimica Acta 317:282-289

Lee CK (2001) Effect of heating on the surface roughness and pore connectivity of TiO_2: Fractal and percolation analysis. J Chem Engin Japan 34:724-730

Lien HL, Zhang WX (1999) Transformation of chlorinated methanes by nanoscale iron particles. J Environ Engin-ASCE 125:1042-1047

Liu C, Huang PM (1999) Atomic force microscopy and surface characteristics of iron oxides formed in citrate solutions. Soil Sci Soc Am J 63:65-72

Lof RW, Vanveenendaal MA, Koopmans B, Heessels A, Jonkman HT, Sawatzky GA (1992) Correlation-effects in solid C-60. Int'l J Modern Phys B 6:3915-3921

Marsh CA, Backx G, Ernst MH (1997) Static and dynamic properties of dissipative particle dynamics. Phys Rev E 56:1676-1691

Maurice PA, Lee YJ, Hersman LE (2000) Dissolution of Al-substituted goethites by an aerobic *Pseudomonas mendocina* var. bacteria. Geochim Cosmochim Acta 64:1363-1374

McCoy JM, LaFemina JP (1997) Kinetic Monte Carlo investigation of pit formation at the $CaCO_3$ $(10\bar{1}4)$ surface-water interface. Surface Sci 373:288-299

Meakin P (1998) Fractals, Scaling, and Growth far from Equilibrium. Cambridge: Cambridge University Press

Meakin P (1988) Fractal aggregates. Adv in Colloid Interface Sci 28:249-331

Mintmire JW, White CT (1998) Universal density of states for carbon nanotubes. Phys Rev Lett 81: 2506-2509

Moktadir Z, Sato K (2000) Wavelet characterization of the submicron surface roughness of anisotropically etched silicon. Surface Sci 470:L57-L62

Monaghan JJ (2000) SPH without a tensile instability. J Comput Phys 159:290-311

Nakano A (1999) Rigid-body based multiple time-scale molecular dynamics simulation of nanophase materials. Int'l J High Perform Comput Appl 13:154-162

Orme CA, Noy A, Wierzbicki A, McBride MT, Grantham M, Teng HH, Dove PM, DeYoreo JJ (2001) Formation of chiral morphologies through selective binding of amino acids to calcite surface steps. Nature 411:775-779

Ottino JM, De Roussel P, Hansen S, Khakhar DV (2000) Mixing and dispersion of viscous liquids and powdered solids. Adv Chem Engin 25:105-204

Papaconstantopoulos DA (1986) Handbook of the Band Structure of Elemental Solids. New York: Plenum Press

Patarin J, Kessler H (2000) A "hole" new breakthrough. Recherche 334:18

Penn RL, Banfield JF (1998) Imperfect oriented attachment: Dislocation generation in defect-free nanocrystals. Science 281:969-971

Penn RL, Banfield JF (1999) Morphology development and crystal growth in nonocrystalline aggregates under hydrothermal conditions: Insights from titania. Geochim Cosmochim Acta 63:1549-1557

Penn RL, Oskam G, Strathmann GJ, Searson PC, Stone AT, Veblen DR (2001a) Epitaxial assembly in aged colloids. J Phys Chem B 105:2177-2182

Penn RL, Stone AT, Veblen DR (2001b) Defects and disorder: Probing the surface chemistry of heterogenite (CoOOH) by dissolution using hydroquinone and iminodiacetic acid. J Phys Chem B 105:4690-4697

Plimpton S (1995) Fast parallel algorithms for short-range molecular-dynamics. J Comput Phys 117:1-19

Rafii-Tabar H (2000) Modeling the nano-scale phenomena in condensed matter physics via computer-based numerical simulations. Phys Reports-Rev Section Phys Lett 325:240-310

Rescigno TN, Baertschy M, Issacs WA, and McCurdy CW (1999) Collisional breakup in a quantum system of three charged particles. Science 286:2474-2479

Resnikoff HL, Welss Jr RO (1998) Wavelet Analysis: The Scalable Structure of Information. New York: Springer-Verlag

Rudd RE, Broughton JQ (1998) Coarse-grained molecular dynamics and the atomic limit of finite elements. Phys Rev B 58:R5893-R5896

Rudd RE, Broughton JQ (2000) Concurrent coupling of length scales in solid state systems. Physica Status Solidi B-Basic Res 217:251-291

Rustad JR, Dixon DA, Felmy AR (2000) Intrinsic acidity of aluminum, chromium (III) and iron (III) mu(3)-hydroxo functional groups from *ab initio* electronic structure calculations. Geochim Cosmochim Acta 64:1675-1680

Schelling PK, Yu N, Halley JW (1998) Self-consistent tight-binding atomic-relaxation model of titanium dioxide. Phys Rev B 58:1279-1293.

Schwertmann U, Cornell RM (1991) Iron Oxides in the Laboratory. Weinheim: Wiley-VCH

Schwertmann U, Friedl J, Stanjek H (1999) From Fe(III) ions to ferrihydrite and then to hematite. J Colloid Interface Sci 209:215-223

Simons M, Hager BH (1997) Localization of the gravity field and the signature of glacial rebound. Nature 390:500-504

Slater JC, Koster GF (1954) Simplified LCAO method for the periodic potential problem. Phys Rev 94:1498-1524

Smith W (1991) Molecular-dynamics on hypercube parallel computers. Computer Phys Commun 62: 229-248

Straatsma TP, Philippopoulos M, McCammon JA (2000) NWChem: Exploiting parallelism in molecular simulations. Computer Phys Commun 128:377-385

Swope WC, Andersen HC (1990) 10^6-Particle molecular-dynamics study of homogeneous nucleation of crystals in a supercooled atomic liquid. Phys Rev B 41:7042-7054

Tadmor EB, Phillips R, Ortiz M (2000) Hierarchical modeling in the mechanics of materials. Int'l J Solids Structures 37:379-389

Tomasi J, Persico M (1999) Molecular-interactions in solution—an overview of methods based on continuous distributions of the solvent. Chem Rev 94:2027-2094

Toubin M, Dumont C, Verrecchia EP, Laligant O, Diou A, Truchetet F, Abidi MA (1999) Multi-scale analysis of shell growth increments using wavelet transform Computers Geosci 25:877-885

Turcotte DL (1997) Fractals and Chaos in Geology and Geophysics. Cambridge: Cambridge University Press

Turcotte DL, Schubert G (1982) Geodynamics: Applications of Continuum Physics to Geological Problems. New York: John Wiley & Sons

Varnek AA, Dietrich B, Wipff G, Lehn JM, Boldyreva EV (2000) Supramolecular chemistry—Computer-assisted instruction in undergraduate and graduate chemistry course. J Chem Educ 77:222-226

Vecsey L, Matyska C (2001) Wavelet spectra and chaos in thermal convection modeling. Geophys Res Lett 28:395-398

Voter AF (1986) Classically exact overlayer dynamics—Diffusion of rhodium clusters on Rh(100). Phys Rev B 34:6819-6829

Wasserman E, Rustad JR, Felmy AR, Hay BP, Halley JW (1997) Ewald methods for polarizable surfaces with application to hydroxylation and hydrogen bonding on the (012) and (001)surfaces of alpha-Fe_2O_3. Surface Sci 38:217-239

Wehrli B (1989) Monte-Carlo simulations of surface morphologies during mineral dissolution. J Colloid Interface Sci 132:230-242

Weidler PG, Degovics G, Laggner P (1998) Surface roughness created by acidic dissolution of synthetic goethite monitored with SAXS and N-2-adsorption isotherms. J Colloid Interface Sci 197:1-8

Xagas AP, Androulaki E, Hiskia A, Falaras P (1999) Preparation, fractal surface morphology and photocatalytic properties of TiO_2 film. Thin Solid Films 357:173-178

Yabusaki S, Cantrell K, Sass B, Steefel C (2001) Multicomponent reactive transport in an *in situ* zero-valent iron cell. Environ Sci Techn 35:1493-1503

Yuen DA, Vincent AP, Bergeron SY, Dubuffet F, Ten AA, Steinbach VC, Starin L (2000) Crossing of scales and non-linearities in geophysical processes. *In* Problems in Geophysics for the New Millenium. Boschi EV, Ekstrom G, Morelli A (eds) p 403-462. Bologna, Italy: Editrice Compositori

Yuen DA, Vincent AP, Kido M, Vecsey L (2001) Geophysical applications of multidimensional filtering with wavelets. Pure Appl Geophys (in press)

Zachara JM, Fredrickson JK, Li SM, Kennedy DW, Smith SC, Gassman PL (1998) Bacterial reduction of crystalline Fe^{3+} oxides in single-phase suspensions and subsurface materials. Am Mineral 83: 1426-1443

7 Magnetism of Earth, Planetary, and Environmental Nanomaterials

Denis G. Rancourt

Department of Physics
University of Ottawa
Ottawa, Ontario, Canada K1N 6N5

INTRODUCTION

Superparamagnetism, the archetypal small particle phenomenon, was first described by Louis Néel (1949a,b) in the context of rock magnetism. Key features of the underlying micromagnetism had been described by Stoner and Wohlfarth (1948), in the context of industrial magnetic alloys. The physical theory of superparamagnetism was further developed by William Fuller Brown (1963, 1979; see Rubens 1979). Many recent and ongoing additional advances are motivated by the potential for improved nanotechnological applications. One of my main aims in writing this chapter is to help bridge the gap that presently exists between the still largely phenomenological methods of rock and mineral magnetism and rapidly developing fundamental advances in the theory of micromagnetism, the theory of magnetic measurements applied to nanomagnets, and the study of synthetic model systems of ensembles of interacting magnetic nanoparticles. It is hoped that the reader will acquire appreciations of both the underlying atomistic theory of small particle magnetism and the wealth of phenomena involving the magnetism of natural materials, that often have nanoscale structural or chemical features, including their sizes. The reward for taking a more fundamental approach is that magnetism becomes one of the most sensitive probes of such structural and chemical features of complex natural solids and composites.

Magnetism in the Earth sciences

Magnetism is many things in the Earth. Three large areas can be distinguished: (1) geomagnetism, where one is primarily concerned with paleorecords of geomagnetic fields (inclination, intensity, reversals), models of the geodynamo and geomagnetic field generation, or the use of geomagnetic fields for prospecting or probing planetary interiors, (2) magnetic geology (magnetic petrology, magnetic fabric, environmental magnetism), where one is primarily concerned with magnetic rock and mineral records that are used to study petrogenesis, geotectonic activity, weathering, diagenesis, sediment transport, etc., and (3) mineral magnetism, where one is primarily concerned with the magnetic properties of natural samples in order to deduce the underlying mineralogy or individual mineral characteristics. Successes in the first two areas largely depend on the latter area of mineral magnetism, that is the subject of the present chapter. We wish to know the extent to which measured magnetic properties can be used to provide mineralogical identification, quantification, and characterization, of either whole samples or separated solid phase fractions of interest and even down to a single nanoparticle.

Relation to other books and reviews

The book by O'Reilly (1984) is a wonderful first attempt at a general textbook on mineral magnetism, although the outlook is slanted towards rock magnetic methods. The early rock magnetism review by Stacey (1963) has a strong mineral magnetic perspective. Banerjee's review (1991) has a mineral magnetism perspective but does

1529-6466/00/0044-0007$10.00

not deal with particle size effects. Frost (1991) has written a nice survey of magnetic petrology. Researchers such as Rochette and Fillion (1988) and Richter and van der Pluijin (1994) have endeavored to bring novel mineral magnetic methods into rock magnetism. Dunlop and Özdemir (1997) have written an authoritative recent textbook on traditional rock magnetism, that complements the classic text by Nagata (1961) and other important texts (e.g., Stacey and Banerjee 1974, Collinson 1983) and reviews (e.g., Clark 1983). Tarling and Hrouda (1993) concentrate on magnetic fabric applications and good examples in this area can be found in the collection assembled by Benn (1999). Thompson and Oldfield (1986) concentrate on environmental magnetism, Creer et al.. (1983) on baked clays and recent sediments, and Opdyke and Channell (1996) on magnetic stratigraphy, all using mainly rock magnetic methods. Maher and Thompson (1999) have surveyed the use of magnetism in the study of paleoclimates via the sedimentary record. Worm (1998) has reviewed the application of classic alternating field magnetic susceptibility measurements to soils, in relation to pedogenic nanoparticles. There are several books on the theoretical and experimental aspects of geomagnetism and paleomagnetism (e.g., Parkinson 1983, Vacquier 1972, Rikitake and Honkura 1986, Lowes 1989, Jacobs 1991, Backus et al. 1996, Merrill et al. 1998, McElhinny and McFadden 1999, Campbell 2000). A nice recent example of geomagnetic research is provided by Gee et al.. (2000). Dormann et al. (1997) have given a comprehensive review of magnetic relaxation effects arising from superparamagnetism. Dormann (1981) had provided an earlier review. Pankhurst and Pollard (1993) have reviewed Fe oxide nanoparticles. Himpsel et al. (1998) have provided a review of magnetic nanostructures that emphasises technological applications and the underlying relations to electronic structure. The present chapter is the first review to concentrate on the fundamental foundations of both mineral magnetism and the relevant measurement methods, especially in relation to magnetic nanoparticles and nanomaterials. This should serve to illustrate both the opportunities for development of the methods and interpretations and to provide some insight into the magnetic phenomena themselves.

Organization and focus of this chapter

Following a brief interdisciplinary look at magnetic nanoparticles and a brief overview of the magnetism of Earth's crust and surface environments, the core of this chapter is broadly organized into the usual divisions of measurement methods, underlying theory of both the phenomena themselves and the measurement methods, and applications to and interpretations of natural phenomena. As in all observational sciences, however, these three areas are highly interdependent and must not be treated as separate topics. Interpretation of the raw measured data relies on a chosen theoretical framework that in turn is judged appropriate to describe the expected phenomenon, etc. Self-consistency is the result of much trial and error and is only attained once the phenomenon is judged to be understood, such that routine applications can be devised. The mineral magnetism of nanomaterials is a developing area of intense present research where the interplays between measurement, theory, and interpretation must be considered with care. For example, characteristic measurement times may be comparable to intrinsic sample property fluctuation times (e.g., SP supermoment fluctuations compared to the measurement frequency in an alternating field susceptibility measurement) such that calculations that assume thermodynamic equilibrium become inapplicable but stochastic resonance may become relevant. I have made a special effort to illustrate these inter-relationships throughout this chapter, in order to do justice to the area and to enhance the general reader's appreciation of the subject.

[Text continued on page 221.]

Symbols and acronyms

1NN	first nearest neighbour
2NN	second nearest neighbour
a	lattice parameter or particle's long ellipsoidal semi-axis
a-HFO	abiotic hydrous ferric oxide
$\langle A\rangle_a$	type-a average of A
$\langle A\rangle(T,H)$	thermal average of A as a function of temperature and field
$\{A_i\}$	set of all values of A_i (for all cations-i in sample or particle)
$\{A_{ij}\}$	set of all values of A_{ij} (for all cation pairs-ij in sample or particle)
AF	antiferromagnetic
b	lattice parameter or particle's short ellipsoidal semi-axis
b-HFO	biotic hydrous ferric oxide
BCC	body centered cubic
χ	magnetic susceptibility ($= \partial M/\partial H$)
χ_o	initial magnetic susceptibility
χ_3	cubic magnetic susceptibility
C	Curie constant (of paramagnetic susceptibility)
CBED	convergent beam electron diffraction
CMEs	collective magnetic excitations
d	particle size (diameter or mean diameter)
d_{QT}	quantum-tunnelling/quantum-blocked transition size
d_{SD}	single-domain/multi-domain transition size
d_{SP}	superparamagnetic/blocked transition size
d_{SR}	characteristic surface region size or thickness
DM	diamagnetic
ε_n	subsystem (cation) eigenenergy in subsystem eigenstate n
e	eccentricity of particle's assumed ellipsoid of revolution shape
E_b	supermoment reversal energy barrier height or barrier function of θ
E_d	dipolar anisotropy energy barrier function
E_n	eigenenergy of system in eigenstate n
E_s	surface anisotropy energy barrier function
E_v	magneto-crystalline anisotropy energy barrier function
EDS	energy dispersive spectroscopy
EELS	electron energy loss spectroscopy
ESC	electronic structure calculation (*ab initio* calculation of electron densities)
ESR	electron spin resonance
f_o	pre-exponential term or attempt frequency in expression for $1/\tau$
FC	field cooling
FCC	face centered cubic
FI	ferrimagnetic
FW	field warming
GBIC	graphite bi-intercalation compound
GIC	graphite intercalation compound
GSC	Geological Survey of Canada
$\mathcal{H}$	Hamiltonian of (total) system (entire solid or particle)

$\mathcal{H}_i$	Hamiltonian of subsystem-i (cation-i)
H	magnetic field
H_o	critical field of induced magnetic moment reversal at T = 0 K
H_{int}	interaction field (local or uniform, static or time-dependent)
$H_{applied}$	applied magnetic field
HFD	hyperfine field distribution
HFO	hydrous ferric oxide (solid precipitate)
HM	high moment (state or phase)
INS	inelastic neutron scattering
J, J_{ij}, $J_{ij}(r_{ij})$	magnetic exchange parameter (energy per interacting pair of moments)
k_B	Boltzmann's constant
k_s	surface anisotropy energy barrier per surface (in J/m^2)
K	single-ion magneto-crystalline anisotropy constant (energy per ion)
K_i	magneto-crystalline anisotropy constant of cation-i
K_v	magneto-crystalline anisotropy energy per volume of the material (= nK/v)
ln(A)	natural logarithm of A
L	orbital angular momentum vector operator (of a cation)
LHT	liquid helium temperature (4.2 K)
LM	low moment (state or phase)
LNT	liquid nitrogen temperature (77 K)
LSSE	Lake Sediment Structure and Evolution (Group)
μ	magnetic moment (atomic or particle or sample)
μ_i	magnetic moment at site-i or on nanoparticle-i
μ_S	magnetic supermoment
μ_{Fe}	atomic magnetic moment on Fe atom
μ_B	Bohr magneton
μ_N	nuclear magneton
m	number of cation moments per net supermoment ($m = \mu_S/\mu = Mv/\mu$)
M	sample or sublattice magnetization ($= \mu/V$)
M(H)	magnetization as a function of field, at constant temperature
M(T)	magnetization as a function of temperature, at constant field
M(t)	magnetization as a function of time, at constant field and temperature
MC	Monte Carlo (calculation method)
MD	multi-domain or molecular dynamics
MFT	mean field theory
MITE	Metals in the Environment (Project)
MQT	macroscopic quantum tunnelling (of the supermoment orientation)
n	number of moment-bearing cations in a nanoparticle or eigenstate index
$\langle n\|A\|n\rangle$	diagonal matrix element of operator A, using eigenstate $\|n\rangle$
NC	non-collinear
NN	nearest neighbour
$P(H_o)$	probability density distribution of critical reversal fields
$P(m,E_b,H_{int},\ldots)$	joint probability density distribution of m, E_b, H_{int}, and other parameters
P(v)	probability density distribution of particle volumes
PM	paramagnetic
ρ	mass density

RT	room temperature (22 °C)
RTM	reaction transport model
s-Fe	surface complexed or sorbed Fe
S	dimensionless spin angular momentum quantum number (of a cation)
S	dimensionless spin angular momentum vector operator (of a cation)
S, S(T,H)	magnetic viscosity
SAED	selected area electron diffraction
SANS	small angle neutron scattering
SD	single domain
SF	superferromagnetic
SG	spin glass
SP	superparamagnetic
SQUID	superconducting quantum interference device (magnetometer)
SR	surface region (of a nanoparticle)
τ	superparamagnetic reversal fluctuation time (zero field)
τ_+	superparamagnetic dwell time for supermoment mostly parallel to field
τ_-	superparamagnetic dwell time for supermoment mostly antiparallel to field
τ_m	measurement or observation time
θ_{CW}	Curie-Weiss temperature
$\theta_{\mu a}$	angle between supermoment and particle's major axis of elongation
$\theta_{\mu K}$	angle between supermoment and particle's anisotropy axis direction
T	temperature or Tesla
T_{0B}	effective Curie-Weiss temperature due to distribution of barrier energies
T_{0i}	effective Curie-Weiss temperature due to inter-particle interactions
T_C	Curie point or Curie temperature (of a ferromagnet)
T_{oB}	magnetic ordering temperature of the bulk material
T_{peak}	temperature of peak in ZFQ-FW magnetization curve
T_{SP}	superparamagnetic/blocked transition temperature (depends on τ_m)
TEM	transmission electron microscopy
TRM	thermoremanent magnetization
$\mathbf{u}_r$	dimensionless unit vector pointing along a vector **r**
$\mathbf{u}_i$	dimensionless unit vector pointing along axis-i
$\mathbf{u}_{ij}$	dimensionless unit vector pointing along the straight line between i and j
U_{dd}	dipolar interaction energy (per pair of interacting moments)
v	volume of a nanoparticle
v_{SP}	transition nanoparticle volume between SP and blocked states
VSM	vibrating sample magnetometer
WF	weak ferromagnetic (or canted antiferromagnetic)
Z	partition function
ZFQ	zero field quench

Note: Boldface type is used to represent vector quantities.

This chapter is not a comprehensive review of the many studies concerned with natural magnetic nanoparticles in the Earth, planetary and environmental sciences. Instead, I concentrate on giving a broad overview and key examples and attempt to motivate a deeper than usual examination of forefront fundamental developments. I stress that the magnetism of nanomaterials and nanoparticles is presently an area of intense and rapid development that touches many fields including technological applications, biology, medicine, chemistry, materials science, condensed matter physics, environmental science, planetary science, information theory, etc. (see below), and that I will primarily focus on the development of mineral magnetism for advanced applications in the Earth, planetary, and environmental sciences.

MAGNETIC NANOPARTICLES EVERYWHERE

In our brains, the animals, space, everywhere

The human brain contains over 10^8 magnetic nanoparticles of magnetite-maghemite per gram of tissue that may be responsible for a variety of biological effects (Kirschvink et al. 1992). The human mind struggles with complex applications involving image processing or data classification and mining and has recently discovered an underlying similarity between superparamagnetism and data clustering, that allows methods of statistical physics developed for superparamagnetism to be applied to great advantage in information handling (Rose 1990, Domany 1999, Domany et al. 1999). This same information is stored on magnetic media composed of magnetic nanoparticles and the need for larger and larger storage densities drives yet more advanced theoretical models of the supermoment and its fluctuations and field driven reversals. Experimentalists, in turn, devise more and more clever synthetic model nanoparticle systems to test and challenge theory and force more and more realistic features into the calculations (see below).

Meanwhile, Earth scientists, planetary scientists, environmental scientists, and biologists keep finding more and more unusual magnetic nanoparticles in the strangest places: as geomagnetic navigational aids in bacteria, eukaryotic algae, and the bodies of higher animals (Kirschvink et al. 1985, Kirschvink 1989, Wiltschko and Wiltschko 1995) such as homing pigeons (e.g., Hanzlik et al. 2000), migratory birds, ants, bees, salmon, tuna, sharks, rays, salamanders, newts, mice, cetaceans, etc., as the ferrihydrite-like mineral cores of the most common iron storage protein ferritin (e.g., St-Pierre et al. 1989), present in almost every cell of plants and animals including humans, as keystone crystals in the cells of hornet combs (Stokroos et al. 2001), as bacterial micro-fossils (e.g., Chang and Kirschvink 1985, Petersen et al. 1986), precipitated to bacterial cell walls (e.g., Ferris 1997, Fortin et al. 1997, Watson et al. 2000), in lunar samples and as common products of space weathering (e.g., Pieters et al. 2000), in Martian meteorites where they may be of biogenic or inorganic origin (e.g., Bradley et al. 1998, Golden et al. 2001), in Martian soil and its analogues (e.g., Morris et al. 1998, 2001), in interstellar space (Goodman and Whittet 1995), at the Cretaceous/Tertiary (K/T) boundary on Earth where they can be either produced in molten impact droplets or oxidized remnants of the meteoritic projectile itself (e.g., Kyte and Bohor 1995, Kyte and Bostwik 1995, Robin 1996, Gayraud et al. 1996, Kyte 1998), at the oxic/anoxic boundary in lacustrine sediments (e.g., Tarduno 1995), etc. And if these researchers become ill from overwork and the medical diagnosis becomes difficult, chances are they will be injected with magnetic nanoparticles as magnetic resonance imaging (MRI) contrast agents (e.g., Roch et al. 1999, Grüttner and Teller 1999, Bonnemain 1998). If these patients are among the few to develop side effects (e.g., Sharma et al. 1999) and have the occasion to take some time off, they may notice while gazing at the stars one night (with polarizing sun glasses!) a slight polarization of the background starlight that may be due to superparamagnetic (SP) particles (Goodman and Whittet 1995).

Applications of magnetic nanoparticles

Back on Earth, in nature, most magnetic nanoparticles are iron oxides and oxyhydroxides. Pankhurst and Pollard (1993) have reviewed iron oxide nanoparticles and some of their applications. The applications range from model systems to study stochastic resonance (Raikher and Stepanov 1995a, Ricci and Scherer 1997), to synthetic industrial catalysts (e.g., Lopez et al. 1997), to magnetic colloids and ferrofluids for seals, bearings, and dampers, such as the ones in our cars (Rosensweig 1985, Cabuil 2000, Popplewell and Sakhnini 1995), to magneto-optic recording devices (Suzuki 1996), to magnetic force microscopy (MFM) tips (Liou et al. 1997, Hopkins et al. 1996), to giant magneto-resistive devices (Lucinski et al. 1996, Wiser 1996, Xu et al. 1997, Altbir et al. 1998), to field sensing technology (Cowburn et al. 2000), to molecular biology (Grüttner and Teller 1999), to cell biology and immunomagnetic methods (Yeh et al. 1993, Sestier et al. 1998, Sestier and Sabolovic 1998), to cell separation (e.g., Honda et al. 1998), to advanced medical applications (Tiefenauer et al. 1993, Jung et al. 1999), to environmental remediation via synthetic magnetically separable sorbants (e.g., Safarik and Safarikova 1997, Broomberg et al. 1999), in addition to the areas mentioned above. There are also associated technologies, such as magnetic separation of nanoparticles (Zarutskaya and Shapiro 2000). And, of course, one must not leave out the largest single client of magnetic nanoparticle use: the magnetic recording industry (Lodder 1995, Brug et al. 1996, Kryder 1996, Onodera et al. 1996). Here, corporate giants have taken charge of the problem, with claims like "The superparamagnetism limit and what IBM is going to do about it" (e.g., Comello 1998, Soltis 2001).

Since the time of Stoner and Wohlfarth (1948), we have known that the first great application advantage of magnetic nanoparticles is that they are single domain and therefore have large coercivities, allowing them to individually retain their magnetization directions. And since the time of Néel (1949a,b) we have known that, as the particle size is decreased, every particle eventually becomes SP and looses its capacity to retain a remanence magnetization. Higher and higher information storage densities, that have doubled every year since the late 1990s, mean smaller and smaller particles, hence is born the superparamagnetism problem or limit, presently estimated at ~100 Gbits/in^2 (Comello 1998, Soltis 2001). Engineers want to suppress it or avoid it while natural scientists want simply to use it as a means to learn about the particles, hence is born mineral magnetism, with its first client that is rock magnetism.

Towards function and mechanisms

There must be at least as many applications and natural phenomena related to magnetic nanoparticles left to be discovered. The mechanisms of magnetic biosensitivity are mostly unknown (Deutschlander et al. 1999, Wiltschko and Wiltschko 1995). The possibility of magnetically sensitive biological chemical reactions has only recently been demonstrated (Weaver et al. 2000) and such reactions may, in turn, interact with biogenic nanoparticle assemblies. Function is as much a puzzle in environmental science where bacteria-mineral interactions are just beginning to be explored and where most reactive authigenic, biogenic, pedogenic, and diagenetic solid phases are nanoparticles and nanomaterials. Beyond observations of particles, uncovering function is the goal. Elucidation of function requires detailed characterization that, in turn, requires advanced characterization tools that are adapted to the subject. Mineral magnetism, coupled with a fundamental understanding of both the measurement process and nanomagnetic phenomena, is such a tool that offers great promise. It has already given remarkably intricate views of these complex beasts that are magnetic nanomaterials.

MAGNETISM OF THE CRUST AND SURFACE ENVIRONMENTS

Diamagnetic and paramagnetic ions

In terms of both abundance in the Earth's crust (and on planetary surfaces) and numbers of species, most minerals are ionic and covalent in character, rather than metallic. Their magnetism, therefore, is appropriately described in terms of the magnetism of their cations and anions, rather than in terms of band magnetism of conduction electrons. Prominent exceptions are the Fe-Ni phases found in meteorites and that are relevant to planetary cores. Even in metallic minerals, however, the magnetism of the closed shell cation cores can, to a good first approximation, be treated independently from the conduction electron magnetism.

The magnetism of constituent ions in solids is mostly due to electron spin and may have an electron orbital component as well (Ashcroft and Mermin 1976). Closed shell ions have paired spin up and spin down electrons in filled orbitals and zero net orbital angular momentum. As a result, closed shell ions (most ions in the Earth's crust) are strictly non-magnetic, in that their net spontaneous (i.e., in the absence of an applied magnetic field) ionic magnetic moments are strictly zero. Such ions and substances that contain only such ions are said to be diamagnetic (DM). When a magnetic field, H, is applied to a DM ion, a relatively small magnetic moment, μ, is induced, in proportion to the magnitude of and in a direction opposite to the applied field. This negative magnetic susceptibility ($\chi \sim \partial\mu/\partial H < 0$) is taken to be a diagnostic feature of non-metallic diamagnetism. The only other known type of diamagnetism is the strong (Meisner effect) diamagnetism associated with superconductivity. This feature is considered an experimental proof of the superconducting nature of a material.

Ions having partially filled orbitals will have net electron spin magnetic moments and may, in addition, have net electron orbital magnetic moments. Such ions are referred to as paramagnetic (PM). These are the moment-bearing atomic entities of ionic and covalent magnetic minerals. Materials that contain PM ions and in which the PM ions do not interact magnetically, other than via the usual classical and relatively weak dipole-dipole interactions, are also referred to as PM. The corresponding thermally induced disordered state, in which the vector orientations of the PM ionic moments in a given solid phase fluctuate randomly such as to produce zero time averages of the local moments on all the ionic PM centres, except to the extent that an applied field would induce a non-zero value, is referred to as the PM state or phase and any material that is in a PM state at a given temperature of interest can also be referred to as PM. An applied magnetic field will tend to align the PM ionic moments in its direction (as is the case with classical magnetic dipole moments), leading to a positive PM susceptibility that is much larger in magnitude than ionic DM susceptibilities. Strong inter-moment quantum mechanical interactions between ionic PM moments give rise to a large array of possible spontaneous ordered magnetic structures, in which the local time average moments are not zero in zero applied field, that are highly sensitive to nanofeatures of the material and that have larger or smaller than PM susceptibilities, as described below.

Magnetism from crustal ions in surface minerals

Table 1 gives the usual elemental abundances of the Earth's crust (Klein and Hurlbut 1999), where I have added the various common ionic valence states, corresponding electron configurations, and resulting expected net spin moments (Ashcroft and Mermin 1976), in units of Bohr magnetons (μ_B). The great majority of abundant ionic species are DM, with zero net moments, and Fe is the main moment-bearing element in crustal rocks. For this reason, magnetic measurements can be considered a means to specifically examine the iron mineralogy of a given natural sample, much as would be the case with

^{57}Fe Mössbauer spectroscopy applied to a whole rock sample. On the other hand, a sample that would have the elemental abundances given in Table 1 but in which all of the Fe were in PM mineral species could have its magnetic response to an applied field dominated by Mn if the Mn was in mineral species that have certain ordered magnetic structures. This is relevant to early diagenesis in marine and lacustrine environments where authigenic nanophase Mn oxyhydroxides play an important role. Similarly, particular mineral assemblages that involve V, Cr, or the magnetic rare earth elements (mostly not shown in Table 1) could exhibit magnetic responses that are due to those elements rather than Fe. In addition, even in the most common cases where the magnetic susceptibility is dominated by Fe signals from several Fe-bearing fractions, magnetic phase transitions, between different ordered magnetic structures or between an ordered phase and the PM phase, can be observed, as temperature and field are varied, that are characteristic of specific majority and minority mineral species, thereby allowing their quantification and characterization.

Table 1. Electronic magnetism of crustal ions.

element	*weight % of crust*	*ion*	*electrons*	*spin magnetic moment,* μ_B
O	46.60	O^{2-}	[Ne] (=$1s^2 2s^2 2p^6$)	0
Si	27.72	Si^{4+}	[Ne]	0
Al	8.13	Al^{3+}	[Ne]	0
Fe	5.00	Fe^{2+}	[Ar]$3d^6$	**4**
		Fe^{3+}	[Ar]$3d^5$	**5**
Ca	3.63	Ca^{2+}	[Ar]	0
Na	2.83	Na^{+}	[Ne]	0
K	2.59	K^{+}	[Ar]	0
Mg	2.09	Mg^{2+}	[Ne]	0
Ti	0.44	Ti^{4+}	[Ar]	0
H	0.14	H^{+}	0	0
P	0.1	P^{5+}	[Ne]	0
Mn	0.09	Mn^{2+}	[Ar]$3d^5$	**5**
		Mn^{3+}	[Ar]$3d^4$	**4**
		Mn^{4+}	[Ar]$3d^3$	**3**
Ba	0.04	Ba^{2+}	[Xe]	0
F	625 ppm	F^{-}	[Ne]	0
Sr	375 ppm	Sr^{2+}	[Kr]	0
S	260 ppm	S^{6+}	[Ne]	0
C	200 ppm	C^{4+}	[He]	0
Zr	165 ppm	Zr^{4+}	[Kr]	0
V	135 ppm	V^{2+}	[Ar]$3d^3$	**3**
		V^{3+}	[Ar]$3d^2$	**2**
		V^{4+}	[Ar]$3d^1$	**1**
Cl	130 ppm	Cl^{-}	[Ar]	0
Cr	100 ppm	Cr^{2+}	[Ar]$3d^4$	**4**
		Cr^{3+}	[Ar]$3d^3$	**3**
Rb	90 ppm	Rb^{+}	[Kr]	0
Ni	75 ppm	Ni^{2+}	[Ar]$3d^8$	**2**
Zn	70 ppm	Zn^{2+}	[Ar]$3d^{10}$	0
Ce	60 ppm	Ce^{3+}	[Xe]$4f^1$	**1**
Cu	55 ppm	Cu^{2+}	[Ar]$3d^9$	**1**

Magnetism from crustal and surface mineralogy

This brings us to examine crustal magnetism from the perspective of crustal mineralogy, which is an application of reverse magnetic petrology (Frost 1991). The plagioclase feldspars, the alkali feldspars, and quartz, which together make up 63% of the volume of the crust (Klein and Hurlbut 1999), contain no Fe and consist only of DM minerals containing only DM ions. Clearly, magnetic measurements will not contribute to the study of these most common rock forming minerals, although high resolution magnetometry of DM response can be used to investigate fine aspects of bonding and electronic structure (e.g., Uyeda 1993). The other large classes of abundant minerals (pyroxenes, micas, amphiboles, clay minerals and chlorites, garnets, olivines, carbonates, oxides) generally contain Fe and other PM cations and have important representative end-member species that are Fe-rich. In particular, the next most abundant group, the pyroxenes, with 11% of the crustal volume, represent a large compartment of crustal Fe, predominantly in the form of the enstatite-ferrosilite solid solution, $(Mg,Fe)SiO_3$, and augite, $(Ca,Na)(Mg,Fe,Al)(Si,Al)_2O_6$. The amphiboles and micas represent the next two largest compartments of crustal and petrogenic Fe. Together, the pyroxenes, amphiboles, micas, olivines, and garnets make up 29% of the crustal volume (Ronov and Yaroshevsky 1969) and contain ~90% of the crustal Fe that is tied to petrogenic mineral species. All of the latter Fe is contained in minerals that are PM at room temperature (RT = 22°C) and that generally remain PM down to liquid nitrogen temperature (LNT = 77 K). The remaining ~10% of crustal Fe tied to petrogenic minerals is mostly contained in spinel group Fe oxides, such as magnetite, Fe_3O_4, that comprise 1.5% of the crustal volume. The bulk Fe-rich spinels (magnetite-titanomagnetite-maghemite system) have magnetic ordering temperatures (below which spontaneous ordered magnetic structures occur) that are well above RT and they occur in a broad range of crystallite sizes. As a result, despite the relatively small size of this petrogenic compartment for Fe, the name of the game in rock magnetism is magnetite and its size-dependent magnetic properties (Dunlop 1990). Only a substance with a sufficiently high magnetic ordering temperature can effectively register and preserve remanent geosignals and only a mineral with a magnetic ordering temperature larger than RT can be detected above the PM signals of the rock or sediment matrix, by the usual RT and high temperature measurement protocols.

But rock magnetism and magnetic petrology are not our only concerns here. Approximately as much Fe as is contained in the petrogenic spinel compartment occurs in non-petrogenic minerals and solid phases that reside in the surface environments (soils and sediments). The Fe-bearing phases can be authigenic, biogenic, pedogenic, diagenetic, etc., and have acquired their Fe by weathering and dissolution of the petrogenic Fe phases discussed above, including magnetite. As a result, the magnetism of surface samples is often not dominated by petrogenic magnetite but is instead largely determined by various abiotic and biotic (Banfield and Nealson 1997, Ferris 1997) precipitation reactions that invariably produce nanophase Fe oxides and oxyhydroxides, such as ferrihydrite (Jambor and Dutrizac 1998, Cornell and Schwertmann 1996), in addition to a broad range of nanophase and poorly crystalline materials that are Fe-poor or DM, such as allophane and amorphous silica (Dixon and Weed 1989). Pedogenic and early diagenetic processes also often create and maintain well defined horizons of magnetic (Fe-rich or Mn-rich) phases, such that specific sections of a depth profile can have a magnetism that is overwhelmingly determined by these processes, rather than the geologic nature of the parent material.

Given the importance of spontaneous magnetic order in allowing rock magnetism applications and in determining magnetic susceptibility magnitudes, I give some magnetic ordering and transition temperatures of bulk end-member materials in Table 2.

Table 2. Magnetic ordering and transition temperatures of some bulk end-members.

Material	*Transition*	*Temperature*[a] *(K)*		*Comments*
magnetite	FI/PM	T_C	853(5)	Table 37, Krupicka & Novak (1982)
	Verwey	T_V	119(1)	p. 264, Krupicka & Novak (1982)
hematite	WF/PM	T_N	948	p. 172, O'Reilly (1984)
	Morin	T_M	260	Rancourt et al (1998)
ilmenite	AF/PM	T_N	55	p. 179, O'Reilly (1984)
ferrihydrite[b]	AF/PM	T_N	350(2)	Seehra et al (2000)
annite[c]	meta/PM	T_N	58(1)	Rancourt et al (1994)

a – C, Curie point; V, Verwey transition; N, Néel temperature; M, Morin transition
b – synthetic two-line ferrihydrite
c – synthetic annite, Fe-end-member of phlogopite-biotite-annite series

Having described the ionic and mineral magnetic species that one should expect to encounter on Earth and the interplay between chemistry and mineralogy that determines the strength of a magnetic response (in the sequence DM << PM << magnetically ordered), we end this section by a few observations that bring us back to our main theme of mineral magnetism. First, as mentioned above, applying mineral magnetic methods (next section) to bulk natural samples or separated fractions of bulk samples should be a powerful tool in elucidating the Fe mineralogy, and in some cases the Mn mineralogy. It has surprised me, as a relative newcomer to the Earth sciences, that this has not been developed as a systematic and widespread approach. Hopefully this chapter will catalyze work along these lines. Second, the cations that are magnetic (i.e., PM rather than DM) are also those, in general, that have multiple valence states (Table 1), for obvious chemical reasons. Such multiple valence states are important in the Earth sciences. They allow various surface reactions such as oxidative precipitation and reductive dissolution in sediment profiles, and several biomediated reactions. They record redox conditions of the melt during petrogenesis. They cause characteristic rare earth element (REE) anomalies that are diagnostic of weathering and pedogenic activity. For these and other reasons, it is important to be able to accurately quantify valence state populations in minerals of interest. Table 1 suggests that magnetic measurements may be an accurate and convenient way to do this. This is indeed the case and it has been exploited in magneto-chemistry (Carlin 1986) but not in the Earth sciences.

MEASUREMENT METHODS FOR MINERAL MAGNETISM

Several methods either directly measure the spontaneous or field-induced macroscopic magnetic moment on a sample or the magnitudes and orientations of microscopic local ionic magnetic moments or are particularly sensitive to various properties of moment-bearing cations or provide some physical property that has a strong interplay with a sample's magnetism. These methods can be classified under the usual categories of microscopy, spectroscopy, diffraction, and bulk properties and are far too numerous to even name here. See for example the various monographs on measurement methods for solid materials (American Society for Metals 1986, Flewitt and Wild 1994, Hawthorne 1988, Lifshin 1992, 1994; Cahn and Lifshin 1993, Sibilia 1996), a recent series of articles on specifically magnetic methods (McVitie and Chapman 1995, Qiu and Bader 1995, Smith and Padmore 1995), and the recent article by Pietzsch et al. (2001) who describe the most advanced magnetic nanoscopic imaging methods. New methods are being devised continuously, involving both large facilities providing synchrotron radiation and specialized table top apparatus. I will concentrate only on the main well established methods that are likely to be used in most materials magnetism laboratories and that are applicable to broad classes of materials and to various sample types. As is often the case, the usefulness of the latter methods is limited more by the researcher's mastery of the underlying theory than by access to the instruments.

Constant field (dc) magnetometry

The first measurement method that is the main tool in mineral and materials magnetism is constant (dc) field magnetometry. The constant field magnetometer allows one to measure the net magnetic moment on a powder or oriented single crystal or oriented textured sample, at a given measurement temperature and at a given value of a constant and uniform applied magnetic field. The measured magnetization, M, is the sample moment divided by the sample volume. One can also define a measured moment per weight by dividing M by the sample's mass density, ρ. Two popular and comparable instruments for bulk samples of nanomaterials are the vibrating sample magnetometer (VSM) and the superconducting quantum interference device (SQUID) magnetometer. Typically, the measurement temperature can be varied between 4.2 K (or as low as 1.5 K by pumping on the liquid helium reservoir) and RT (or up to ~1000°C with a furnace attachment). The field can be varied through both polarities up to some maximum magnitude that is typically 1 T (10 kG) for an electromagnet or up to 10 T or more for a superconducting magnet. Measurement programs usually involve temperature cycles at fixed field values and field cycles at fixed temperatures. The most common fixed temperatures for field cycles (called hysteresis cycles) are liquid helium temperature (LHT = 4.2 K), LNT, and RT.

The measurements are often sensitive to the magnetic and thermal history of the sample. Remanence may be induced by quenching in a large field, followed by turning off the field and measurement as a function of increasing temperature, at some set heating rate. The resulting curve is called a thermoremanent magnetization (TRM) curve. In rock magnetism, one typically starts at RT, from the natural remanence or from an applied field induced remanence, and one measures as the sample is heated. A common cycle in materials science is to first perform a zero field quench (ZFQ), followed by turning on the applied field to then measure the field warming (FW) curve as a function of increasing temperature, followed by a field cooling (FC) curve back down to the initial cryogenic temperature. In all such measurement cycles, it is important to remember that the measurement in general depends on the heating and cooling rates, waiting times during actual measurements, and any waiting times during magnetic field applications. The latter effects arise from all the physical processes that occur on time scales that are comparable to or larger than the measurement time. Another important measurement schedule is to measure the sample's magnetic moment as a function of time, at fixed temperature and applied field, either following a specific treatment or not. Such measurements performed as functions of time are referred to as relaxation measurements and allow one to obtain the magnetic viscosity, S, defined as $S(T,H) = dM(T,H)/d\ln(t)$, the logarithmic time derivative of M. S is often a constant, as predicted by domain wall motion theory (Street and Woolley 1949) and under common assumptions for ensembles of SP particles (Néel 1949a), but also often depends on time, especially with interacting small particle systems (see below).

The measured sample's magnetic moment is the net spontaneous or induced moment arising from a vector sum of all the microscopic DM, PM, electronic, and nuclear contributions. Nuclei can have permanent magnetic moments, although they are much smaller than electronic moments, $\mu_N/\mu_B \sim 10^{-3}$, and have much weaker interactions with each other and with electronic moments. For our purposes, the sample's magnetic moment is essentially a vector sum of all the PM moments of moment-bearing cations in the sample, since the nuclear and DM contributions are much smaller and can be reliably accounted for and we are ignoring metallic and superconducting samples. The vector sum of interest is actually a vector sum of local thermal averages of cation moments, since each cation moment has rapid fluctuations in orientation and an average orientation and value that depend both on its local environment and the local field that it is subjected to. We must therefore distinguish between the cation moment magnitude, μ_i, of cation at site-i in the material and the local thermal average moment at site-i, $<\boldsymbol{\mu}_i>_T$, where $< >_a$ signifies averaging of type-a and bold face type is used to represent a vector quantity. Table 1 gives nominal values of the spin component of the ionic moment magnitude, that can also contain an electronic orbital component and, in general, a component from the local polarization of conduction electrons.

With measurements that have characteristic measurement times that are comparable to or shorter than the characteristic cation moment fluctuation time (typically, 10^{-6} to 10^{-12} seconds, depending on the system), the local average cation moments are not directly relevant in interpreting the measurements. This is clearly not the case with classic magnetometry, where the measurement time (per sample moment evaluation) is typically 10 to 0.1 seconds. The power of magnetometry comes from the fact that the individual average moments and their vector sums in a given sample of known structure can be calculated from first principles (quantum mechanics and statistical mechanics) and compared to the intricate field and temperature dependent curves that are collected. As it turns out, various sample features (type of inter-moment interactions, domain structure, size, shape, chemical order, etc.) give rise to dramatically different and often characteristic features in the measured and calculated curves. We will therefore spend some time developing the underlying theoretical framework.

Recent examples of key theoretical and laboratory studies of constant field magnetometry applied to nanoparticles can be divided into the following categories: M(T) temperature cycles (El-Hilo et al. 1992a, Vincent et al. 1996, Sappey et al. 1997a, Friedman et al. 1997, Vaz et al. 1997, Spinu and Stancu 1998, Stancu and Spinu 1998, Ezzir et al. 1999, Chantrell et al. 1999, 2000; Bodker et al. 2000), M(H) field hysteresis cycles (Jiles and Atherton 1984, Victora 1989, Ignatchenko and Mironov 1991, Roberts et al. 1995, Zhu 1995, Tauxe et al. 1996, Vaz et al. 1997, Spinu and Stancu 1998, Stancu and Spinu 1998, Nowak and Hinzke 1999, Basso et al. 2000, Schmidt and Ram 2001), and M(t) viscosity or relaxation measurements (Mullins and Tite 1973, Aharoni 1992, Barbara et al. 1994, Balcells et al. 1997, Basso et al. 2000). In all cases, the starting point should be the remarkable early work of Néel (1949a).

Alternating field magnetometry (ac susceptometry)

The other most used classic magnetic measurement method is alternating field (ac) magnetometry. The derivative of a measured sample moment with respect to applied field magnitude, H, divided by the sample volume, is called the susceptibility ($\chi = \partial M/\partial H$) and it is called the initial susceptibility (χ_o) when evaluated at zero applied field. Alternating field magnetometry is, under certain circumstances, a method of directly measuring the susceptibility. One applies a sinusoidally varying small amplitude field (the driving field) to the sample held at some constant measuring temperature and in a constant and uniform field (often zero or unscreened Earth's field). One then measures the amplitude of the resulting sinusoidally varying component of the sample moment that is induced by the driving field and its phase shift with respect to the driving field. This measurement is particularly easy to implement experimentally, using lock-in amplifier technology. At ambient temperatures and without applied constant fields, it is the basis of many routine laboratory and in the field measurements in rock and environmental magnetism. In advanced mineral magnetism applications, the experimenter not only controls measurement temperature and the magnitude of the applied constant field (typical ranges 1.5-1500 K and 0-10 T, as in constant field magnetometry) but also: the orientation of the driving field with respect to the constant field, the amplitude of the driving field (typically 10^{-3} to 1 mT), and, most importantly, the frequency of the driving field (typically 1 to 10^4 Hz). In many applications, the measurement frequency is slow compared to the internal equilibration times such that there is no phase shift and the induced moment instantaneously follows the driving field. In this case, the susceptibility is given by dividing the volume normalized induced moment amplitude by the driving field amplitude and is the same as would be obtained from the slope of the M versus H curve. In many other applications, the measurement time is comparable to or shorter than a relevant characteristic time in the sample, a phase shift is observed, and one must treat the measurement as a spectroscopic measurement of response versus probing frequency. This gives much information about the sample and its magnetism that cannot be obtained by constant field magnetometry or easily by any other method. For example, one can extract the distribution of magnetic particle sizes or the distribution of excitation volumes related to domain wall motion or the temperature and field dependencies of supermoment fluctuation times, etc., depending on the system.

The susceptibility described above is also called the linear susceptibility because, in a Taylor expansion of M(H) about the applied field value, it is the coefficient of the term that is linear in H. In this way, higher order susceptibilities can be defined and measured. The higher order (non-linear) susceptibilities can provide information that is not contained in the linear term alone. This information has turned out to be vital in systems of interacting clusters or particles, such as ensembles of magnetic nanoparticles, canonical spin glasses and cluster spin glass systems (see below).

Examples of key theoretical and laboratory studies of alternating field magnetometry applied to nanoparticles are provided by Néel (1949a), Mullins and Tite (1973), Gittleman et al. (1974), Raikher and Shliomis (1975), Aharoni (1992), Sadykov and Isavnin (1996), Djurberg et al. (1997), Garcia-Palacios and Lazaro (1997), Jonsson et al. (1997), Raikher and Stepanov (1997), and Svedlindh et al. (1997). Much of the theory applies equivalently to electron spin resonance (see below) which is an ac susceptibility measurement at much higher driving frequency in which electronic transitions between spin levels are also probed. Néel (1949a) gives a particularly lucid description of alternating field measurements of nanoparticles and of the theoretical relationships with constant field magnetometry measurements.

Neutron diffraction

The next most important method has played a pivotal role in the development of our understanding of magnetism in materials. It has been as important in magnetism as X-ray crystallography has been in mineralogy. Neutron diffraction not only probes nuclear positions but also magnetic structures, by virtue of the neutron's intrinsic spin magnetic moment and its interaction with the atomic magnetic moments of materials. The magnetic Bragg peaks can be analysed to give both the magnitudes and orientations of the cationic moments in the magnetic unit cell. Such measurements provided the first experimental proofs of Néel's proposal of unusual magnetic structures, now well known as antiferromagnetic (AF) and ferrimagnetic (FI) structures (next section). Neutron diffraction is the main method for obtaining magnetic structures and these structures are as diverse as are crystal structures in X-ray diffraction.

Because the measurement time in neutron diffraction (the time that an individual neutron interacts with the diffracting region) is very short (~10^{-13} s), the measurement is faster that most SP reversal events such that a true intra-particle average magnetization can be measured. In this way, the true magnetic ordering temperature of superparramagnetic ferrihydrite was recently measured for the first time and its AF magnetic structure was demonstrated (Seehra et al. 2000). In another study, Lin et al. (1995) used polarized neutrons to demonstrate the presence of noncollinear surface moment disorder in synthetic colloidal $CoFe_2O_4$ particles. Small angle neutron scattering (SANS) allows one to probe magnetic correlations on length scales in the range 1 to 10^3 nm, which is ideal for nanomaterials (Bacri et al. 1993, Upadhyay et al. 1993). One can also perform magnetic inelastic neutron scattering (INS) which probes the densities of state of the magnetic excitations, much as nuclear INS probes the phonon densities of state. Using quasi-elastic neutron scattering, Gazeau et al. (1997) were recently able to resolve the magnetic ion and supermoment excitations in 4 nm synthetic maghemite nanoparticles.

Mössbauer spectroscopy

Another important magnetic measurement method, that is particularly relevant to the Earth sciences because of the importance of Fe, is ^{57}Fe Mössbauer spectroscopy (Rancourt 1998). This hyperfine nuclear spectroscopy is a probe of atomic magnetism by virtue of the fact that the nuclear hyperfine magnetic field has a large component that is strongly coupled to the local cationic moment on the probe nucleus. Indeed, normally its largest component is the contact electron spin term, that arises and is proportional to the total s-electron spin polarization at the nucleus, which, in turn, in insulators, is predominantly caused by s-d spin-spin interactions between the filled s shells and the partially filled 3d shell (or net cation moment). Other hyperfine parameters are also measured that give local structural and chemical information related to the immediate coordinating environment of the probe. In the Earth sciences, the method is commonly used to measure crystallographic or coordination site populations of Fe^{2+} and Fe^{3+} or site- and valence-state-specific distributions of local distortion environments, in Fe-bearing minerals (Rancourt 1994a,b; Rancourt et al. 1994, Lalonde et al. 1998).

The measurement time in ^{57}Fe Mössbauer spectroscopy is the time during which the γ-ray interacts with the ^{57}Fe nucleus to cause a transition between the ground state and the first excited state. This time is equal to the life time of the first excited state, $\sim 10^{-8}$ seconds. The so-called static limit, in which the usual multiplets (singlets, doublets, sextets, and octets) of Lorentzian lines are observed, occurs when all the characteristic fluctuation times of the relevant hyperfine interactions are much smaller or much larger than the measurement time. Since the usual fluctuations in bulk materials (electron residence times << atomic vibrational times << atomic magnetic moment orientation fluctuation time) are much faster than the measurement time, the static limit is quite common. The so-called dynamic or relaxation case, on the other hand, is not uncommon. Characteristic fluctuation times in a broad range centered on $\sim 10^{-8}$ seconds occur in many circumstances, including: at temperatures near structural or magnetic or other phase transition temperatures, when atomic diffusion hopping times are in this range, when valence fluctuations ($Fe^{3+} + e^- = Fe^{2+}$) occur with these characteristic times, and many cases where structural constraints, such as reduced dimensionality of the bonding network or a particular microstructure, are associated with relatively slow magnetic excitations. This is particularly true with nanophase materials and nanoparticles, where (non-Lorentzian) relaxation lineshapes are as common as their bulk Lorentzian multiplet counterparts. One must therefore analyse the data in the context of a general lineshape model that allows the full continuum of possible characteristic times. This can directly give measured microscopic times (in the range 10^{-6} to 10^{-10} seconds) that cannot easily be obtained by other methods (Rancourt 1988, 1998).

Mössbauer spectroscopy is also able to give local moment orientations, with respect to the crystalline lattice, or the correlations between moment orientations and local distortion axis orientations in a chemically disordered or amorphous material. This arises from the interplay between the structural (electric field gradient) hyperfine parameters and the magnetic hyperfine parameters. In this way, the spin flop Morin transition of hematite, for example, is easily detected and characterized (e.g., Dang et al. 1998). The noncollinear magnetic structures of nanoparticles can also be characterized.

Finally, because of the complexity of magnetic nanomaterials and because of the intrinsic difficulties related to spectral modeling, there are probably many incorrect interpretations of the Mössbauer spectra of nanoparticles (as there already are of bulk complex minerals). I discuss this further below where I argue that Mössbauer spectroscopy will not attain its full potential in this area until a significant shift in paradigm is achieved.

Electron spin resonance

One final method that is worth mentioning in this brief and selective overview is electron spin resonance (ESR). This is a resonance absorption spectroscopy, involving transitions between the applied-field Zeeman split ground state level of a PM center. Selection rules are such that not all cation species of a given PM element (e.g., Fe, Mn, V, Cr, Table 1) are detected, thereby providing a sensitive probe of valence state. Again, thanks to hyperfine and fine structure (here on electronic energy levels rather than on nuclear levels), the various ESR active cations and their local bonding environments can be resolved and quantified. The method is characterized by a remarkable sensitivity, easily allowing dilute solutions and rare impurities to be studied. For example, PM centers corresponding to electrons trapped in radiation-induced defects in quartz are routinely used to date sand (Grun 2000) and a similar method allows one to date fossil teeth (Oka et al. 1997).

The sensitivity of ESR is, in a sense, a weakness because in samples with large concentrations of PM cations (whether ESR active or not) inter-moment interactions cause resonance line broadening and associated loss of hyperfine and fine structures. On the other hand, the latter broadening can be used to study concentrated magnetic materials since it has a large dependence on the degree and type of spontaneous magnetic order. It is also sensitive to the superparamagnetism of magnetic nanoparticles (Dubowik and Baszynski 1968, Sharma and Waldner 1977, de Biasi and Devezas 1978, Griscom et al. 1979, Sharma and Baiker 1981, Raikher and Stepanov 1995b, Berger et al. 1997, 1998; Gazeau et al. 1998). In a natural sample that combines concentrated and dilute magnetic materials, the resulting ESR signal is a simple linear superposition of all material-specific signals, thereby allowing each to be studied and quantified separately. I expect ESR to be rediscovered in the near future, as a valuable probe in environmental science.

TYPES OF MAGNETIC ORDER AND UNDERLYING MICROSCOPIC INTERACTIONS

Intra-atomic interactions and moment formation

Strong intra-atomic electron-electron interactions, in the presence of the central forces due to the charged nucleus and as perturbed by the crystalline environment, are such that a certain net atomic (i.e., cationic) magnetic moment will result, depending on the total number of atomic electrons. This leads to the DM and PM cations discussed above in the context of Table 1. For a given number of atomic electrons, these intra-atomic forces are sufficiently strong to establish a permanent magnetic moment magnitude having a value that is little affected by either applied magnetic fields, up to the largest fields that are present in nature or the laboratory, or ambient temperatures, to far beyond melting and vaporization temperatures. As a result, we treat the cationic moments as permanent vector magnetic dipole moments having fixed magnitudes (represented by vector operators in quantum descriptions of solid state magnetism). On the other hand, nothing intra-atomic imposes a preferred orientation of the cationic moment. Therefore, in the absence of extra-atomic aligning influences the cationic moment experiences random fluctuations in orientations, induced by the ambient bath of relevant thermal excitations, leading to a thermal or time average of exactly zero ($\langle \boldsymbol{\mu}_i \rangle_T = \langle \boldsymbol{\mu}_i \rangle_t = \mathbf{0}$). This is the PM state of a material having non-interacting PM moments. An instantaneous snap shot of the PM state (in zero applied field) would have each cationic moment frozen in some random direction, such that the spatial average over all instantaneous moment vectors in the large sample is also zero ($\langle \boldsymbol{\mu} \rangle_s = (1/N)\, \Sigma_i\, \boldsymbol{\mu}_i = \mathbf{0}$).

Inter-atomic exchange interactions

The atoms touch and participate in bonding. As a result, there will often be inter-moment interactions that tend to align neighboring cationic moments. These interactions are of quantum mechanical origin and are much stronger that the classical dipole-dipole interactions that must act between all moments (and bar magnets). They are called magnetic exchange interactions and they are said to be ferromagnetic (FM) if they tend to make the interacting moments parallel (↑↑) and AF if they tend to make the interacting moments antiparallel (↓↑). Because they arise from orbital overlap and electron sharing, they tend to be short range, unlike the long range dipole-dipole forces that are a direct consequence of electromagnetism. Exchange interactions are sufficiently strong to align cationic moments against the randomizing influence of temperature and are the cause of magnetic order and the spontaneous magnetic structures mentioned in the previous sections.

When the exchange directly arises from the overlap of nearest neighbor (NN) atoms it is called direct exchange. The exchange interaction between moment-bearing (PM) cations can also be propagated via a shared coordinating (DM) anion, such as O^{2-}, OH^-, Cl^-, and F^-. This is called superexchange and is the relevant exchange interaction in oxides, hydroxides, and silicates. Note that one consequence of superexchange is that the magnetism of a mineral can be dramatically affected by its anion chemistry, without changing anything else. In metals, the exchange interaction can be propagated via the conduction electrons. This type of exchange is usually referred to as Ruderman-Kittel-Kasuya-Yosida (RKKY) exchange and it can be of longer range than just NN. The exchange interactions in meteoritic Fe-Ni alloys (e.g., kamacite, taenite) are visualized as a combination of direct and conduction electron mediated exchange.

Magnetic order-disorder transitions

The magnetic structures (or spin structures or moment configurations) that are caused by the exchange interactions and, to a lesser extent, by dipole-dipole interactions and spin-orbit coupling effects that I describe below, are remarkably varied and can involve magnetic unit cells many times larger than the crystallographic unit cells or magnetic structures that are incommensurate with the underlying crystalline frameworks. I next describe the more common magnetic structures, including those relevant to magnetic minerals and nanoparticles. Examples are illustrated in Figure 1. These structures typically occur at low temperatures and persist up to some magnetic ordering temperature that is characteristic of the material (but is affected by particle size), above which the system is in a PM state (Table 2). As temperature is increased towards the magnetic ordering temperature, the magnetic structure is usually preserved (or a

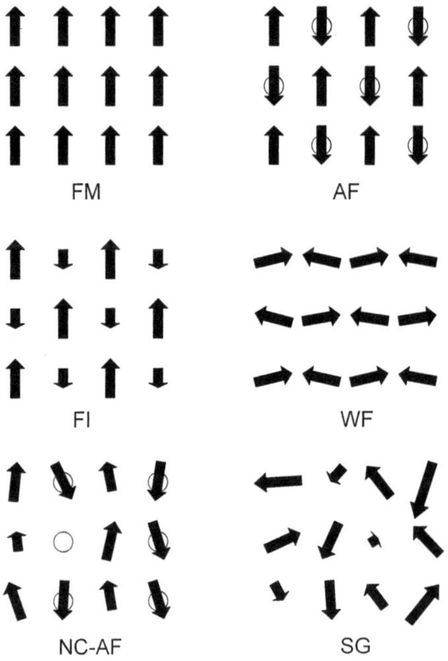

Figure 1.
Classic spin structures corresponding to:
- ferromagnetic (FM)
- anti-ferromagnetic (AF)
- ferrimagnetic (FI)
- weak ferromagnetic (WF)
- noncollinear antiferromagnetic (NC-AF)
- spin glass (SG)

arrangements of the thermal average cation magnetic moments. In the AF and NC-AF structures, the down sub-lattices are identified by open circles. Only the simplest two-sublattice cases are illustrated. In the NC-AF example, a cation vacancy (or substitution with a DM cation) on the down sublattice is illustrated, which gives rise to a net uncompensated sample moment.

transition to another magnetic structure can occur at a magnetic structure transition temperature) and each local average moment generally decreases continuously, while maintaining its orientation, up to the magnetic ordering temperature, above which it is zero, in the absence of an applied field.

Collinear and noncollinear ferromagnetism

When the magnetic structure corresponds to all the moments being aligned in the same direction, the ordered magnetic structure is said to be FM and we speak of ferromagnetism (Fig. 1). The magnetic ordering temperature in this case is called the Curie point or Curie temperature. This is the magnetic structure of the FM first row transition metals: body centered cubic (BCC) α-Fe, hexagonal Co, and face centered cubic (FCC) Ni. In a FM material, all moment-bearing cations need not be the same but all moments (of different magnitudes and different local thermal averages) must point in the same direction, to within allowed small site to site deviations of the average direction. When there are no such deviations, one can explicitly refer to a collinear ferromagnet. Or, if one wishes to stress the importance of small deviations that occur in a particular material, one can use the term noncollinear (NC) ferromagnetism. NC-FM structures occur in the presence of substitutional (chemical) disorder and positional (strain or structural) disorder. They appear to be common magnetic structures in FM alloys, especially amorphous FM alloys (Coey and Readman 1973a, Coey 1978, Ferchmin and Kobe 1983). An important defect that can lead to NC magnetic structures is the surface of a nanoparticle, via the mechanism of surface pinning that I describe below. Also, some FM alloys have a relatively small population of cationic moments that point opposite to the main FM direction. This is the case with the well known (synthetic and natural) Fe-rich FCC Fe-Ni alloys (e.g., Invar, taenite), where it has been referred to as latent antiferromagnetism.

Collinear and noncollinear antiferromagnetism

The magnetic structure that is most common in the oxides arises from the fact that the superexchange bonds are AF in these materials and is referred to as the AF structure (Fig. 1). Here, the identical moment-bearing cations are typically divided into two interpenetrating sublattices that have opposite moment orientations. In Figure 1 the lattice sites of the sublattice with down moments are identified by open circles in the AF structure. The other sublattice is populated with up moments. As a result of this arrangement, the net spontaneous macroscopic moment on the sample and the corresponding magnetization are zero, just as they would be in DM or PM materials. Historically, this made the recognition of antiferromagnetism difficult. The magnetic ordering temperature of an AF material is called the Néel temperature.

As is the case with the FM state, the ideal AF structure that occurs in bulk ordered crystalline compounds has perfectly collinear moments, all aligned along the same axis, but NC-AF structures typically occur in nanophase, chemically disordered, and poorly crystalline materials. The NC-AF structure is illustrated in Figure 1. The down sublattice is again identified by open circles. The thermal average moment sizes (lengths of arrows, Fig. 1) vary from site to site, as does the degree of noncollinearity. In this example (NC-AF, Fig. 1), a vacancy of one of the down sublattice moments is also illustrated. Such a magnetic vacancy, can either be due to a true cation vacancy or to a chemical substitution of a PM cation by a DM cation (e.g., Al^{3+} for Fe^{3+}). The type of NC-AF structure illustrated in Figure 1 is the relevant starting point for describing important environmental nanomaterials such as the hydrous ferric oxides (HFOs) and the manganese hydroxides. It is a key feature of nanophase AF materials that the two magnetic sublattices on a given particle are not usually equally populated. This gives rise to a residual net magnetic moment per particle (i.e., the supermoment) that is much larger than one would expect from an AF structure.

Ferrimagnetism

The magnetic structure of the most prominent natural magnetically ordered material, magnetite, is a ferrimagnetic (FI) structure (Fig. 1). FI magnetic structures are similar to AF structures except that the moments on the different sublattices do not cancel each other completely. For example, one of the two sublattices (the down sublattice in Fig. 1) is populated by a cation species having a smaller moment magnitude than the moment magnitude of the cation species on the other sublattice. As a result, there is a large net magnetic moment per sample and large corresponding magnetization, in the direction of the moments of the dominant sublattice. Only FM and FI materials are "magnetic" in the sense that only these two magnetic structures give rise to net magnetizations that are sufficiently large for the material to be noticeably affected by an ordinary bar magnet and only these materials are used to make permanent magnets for various applications.

Whereas in ideal FM and AF compounds having only one type of magnetic cation on one type of crystal site the evolution with changing temperature of the average local moment is the same for all moments, in FI materials the temperature dependence of one sublattice magnetization (i.e., net average magnetic moment on a given sublattice divided by sample volume) can be quite different from that on the another sublattice. As a result, compensation points can occur at temperatures where a dominant sublattice becomes subordinate and vice versa (Néel 1955, Kahn 1999). In magnetometry experiments, this can lead to unusual situations where large measured sample moments occur in the direction opposite to the applied field, as temperature is varied.

As with FM and AF materials, the FI structure can be noncollinear and can include imperfections such as vacancies, chemical disorder of the magnetic cations, exchange bond disorder from superexchange anion substitutions, and distributions of exchange values from strain distributions of inter-ionic distances. This, in particular, is the case with natural magnetite-titanomagnetite-maghemite materials.

Weak ferromagnetism, canted antiferromagnetism

At ambient temperature, the stable surface environment end-product of Fe dissolution and oxidation and the principle iron ore, hematite, α-Fe_2O_3, has a magnetic structure of the weak ferromagnetic (WF) or canted antiferromagnetic type that is illustrated in Figure 1. This structure consists of an AF starting point, with exact cancellation of the two sublattice magnetizations, in which a small moment canting of both sublattices leads to a small non-compensated net residual moment. In Figure 1, the underlying AF structure has horizontal moments pointing to the left in one sublattice and to the right in the other and the canting involves counter rotations of the two sublattice magnetizations such that a net up-pointing vertical moment results.

From the perspective of the microscopic theory of magnetism, WF magnetic structures are quite unusual in that the canting arises from the combined effects of superexchange and spin-orbit coupling to the crystalline lattice (Dzyaloshinsky 1958, Moriya 1960). This coherent canting must be distinguished from the local noncollinear effects that arise from disorder. In following the magnetization versus temperature and applied field, it is expected that both the local average moment value and the canting (of the average moment) change.

A material such as hematite, that can host a WF magnetic structure, also is able to host a classic AF structure with magnetic sublattices along a different crystalline axis. A spin flop transition, known as the Morin transition in hematite, can occur where the AF axis abruptly changes from one crystal orientation to another, at a certain transition temperature. Such spin flop transitions are sensitive to sample features such as impurity chemistry and particle size and shape, as discussed below (Dang et al. 1998).

Metamagnetism of layered materials

Materials with layered crystallographic structures, such as layer silicates, layered double hydroxides, layered chlorides, and graphite intercalation compounds (GICs), often have layered magnetic structures arising from intra-layer exchange interactions being much stronger than inter-layer exchange interactions. Indeed, often there are no superexchange paths between the moment-bearing cations in different layers and there are no exchange interactions between the layers. In many cases the intra-layer superexchange is FM and the layer plane is a magnetic easy plane, as determined by magneto-crystalline effects. The magnetic structure that then results is that of annite below its magnetic ordering temperature of 58 K (Rancourt et al. 1994), in which planes of ferromagnetically aligned moments lie in the ab-plane and are stacked antiferromagnetically along the c-axis. In general, the in-plane ferromagnetism is due to FM superexchange and the AF stacking is due to some combination of dipole-dipole coupling and a relatively weak AF inter-plane exchange. Because the inter-plane coupling is weak, there often occurs an applied field-induced spin flop transition, at attainable in-plane laboratory fields, in which the magnetic structure changes from AF stacking to FM stacking, along the field direction. The term metamagnetism is used in relation to the existence of this spin flop feature, in materials that have in-plane ferromagnetically aligned moments and AF stacking of the FM planes in zero field. In Fe-rich biotites, the LHT spin flop field is typically 0.1–0.2 T (Rancourt et al. 1994).

Spin glasses, cluster glasses, and multi-configuration states

There are many other magnetic structures, in addition to the above described classical FM, AF, FI, WF, and metamagnetic structures, that include: decoupled sublattices having various relative sublattice magnetization orientations and different sublattice-specific magnetic ordering temperatures, helical spin structures, and static spin density waves. Elemental Cr, for example, has a particularly complicated ordered magnetism and helical structures are common in rare earth element compounds. Non-classical magnetic structures may also be present in many important Fe-bearing rock forming minerals that have not been extensively studied to low temperatures. At present, such structures can be considered rare in Earth materials. One exotic magnetic structure, however, has often been suggested to occur at low temperatures in minerals and should be considered here: the spin glass (SG) magnetic structure (Binder and Young 1986, Chowdhury 1986, Fischer and Hertz 1991).

I stress that this magnetic structure represents a theoretical end-point that, in my opinion, has never been conclusively observed in a real material, synthetic or natural. The SG state is represented in Figure 1 and has the following defining ingredients: (1) random orientations of the average local moment vectors, and (2) some statistical distribution of average local moment values. That is, the SG state is a static configuration of average moments that is similar to an instantaneous snap shot of the ideal PM state, except that in a crystalline PM compound with only one moment-bearing cation species all local moments have equal values. The SG state can equally well be postulated to occur in amorphous or crystalline materials. It is believed to be the result of magnetic frustration, in which the moment configuration is established by competing exchange interactions. The required frustration has been described as occurring either from a distribution of exchange values, including both negative (AF) and positive (FM) values, or from a geometric effect called magnetic topological frustration whereby the exchange bond preferences cannot all be satisfied simultaneously because of the particular exchange bond network architecture.

Crystalline and amorphous materials that contain small concentrations of one or several moment-bearing cations, such as Fe-poor silicates, normally display a low temperature alternating field magnetometry peak or cusp having a magnitude and position that are sensitive to the driving field frequency and to the size of the constant applied field. An analogous cusp occurs in the low-field ZFQ-FW constant field magnetization curve, having a magnitude and position that is sensitive to the applied field magnitude. These systems are generally believed to have SG order below the cusp temperature and the cusp temperature is often associated with a SG transition temperature, at which the system is assumed to become PM on increasing the temperature. It is also possible to interpret these phenomena in terms of the response of correlated clusters of moments, without reference to SG formalism (e.g., Continento and Malozemoff 1986). Indeed, systems of FM, AF, and FI nanoparticles display very similar alternating and constant field responses (discussed below).

The SG picture has a life of its own and when disordered multi-component FM materials are observed to have unusual features at temperatures below their Curie points, researchers often attempt to describe their observations in terms of a reentrant SG phase. Such interpretations should be considered tentative and have not provided much insight into the mechanism(s) for the unusual behaviors. A true understanding will emerge as microscopic models include more and more realistic features and directly deal with simulating the actual measurements, rather than via attempts to map an enticing simple theoretical picture onto complex real materials. Realistic features that are not usually taken into account include: the effects of dipolar interactions (last subsection), domain wall motion, chemical clustering on several length scales, coexistence of several moment-bearing cation species, and coexisting correlated distributions of local ionic anisotropy (next subsection), exchange strengths and signs, and cation species and positions. Nonetheless, the extensive theoretical musings related to SG magnetism seem to have had applications in several other areas, such as biology, neural networks, and machine learning (e.g., Chowdhury 1986, Fischer and Hertz 1991, Stein 1992). This stems from an underlying theoretical difficulty of dealing with systems that have large numbers of energetically equivalent state configurations.

Systems of randomly oriented magnetic nanoparticles randomly dispersed in a supporting medium or matrix and that interact via dipole-dipole forces (last subsection) are systems having several energetically equivalent supermoment orientational states, at given temperatures and applied fields. As such, it is relevant to compare their magnetic behaviors with both the observed behaviors of canonical SG systems (dilute magnetic alloys such as MnCu) and the theoretical predictions from overly simple SG models. This has lead to a productive examination of the effects of dipolar and other inter-particle interactions in synthetic nanoparticle model systems that is reviewed below. Hopefully, this will in turn motivate the development of more realistic theoretical models of disordered dipolar systems.

Spin-orbit coupling and magneto-crystalline energy

Spin-orbit coupling is the intra-ionic coupling between an ionic shell's electronic spin angular momentum and its electronic orbital angular momentum. Since the ionic moment is largely due to electronic spin and since the orbitals participate in bonding, the spin-orbit interaction effectively couples the ionic magnetic moment to the underlying crystalline lattice or amorphous structure. For this reason, there can be preferred or easy directions along which the moments prefer to align. These directions have the symmetry of the crystallographic site or local bonding environment such that they cannot be the cause of magnetic order in a given vector direction but only define easy and hard magnetic axes. The energy required to turn an ionic moment from an easy direction to a hard direction is called the local magneto-crystalline energy or magneto-crystalline constant, K. This is the phenomenon that binds a given magnetic structure to the underlying crystal structure.

In most cases of interest K is much smaller than the NN magnetic exchange constant, J, such that an applied field will typically cause coherent rotations of exchange correlated moments rather than moment flips that would break exchange bonds. For example, when a sufficiently large field is applied to a classic AF structure along the easy axis of sublattice magnetizations, the system responds by a spin flop in which both sublattices go to a hard direction, perpendicular to the field, while preserving their AF relation and allowing a small WF rotation to produce an induced moment along the field. The field at which this occurs is called the spin flop field. It is determined by K, J, and the magnitude of the ionic moment, and it is temperature dependent. In a fixed sample of FM material, K determines how easy it is for an applied field to cause a coherent rotation of the FM magnetization along the field direction. K also plays a key role in determining the domain wall width and the ease with which domain walls can be moved by a driving applied field (next subsection). FM materials with small K are called shoft (Permalloys are good examples) and FM materials with large K are called hard. Hard materials are required for permanent magnets. Soft materials are required for transformer cores.

Electronic shells that have all orbitals equally populated, such as filled and half filled shells, have zero net total angular momentum. As a result, they have no spin-orbit coupling of the moment to the underlying crystal structure. This is a major difference between ferric and ferrous cations: Fe^{3+} ($3d^5$) has K ~ 0 whereas Fe^{2+} has a significant K, that in the flattened octahedral environment of micas, for example, causes a net preference for in-plane alignment and a larger in-plane PM susceptibility. In nanophase materials, near-surface distortions can cause deviant magneto-crystalline axes and altered K values, leading to surface moment pinning and noncollinearity.

Dipole-dipole interactions and magnetic domains

Any two magnetic dipole moments, $\boldsymbol{\mu}_1$ and $\boldsymbol{\mu}_2$, whether macroscopic or microscopic, have a dipole-dipole interaction energy that is given by the classical electromagnetic expression:

$$U_{dd} = (1/r^3)\,[\,\boldsymbol{\mu}_1 \cdot \boldsymbol{\mu}_2 - 3\,(\boldsymbol{\mu}_1 \cdot \mathbf{u}_r)\,(\boldsymbol{\mu}_2 \cdot \mathbf{u}_r)\,] \quad (1)$$

where r is the distance between the two moments and $\mathbf{u}_r$ is a unit vector that points along the straight line joining the two moments. The strength of this interaction falls off as $1/r^3$ and it does not have a limited range as is the case with exchange interactions. In astrophysics, for example, these forces may play a significant role in plantesimal accretion (Nuth and Wilkinson 1995). For ionic moments and typical inter-ionic separations in solids, U_{dd} is much smaller than typical exchange bond energies. As a result, dipole-dipole interactions play a minor role in establishing the magnetic structures described above. However, dipolar interactions occur between all pairs of moments and extend over large distances such that the dipolar fields due to large regions of ordered moments do become quite large. The dipolar field of a bar magnet is an example of this phenomenon.

Because of the first term in Equation (1), dipolar forces tend to advantage antiparallel configurations but in a FM or FI material, for example, the stronger exchange forces must prevail at short distances. The compromise (i.e., the situation that minimizes the system's total free energy) involves large regions of ordered moments that organize their respective magnetization directions in antiparallel and canted arrangements that mostly cancel any net sample moment. The regions of given directions of magnetic order are called domains and a macroscopic sample in the absence of an applied field will have a domain structure that results in no net sample moment, as is the case with a piece of demagnetized iron. The magnetic structures (FM, AF, FI, etc.) described above are intra-domain structures whereas the sample also has a magnetic microstructure or domain structure.

The boundaries between domains are called domain walls. These walls must involve a rotation of ionic moments from one domain orientation to the other and therefore constitute high energy defects. The domain wall thickness is also determined by free energy minimization and arises from a balance between exchange energy J and magneto-crystalline energy K. The domain structure of a macroscopic sample is difficult to calculate because of the long range nature of dipole-dipole forces and it depends on the size and shape of the sample. It is also sensitive to any applied field, including stray fields from nearby FM or FI samples.

In magnetizing a FM or FI material, it is initially easier, as applied field is increased, to move domain walls in such a way as to favor domains that are more aligned with the applied field. In a bulk material, this will generally be the dominant process in the initial rise of a hysteresis curve. Only domain wall pinning by inclusions and defects retards this process. The next step is typically to work against the magneto-crystalline energy in order to coherently rotate the magnetization of the surviving domain along the field direction, until saturation is reached. In a FI material, a spin flop with continued further rotation or a spin flip of the subordinate sublattice can occur at higher fields.

An AF sample may also have domains, that are defined in terms of how the up and down sublattices are assigned to the two underlying crystallographic sublattices: as up/down = A/B or as down/up = A/B, for example. In one dimension, this situation with one domain wall in the middle would be ↑↓↑↓↓↑↓↑. Alternatively, AF domains can be defined with the AF sublattices in a given domain along any of the equivalent n-fold magnetic easy axes. These domain walls in AF materials are not imposed by dipole-dipole forces and do not represent equilibrium situations. Instead, they can be considered quenched-in defects that increase the free energy. In small AF particles, this type of domain wall can cause significant residual sample moments or can participate in reducing residual moments. Analogous domain walls tend not to occur in FI materials, unless they are accompanied by chemical ordering domain walls, of the two moment-bearing cation species on the underlying crystallographic lattice sites, as: XOXOOXOX = ABABABAB.

FM and FI nanoparticles are always single domain particles because calculated domain sizes (and indeed even domain wall widths, often) are larger than the nanoscale (Stoner and Wohlfarth 1948). Microcrystals, however, can be multi-domain and it is of great concern in rock magnetism whether magnetite particles are single or multi-domain because this dramatically changes their relevant magnetic characteristics (Dunlop 1990). With single-domain particles, the magnetic and crystallographic domains are defined by the particle itself, although particles can also be polycrystalline. In any case, dipole-dipole forces will be an important inter-particle interaction that affects the applied field response of an ensemble of magnetic particles. The other main inter-particle interaction consists of exchange bridges, via inter-particle chemical bond bridges. I review this aspect of magnetic nanoparticle systems below.

FROM BULK TO NANOPARTICLE VIA SUPERPARRAMAGNETISM

Phenomena induced by small size and sequence of critical particle sizes

Having described the exchange, magneto-crystalline, and dipole-dipole interactions and their roles in determining the magnetic nanoscale and mesoscale structures and magnetic responses to applied fields of bulk samples, I now examine how particle size affects magnetic phenomena. What happens to a sample's magnetism as one continuously decreases the sample size, from macroscopic bulk dimensions down to an individual PM cation with its inner sphere coordination shell in solution?

Magnetic domain structure. Let us start with a bulk single crystal sample of FM or FI or AF material and let us consider that we are at a temperature, T, that is much lower than the magnetic ordering temperature, T_{oB}, of the bulk material (e.g., RT in magnetite or hematite is such a temperature, Table 1). Initially, we also assume that we are in thermodynamic equilibrium at each step in the size reduction process. The system first finds itself with an ordered magnetic structure, with well defined average moments on each PM cation site. The FM and FI varieties also have equilibrium domain structures, that are consistent with any applied field and the sample's shape and orientation in the applied field. As the particle size decreases and becomes compable to equilibrium domain sizes of the bulk material (10^6 to 1 nm, depending on the intrinsic material properties), the domain configuration or domain microstructure must significantly modify and adjust itself to the new equilibrium configuration of each new size. Here particle shape also plays an important role and the surface acts as a significant perturbation. Domain wall pinning by various intra-particle and surface defects will play an important role in that domain walls will take advantage of regions that are comparable in size to or larger than the domain wall thickness (10^2 to 10^{-1} nm, depending on the material) and that lower their energies.

Multi-domain/single-domain transition. At a first critical size, d_{SD} ~ equilibrium domain size, as particle size continues to decrease, the particle will become single domain (SD), rather than multi-domain (MD). This is an important boundary that dramatically affects the particle's magnetic properties. The particle no longer sustains a domain wall and now consists of an essentially uniformly magnetized grain carrying a net magnetic supermoment frozen along one of its magneto-crystalline easy directions. At $d > d_{SD}$, one could magnetize or demagnetize the particle under changing applied field simply by moving the domain wall(s), a relatively low barrier energy process. Now, the only way, at constant temperature, to change the magnetization of a sample of fixed SD particles is for the applied field to work against the magneto-crystalline barrier by causing a uniform rotation of the strongly exchange coupled moments, that is, of the supermoment. This is of course a simplified picture (the magnetization and the rotation are never perfectly uniform, as discussed below) but one that is often close enough to reality to be extremely useful. It is due to Stoner and Wohlfarth (1948).

Single-domain/superparamagnetic transition. As we continue to decrease the particle size, the SD state is stable down to the next critical size, d_{SP}, the size below which the particle becomes SP. This critical size depends strongly on both temperature and the measurement or observation time, τ_m. It is the size below which spontaneous thermally driven fluctuations of the supermoment between the particle's easy directions occur with a large enough probability to be observed within a chosen measurement or observation time. In the limit of an infinite observation time, all SD particles are SP and $d_{SP} = d_{SD}$. That is, all SD particles have true equilibrium thermal average moments of exactly zero, in zero applied field. This is an academic point but one that has large consequences in terms of how the theoretical calculations are done: one cannot simply use the methods of equilibrium thermodynamics and statistical mechanics. In rock magnetism, d_{SP} is normally given for RT and assuming the measurement time of constant field magnetometry. It is adjusted for the relevant measurement time when using alternating field magnetometry.

There has been much theoretical work, starting with Néel (1949a,b) and reviewed below, in developing valid expressions for the characteristic SP fluctuation time, τ. This is the average time one must wait before the supermoment spontaneously jumps from an initial easy direction orientation to a different easy direction orientation. The energy barrier that the supermoment must cross in doing this is denoted E_b. The simplest case of uniaxial magneto-crystalline anisotropy has been treated most frequently. In this case $E_b = nK$, where n is the number of PM cations in the particle and K is the ionic anisotropy constant. There can also be a dipolar contribution to E_b that depends on the sample shape and its magnetization. For uniaxial anisotropy and in zero applied field, the theoretical consenssus is that τ is given by:

$$1/\tau = f_o \exp(-E_b/k_B T) \tag{2}$$

where k_B is Boltzmann's constant and f_o is called the pre-exponential factor or attempt frequency. Temperature of course is in Kelvin. Various calculations obtain similar but different expressions for f_o which depends somewhat on temperature and on various material properties of the particle. Typical calculated, measured, and estimated values of the attempt frequency are in the range 10^6 to 10^{13} Hz. As I discuss below, the validity of Equation (2), first developed by Néel, has now been amply demonstrated experimentally on individual nanoparticles, not just macroscopic ensembles of particles.

One notes that d_{SP} is, in cases where Equation (2) applies, given by satisfying the equation $1/\tau_m = f_o \exp(-E_b/k_B T)$, where both f_o and E_b depend on the particle size. To a good approximation, f_o can be taken as a constant and $E_b = vK_v$ where v is the particle volume and $K_v = nK/v$ is the (known) total anisotropy energy constant per unit volume of the material. This gives:

$$v_{SP} = (\pi/6)d_{SP}^3 = (k_B T/K_v)\ln(f_o\tau_m) \quad (3)$$

which leaves the most difficult task of correctly evaluating the attempt frequency f_o in order to predict a correct d_{SP}. Clearly, however, temperature plays the dominant role and it is possible to block all SP fluctuations by going to sufficiently low temperatures. This is one reason that advanced mineral magnetic methods include variable temperature measurements over broad ranges from cryogenic to high temperatures. In such measurements, the concept of a SP blocking temperature, T_{SP}, is a useful one. Below this temperature, the SP supermoment fluctuations are observed to be blocked on the timescale of the measurement. T_{SP} is obtained by solving Equation (3) for temperature, given the particle volume: $T_{SP} = (E_b/k_B)\ln(f_o\tau_m)$.

Surface region size. Now, let us continue decreasing the particle size. One will next encounter a third characteristic size, d_{SR}, the surface region (SR) thermal local moment disorder size. At this point, the particle size is equal to the SR thickness of the equilibrium surface layer that has thermal average local ionic moments that are significantly affected (usually decreased but can be increased) by the surface in a semi-infinite slab of the material. This SR of modified average moments arises because the moment-bearing cations are coordinated via exchange interactions to fewer neighbors when at the surface than when in the bulk. That is, the magnetic interaction coordination is smaller at the surface than in the bulk. The ideal SR thickness depends on temperature, the strengths of the exchange interactions, and the crystallographic orientation of the surface (Hohenburg et al. 1975). In real particles, the SR thickness is also sensitive to: particle shape, surface roughness, surface complexed impurities, surface chemical bond relaxation and reconstruction, surface defects such as vacancies, SR compositional inhomogeneities, and particle shape orientation on the underlying crystal lattice. To understand this SR effect, consider that magnetic order is, in general, very sensitive to the degree of magnetic interaction coordination. Indeed, it can be rigorously proven that, with the usual isotropic short range exchange interactions, magnetic order cannot be sustained at any non-zero temperature in one or two dimensions, irrespective of the finite exchange strength (Mermin and Wagner 1966, Hohenburg 1967). In a finite one dimensional system (an extreme case of an elongated magnetic nanoparticle), magnetic order cannot be sustained at any temperature, irrespective of the exchange and magneto-crystalline anisotropy strengths (de Jongh and Miedema 1974, Steiner et al. 1976). When $d < d_{SR}$, the concept of a well defined supermoment breaks down, as does that of SP fluctuations. One is back to modelling individual ionic moments using equilibrium statistical physics methods (on finite systems), a task that is considerably simpler than dealing with superparamagnetism. Reviews of SR effects and their importance in dealing with real SP particles have recently been given by Kodama (1999) and Berkowitz et al. (1999). Monodisperse synthetic nanoparticles are now sufficiently small for these effects to become important (see below).

Quantum tunnelling. If we continue to decrease the particle size, we encounter a fourth characteristic size, d_{QT}, the quantum tunnelling size, that is relevant at low temperatures where thermal fluctuations become negligible. This is the size at which, at zero or near zero temperatures, the quantum tunnelling barrier is low or narrow enough to allow macroscopic quantum tunnelling (MQT) of the supermoment between its easy axis orientations, within a set measurement or observation time. Such tunnelling has recently been reported in, for example, molecular clusters of twelve PM cations (Friedman et al. 1996) and magnetic proteins (Awschalom et al. 1992). We review some of the evidence for MQT below.

Molecular clusters and N-mers. One eventually reaches the smallest sizes, where it may be more relevant to speak of polynuclear molecules and small polymer networks than of nanoparticles. Here we think of molecular clusters of only ten or so PM cations, down to the bi-nuclear molecule. Crystals of molecular clusters can be elegant model systems of identical magnetic nanoparticles (e.g., Barra et al. 1996). When one reaches these sizes, any change in cluster size (i.e., by as little as one PM cation or one coordinating anion) can cause dramatic changes in the magnetic properties. For example, with AF coupling, clusters with odd numbers of PM cations have large residual supermoments whereas clusters with even numbers of PM cations do not. Similarly, substitution of a single coordinating (i.e., superexchange) anion can change a key magnetic exchange bond from being AF to being FM or vice versa. Also, with such small clusters, the magnetic exchange bond topology becomes critical and, as with low dimensional materials, the concept of a bulk magnetic ordering temperature becomes irrelevant. This is because soliton and single-ion magnetic exitations become the energetically preferred excitations, even at the lowest temperatures, relative to cooperative magnetic excitations such as supermoment and SP fluctuations (e.g., Rancourt 1986). In this limit, therefore, depending on temperature, magneto-crystalline anisotropy, exchange strength, and exchange bond topology, the concept of a supermoment looses its usefulness because fluctuations in the supermoment magnitude become comparable to the fluctuations in supermoment direction. Often, the magnetism of molecular clusters is better described in terms of exchange-modified paramagnetism than in term of superparamagnetism (Carlin 1986).

Single coordinated cation. The smallest size of interest is the single PM cation and its inner sphere coordination shell of anions. These are the $Fe^{3+}(OH_2,OH^-)_6$ and $Fe^{2+}(OH_2,OH^-)_6$ octahedral cations of aqueous chemistry, for example. One encounters the usual crystal field stabilisation energetics and a PM fluid that has an ideal Curie law paramagnetism. Frozen and unfrozen solutions have been studied by ESR, magnetometry, and Mössbauer spectroscopy. These ionic clusters are mainly of interest to us here because they are the relevant starting points in various synthesis reactions of magnetic nanoparticles and they are the precursor forms that will surface complex to nanoparticles of interest. In this regard, it is important to note that geochemists often incorrectly assume the free metal ion model; that the equilibrium cation species in solution will be the only species that surface complex or that participate in nucleation and growth of new nanoparticles and that, therefore, the solution conditions (pH, Eh) completely determine the valence states and coordinations of intra-particle and surface cations. In fact, the valence states and coordinations of intra-particle and surface cations must be consistent with intra-particle and surface stereochemical constraints and can be quite different from those predicted by the free metal ion model, as they are near cell membranes (Simkiss and Taylor 2001). For example, tridentate complexation of a surface phosphate group can stabilize a surface ferrous cation on a hydrous ferric oxide nanoparticle, via a crystal field stabilization energy that is greater than the electron transfer energy from the environment, under conditions where only a ferric oxide solid phase is thermodynamically stable (Rancourt et al. 1999a, 2001a).

The above sequence of characteristic sizes is meant as a representative scenario only, not a rigorous path that all materials must follow when particle size is decreased. The same is true of the thermal path from high temperatures, above T_{oB}, to 0 K where zero point motion and quantum tunnelling persist. For example, goethite has a T_{oB} just above RT and nanoparticles of goethite probably have SP fluctuations starting at T_{oB} itself, where the single ion and supermoment excitations coexist and cannot be treated separately as with most other Fe oxide nanoparticles. This is probably the reason that goethite exhibits such unusual Mössbauer specira compared to other Fe oxide nanoparticle systems (Morup et al. 1983, Bocquet 1996).

Magneto-sensitive features of magnetic nanoparticles

In this subsection, I describe the features of real nanoparticle systems that need to be added to the ideal starting point of a uniform magnetization on a regular crystal lattice, both the magnetization and the lattice being unperturbed by the particle surface, relative to the magnetism and structure of the bulk material. These crystal-chemical features significantly affect the magnetism and are therefore probed by magnetic measurements. Most of these effects were anticipated and described by Néel (1949a, 1953, 1954, 1961a,b,c,d).

SR thermal effects. I have already described the SR average thermal local moment effect that was characterized by d_{SR}. If the bulk magnetic bond structure is such that there is some degree of exchange bond frustration, where, for example, NN bonds would be AF whereas next or second nearest neighbor (2NN) bonds would be FM, and if the surface preferentially removes, for example, more 2NN bonds than NN bonds, then some or all of the local thermal average ionic moments in the SR can be larger than the local thermal average ionic moments in the middle region of the particle. The different local thermal average ionic moments will also have different applied field responses or local susceptibilities. To some extent, all moments in a particle are affected by this phenomenon and the magnetic ordering transition itself is affected compared to its occurrence at T_{oB} in the bulk material. Indeed, the normally sharp transition is affected by what are known as finite size effects. The transition does not exhibit the usual divergences of thermodynamic properties at T_{oB} and critical exponents of the transition region are different (e.g., Fisher and Privman 1985). For sufficiently small particles ($d \sim d_{SR}$) the perturbation extends to temperatures far below T_{oB} and the entire intrinsic magnetization curve is affected down to 0 K.

Surface magneto-crystalline anisotropy and SR anisotropy disorder. Regarding the surface, another important effect is that the local ionic magneto-crystalline anisotropies (K) in the SR can be significantly different in magnitude (including sign), symmetry, and orientation, compared to their values in the middle regions of particles. This is due to the different ionic coordination environments near the surface and was first described by Néel (1954a,b). Indeed, using a classical approach and the continuum limit, Néel showed that a surface magneto-crystalline anisotropy, distinct from the dipolar anisotropy described below, must occur that produces a local easy direction that is either perpendicular to the surface or in the plane tangent to the surface, depending on the sign of the surface anisotropy density k_S (in J/m^2). The density k_S was shown to depend on the crystal chemistry of the material and the crystalline orientation of the surface and to be significant for nanoparticles. Modern ab initio electronic structure calculations are now able to calculate such effects at the atomic and intra-atomic levels (e.g., Pastor et al. 1995, Nordström and Singh 1996). In discrete atomic models, the local anisotropy is not simply related to the surface normal but instead depends on the local coordination environment. A similar effect arises from chemical disorder in an alloy or solid solution, where K is locally defined by the chemical environment at a cation site. This means that, in addition to the above described d_{SR} thermal effect, the local thermal average moments in the SR will not only have different magnitudes than in the middle region but may also have different local easy directions and local orientations. The surface moment orientations (i.e., misalignments or non-collinearities) will result from an interplay and compromise between thermal agitation, exchange interactions, and local magneto-crystalline anisotropies. Such misaligned moments have been called surface pinned spins. They contribute to the particle's overall supermoment, have temperature and field dependent misalignments, and represent a main mechanism for large residual high field susceptibilities, above the saturation fields that give technical saturations in which all supermoments and middle region local moments are aligned at low temperatures. These phenomena have recently been elegantly modelled and described by Kodama and co-workers (Kodama 1999, Kodama and Berkowitz 1999, Berkowitz et al. 1999). These researchers also point out that the associated phenomena have probably been frequently misinterpreted as quantum tunnelling effects.

Incomplete sublattice cancellation and residual moments. Another important finite size effect, again first described theoretically by Néel (1961a), is relevant with AF and FI nanoparticles: In a two sublattice material, there probably will not be equal numbers of cations on each sublattice. This implies that not all moments will be compensated, as they would be in the AF or FI bulk materials. With AF nanoparticles, this will usually be the main mechanism to create a large supermoment, as is the case with ferrihydrite, ferritin, and hematite nanoparticles (Rancourt et al. 1985, Makhlouf et al. 1997a, Dang et al. 1998, Rancourt et al. 2001b). An elegant example of this effect, where exchange bond frustration also plays an important role, is provided in the case of a synthetic two dimensional nanoparticle system (Rancourt et al. 1986). The resulting residual supermoment that resides on an AF nanoparticle does not have a simple dependence on particle size and depends on the details of the material, including: vacancies, cation substitutions, shape, surface complexation, etc. Néel (1961a) proposed simple situations where the residual moment would go as $n^{1/2}$ or $n^{1/3}$ or $n^{2/3}$, depending on the underlying conditions.

Dipolar shape anisotropy barrier. Given the nature of dipole-dipole forces (Eqn. 1), a uniformly magnetized FM or FI SD particle having a non-spherical shape will tend to be magnetized along the particle's axis of greatest elongation. That is, the dipolar contribution to the energy is minimized for a magnetization axis that is parallel to the main elongation axis. It is simple to see that one consequence of Equation (1) is that, for the same local ionic moments having the same inter-ionic separations, the FM configuration → → (or → → →, etc.) has a smaller dipole-dipole interaction energy than that of the FM configuration ↑ ↑ (or ↑ ↑ ↑, etc.). Elongated or acicular SD particles are common so this is an important effect and the shape dependent dipolar contribution to E_b can be significant. For AF particles that have vanishingly small supermoments, this effect is relatively negligible because many of the long range terms in the total energy sum cancel. It is important to note that the magneto-crystalline anisotropy easy axis is coupled to the underlying crystal lattice and that it may not, therefore, be aligned with the elongation axis. The resulting easy axis directions will represent a compromise between the middle region (bulk) K axes, the SR K axes, and the dipolar anisotropy. For a uniformly magnetized particle of magnetization $M = m\mu/v = \mu_S/v$, where m is the net number of cationic moments μ per supermoment μ_S, having the shape of an ellipsoid of revolution with long (rotation) semi-axis *a*, short semi-axis *b*, and small eccentricity *e*, the dipolar anisotropy energy is given by (Néel 1954):

$$E_d = -(16\pi^2/30)\, a\, b^2\, e^2\, M^2 \cos^2\theta_{\mu a} = -(4\pi e^2/10)\,(\mu_S^2/v)\cos^2\theta_{\mu a} \tag{4}$$

where $\theta_{\mu a}$ is the angle between the supermoment and and the long axis. This continuum limit result applies irrespective of the underlying crystalline orientation. By comparison, the net surface magneto-crystalline anisotropy energy for such a particle is given by (Néel 1954):

$$E_s = -(16\pi/15)\, a\, b\, e^2\, k_S \cos^2\theta_{\mu a}. \tag{5}$$

For nanoparticles, these two shape and size dependent energies are significant and comparable in magnitude. Depending on the sign of k_S, they can add or partially cancel.

They are independent of and in addition to the net volume magneto-crystalline energy, E_v, which, for the uniaxial case, is given by:

$$E_v = -nK \cos^2\theta_{\mu K} \quad (6)$$

where $\theta_{\mu K}$ is the angle between the supermoment and the crystalline K axis. K, like k_S, can also be positive or negative. In the simple case where the K axis and the ellipsoid long axis *a* are parallel, the total uniaxial barrier energy of Equation (2) is

$$E_b = E_d + E_s + E_v.$$

Of course, in general, the magnetization is not uniform, the particle is not an ellipsoid of rotation, the volume magneto-crystalline anisotropy is not uniaxial nor is it aligned with the particle shape, there are local orientational and magnitude variations of both SR and middle region magneto-crystalline anisotropies, etc.

Intrinsic and quenched-in nanoparticle defects. Even if a nanoparticle could be made by simply cutting out a piece of desired shape and size from a perfect single crystal of material, it would then relax and spontaneously change itself to lower its free energy, via all mechanisms available to it within its life or annealing time. First and foremost, an ionic nanoparticle will not generally be stoichiometric and this problem will require significant SR adjustments to achieve electrical and valence stability. In additon, the SR bond lengths relax, relative to the bond lengths in the bulk, as a direct consequence of lower surface atom coordinations. Indeed, there is a continuous radial distribution of bond lengths, from bulk-like middle region values to SR values. This distribution is affected by the surface conditions. Surface modifications will include: reconstruction, complexation of impurities from the environment, adsorption of neutral molecules such as water molecules that complete ligand field stabilized inner shell coordinations of surface cations, and loss or gain of protons depending on solution pH. The changes in bond length are small (10^{-2} to 10^{-4} nm or less) but magnetic exchange interactions are very sensitive to bond lengths and bond angles and can even change signs under moderate structural perturbations. As a result, magnetic probes can often uncover changes, relative to the bulk material, that are difficult to detect by any other method. For example, the particle size dependent moment rotations seen by ^{57}Fe Mössbauer spectroscopy in the WF state of hematite (Dang et al. 1998, and references therein).

The intrinsic size-induced changes in bond lengths are accompanied by changes in atomic vibrational amplitudes, such that vibrational entropy per atom and effective Debye or Einstein temperatures will be different than those of the bulk. The Debye-Waller factors of measurements that are affected by lattice vibrations (e.g., X-ray diffraction, Mössbauer spectroscopy) will also be different than in the bulk. In addition, the surface will cause atomic vibrations of atoms near the surface of the particle to be anisotropic, even if they were isotropic in the bulk, typically with larger vibrational amplitudes perpendicular to the surface. With magneto-volume coupling, the surface-perturbed magnetic structure also contributes to changing the vibrational characteristics, relative to those of the bulk material.

Regarding both vibrational (phonon) and magnetic (magnon) excitations, the finite particle size imposes an upper wavelength limit. This can be detected in measurements that probe the excitation spectra up to large wavelengths (e.g., inelastic neutron scattering) and it affects the thermodynamic properties of the nanoparticles.

On the other hand, natural nanoparticles are usually nothing like clean crystalline surfaces that simply relax their structure and admit complexing cations on well defined bonding sites. They are often formed by rapid precipitation and represent metastable phases with high densities of quenched-in defects. These defects include: vacancies, often charged balanced by O^{2-}/OH^- substitutions, cation substitutions such as Al^{3+} for Fe^{3+} and substitutions of cations with different valences using various charge balancing schemes, strong surface complexation involving ubiquitous elements such as C and Si, an interface region with a supporting material or matrix phase, and crystal grain boundaries both between grains of the same phase and between grains of different phases.

Exchange anisotropy in multi-phase particles. An example of the latter situation is a metal nanoparticle that has a metal oxide layer or a magnetite nanoparticle that has an oxidized maghemite layer. The interfaces between phases can be disordered or crystallographically continuous or some combination of these situations. The interfaces can also be atomically sharp or continuous. For example, a magnetite/maghemite particle could have a continuous radial gradient of oxidation, from the middle region to the SR. When both phases of a two-phase particle are magnetic, such as the case of magnetite/maghemite particles, there may be continuous magnetic exchange interactions across the phase boundary. In this case, at temperatures between the blocking temperatures of the two phases, a phenomenon known as exchange anisotropy occurs, whereby the blocked component exerts a significant exchange force on the supermoment of the unblocked phase. This exchange force can be represented as a mean magnetic field that is applied to the SP component, thereby polarizing it in some preferred direction. This direction will have established itself at the higher blocking temperature where the blocked phased became blocked, depending on the ambient applied or interaction fields present at that time.

Inter-particle interactions. The various forms of magnetic and non-magnetic inter-particle interactions, in an ensemble of, say, identical nanoparticles, can dramatically affect the magnetic behavior of the macroscopic ensemble or sample. It has only recently become possible to study individual isolated nanoparticles (see below) so that most studies have had to struggle with resolving individual particle properties and properties that arise from inter-particle interactions. In a given sample, such as a rock or a consolidated or frozen sediment or a frozen ferrofluid, etc., there will be inter-particle interactions that depend on such factors as: the concentration of magnetic particles, the spatial distribution of particles, the nature of the matrix phase, and the nature of the particles. These factors have recently been given significant attention in controlled synthetic model systems (see below). If the particles make contact, there may be chemical bonds between them and relatively strong inter-particle exchange interactions via exchange bond bridges. The geometry of these bridges will predominantly determine the strength of the corresponding inter-particle exchange interactions (Rancourt 1987, p. 91-92). If the particles are not connected via exchange interactions, then the remaining relevant magnetic inter-particle interaction is the dipolar interaction: Equation (1) where each supermoment is considered a fundamental interacting moment and the interaction sum is carried out on all pairs of magnetic particles in the sample. As a result of these long range interactions, the magnetic properties depend on the shape of the macroscopic sample, assumed to contain a uniform distribution of particles. For this reason experimenters must take special precautions to make proper demagnetizing field corrections or calculations.

I have already mentioned that inter-ionic exchange interactions depend on inter-ionic separation (bond length). This gives rise to magneto-volume effects, where bond lengths adjust to the degree of local magnetic order, that is, to the magnitudes of local average ionic moments and to their relative orientations (Grossmann and Rancourt 1996). In addition, spin-orbit coupling induces a phenomenon known as magneto-elastic effects, where inter-ionic separation (bond length) is affected by the angle that an average local ionic moment makes with its local magneto-crystalline easy axis. That is, the effective ionic radius is a function of the angle between the local moment and the local magneto-crystalline axis. These magneto-volume and magneto-elastic effects are generally small in terms of the bond length changes but their influence on the magnetic properties can be large. In particular, if the particles are imbedded in a rigid matrix, then strain mediated interactions can become important (Jones and Srivastava 1989). Unlike the exchange and dipolar inter-particle interactions (and unlike exchange anisotropy described above), the later strain mediated inter-particle interactions cannot be modelled by mean interaction fields (i.e., by effective magnetic fields arising from the interactions) because they cannot induce a local preferred direction, although they could, for example, change the orientations of easy axes. Here one must distinguish the direction of a vector from the orientation of an axis. This has important consequences in interpreting measurements because an external or effective applied field will induce different SP dwell times on a given particle (τ_+ and τ_- given by an equation similar to Equation (2), that has a field dependence, more below) whereas magneto-crystalline and strain effects cannot.

More on magneto-elastic and magneto-volume effects. The SP supermoment fluctuations themselves arise from thermal lattice vibrations via magneto-elastic coupling (Néel 1949a,b). The supermoment fluctuations are intimately tied to lattice vibrations, contractions, and expansions. This is the source of the matrix mediated strain interactions described above but it also implies that, even in an isolated SP particle, there will be whole-particle volume fluctuations occurring in phase with the SP supermoment fluctuations that will have the same characteristic time τ, given by Equation (2). It should be possible to detect a sound of characteristic SP frequency $1/\tau$ emitted from a suitably prepared sample of imbedded nanoparticles and this may have applications as a means to characterize ensembles of nanoparticles where the acoustic noise spectrum would contain information about the distributions of τ values and their field and temperature dependencies. Gunther and Mohie-Eldin (1994) have suggested that such magneto-strictive SP lattice fluctuations should produce measurable effects in the Mössbauer spectra of SP nanoparticles.

Distributions of everything. The latter comment on distributions of τ values brings us to our next topic, the fact that most samples containing nanoparticles contain many different nanoparticles, with entire distributions of particle sizes and shapes, particle compositions and structures, matrix media, etc. Natural and synthetic assemblies of nanoparticles are complex, mainly because there are correlated distributions of all the physico-chemical properties of the nanoparticles themselves, not to mention the supporting medium or matrix. As a result, most measured properties cannot be understood on the basis of the properties of individual nanoparticles alone. For example, Equation (2) leads to a predicted exponential time dependence of the sample magnetization, at constant temperature and applied field, of the form

$$M(t) = M(\infty) + [M(0) - M(\infty)] \exp(-t/\tau) \tag{7}$$

whereas measured time dependences are most often logarithmic, with near-constant viscosities. This striking discrepancy can be shown to arise from the distribution of nanoparticle volumes, via the volume's effect on both τ and the supermoment magnitude (Néel 1949a). Indeed, the distribution of particle volumes or sizes is most often the single most important distribution to affect macroscopic sample properties, mainly because of the particle volume effect on τ, via the barrier energy E_b (Eqn. 2). As a result, magnetic measurements allow detailed quantitative determinations of particle size distributions. On the other hand, the barrier energy has significant contributions that are not simply proportional to particle volume (e.g., Eqns. 4-5) and have strong dependencies on particle shape and supermoment formation mechanism. The pre-exponential factor in Equation

(2), f_o, also has a complex dependence on particle properties and its value will be strongly correlated to the value of E_b.

The above discussion assumes non-interacting (well dispersed) nanoparticles but, in non-dilute samples of nanoparticles, interactions are the other major complicating factor, in addition to the particle volume distribution. Because of inter-particle interactions (both dipolar and exchange bridges), the distributions of spatial positions and orientations of the particles can significantly affect the measured properties. For example, the nano-particles may form aggregates or clusters, with exchange interaction bridges between the nanoparticles in a cluster and dipolar interactions within and between clusters. The distributions of cluster sizes and structures and their separations then become important. Such clusters may act as large magnetic particles and dramatically affect the observed behavior.

As a result of magnetic inter-particle interactions, each nanoparticle will feel a net local interaction field that can be modelled by a time-dependent local applied field, $H_{int}(t)$. If the time variation of the local interaction field is either very fast or very slow compared to the relevant characteristic times (e.g., τ) of the particle, then $H_{int}(t)$ can in turn be modelled as a static local field, that will, of course, depend on temperature and macroscopic applied field. The distributions of particle positions, orientations, and supermoments will determine the distribution of local interaction fields. These interaction fields are present in zero applied field and dramatically affect the behaviors of the individual nanoparticles and, consequently, of the sample as a whole. They achieve this in two important ways. First, they change the equilibrium magnetic properties of the sample, giving rise, for example, to superferromagnetic ordering or interaction Curie-Weiss behaviors (see below). Second, and possibly more importantly, they affect dynamic response, via their influence on SP dwell times.

The latter point has not been sufficiently recognized, despite Néel's (1949a) clear original exposition, and deserves more explanation. Equation (2) is valid only in zero applied field. At T = 0 K, a critical applied field, H_o, equal to the coercivity at this temperature for our uniaxial particle, is given by

$$H_o = 2\,E_b/\mu_S = 2\,E_b/m\mu. \qquad (8)$$

where $E_b = E_d + E_s + E_v$ (Eqns. 4-6). Applied fields greater than this magnitude will remove the barrier and completely saturate the sample magnetization, within the limits of equilibrium thermal averaging at the given temperature. At smaller fields, $H < H_o$, a barrier between the two easy directions persists but it is of different heights depending on whether the supermoment is predominantly parallel or antiparallel to the field direction. As a result, one defines two different SP dwell times: τ_+ is the average time one must wait for the supermoment to spontaneously flip to the other (i.e., minus) easy direction if it is initially in the plus easy direction (i.e., the easy direction that is predominantly aligned with H, for a given orientation of the particle) and τ_- is the average time one must wait for the supermoment to spontaneously flip to the plus easy direction if it is initially in the minus easy direction. The action of the field is such that $\tau_+ > \tau_-$. Both are field-dependent and one defines the field-independent (for $H << H_o$) average dwell time, τ, as

$$1/\tau = 1/\tau_+(H) + 1/\tau_-(H) \qquad (9)$$

which, to first order in H, is given by Equation (2). Derivations of correct microscopic expressions, analogous to Equation (2), for $\tau_+(H)$ and $\tau_-(H)$ in an ideal uniaxial particle are non-trivial and have involved much fundamental work (see below). For example, the expressions of Néel, Brown, and recent workers differ somewhat, depending on details of the assumptions and approximations made.

The main points are that there are at least two characteristic microscopic dwell times, at the level of a single nanoparticle, that they are field dependent, and that they are distributed, at the level of the sample, by two mechanisms: via the distributions of particle properties (E_b, μ_S) and via the distribution of interaction fields (H_{int}). Such is the added complexity due to inter-particle interactions.

Measurement time complexity. Yet another level of issues must be addressed when one considers whether the microscopic dwell times introduced above are much smaller than, comparable to, or much larger than the measurement time, that, in turn, is highly dependent on the type of measurement (see above). The simplest situation arises when all the microscopic dwell times are much smaller than the measurement time. In this case, one observes equilibrium values. For example, the average supermoment on a given particle is

$$<\mu_S>(T, H) = ([\tau_+(T,H) - \tau_-(T,H)]/[\tau_+(T,H) + \tau_-(T,H)])\ m <\mu>(T,H) \quad (10)$$

where $<A>(T,H)$ denotes the equilibrium thermal average of A, evaluated at temperature T and field $H = H_{applied} + H_{int}$. For a sample with N magnetic particles and of volume V, one then has, a sample magnetization given by

$$M(T,H) = (N/V) \int P(m,E_b,H_{int},...) <\mu_S>(T, H)\ dm\ dE_b\ dH_{int} \ldots \quad (11)$$

where the integral is over all distributed parameters and $P(m,E_b,H_{int},...)$ is the joint probability distribution of all distributed parameters. An expression such as Equation (11) is relatively simple to evaluate and interpret.

If the microscopic dwell times are all much larger than the measurement time, then one probes the system as a static distribution of its parameters in order to deduce the state in which the system was prepared, by its previous temperature, applied field, and structuro-chemical history. For example, this would correspond to a remanence magnetization measurement, in the absence of time or relaxation effects. Alternatively, one can consider that all the particles in the sample that have microscopic dwell times much larger than the measurement time form a subgroup or subsystem that has reliably preserved a subsystem-specific remanence signal. Since dwell times are highly temperature dependent (Eqn. 2), partial thermoremanence measurements are a powerful tool to reconstruct a rock's thermomagnetic history.

If the sample contains very broad distributions of microscopic dwell times, arising from broad distributions of E_b, H_{int}, etc., with many dwell times that are much larger than and much smaller than the measurement time, and if the measurement is one that is similarly sensitive to all signals from the sample, then one can ignore those particles having dwell times that are comparable to the measurement time and simply divide the sample into two groups having $\tau > \tau_m$ and $\tau < \tau_m$, in the spirit of Equation (3). One then proceeds to use static configuration calculations on the first group and equilibrium calculations on the second group. This approximation was first introduced by Néel (1949a) and is widely used in many fields related to magnetic nanoparticles. It is often a very good approximation for constant field magnetometry with natural geologic or environmental samples, where very broad distributions of magnetic particle sizes do occur. Its validity depends on both the nature of the sample and the characteristics of the measurement technique, since different measurement methods have different measurement response time windows (centered on τ_m).

If all of the particles or a significant fraction of all the particles in the sample have dwell times that are comparable to the measurement time, then one has the most complex situation. In particular, we stress that the measurement may be very sensitive to the fact that $\tau_+ \neq \tau_-$ and to the actual values of both τ_+ and τ_-, not just the average τ. Mössbauer spectroscopy is an example of a measurement method where this is important and has not generally been recognized (Rancourt and Daniels 1984). We discuss this in more depth below.

More measurement complexity. In addition to the problem of measurement time, each measurement method normally has a non-flat response or sensitivity function. By this I mean that the measurement of a macroscopic sample of nanoparticles will, for example, often involve a higher sensitivity to a particular subgroup of particles in the sample. Examples are as follows: alternating field susceptibility resonance damping (phase shift) being predominantly caused by those particles having $1/\tau$ comparable to the probing field frequency, Mössbauer recoilless fractions being smaller for the smaller loosely bound particles, high frequency alternating field susceptibility being predominantly sensitive to supermoment fluctuations about a given easy axis rather than to SP fluctuations, magnetometry being more sensitive to those particles with larger supermoments, ESR not detecting contributions that are broader than the experimental field (i.e., resonance) range, etc. One must guard against assuming a flat response function and assigning the observed behavior from the given measurement method to an "average particle" that has a behavior representative of all the particles in the sample.

Size and shape versus crystal chemical effects. Another aspect of the complexity of real nanoparticles is that there can often be strong correlations between quite different structuro-chemical features of the nanoparticles, thereby leading the experimenter to identify incorrect causal relationships between measured characteristics (e.g., size, shape, composition) and measured behaviors (e.g., phase transitions, changes in physical properties). The strong correlations between different particle characteristics arise because, for a given suite of natural or synthetic samples, the main relevant synthesis or treatment parameters (temperature, pH, oxygen fugacity, kinetic supply of reactants, bulk composition, annealing or aging time, etc.) simultaneously control several resulting nanoparticle characteristics (size, shape, composition, degree of agglomeration or sintering, degree of oxidation, surface conditions, degree of chemical order, etc.) and their distributions. This is very important because the goal of much research on nanoparticles is often to identify the role of particle size, as a key controlling parameter. There is an unfortunate tendency for assigning causal relationships between those particular particle characteristics and sample behaviors that the experimenter has happened to measure, although the causal link may actually be with characteristics that were not measured. The best defences against such errors is to measure as many different sample characteristics as possible on the same samples (Waychunas, this volume). We treat one classic example below, where many authors have attributed changes in the Morin transition of hematite to particle size in cases where these changes are due to chemically induced lattice deformations.

Ferrofluids. Ferrofluids are an important class of magnetic nanomaterials that is not of primary interest in this chapter. Nonetheless, it is useful to point out some key differences between magnetic fluids and solid phase magnetic nanomaterials. There are three main differences regarding microscopic magnetic phenomena: (1) regarding supermoment relaxation, ferrofluids have a Brownian relaxation component in addition to Néel superparamagnetism, (2) in relation to applied field response, in a ferrofluid the particle itself can rotate, thereby providing an independent induced magnetization mechanism that depends on fluid viscosity, particle surface properties, and magneto-crystalline coupling strength, and (3) particle aggregation and spatial correlations are largely determined by magnetic inter-particle interactions and change in response to applied fields. Some of these features are relevant to aquatic sediments, industrial sedimentation and holding pools, and to the filtering or separation of magnetic precipitates in environmental and industrial applications.

MICROSCOPIC AND MESOSCOPIC CALCULATIONS OF MAGNETISM IN MATERIALS

The remarkable situation in which we find ourselves in modern materials science is that physics has for some time been sufficiently developed, in terms of fundamental quantum mechanics and statistical mechanics, that complete and exact *ab initio* calculations of materials properties can, in principle, be performed for any property and any material. The term "*ab initio*" in this context means without any adjustable or phenomenological or calibration parameters being required or provided. One simply puts the required nuclei and electrons in a box and one applies theory to obtain the outcome of a specified measurement. The recipe for doing this is known but the execution can be tedious to the point of being impossible. The name of the game, therefore, has been to devise approximations and methods that make the actual calculations doable with limited computer resources. Thanks to increased computer power, the various approximations can be tested and surpassed and more and more complex materials can be modelled. This section provides a brief overview of the theoretical methods of solid state magnetism and of nanomaterial magnetism in particular.

Methods of calculation in solid state magnetism

Electronic structure calculations. The most fundamental approach is *ab initio* electron structure calculations (ESCs) (e.g., Sutton 1993, Szabo and Ostlund 1996). In this method, one specifies the nuclei and their positions and one uses quantum mechanics to deduce how the electrons will organize themselves about these nuclei. All the properties due to electrons, including chemical bonding, valence states of ions, and magnetism, are calculated from first principles. Such calculations are normally performed assuming a temperature of T = 0 K. The characteristic temperature at which electronic structure deviates from its T = 0 K configuration, known as the Fermi temperature, is normally very high (~10^4 K) such that calculated ESCs are normally valid at all temperatures of interest, in terms of the electron distributions (i.e., bonding configurations, bond strengths, bond types, relative bond lengths, ionic magnetic moment magnitudes, magnetic exchange bond strengths, fine and hyperfine parameters, etc.). ESCs represent the methods of choice for understanding crystal chemistry at a fundamental level and for predicting the structures of unknown materials of given compositions and subjected to any pressure. They can be used to investigate dramatic changes in electronic structure and the associated magnetism that can occur as one varies the composition and/or the lattice parameters. A good example of the latter application is provided by the study of low-moment (LM) and high-moment (HM) meteoritic and synthetic taenite phases (Rancourt et al. 1999b, Lagarec et al. 2001). They can equally well be used on bulk materials, molecules, clusters, and nanomaterials.

At present, there has been only limited success at developing ESCs for non-zero temperatures. This means that ESCs can only be used for obtaining ground state configurations (e.g., T = 0 K magnetic moment structure or T = 0 K chemical ordering energy) and the T = 0 K properties of imposed structures that may only be stable at non-zero temperatures. ESCs are not presently available for treating non-zero temperature behavior, including magnetic and chemical order-disorder transitions and vibrational, electronic, and magnetic excitations. One can perform ESCs for any set of nuclear positions and, since electron density equilibration times (~10^{-18} s) are much shorter than any other relevant time such as that associated with nuclear motion, thereby examine chemical reactions, atomic diffusion, precipitation and coagulation, etc., but always from a perspective that ignores the low energy excitations, compared to chemical bond energies. The low energy excitations that are ignored are: atomic and collective vibrations, valence fluctuations, magnetic moment orientation excitations (including ionic and SP varieties), entropic defect generation and diffusion, etc. When chemical bonding is a given and it is the latter phenomena that are of interest, then present ESCs need to be abandoned for more phenomenological calculations that take the electronic structure as fixed and treat the excitations of interest, to various degrees of exactness.

Magnetic subsystem. Next we briefly describe the most common such methods that are used in magnetism, to specifically study magnetic phenomena that are due to the responses of and interactions between permanent ionic moments in solids, to the extent that the magnetic subsystem can be decoupled from the underlying chemical, electronic, vibrational, etc. subsystems. As it turns out, the magnetic subsystem of a solid can usually be conveniently decoupled from the other degrees of freedom for calculation, thanks to significantly different time scales for the different excitation subsystems (e.g., Grossmann and Rancourt 1996). In these calculations, the ionic magnetic moments of the PM cations become the fundamental microscopic entities that replace the electrons of ESCs. The latter ionic moments are taken to be "permanent" in that their magnitudes are assumed not to vary with temperature. That is, ground state ionic configurations are assumed with set dimensionless total ionic angular momentum and spin quantum operators, **L** and **S**, respectively. Inter-moment exchange interactions, J_{ij}, between PM cations i and j, are also assumed to be fixed and given, as are the local ionic magneto-crystalline anisotropy constants, K_i, for cations i. The reader now appreciates that the cation-specific physical parameters **L**, **S**, J_{ij}, and K_i are determined by the electronic structure and normally have temperature variations, to the extent that the electronic structure itself does. The operational link between ESCs and the methods described below is that ESCs can be used to calculate the T = 0 K values of **L**, **S**, J_{ij}, and K_i rather than treating the latter as adjustable phenomenological parameters.

Exact analytic solutions. As usual in solid state magnetism, the recipe for a formally exact solution to the magnetic subsystem problem described above is known. For simplicity, we take the angular momenta to be quenched (**L** = **0**), as often occurs from bonding, and the Hamiltonian of the magnetic subsystem is written

$$\mathcal{H} = - \Sigma_{ij} J_{ij} \mathbf{S}_i \cdot \mathbf{S}_j + \Sigma_i K_i (\mathbf{u}_i \cdot \mathbf{S}_i)^2 + g\mu_B \Sigma_i \mathbf{S}_i \cdot \mathbf{H} \quad (12)$$

where bold symbols are vectors or vector operators. The first sum is over all interaction pairs of PM cations in the sample and represents the total exchange bond energy. The negative sign in front of the first term is such that positive exchange constants correspond to FM exchange bonds whereas negative constants correspond to AF bonds. The second sum is over all cations and represents the total magneto-crystalline energy for an assumed local uniaxial symmetry where $\mathbf{u}_i$ is a dimensionless unit vector parallel to the local symmetry axis. The third term is also over all cations and represents the total energy of interaction between an applied field and the cation moments. A given cation moment has magnitude $\mu_i = g\mu_B S_i$, where g is the electronic g-factor and S_i is a dimensionless half integer quantum number (in Table 1, g = 2). The quantum operator $\mathcal{H}$ acts on a $(2S+1)^N$ dimensional state space, where N is the total number of cations in the sample. One solves the associated eigenvalue problem for the $(2S+1)^N$ energy eigenvalues, E_n, and eigenstates, $|n\rangle$, and one uses these in the usual Boltzmann statistics to calculate the equilibrium value of any physical property A as

$$\langle A \rangle (T,H) = (1/Z) \Sigma_n \langle n|A|n \rangle \exp(-E_n/k_B T) \quad (13)$$

where $Z = \Sigma_n \exp(-E_n/k_BT)$ is the partition function and the sums are over all eigenstates. Non-equilibrium properties and transition probabilities can also be calculated.

One notes that the problem is separated into two parts: a purely quantum part of obtaining the energy eigenvalues and eigenfunctions and matrix elements $\langle n|A|n\rangle$ and a purely statistical mechanics part of evaluating the Boltzmann sum and partition function. The problem is solved, in principle. In practice, exact analytic solutions have been obtained in only a few cases corresponding to simplified (low dimensional) bulk materials with simplified exchange and magneto-crystalline interactions. Nonetheless, this has represented a fruitful area of interaction between theory and synthetic low dimensional model systems. Most importantly in the present context, available computers now allow one to treat clusters and small nanoparticles directly by these exact methods and to add realistic additional features such as magneto-elastic and magneto-volume coupling to the underlying lattice or chemical effects.

Other analytic methods. Several analytic methods have been developed to treat the statistical mechanical part of the above exact problem, since the quantum part is often known (Van Vleck 1959). These include: near-critical point renormalization group methods, temperature expansion methods for limited temperature regions, and various rigorous proofs relating to ground state or high temperature properties and to relations between critical exponents, near magnetic phase transitions. These results have often been extensively verified experimentally. Each such new result or method is considered a milestone in statistical physics, and magnetism is one of the major model system testing grounds used by statistical physicists.

Approximate analytic methods. If one is willing to make certain physically reasonable approximations, one can often obtain powerful and elegant closed form analytic solutions to the problem of evaluating expressions such as Equation (13). Historically, the most important of these methods is called mean field theory (MFT), as first developed by Weiss (1948, Pathria 1988) and used to great advantage by Néel and others. The basic simplifying assumption consists in replacing the cation moments that interact with a given central moment with their thermal averages in Equation (12). That is, we replace the operator $\mathbf{S}_j$ in Equation (12) by the scalar $<\mathbf{S}_j>(T,H)$, given by Equation (13) where $A = \mathbf{S}_j$. If all thermal average local moments are equal, as in an all-sites-equivalent ferromagnet for example, then the inter-cation terms in the interaction sum drop out and the problem goes from being a $(2S+1)^N$ dimensional one to being 2S+1 dimensional. The relevant Hamiltonian becomes

$$\mathcal{H}_i = -\Sigma_j J_{ij}\, \mathbf{S}_i \cdot <\mathbf{S}_j> + K_i\, (\mathbf{u}_i \cdot \mathbf{S}_i)^2 + g\mu_B\, \mathbf{S}_i \cdot \mathbf{H} \tag{14}$$

and the problem is reduced to a single ion problem. The local interaction sum is over all j-cations that interact with the central i-cation. We denote the single ion eigenenergies as ε_n and, in the case of the all-sites-equivalent ferromagnet, one has

$$<\mathbf{S}>(T,H) = (1/Z)\, \Sigma_n \langle n|\mathbf{S}|n\rangle \exp(-\varepsilon_n/k_BT) \tag{15}$$

where the sum has only 2S+1 terms (e.g., for Fe^{3+}, S = 5/2, Table 1) and all the eigenvalues, eigenstates, and matrix elements are known. Equation (15) is solved self-consistently for $<\mathbf{S}>(T,H)$ and the sample magnetization density is simply $M(T,H) = (N/V)<\mathbf{S}>(T,H)$. For $K_i = 0$, the calculation predicts a Curie point occurring at

$$T_C = [S(S+1)/3k_B]\, \Sigma_j J_{ij} \tag{16}$$

where the sum is over all pair-wise exchange bonds with the central i-cation. For example, with NN only exchange bonds and z NNs in the structure, $T_C = zJ_{NN}S(S+1)/3k_B$. Otherwise, T_C depends on the non-zero value of K_i and the induced magnetization depends on the relative orientations of **H** and $\mathbf{u}_i$. The calculation also predicts the usual high temperature ($T > T_C$) Curie-Weiss behavior of the initial susceptibility:

$$\chi_o = C/(T - \theta_{CW}) \tag{17}$$

with Curie constant $C = (N/V)\,(g\mu_B)^2\, S(S+1)/3k_B$, as given by an exact calculation for a paramagnet where $\chi_o = C/T$, and Curie-Weiss temperature $\theta_{CW} = T_C$. In reality, one normally has $\theta_{CW} \sim T_C$. These are remarkably good predictions for such a simple model.

For a two-sublattice antiferromagnet, one simply has two coupled single-ion equations such as Equation (15) that are simultaneously solved selfconsistently for the two sublattice average moments. A p-sublattice material gives rise to p coupled equations and presents no special difficulty. The MFT method has been generalized to random, disordered, and amorphous magnetic materials by Rancourt et al. (1993).

MFT can also be made as exact as computer resources permit by applying a Bethe-Peierls cluster approach (Pathria 1988). In this adaptation of the model, one considers an entire cluster of PM cations around a central cation of interest in the bulk material and one exactly solves the intra-cluster system, using the exact analytic approach described above, by replacing all extra-cluster cation moments that have exchange interactions with intra-cluster moments by their respective thermal averages. This calculation for the entire bulk material becomes more and more exact as the assumed cluster size is increased. In practice, it is often found that using relatively small clusters with only 1NN or 2NN cations provides a remarkably good approximation that is significantly better than MFT alone in its simplest form (e.g., Chamberlin 2000).

The relevance of MFT and related methods to magnetic nanoparticles is twofold. First, MFT provides valuable insight regarding the depth-wise profile of thermal average moments starting at the surface of a semi-infinite solid. In this application, each successive surface layer is treated as a separate sublattice. This allows a fair estimate of the SR depth, d_{SR}, defined above and its temperature, surface structure, and applied field dependencies. Second, although MFT is not well suited to describing the magnetism of a given nanoparticle compared to the recommended exact analytic method, it is a valuable method for treating inter-particle interactions. In this application of MFT (e.g., Rancourt 1985), one treats each supermoment as an elementary cation moment in a bulk magnetic material. That is, one models the interactions of a given supermoment (on a given nanoparticle) with all other supermoments in the sample by replacing the supermoments of interaction by their thermal averages. In this way, the central supermoment of interest sees an effective interaction field of interaction that is due to the thermal averages of all the other supermoments. This mean field is the H_{int} field introduced above in the case where it is not time dependent. The same is true in the application of MFT to bulk magnetic materials. The mean field or molecular field of MFT does not need to be calculated explicitly in obtaining the various thermal average properties but it is a useful representation of the average effects of interactions. For a bulk magnetic material, the vector mean field at site-i is given by

$$\mathbf{H}_{int}(1) = (1/g\mu_B)\,\Sigma_j\, J_{ij} <\mathbf{S}_j> \tag{18}$$

where the sum is over all interaction pairs with the central cation. In a sample of nanoparticles interacting by dipolar interactions (Eqn. 1), the static vector mean field of interaction seen by particle-i at position vector $\mathbf{r}_i$ is given by

$$\mathbf{H}_{int}(\mathbf{r}_i) = \Sigma_j (1/r_{ij})^3 [<\boldsymbol{\mu}_j> - 3 (<\boldsymbol{\mu}_j> \cdot \mathbf{u}_{ij}) \mathbf{u}_{ij}] \quad (19)$$

where r_{ij} is the distance between central supermoment-i and supermoment-j, $\mathbf{u}_{ij}$ is a dimensionless unit vector pointing in a direction parallel to the straight line separating supermoment-i and supermoment-j, and the sum is over all $j \neq i$, to include all supermoments interacting with supermoment-i. One notes that the interaction mean field depends on relative positions of particles, position of the particle considered within the sample, orientations and magnitudes of all the average supermoments of interacting particles, number density of nanoparticles, etc. These factors give rise to the distributions of interaction fields described above.

Numerical simulation methods. The two most widespread numerical simulation methods in materials science are Monte Carlo (MC) simulation and molecular dynamics (MD). MC simulation consists in evaluating the integrals and sum of statistical mechanics, such as the sums in Equation (13) for a given equilibrium physical property, by a trial and error statistical method using computer generated random trials rather than analytically or by direct summation (Binder and Heermann 1992). It has been used extensively to evaluate the properties of both ferrofluids and solid nanoparticle systems, including inter-particle interactions and particle aggregation effects (e.g., Menear et al. 1984, Ferré et al. 1995, Dimitrov and Wysin 1996, Ribas and Labarta 1996a,b; Kechrakos and Trohidou 1998a,b; Chantrell et al. 1999, 2000; Psenichnikov and Mekhonoshin 2000). Cluster MC algorithms (Wang and Swendsen 1990) are particularly relevant to magnetic nanoparticles. MC simulation has also recently been extended to be applicable to the problem of SP fluctuations on a single nanoparticle (Nowak et al. 2000). In the latter work, a time step calibration allows one to go beyond the usual extraction of equilibrium properties only and to model the microscopic SP fluctuations themselves. It can also be used to study the SP rotation mechanism on a single particle (Nowak and Hinzke 1999), beyond the simple uniform coherent rotation picture of Stoner and Wohlfath (1948) that most workers assume.

MD consists in modelling the atomic vibrations and movements using Newton's classical laws of motion, by including the relevant microscopic inter-atomic interactions. MC can also be used to model atomic displacements but only by discontinuous discrete steps and only to obtain equilibrium average positions and without including kinetic or real time effects or true vibrational effects, as is done with MD. MD is useful when one is primarily concerned with atomic motion and vibrational energy, such as to model chemical intercalation reactions, chemical mixing and reactions, surface adsorption phenomena, diffusion through micro- and nanopores, chemical order-disorder phenomena, phase separation, solid-liquid-gas phase transitions, etc. The chemical binding forces can be modelled via phenomenological microscopic forces or full ESCs can be performed at discrete times, since the electronic structure response time is much smaller than the MD atomic displacement times. MD has not been used to model the atomic vibrational part of magnetic nanoparticles and has not been applied to solid state magnetism in general, until recently. SP fluctuations themselves are due to thermal atomic vibrations and magneto-elastic, magneto-volume, and magneto-shape couplings, such that it would be important to use MD as part of complete and realistic simulations of magnetic nanoparticles. Grossmann and Rancourt (1996) have made a first step along these lines by showing how MD and MC can be combined to model the atomic motion and magnetic components, respectively, of a magneto-volume active bulk material. The method relies on the fact that, most often, atomic vibrational times ($\sim 10^{-13}$ s) are much shorter than ionic magnetic moment fluctuation times ($\sim 10^{-6}$ to 10^{-12} s) such that average atomic positions can be used to calculate magnetic exchange strengths and equilibrium spin structures and instantaneous moment configurations can be used to calculate inter-atomic potentials. Supermoment fluctuations are in turn much slower than ionic moment fluctuations such that a similar approach should be fruitful in modelling magnetic nanoparticles.

Calculations of superparamagnetism and inter-particle interactions

Uniaxial particles and fixed supermoment magnitudes. Starting with Néel (1949a,b) and Brown (1963), the great majority of researchers who have developed our theoretical understanding of SP fluctuations have treated only the simplest and most common case of uniaxial particles having well defined supermoments of fixed magnitudes. That is, particles having axial symmetries of their barrier energy profiles, such as those predicted for axial magneto-crystalline anisotropy (Eqn. 6), surface magneto-crystalline anisotropy of an ellipsoid of revolution (Eqn. 5), and dipolar anisotropy of a uniformly magnetized ellipsoid of revolution (Eqn. 4), and in which the supermoments have robust magnitudes arising from strong inter-cation exchange interactions. All researchers agree that the SP fluctuation time of an isolated uniaxial nanoparticle in zero applied field is given by an expression analogous to Equation (2) but there are several different expressions for the pre-exponential factor f_o, depending on the physical couplings that are included and the particular approximations that are used. In this regard, Yelon and Movaghar (1990) have made some interesting general comments. There are also important variations in the expressions obtained by various authors for $\tau_+(H)$ and $\tau_-(H)$ of Equation (9). It is beyond the scope of this chapter to review the details of these developments but the reader should know that much recent fundamental work has been done in this very active area and that the early results have been but in a broader context that allows one to understand the different approaches and approximations. Key papers and reviews addressing the underlying theory and calculations of τ, $\tau_+(H)$, and $\tau_-(H)$ include those of: Aharony (1969, 1992), Bessais et al. (1992), Coffey et al. (1993, 1994a,b,c, 1995a,b,c, 1996, 1998a,b), Cregg et al. (1994), Garanin (1996, 1997), Garanin et al. (2000), Garcia-Palacios and Svedlindh (2001), Jones and Srivastava (1989), Kalmykov and Titov (1999), Kalmykov (2000), Nowak et al. (2000), Pfannes et al. (2000), Popkov et al. (1999), Raikher and Stepanov (1995a), and Srivastava and Jones (1988). It is a non-trivial problem that consists in calculating the kinetics of a complex magnetic system, not just equilibrium properties.

Beyond isolated uniaxial particles. Several researchers have attempted to extend the fundamental envelope beyond isolated uniaxial particles with pre-defined supermoments. Raikher and Shliomis (1994) have reviewed the application of fundamental results to magnetic fluids. In addition to the important MC work cited above, other researchers have also attempted to calculate inter-particle interactions effects. Examples are provided by Kneller (1968) and by Song and Roshko (2000). Kalmykov and Titov (2000) and Coffey et al. (1998a) have treated non-axially symmetric particles. Chang et al. (1997) have described a way to go beyond the fixed supermoment two-level approximation. Victora (1989) has made an important link to measured properties by developing the concept of the time-dependent switching field. This is along the lines of Garanin's (1999) integrated relaxation time. Some authors have made first important steps in going beyond the coherent supermoment rotation model. New effects include domain wall nucleation (Broz et al. 1990) and non-uniform rotation (Braun 1993, Nowak and Hinzke 1999). Dormann et al. (1987) described the field dependence of the blocking temperature, thereby making another useful link to measurements.

Towards natural nanoparticle systems. I have already described the main physico-chemical features of real nanoparticle systems. Much theoretical work is still needed to bring the kind of fundamental understanding that has been achieved for the uniaxial fixed-magnitude supermoment model towards including these real features. Ongoing theoretical work is needed in three areas of active present research. First, as described above, to continue expanding from the uniaxial fixed-magnitude supermoment picture towards piece-wise additions of more and more real features. Second, to continue developing our theoretical understanding of the measurement methods themselves, as discussed further above and below. Third, to develop methods that combine the various results from the latter two areas, using known approaches, in order to model and analyse the behaviors of real samples. I review advances in the third area, that attempts to make links between fundamental theoretical results and observed behaviors, below.

INTERPLAYS BETWEEN MAGNETISM AND OTHER SAMPLE FEATURES

In this brief section I attempt to make a list of the main physico-chemical features of a material that can have strong interplays with the magnetism of the material. With nanoparticles, the strength and nature of such an interplay or coupling may vary dramatically with size. The size itself, independently of anything else, is such a feature because of the SR and finite size effects on magnetic order already described above. Simply put, spontaneous magnetic order (i.e., spontaneous existence of some array of non-zero thermal average local ionic magnetic moments) is more difficult to establish in a small particle than in a bulk, effectively infinite, material. With nanoparticles, shape itself, independently of any other factor, is also such a feature because of the dipolar (Eqn. 4) and surface anisotropy (Eqn. 5) effects described above. Shape can also play a role in producing sample texture (i.e., non-randomness of particle orientations) which in turn affects inter-particle magnetic interactions.

Chemistry and structure via $\{\boldsymbol{\mu}_i, \mathbf{K}_i, J_{ij}(\mathbf{r}_{ij})\}$

The other main features that couple strongly with magnetism, excluding exotic effects involving superconductivity, all fall into two categories: chemistry and structure. Indeed, mineral magnetic studies are geared towards extracting crystal chemical information from magnetic measurements and therefore rely on the following coupling mechanisms. In terms of its magnetism, a local moment system is completely specified by: the set $\{\mathbf{r}_i\}$ of the vector positions of the PM cations, the set $\{\boldsymbol{\mu}_i\}$ of local magnetic moment magnitudes on each of the PM cations, the set $\{\mathbf{K}_i\}$ of local magneto-crystalline anisotropy constants and directions, and the set $\{J_{ij}\}$ of all inter-cation pair-wise magnetic exchange bonds. The electronic origins of these parameters are such that J_{ij} has a small dependence on inter-cation distance as $J_{ij}(r_{ij})$ and $\mathbf{K}_i$ has a small dependence on the positions and identities of neighboring cations. Indeed, excluding the relatively weak inter-cation dipolar interactions, the only dependence on $\{\mathbf{r}_i\}$ is via $J_{ij}(r_{ij})$ and $\mathbf{K}_i$. In the present context, one therefore asks how structure and chemistry affect or determine the set $\{\boldsymbol{\mu}_i, \mathbf{K}_i, J_{ij}(r_{ij})\}$ and how this in turn determines all the magnetic properties. The identity of cation-i is the main factor determining both $\boldsymbol{\mu}_i$ and $\mathbf{K}_i$ and the identities of the interacting cations (i and j) and of the related superexchange anions are the main factors determining J_{ij}.

Chemical coupling to magnetism in nanoparticles

Bond and site disorder. Thermodynamic aspects of the interplay between chemical order and magnetism have been reviewed by Burton (1991), without consideration of size effects. Changing the degree of chemical order in a material changes the magnetic exchange bond topology of the network of interacting PM cations, thereby directly affecting the magnetism. For example, in a MFT description, the degree of chemical order determines the set $\{J_{ij}\}$, the sums in Equations (16) and (18), and all the magnetic properties. With complete solid solution randomness, one has a random set $\{J_{ij}\}$ and one speaks of magnetic exchange bond disorder. If the exchange bonds are superexchange bonds, then the degree of anion order (or substitution) does indeed directly control the bond set $\{J_{ij}\}$ and the degree of exchange bond disorder. If, on the other hand, it is the cations that are distributed (or substituted), then in addition to exchange bond disorder there is also cation site disorder. That is, the magnetism is affected by correlated distributions of both exchange bonds and cation species that may have different magnetic moments and different local magneto-crystalline anisotropies. Pure site dilution occurs when one type of PM cation is substituted by a DM cation species (the magnetic equivalent of a vacancy), without any change in the anion network. Pure bond dilution is said to occur when exchange bonds are removed, without removing moment-bearing cations.

Magnetism affects chemistry. Given that chemical bond energies are much larger ($U_{ij} \sim 10^4$ K) than magnetic exchange bond energies ($J_{ij} \sim 1$ to 10^3 K), one might expect that chemical order can affect magnetic properties but not vice versa. In fact, the chemical bond differences that drive chemical order are comparable to magnetic exchange bond energies such that magnetic degrees of freedom can significantly affect chemical order-disorder effects and even induce chemical order in situations where chemical bonds alone would lead to randomness or even phase separation (Dang and Rancourt 1996). This, in turn, implies that the magnetic SR effect could induce a radial gradient of chemical order in a magnetic nanoparticles, in addition to the surface-induced purely chemical effects that could produce such gradients. In general, we must expect radial gradients of chemical order, defect densities, and composition that must have strong interplays with magnetic radial gradients. This is true of both equilibrium properties and quenched-in defects and configurations that do not evolve because of the slowness of diffusion at the temperatures of interest.

Surface diffusion faster than bulk diffusion. In the latter regard, it is important to keep in mind that lateral surface diffusion times are typically two orders of magnitude smaller than bulk inter-site diffusion or hopping times. This implies that when a particle is annealed it will repair its surface before it can repair its middle region. Since magnetism is very sensitive to surface conditions, one can observe significant changes in magnetic response as a function of annealing temperature (or annealing time) at temperatures far below chemical reaction or sintering temperatures.

Structural coupling to magnetism in nanoparticles

Low magnetic dimensionality effects. Magnetic dimensionality of a bulk material was described above, in the context of SR effects, as a structural feature that dramatically affects magnetism. By low dimensionality we mean that the J_{ij}s are such that they form chains or ribbons or planes instead of three-dimensional frameworks. Indeed, low dimensionality can suppress the occurrence of magnetic order, even with exchange strengths that would normally give large Curie or Néel points. This is because the interactive feedback provided by fewer interacting neighbors is not enough to counter thermal agitation and because novel excitations occur that are not quenched by low temperatures. In addition, the behavior of finite low dimensional systems can be significantly different from the behaviors of their infinite low dimensional counterparts. For example, a short AF chain with an odd number of PM cations gives rise to a large initial susceptibility at low temperatures whereas the infinite chain does not. For these reasons, nanoparticles made of materials that have low dimensional exchange bond networks have magnetic properties that are particularly sensitive to particle size (e.g., Rancourt et al. 1986).

Magnetic frustration effects. A phenomenon known as magnetic frustration can effectively lower the dimensionality or connectedness of an exchange bond network. For example, consider a chain of moments that interact via 1NN and 2NN exchange interactions only. Each cation has two 1NN bonds and two 2NN bonds and if the 1NN exchange bonds are FM whereas the 2NN bonds are AF and of equal magnitude ($J_{1NN} = -J_{2NN}$) then each cation is completely frustrated regarding its orientation that would minimize the magnetic energy. As a result, the cations respond to applied magnetic fields much as isolated cations would in a PM material, despite the presence of strong exchange interactions. In this example, the dimensionality has been effectively reduced from 1 to 0. If a nanoparticle is made of such a material that has magnetic frustration in its bulk structure, then the magnetism of the particle can be very sensitive to size because cations at or near the surface will have missing bonds and may be less frustrated than they would be in the bulk. As a result, the SR can be more strongly magnetic than the bulk of the particle. That is, thermal average cation moments near the surface can be larger than in the middle region of the particle. Similarly, chains and planes of cations that were frustration isolated in the bulk may become exchange coupled via an uncompensated edge effect in the nanoparticle. In the above discussion, frustration referred to the local moment experiencing a mean field of interaction of zero and not receiving conclusive instructions from its neighbors regarding the orientation that it should adopt. I call this site frustration. Frustration can also refer to magnetic exchange bonds that are not energetically satisfied (i.e., do not participate in lowering the energy of the system) in a magnetically ordered structure because the orientations of the relevant moments are determined by other exchange bonds. I call this exchange bond frustration.

Chemical origin of positional disorder. The degree of positional disorder or the degree of crystallinity or amorphousness, for a given composition or fixed compositional distribution, in a nanoparticle is an important structural feature that couples to the magnetism via $\{\mathbf{K}_i, J_{ij}\}$. Much of the cation positional disorder comes from quenched-in disordered anion substitutions, such as (OH^-,OH_2) in authigenic oxyhydroxides. Much position disorder must also arise from cation chemical disorder, such as $(Fe,Al)^{3+}$ in alumino-oxides, in compliance with cation-anion bond length constraints and associated local anion types. Of course, positional disorder does not occur alone and the combination of several chemical and positional effects will lead to such striking nanoparticle phenomena as anticipated by Néel and recently described by Kodama (1999) and Berkowitz et al. (1999).

MAGNETIC AND RELATED TRANSITIONS AFFECTED BY PARTICLE SIZE

In this section, I briefly survey the different kinds of magnetic and related transitions that should or may be affected by particle size and shape. Transitions occur at special compensation or cooperative build-up points (critical temperatures, fields, compositions, or pressures, and at multi-critical points) where several factors conspire to give an abrupt change in physical state of the system. As such, it is expected that transitions will be particularly sensitive to particle size, not to mention intra-particle and surface conditions...

Defining the size effect question

When we ask whether a certain transition or property is affected by particle size, we generally mean "Would reducing the size of a bulk sample, without otherwise changing the material, eventually produces changes that are directly attributable to size?" The problem with this question is that it cannot be answered by experiment, for two main reasons. First, even with a pure and initially undisturbed piece of bulk material, the bond structure, chemical distributions, particle shape and size, etc., will relax, anneal, reconstruct, redistribute, etc., to best minimize the free energy of the particle in its environment within the time available to it. Second, nanoparticles are not produced by cutting out perfect pieces of bulk material and are often nothing like their closest bulk counterparts, as amply discussed above. The classic examples are ferrihydrites, ferritin cores, and hydrous ferric oxides (HFOs) that exist only as nanoparticles and do not have known bulk counterparts. On the other hand, the question can be posed and answered by theoretical calculations or simulations and such studies can be useful in interpreting the results of experiments. It is in this context and perspective that I pose the question.

Classic order-disorder, spin flops, electronic localization, and percolation

The classic magnetic order-disorder transitions at which a PM cation-bearing material goes from a PM state to a FM or FI or AF or WF state, etc., at associated Curie or Néel or equivalent transition points, as temperature is decreased, are affected by particle size as already described above. Spin flop transitions, such as the Morin transition of hematite, should also be affected by particle size for sufficiently small sizes and this is reviewed below for the case of hematite. Electron localization or cooperative valence fluctuation transitions, such as the Verwey transition of magnetite, should be affected by particle size for sufficiently small particle sizes but there have been no systematic studies of this or theoretical estimates of the size effect. Magnetic percolation transitions refer to the percolation of infinite clusters of exchange coupled cations in bulk dilute magnetic materials. The concept of percolation (Essam 1980) has relevance to the Verwey transition in impurity-bearing magnetite (Marin et al. 1990, Aragon et al. 1993) and to dynamic and relaxation effects in solid solution magnetic materials (e.g., Chamberlin and Haines 1990). In small particles the concept must be revised in that the effective critical composition becomes the composition at which the mean cluster size of the bulk analogue becomes equal to the largest cluster that the particle can sustain. One should expect, therefore, that all properties that critically depend on percolation will be significantly affected by particle size and will exhibit a critical size anomaly or saturation point.

Frustration and magneto-strain effects

I have already described how magnetic frustration that directly results from the geometrical arrangements of exchange bonds, such as three equal AF exchange bonds between identical PM cations at the corners of an equilateral triangle that cannot all be simultaneously satisfied, can suppress a magnetic ordering temperature that would other wise occur given the large magnitudes of the exchange interactions. It is easy to see how the surface of a nanoparticle could reduce frustration in the SR, thereby effectively allowing a higher degree of magnetic order to occur than would occur in the bulk. Another factor must be considered also. Perfect site frustration does not normally persist to low temperatures because the inter-cation separation dependence of J_{ij}, explicitly expressed as $J_{ij}(r_{ij})$, often will drive a spontaneous distortion of the lattice that will remove the site frustration and allow establishment of magnetic order, in the presence of residual bond frustration. Such cooperative magneto-strain transitions are expected to be sensitive to particle size since both strain and magnetic order exhibit surface-induced intrinsic radial distributions.

Magneto-volume effects

Magneto-volume effects refer to the balance between chemical bond energy inter-atomic separation dependence and magnetic exchange energy inter-moment separation dependence that, along with vibrational energy inter-atomic separation dependence, produces a resulting equilibrium distribution of inter-atomic separations in the material. Rancourt and Dang (1996) and Lagarec and Rancourt (2000) have developed the theory of magneto-volume effects in local moment systems with some magnetic exchange bond frustration and varying degrees of chemical order. Clearly, all relevant aspects of the problem (i.e., nature of the chemical bonds, separation dependence of the exchange parameters, vibrational amplitudes and energies, presence of exchange bond frustration, degree of chemical order, etc.) and their interrelations are affected by particle size and the presence of the surface. As a result, a nanoparticle can be significantly expanded or contracted relative to its bulk counterpart and in a non-uniform way, depending on the details of the inter-atomic interactions and the size and shape of the particle. Matrix effects, including differential thermal expansion, can also play an important role via positive or negative effective pressures on the nanoparticle.

Exotic effects and transitions

Other important transitions and cooperative solid state effects that have a coupling to magnetism and that would be sensitive to particle size include: superconductivity, Jahn-Teller effects, photo-magnetic effects as in magnetic semiconductors, radiation-induced defect susceptibility and associated PM centres, and magnetic moment formation or stabilization effects. In the latter case, one departs from our simple ionic picture of fixed cation magnetic moment magnitudes and covalent-type superexchange interactions to venture into the metallic domain where conduction electrons may or may not form local magnetic moments corresponding to non-integer values of electron spins. Some metals, such as body centered cubic iron (α-Fe, $\mu_{Fe} \sim 2.2\ \mu_B$, $T_C = 1043$ K), have stable and constant moment magnitudes up to and far beyond their magnetic ordering temperatures (Pindor et al. 1983) whereas others have weak local moments on the verge of stability or exhibit moment loss/formation transitions. Such exotic phenomena have recently been discovered in synthetic Fe-Ni alloys (Lagarec et al. 2001) and are relevant to meteoritic Fe-Ni phases where the nanoscale microstructure probably plays an important role (Rancourt et al. 1999b), as described below. They may also be relevant to geomagnetism?

OVERVIEW OF RECENT DEVELOPMENTS

Main recent developments in magnetic nanoparticle systems

Fundamentals of SP fluctuations. In the calculation methods section, I described the recent and ongoing flurry of activity in the area of fundamental calculations of the SP supermoment reversal fluctuations themselves and their dependencies on applied field, underlying magneto-crystalline anisotropy, etc. Such work is vital in that it links the main microscopic fluctuations that determine sample response and its time evolution to their fundamental causes and the associated particle features and material properties. All areas of science and technology that involve magnetic nanoparticles can only benefit from these fundamental advances.

Addressing inter-particle interactions. Another important recent development has been the explicit considerations of inter-particle dipolar interactions. In solid state magnetism as a whole, dipolar interactions have been largely ignored except for their acknowledged role in causing magnetic domain structures in bulk materials. This is because in bulk materials inter-cation dipole-dipole interactions are several orders of magnitude smaller than typical inter-cation exchange interactions. In nanoparticle systems, however, dipolar interactions are often the main inter-particle interactions and they are amplified by the presence of supermoments.

Detailed analyses using synthetic model systems. A third significant recent development, in terms of moving towards complete understanding of natural magnetic nanoparticle systems, has been the study of synthetic magnetic nanoparticle systems having controlled nanoparticle characteristics and controlled nanoparticle dilution or number density in an inert matrix phase. Typically, such samples are frozen synthetic ferrofluids. These studies have often pushed the limit of analysis by including most needed realistic features, such as the dipolar interactions mentioned above, to model several different types of magnetic measurements and have provided interesting debates and clever insights.

Other continued recent advances include the explicit recognition and analysis of intra-particle complexity (e.g., Kodama 1999, Berkowitz et al. 1999) and the development of the measurement methods themselves and their underlying theoretical understanding (e.g., Victora 1989, Kodama and Berkowitz 1999, Pfannes et al. 2000). I present the above main areas of development and other key advances in the sections that follow. Readers should also note the comprehensive review by Dormann et al. (1997), in comparison with an earlier review (Dormann 1981).

Measurements on single magnetic nanoparticles

Some remarkable recent experiments involve measurements of SP fluctuations and time-resolved supermoment dynamics on single magnetic nanoparticles. Probably the first measurements on individual magnetic nanoparticles were performed by Awschalom et al. (1990) where a scanning tunneling microscope was used to deposit nanoparticles directly into the micro pickup coil of a SQUID alternating field susceptometer. The measurements were performed at ultra-low temperatures (~20 mK) where SP fluctuations are quenched out and MQT of the supermoment can occur. The results were found to be "difficult to reconcile ... with the current theoretical picture of magnetic MQT". This first work lead to several more studies using either magnetic force microscopy (MFM) or micro-SQUID methods.

Wernsdorfer et al. (1997a) performed the first magnetization measurements of single nanoparticles in a range of low temperatures, thereby providing direct observations of SP supermoment reversals and quantitative verifications of the Néel-Brown SP model (Eqn. 2) and the Stoner-Wohlfarth reversal mechanism. Measured hysteresis curves and switching fields were like theoretical textbook examples of the predictions arising from Equation (2) and its underlying assumptions. This study provided a first direct evaluation of the attempt frequency pre-factor of Equation (2) which was found to be $1/f_0 \sim 4 \times 10^{-9}$ s for an ellipsoidal single-crystal face centered cubic Co particle having a mean diameter 25±5 nm.

At the lowest temperatures, Wernsdorfer et al. (1997b) found that deviations from the Néel-Brown model of SP fluctuations were in quantitative agreement with magnetic MQT models in the low dissipation regime. They briefly reviewed past attempts to measure magnetic MQT phenomena on individual nanoparticles and discussed the various processes that have been pointed out to cause MQT-like effects. In particular, MQT-like artefacts are more likely in many-particle systems such that MQT observations in bulk or many-particle samples (Awschalom et al. 1992, Gider et al. 1995, 1996; Ibrahim et al. 1995, Tejada et al. 1996, Sappey et al. 1997b) must be analysed with care (Kodama and Berkowitz 1999), although single-crystal samples of molecular clusters (Thomas et al. 1996, Friedman et al. 1996) may not suffer from these difficulties. Theoretical developments in the area on magnetic MQT phenomena are ongoing (e.g., Barbara and Chudnovsky 1990, Stamp 1991, Politi et al. 1995, Hartmann-Boutron et al. 1996, Pfannes 1997).

Bonet et al. (1999) performed the first three dimensional switching field measurements on individual magnetic nanoparticles of barium ferrite in the size range 10-20 nm. They showed conclusively that the total anisotropy energies of such particles of hexagonal symmetry could not be modelled as uniaxial but that a hexagonal anisotropy term needed to be added to the principle uniaxial term. These authors indicated the possibility of exploring the underlying physics of the pre-exponential factor f_0 and its relation to the total anisotropy energy function in individual particles, thereby opening the way to quantitatively test the most recent models of SP fluctuations described above, without the almost unbearable complexities associated with ensembles of nanoparticles and their distributions of properties and inter-particle interactions.

Acremann et al. (2000; Miltat 2000) recently reported picosecond time-resolved imaging of the precessional motion of individual supermoments, in response to sudden applied field changes, measured using a vectorial Kerr effect laser system. Such measurements allow one to compare the classical motion of the supermoment to well established classical models of its dynamics. In this way, the validities, necessities, and magnitudes of various damping and noise terms in the classical equations can be evaluated by comparing predicted and measured time evolutions.

Synthetic model systems of magnetic nanoparticles

Synthesis methods. Synthetic nanoparticles can be made by several quite different methods including: scanning tunnelling microscopy deposition, evaporative deposition and masking, molecular beam epitaxy, arc discharge, electron beam evaporation, gas phase condensation, laser annealing, mechanical alloying methods such as ball milling, particle irradiation-induced phase separation, nuclear implantation methods, and a great variety of wet chemical methods such as those involving ion exchange, molecular intercalation, precipitation, Langmuir-Blodgett films, micelle and reverse micelle syntheses, etc. Other fabrication methods involve growing organisms, in controlled growth media, that produce biogenic particles or, for example, using organic components such as the ferritin polymer shell as templates for particle growth. The wet chemical methods are generally those most accessible to large numbers of researchers and they often more closely mimic natural synthesis routes, in aquatic environments in any case.

Simplified model systems. Wet chemical methods have been developed that provide single crystals or polycrystals of identical nanoparticles or molecular clusters. Organometallic clusters (e.g., Van Ruitenbeek et al. 1994) constitute one example where identical individual euhedral nanocrystals of metallic atoms (typically 10-10^3 atoms per nanocrystal) are surrounded by organic ligands and assembled into macrocrystals or polycrystalline powders. These are ideal for studying size and surface effects in metallic nanosystems. Molecular clusters (e.g., Gatteschi and Sessoli 1996) tend to be smaller (typically 2-20 cations per cluster) and are systems where the metal cation centers are covalently bonded via coordinating anions or organic groups that propagate inter-cation magnetic exchange interactions. Molecular clusters have the remarkable attributes that all clusters in a sample are identical and that they can form single crystals where the relative positions and orientation of the clusters are known. One the other hand, they are unlike many natural systems that have larger nanoparticles, distributions of particle sizes, shapes, and types, and distributions of particle positions and orientations. Reverse micelle synthesis appears to provide the possibility of continuously varying the the particle size in the 4-15 nm range with narrow size distribution widths of ~9% (Liu et al. 2000). Intercalation or ion exchange into zeolites, clay minerals, layered hydroxides, and graphite offer many other possibilities for nanoparticle synthesis.

Model systems of Fe oxides and oxyhydroxides. From an environmental and magnetic nanoparticle perspective and given the importance of Fe (Table 1), it is the Fe oxides and oxyhydroxides that are of greatest interest to us here and, fortunately, these systems have been at the center of much recent work, mainly involving precipitation syntheses, directly aimed at understanding the magnetic properties of realistic model systems. Examples of magnetic nanoparticle studies and reviews involving specific Fe oxides or oxyhydroxides (synthetic, natural, and biogenic) are as follows: (1) magnetite, Fe_3O_4 (Sharma and Waldner 1977, Kirschvink et al. 1989, 1992; Luo et al. 1991, El-Hilo et al. 1992a,b; Popplewell and Sakhnini 1995, Balcells et al. 1997, Hayashi et al. 1997, Malaescu and Marin 2000, Golden et al. 2001, van Lierop and Ryan 2001, Morris et al. 2001), (2) maghemite, γ-Fe_2O_3 (Bacri et al. 1993, Chaput et al. 1993, Parker et al. 1993, Morup and Tronc 1994, Dormann and Fiorani 1995, Jonsson et al. 1995a, 1997, 1998, 2000; Morup et al. 1995, Dormann et al. 1996, 1998a,b; Gazeau et al. 1997, Moskowitz et al. 1997, Sappey et al. 1997a, Svedlindh et al. 1997, Vaz et al. 1997, Fiorani et al. 1999), (3) hematite, α-Fe_2O_3 (Kundig et al. 1966, van der Kraan 1971, Rancourt et al. 1985, Amin and Arajs 1987, Ibrahim et al. 1992, Morup 1994a, Hansen et al. 1997, Dang et al. 1998, Suber et al. 1998, 1999; Vasquez-Mansilla et al. 1999, Bodker and Morup 2000, Bodker et al. 2000, Hansen and Morup 2000, Zysler et al. 2001), (4) ferrihydrite or hydrous ferric oxide (Van der Giessen 1967, Coey and Redman 1973b, Madsen et al. 1986, Lear 1987, Quin et al. 1988, Murad et al. 1988, Cianchi et al. 1992, Pankhurst and Pollard 1992, Pollard et al. 1992, Ibrahim et al. 1994, 1995; Zhao et al. 1996, Mira et 1997, Prozorov et al. 1999, Seehra et al. 2000, Rancourt et al. 2001b), and (5) ferritin cores (Bauminger and Nowik 1989, St-Pierre et al. 1989, 1996; Awschalom 1992, Dickson et al. 1993, Hawkins et al. 1994, Tajada and Zhang 1994, Gider et al. 1995, 1996; Friedman et al. 1997, Makhlouf et al. 1997b, Moskowitz et al. 1997, Sappey et al. 1997b, Tajada et al. 1997, Allen et al. 1998, Gilles et al. 2000). Key aspects of these studies are described below.

Inter-particle interactions and collective behavior

Phenomenological models. The recognition of the importance of inter-particle interactions in describing the field-induced responses of magnetic nanostructured materials and their theoretical description by Preisach (1935) predate the discovery of superparamagnetism. The status of such early mean field models of inter-particle interactions was reviewed by Wohlfarth (1964) and by Kneller and Puschert (1966). Phenomenological Preisach models still form the basis of powerful current approaches for handling the effects of inter-particle interactions (e.g., Spinu and Stancu 1998, Stancu and Spinu 1998, Song and Roshko 2000). These calculations are mainly concerned with modelling measured relaxation and response behaviors using phenomenological parameters, without much concern for the nature of the collective ordered state that can be induced by interactions or the nature of a possible phase transition to such a collective state.

Dipolar interactions and SG behavior. Given the significant progress described above in dealing with the disordered interacting magnetic systems known as SGs, many researches have recently examined inter-particle interaction effects with this perspective, based on classic SG results (e.g., Chalupa 1977, Omari et al. 1983), using synthetic model systems. Several authors have written critical reviews of this recent approach (Morup 1994b, Dormann et al. 1997, Hansen and Morup 1998, Fiorani et al. 1999). Luo et al. (1991) were among the first to take this modern perspective, in describing the behavior of their ferrofluid of 5.0±1.6 nm shielded magnetite particles. This launched significant activity along these lines, using carefully prepared synthetic model systems in which inter-particle interactions were varied by controlling particle dilution and type (El-Hilo et al. 1992a, Morup and Tronc 1994, Morup 1994a, Dormann and Fiorani 1995, Jonsson et al. 1995b, 1998a,b; Ligenza et al. 1995, Morup et al. 1995, Bitoh et al. 1996, Dormann et al. 1996, 1998a,b; Scheinfein 1996, Vincent et al. 1996, Djurberg et al. 1997, Hansen and Morup 1998, Mamiya et al. 1998, Fiorani et al. 1999, Prozorov et al. 1999, Garcia-Palacios and Svedlindh 2000, Kleemann et al. 2001), that had been anticipated by

Morup et al. (1983), Fiorani et al. (1986), Dormann et al. (1988), Rancourt et al. (1990a,b), and others.

The above examinations from the SG perspective have lead to a consensus that dipolar coupled nanoparticles with disordered spatial arrangements have several characteristics in their non-equilibrium responses to applied fields that directly stem from the dipolar interactions and that are typical of multi-state free energy structures. On the other hand, it seems that there may be significant differences with canonical SGs in that divergences associated with a SG transitions do not occur (e.g., Dormann et al. 1988). The latter point is one of debate, as it was with the alloy SG systems. The next question is "What happens if there are inter-particle exchange bridges, in the common cases of dense arrays of nanoparticles?" Kleemann et al. (2001) interpret the behavior of their dense particle system in terms of a superferromagnetic (SF) transition, driven by inter-particle exchange interactions, followed by a reentrant super-SG transition as temperature is lowered and weaker dipolar interactions become relevant.

Exchange bridges and superferromagnetism. This brings us to the concept of superferromagnetism. The effects of inter-particle interactions are clearly seen in the Mössbauer spectra of magnetic nanoparticles, although the actual spectral analysis and interpretation is complicated (more below). Certain Mössbauer spectral features were first attributed to inter-particle interactions and associated phenomena first termed "superferromagnetism" independently by Morup et al. (1983) and by Rancourt and Daniels (1984). Morup et al. attributed variations in hyperfine field distributions (HFDs) to mean interaction fields and proposed the existence of a spontaneous ordered supermoment state, the SF state. Rancourt and Daniels used the term to mean any effect, such as inter-particle interactions and exchange anisotropy, that could locally cause $\tau_+ \neq \tau_-$ (Eqn. 9), thereby changing the fluctuation-dependent lineshape. Both Rancourt (1988) and Morup (1994b) have further explained their positions. Both authors agree that the term superparamagnetism should be reserved for magnetic effects resulting from inter-particle interactions, specifically, a spontaneous ordered state called the SF state. Explicit theoretical calculations of exchange bridge coupled SF systems and quantitative comparisons with magnetic measurements on synthetic model systems were provided by Rancourt (1985, 1987) and Rancourt et al. (1986).

Noteworthy attempts at dealing with nanoparticle complexity

In this section, I describe some classic and recent examples of studies in which the authors have made noteworthy attempts to deal with the complexities of real nanoparticle systems. The difficulties have been outlined above. In these examples, the authors have provided in depth quantitative analyses of data from several types of measurements on several samples, in an effort to identify or demonstrate key needed physico-chemical characteristics of the samples and to show how these affect measured properties. The selection of works is only meant as an illustration of promising approaches. I also include some examples of theoretical work aimed at understanding particular measurements themselves in application to magnetic nanoparticles.

Early classics. I must first emphasise that the early comprehensive studies of Stoner and Wohlfarth (1948) and Néel (1949a) have retained their validity and remarkable insight. Their rigorous and lucid theoretical developments are recommended to anyone wishing to understand the magnetic properties of environmental materials. They lay the foundations for modeling time, temperature, and field dependent magnetic measurements of small particle systems with broad distributions of particle sizes. Another such classic paper is that of Gittleman et al. (1974) who showed how the initial and alternating field susceptibilities could be modeled. The following examples are a subset of the many high quality papers published in the last 10 years or so, excluding the important works already mentioned in the sections above and below.

Molecular cluster system. Barra et al. (1996) studied an eight Fe atom molecular cluster systems by ESR, alternating field susceptometry, and Mössbauer spectroscopy. The system has the great advantages of molecular cluster model systems: There is only one particle size and type, the particles are arranged on a known regular lattice, and the particle or molecular structure is known. Combining alternating field susceptometry with Mössbauer spectroscopy allowed the SP relaxation time to be followed over six orders of magnitude in frequency as temperature was varied. A value of the pre-exponential frequency factor (Eqn. 2) could thereby be obtained as $1/f_o = 1.9 \times 10^{-7}$ s. This may be the first such combination of alternating field susceptometry and Mössbauer spectroscopy to obtain dynamic information in a nanoparticle system but the procedure is well known (e.g., as discussed and reviewed by Rancourt 1987) in one dimensional magnetic soliton systems, that can be considered one dimensional nanoparticle systems. The first use of Mössbauer spectroscopy to measure SP fluctuation frequencies in a nanoparticle system by dynamic lineshape analysis was made by Rancourt et al. (1983, 1985) and described by Rancourt and Daniels (1984).

Magnetic viscosity and distributions. Barbara et al. (1994) made an in depth study of the field and temperature dependencies of the magnetic viscosity S(T,H) of well characterized Ba-ferrite nanoparticles. They expressed S(T,H) = dM(T,H)/dln(t) in terms of the distributions of particles sizes, P(v), and of switching fields, $P(H_o)$, (H_o given by Eqn. 8) and showed that these were the main required distributions. P(v) was imposed by detailed transmission electron microscopy measurements and $P(H_o)$ was obtained from the extrapolated T = 0 K hysteresis cycle. In this way, the intricate behaviors of S(T,H) were quantitatively modeled without any free parameters. This impressive study shows the extent to which the time dependence of the magnetization of non-interacting nanoparticles can be understood in terms of the underlying distributions of relevant particle characteristics.

Alternating field susceptibility and intra-barrier fluctuations. Svedlindh et al. (1997) have presented a comprehensive description of the real and imaginary (i.e., driving frequency linear response and phase shift) parts of the alternating field susceptibilities of non-interacting nanoparticle systems having distributions of axial barrier energies and associated particle size distributions. They have critically reviewed previous work, including the classic paper of Gittleman et al. (1974) and the various methods for obtaining particle size distributions from alternating field susceptometry. They present convincing arguments that the alternating field susceptibility is sensitive to both near anisotropy axis intra-potential well supermoment fluctuations and the SP inter-well reversals described by Equation (2), not just the usual SP fluctuations (Eqn. 2) and an average perpendicular term. They are able to extract the separate frequencies of the two types of fluctuations first described by Néel.

Alternating field susceptibilities and inter-particle interactions. Jönsson et al. (2000) have made an exemplary analysis of the linear and cubic ($M = \chi H + \chi_3 H^3 + \ldots$) alternating field susceptibilities of non-interacting and interacting nanoparticles. Inconsistencies between the linear and cubic susceptibilities of non-interacting particles were interpreted as evidence for a more than axial structure of the barrier potential (Garcia-Palacios and Lazaro 1997) or an inadequate description of the intrinsic SP damping constant (that enters in the theoretical determination of f_o in Eqn. 2). The effects of inter-particle dipolar interactions were clearly recognized and characterized. Similarities with the behaviors of canonical SG systems were demonstrated and the key difference of the lack of a SG ordering divergence was again noted (Fiorani et al. 1986,

1999; Dormann et al. 1988). This is in contrast to an earlier paper by Jonsson et al. (1998) who report the first SG divergence transition in a dipolar interacting nanoparticle system. Jönsson et al. (2000) argued against the approach developed by Dormann and co-workers (see review, Dormann et al. 1997, and Dormann et al. 1998a,b) that has the main effect of inter-particle interactions modeled via an average interaction field that shifts the barrier energy distribution to larger energies thereby increasing the SP relaxation times. A similar approach was developed by Morup and Tronc (1994) and further advanced by Hansen and Morup (1998). Jönsson et al. (2000) argue that the low temperature magnetic relaxation of interacting nanoparticles is "dominated by collective particle dynamics" and "cannot be reduced to that of non-interacting particles shifted to longer time scales".

Constant field magnetometry and distributions. Sappey et al. (1997a) have made a thorough analysis of ZFQ-FW-FC and TRM curves for non-interacting nanoparticles with broad particle size distributions. They demonstrated the dependence of these curves, including the temperature T_{peak} of the peak in a ZFQ-FW curve, on the particle size distribution and showed that anomalous increases in T_{peak} as applied field is increased can arise from an effective field-induced broadening of the energy barrier distribution, rather than from MQT effects as sometimes claimed. The authors introduce a reverse-field TRM measurement procedure that has some advantages over standard methods and provide a lucid discussion of all such constant field magnetometry measurements, in the case of non-interacting (highly dilute) nanoparticles. El-Hilo et al. (1992a) and Vincent et al. (1996) have provided detailed analyses of how inter-particle interactions affect and participate in determining T_{peak} in less dilute systems. See also the review by Dormann et al. (1997) and the tentative mean field discussion by Morup (1994b).

Curie-Weiss behavior and interactions versus barriers. El-Hilo et al. (1992b) have made a thorough examination of the Curie-Weiss behaviors (e.g., Eqn. 17) of the high-temperature ($T > T_{peak}, \theta_{CW}$) constant field initial susceptibility of non-interacting and interacting nanoparticles. Extending the works of Gittleman et al. (1974) and others, they find that distributions of barriers arising from distributions of particle sizes in non-interacting systems give rise to negative effective Curie-Weiss temperatures, T_{0B}, and that interacting systems have effective Curie-Weiss temperatures that combine a negative T_{0B} contribution and a positive contribution, T_{0i}, arising from the inter-particle interactions. In systems where the barrier term dominates at the lower temperatures but becomes smaller at the higher temperatures, one obtains breaks in the χ_0^{-1} versus temperature curves, with separate temperature ranges giving different T_{0B} and T_{0i} intercepts. This illustrates the care that must be taken in interpreting high temperature initial susceptibility results. I would add that at still higher temperatures (or not higher is some systems) one must cross over to the true PM behavior of independent ionic moments and an ionic value of θ_{CW} (e.g., Eqns. 16-17).

High-field non-isotropy effects. Gilles et al. (2000) have made a detailed study of artificial ferritin nanoparticles, combining high field and low field constant field magnetometry and Mössbauer spectroscopy. Novel features include an attempt to give a dynamic lineshape interpretation to Mössbauer spectra in the blocking transition region and a thorough analysis of the non-Langevin high-field behavior of the supermoment susceptibility. They show that at high fields it becomes important to treat the SP fluctuations as occurring between the easy directions of the particles rather than as isotropic fluctuations, as assumed in the classic Langevin derivation. This is expected to be important in all studies involving high and moderate applied fields.

Theory of measurements and processes. Finally, a few theoretical studies can be mentioned, in addition to the contributions cited above, in the present context of dealing with the complexities of real nanoparticle systems. Yelon and Movaghar (1990) made some general deductions, about the relationship between the barrier energy and the pre-exponential factor or attempt frequency in equations of the type of Equation (2), that should be examined further in the context of SP particles. Pfannes et al. (2000) have proposed calculations of the SP fluctuation times that depart from the Néel-Brown coherent rotation picture and explicitly include coupling to the phonon spectrum of the particle. They have made a fundamental link to the Mössbauer spectral interpretation for SP and SF nanoparticles proposed by Rancourt and Daniels (1984) and discussed in the next subsection. Kliava and Berger (1999) have provided an elegant analysis of the ESR spectra of non-interacting nanoparticles with broad distributions of sizes and shapes. As mentioned above, Kodama and Berkowitz (1999) (and as reviewed by Kodama 1999 and Berkowitz et al. 1999) have provided lucid numerical analyses of various intra-particle and SR defects and their effects.

Interpreting the Mössbauer spectra of nanoparticle systems

Pervasiveness of dynamic effects. ^{57}Fe Mössbauer spectroscopy is potentially one of the most powerful methods for studying Fe-bearing nanoparticles because, in addition to being sensitive to several local crystal chemical and magnetic features, it is also sensitive to fluctuations in the local hyperfine interactions in a broad characteristic time window (10^{-6} to 10^{-10} s, or so) centered around the measurement time (Rancourt 1988). This means that SP supermoment reversal times (τ_+ and τ_- of Eqn. 10) can be measured directly, as functions of temperature and applied field and for different samples. The Mössbauer spectra arising from such dynamic or relaxation effects are quite different from the classical multiplets (singlets, doublets, sextets, and octets) that arise from the static limit (see above discussion) and the relevance of this to nanoparticle applications of Mössbauer spectroscopy has unfortunately generally not been recognized.

Dominant spectral analysis paradigm. Possibly the first suggestion that dynamic spectral effects are important in nanoparticles was made by van der Kraan (1971) who cited the early relaxation lineshape models of van der Woude and Dekker (1965a,b) and Wickman et al. (1966). This idea was not followed up until much later and has only very rarely been effectively applied in the analysis of Mössbauer spectra of nanoparticles. Instead, most authors followed the lead provided by the outstanding early work of Kündig et al. (1966) who divided the Mössbauer spectra of samples having broad distributions of particle sizes into two subspectra: a doublet for those particles in the sample that are SP and a sextet for the larger particles that are blocked. This approach neglects the spectral contribution from intermediate size particles ($v \sim v_{SP}$; Eqn. 3) that have $\tau \sim \tau_m$ and transitional type or relaxation type spectra, the possible effects of dwell time values (τ_+ and τ_-) on the sextet subspectrum, and the possible effects of dwell time (τ) on the doublet subspectrum. Nonetheless, this approximation should be valid for sufficiently broad distributions of barrier energies (E_b in Eqn. 2) and sufficiently large values of f_0 (Eqn. 2) for the resulting distributions of SP fluctuation times to mainly contain $\tau >> \tau_m$ and $\tau << \tau_m$ values. It proved to be applicable for many samples and allowed a particle size distribution to be extracted from the temperature dependence of the spectral area ratio of the doublet and sextet subspectra (Kündig et al. 1966). Unfortunately, it also became a spectral interpretation paradigm that until today has not been superseded.

Collective magnetic excitations. In the face of various inadequacies of the above interpretation paradigm, ingenious and physically reasonable additional physical mechanisms were proposed (and widely accepted) to explain the main anomalous features. One anomaly was that the hyperfine magnetic field spitting (see above) at low temperatures was usually found to be somewhat smaller than that of the corresponding bulk material. Morup et al. (1976) proposed that this was due to the thermal reduction of the average supermoment arising from relatively rapid intra-barrier fluctuations. That is, supermoment fluctuations in orientation within a minimum of the potential energy curve (Eqns. 4-6), occuring during the dwell times between supermoment reversals. This phenomenon was termed collective magnetic excitations (CMEs). This mechanism assumes both that the supermoment intra-barrier fluctuations are coherent fluctuations of the supermoment orientation, rather than arising from single-ion or spin wave excitations (that are assumed to be the same as in the corresponding bulk material), and that the local average ^{57}Fe hyperfine field is effectively static and directly proportional to the CME average. The main problems with the first assumption are (1) that the supermoment fluctuations are not simple coherent fluctuations of orientation but include magnitude fluctuations from the relevant single-ion and spin wave fluctuations and (2) the single-ion and spin wave fluctuations may be significantly different from those in the bulk material and these differences can be the dominant causes of local average hyperfine field reduction. The main problem with the second assumption is that the local fluctuations of the hyperfine field arising from all the relevant collective and single-ion excitations, although expected to be very fast compared to both the measurement time and τ (Eqn. 2), as assumed, may not be the only fluctuations to affect the resulting lineshape. Indeed, Rancourt and Daniels (1984) have shown that relaxation lineshapes with SP characteristic times $\tau_+ \neq \tau_-$ can lead to sextet patterns having reduced hyperfine field splittings under the most common circumstances involving ensemblies of nanoparticles. In other words, reduced magnetic hyperfine spittings are often not directly related to fast averaging processes.

Distributions of static hyperfine fields. Another difficulty of the dominant spectral analysis paradigm is that the Mössbauer spectra of nanoparticle samples often have sextet contributions or subspectra that are significantly different from the static sextets expected from blocked particles of the bulk material, having much broader and asymmetric absorption lines. Morup et al. (1983) proposed that this arose from inter-particle interactions, as follows. The interactions in a sample of positionally disordered nanoparticles would give rise to a broad distribution of different average interaction field magnitudes (e.g., Eqn. 19 applied to supermoments) that, in turn, would cause a broad distribution of average supermoments, each supermoment having a thermal average in accordance with the average interaction field that it experiences. The supermoment averages were taken to give rise to a distribution of average local hyperfine fields by assuming an effectively instantaneous coupling of the supermoment and the hyperfine field. Consequently, the sextet components were analysed in terms of distributions of static hyperfine field magnitudes or HFDs. This has become the dominant method for analysing the sextet components of the spectra of nanoparticle samples. The underlying assumptions and the analysis method itself are often referred to as superferromagnetism. The most tenuous underlying assumption here is that all supermoment fluctuations are fast enough for static sextet lineshape components to be used and distributed to account for the observed absorption spectra. The extracted HFDs are extracted to obtain agreement with measured spectra and have not been put on a firm theoretical basis. Indeed, most estimates of f_o (10^6 to 10^{11} Hz) suggest that the usual barrier energy values should give rise to SP fluctuation times that are within the dynamic effect measurement window. Inter-particle interactions are important but they need not give rise to static HFDs.

Anisotropic fluctuation lineshapes. Rancourt and Daniels (1984) showed that the τ_+ and τ_- values that arise from typical interaction fields, with typical values of f_o, give dynamic lineshapes of sextets having broad and asymmetric absorption lines similar to the observed ones in many nanoparticle systems. This suggests that the often extracted HFDs are artefacts of an incorrect spectral analysis. One difficulty that inhibited the application of dynamic effect lineshapes to nanoparticle systems was that isotropic fluctuation lineshape models (with $\tau_+ = \tau_-$) gave predicted lineshapes that were very different from observed spectra. Rancourt and Daniels (1984) demonstrated that anisotropic fluctuation lineshape models (with $\tau_+ \neq \tau_-$) give dramatically different lineshapes than those predicted by isotropic models and that the window of characteristic times inside of which dynamic spectral effects could be detected was much larger than previously thought. See Rancourt (1988), Rancourt (1998), and section 3 of Rancourt and Ping (1991) for further discussion. Rancourt et al. (1983, 1985) were the first to use dynamic effect lineshapes to directly extract SP fluctuation times and interaction field magnitudes from the Mössbauer spectra of nanoparticle samples.

Other dynamic lineshape attempts. Hansen et al. (2000) used an isotropic fluctuation model for the partially collapsed doublet contributions in their Mössbauer spectra while retaining a HFD analysis for the sextet contributions. I would argue, as above, that the sextet contribution is being incorrectly interpreted. Van Lierop and Ryan (2001) used distributions of various particle properties and isotropic fluctuations to model their spectra. I would argue that their restriction to isotropic fluctuations is not justified. Pfannes (1997) made a theoretical analysis that considered only isotropic fluctuations in describing the Mössbauer spectra. Pfannes et al. (2000) later gave a theoretical analysis that stresses the importance of anisotropic fluctuations and questions the validity of CMEs. Gunther et al. (1994) discussed relevant features related to the isotropic fluctuation effects of non-interacting particles. St-Pierre et al. (1987) and Gilles et al. (2000) used an interpretation model that admits anisotropic fluctuations ($\tau_+ \neq \tau_-$) but that assumes τ_+, $\tau_- << \tau_m$ and an associated absence of dynamic lineshape effects. Such assumptions that are used to justify a static analysis approach should always be substantiated when dealing with nanoparticles. Finally, I have already mentioned the interesting work of Barra et al. (1996) who extracted isotropic fluctuation times, for monodisperse non-interacting molecular clusters, that correlated with those extracted from alternating field susceptibility measurements.

Clearly, there is much room for improvement in the analysis of the Mössbauer spectra of nanoparticle systems. One can only hope that the tendency to adopt unjustified simplifying assumptions, in the face of admittedly somewhat overwhelming complexity, will be gradually overcome. A recommended approach is to insist that the same model produce agreement with many different measurements from several measurement methods, including methods having measurement times that are comparable to the expected characteristic times, and to apply as many rigorously valid theoretical constraints as possible.

Needed areas of development

Intra-particle crystal chemistry. Environmental nanoparticles are expected to have intra-particle radial distributions and inhomogeneities involving various coordinating anion substitutions, cation substitutions, vacancies, surface complexed groups, etc. For example, the Al, Mn, and Fe oxyhydroxides are expected to have significant (O^{2-}, OH^-, OH_2) substitutions with various charge balancing mechanisms involving vacancies, cation substitutions, and surface non-stoichiometry. This chemical disorder is expected to modify the bond lengths, bond angles, and most particle properties more than the purely nanocrystalline effects arising directly from small size alone, yet it is rarely measured directly at the single particle scale. Mineral magnetic properties are expected to be very sensitive to these features and first principles electronic structure calculations can be used to predict the relevant magnetic and intra-particle crystal chemical interplay.

Electronic structure calculations. A promising direction is to use first principles

ESCs that are well suited to dealing with magnetic and finite size clusters. The needed codes are rare because most advanced methods either deal with infinite crystalline inorganic materials using periodic boundary conditions or organic molecules that are DM and do not contain transition metals. Particularly promising methods are the spin-polarized molecular orbital calculations in the local density approximation (Grodzicki 1980, Blaes et al. 1987). For example, this method has been applied to the study of biologically relevant polynuclear clusters where intra-cluster magnetic interactions have a direct bearing on charge transfer functionality (Kröckel et al. 1996). Such methods could be used to predict relative stabilities, surface chemical reactions, spectroscopic properties, microscopic and macroscopic magnetic properties, etc., of environmentally relevant nanoparticles.

Dipolar interactions. Inter-particle interactions are important and are often dipolar in nature. These interactions have, to date, not been modeled correctly. Realistic calculations must include both the spatial distributions of average dipolar interaction fields (Eqn. 19) and the temporal fluctuations of the local dipolar interaction field. The latter fluctuations must have characteristic times that are comparable to the SP fluctuation times of the particles since the interaction field is directly caused by the neighboring supermoments. Indeed, for this reason, the concept of an interaction field must be replaced by a proper handling of inter-particle spatial and temporal correlations. This will determine the dynamics in assemblies of interacting particles and has not yet been attempted.

Integrated multi-method studies. As is true in many areas of natural science, there is a need for an integrated approach combining several measurement and theoretical methods and involving systematic studies of both natural and synthetic samples. The high resolution TEM, with its associated methods (EDS, CBED, SAED, EELS) goes a long way in this respect but suffers from the small size of the subsample. It must be combined with whole-sample diffraction, spectroscopy, and bulk property measurements, in order to deal with the macroscopic sample distributions. There is a great need for organized groups of experts to share the same samples and, if possible, to provide reference materials to other groups. Magnetism, in particular, is sensitive to several sample features that are not easily detected by other methods.

Development and application of mineral magnetism. Mineral magnetism, including measurements over broad temperature and applied field ranges and using measurement methods with vastly different measurement times, needs to be developed for and applied to environmental, geochemical, geological, and space nanomaterials, as stressed in this chapter. More controlled syntheses that model the materials of interest are needed. Further development of the underlying theory for the various measurement methods is vital. The case of Mössbauer spectroscopy has been described above. Further developments of the fundamental theory of the physical properties of nanoparticles are needed, with continued inclusion of more and more realistic crystal chemical features. Explicit inclusion of interparticle magnetic interactions is of interest.

Diffraction theory of nanoparticles. Bulk nanoparticle sample powder diffraction, including X-ray diffraction, neutron diffraction, synchrotron-based anomalous scattering, and electron diffraction, needs to be developed to explicitly treat nanomaterials. The standard crystallographic approach to work from the infinite crystal limit towards small-particle size effects is not appropriate for nanoparticles, where it is no longer useful or mathematically justified to divide the diffraction pattern into separate Bragg reflections. Instead, a Debye formula approach can be used in which one explicitly includes positional and chemical disorder and site-specific Debye-Waller factors. Here, the term "site" refers to the position in a given cluster rather than to a crystallographic site. This approach is also valid, using the appropriate ionic magnetic form factors, for modeling the magnetic neutron scattering of nanomaterials.

Mössbauer spectroscopy. The inclusion of anisotropic fluctuations, modeled as $\tau_+ \neq \tau_-$ in uniaxial symmetry, in the presence of applied magnetic fields, exchange anisotropy, or inter-particle interactions, must be used as a starting assumption unless the more restricted assumption that all relevant fluctuations are much faster than the measurement frequency (τ_+, $\tau_- << \tau_m$, in uniaxial symmetry) is justified independently. All other relevant realistic features (distributions of characteristics and properties!) must also be included, by applying as many justified theoretical constraints as possible.

EXAMPLES: CLUSTERS, BUGS, METEORITES, AND LOESS

In this section, I give brief descriptions of a few case studies of magnetic nanomaterials that I have been involved with, with reference to related works.

Two-dimensional nanomagnetism of layer silicates and layered materials

Two-dimensional magnetic materials have one dimension that is reduced to the nanoscale. That is, magnetic exchange interactions exist only within layers and do not propagate between layers. In-plane chemical inhomogeneity can then cause further subdivision to produce nanoscale two-dimensional magnetic clusters. The layer silicates are such materials where Fe^{2+} and Fe^{3+} are almost the only PM cations that are mainly confined to the octahedral sheet, except Fe^{3+} that can occupy tetrahedral sites (Rancourt et al. 1992). In addition, as clay minerals, layer silicates can also approach nanoparticle sizes and may sometimes form from nanomaterial precursor phases such as Si-bearing HFOs (Rancourt et al. 2001b).

Given their platy habits, layer silicates such as biotite play an important role in determining the magnetic fabrics of rocks, both because of their easy-plane paramagnetism and because they often contain stoichiometric magnetite micro-inclusions. The latter petrogenic micro-inclusions are easily detected and quantified by mineral magnetometry and are found to exhibit Verwey transition temperatures of 119 K, that can only occur in stoichiometric magnetite. Layer silicates also form a major class of diagenetic clay minerals (illite, smectite, etc.) that have important associations with diagenetic Fe oxides (see below).

From early work (Ballet and Coey 1982, Beausoleil et al. 1983, Coey and Ghose 1988) it was concluded that layer silicates would not order magnetically above ~10 K for intrinsic reasons due to the presence of chemical disorder in two dimensions. Rancourt et al. (1994) showed this to be incorrect and found magnetic ordering temperatures of 42 K and 58 K in a natural annite sample and in a synthetic annite end-member sample, respectively. Neutron diffraction showed an AF stacking of in-plane ferromagnetically aligned moments. The AF stacking was attributed to dipolar interactions and can be thought of as part of the domain structure of this layered material. The $^{[4]}Fe^{3+}$ and $^{[6]}Fe^{3+}$ moments were found to be magnetically exchange coupled to the octahedral sheet Fe^{2+} FM backbone but to only acquire significant average values at temperatures far below the magnetic ordering temperatures, as clearly seen in the Mössbauer spectra.

Transition metal di- and tri-chloride ($FeCl_2$, $FeCl_3$, $NiCl_2$, $CoCl_2$, $CrCl_3$, $CuCl_2$) GICs and graphite bi-intercalation compounds (GBICs) are ideal model systems for the study of two-dimensional nanomagnetism because most of these compounds intercalate by forming two-dimensional islands with diameters of ~15 nm and corresponding stabilizing charge transfers arising from island edge non-stoichiometry. The magnetism of such compounds and its relation to microstructural and nanostructural details have been reviewed (Rancourt et al. 1986, Rancourt 1987). These systems present many classic examples of effects that are relevant to natural systems, such as: inter-particle interactions via exchange bridging, coexisting nanophase materials exhibiting distinct magnetic ordering temperatures, dipole-dipole mediated magnetic nanostructures, supermoment formation by incomplete AF lattice cancellation, frustration-isolated loose end or SR moments giving rise to large low temperature PM Curie signals, etc., and allow direct comparisons with the pristine Van der Waals layered chlorides.

Abiotic and biotic hydrous ferric oxide and sorbed-Fe on bacterial cell walls

Ferrihydrite is possibly the only accepted mineral species that is intrinsically nanophase (d ~ 5 nm), without any known bulk material counterpart. It is an important environmental material that forms by precipitation, in not too acidic pH environments, whenever Fe(2) meets oxidizing conditions. Its less ordered and smaller variety, known as 2-line ferrihydrite or HFO, is believed to play key roles in controlling the cycling of both nutrients (such as P) and heavy metals in aquatic environments. When formed in the presence of bacteria, this material is often found to be attached to the cell walls and to encase bacterial cells. For this reason, it is generally believed that bacteria play an active role in HFO formation in many environments.

It is also observed in laboratory experiments that bacterial cell walls can directly surface complex Fe and it has been proposed that this may be a first step towards biotic HFO formation (Warren and Ferris 1998). I follow Thibault (2001) in referring to such HFO as biotic (b-HFO) rather than as biogenic or bacteriogenic since the mechanism of its formation is not known. In the absence of bacteria, authigenic HFO can be referred to as abiotic (a-HFO). Rancourt (Rancourt et al. 1999a, 2001a; and as cited by Thibault 2001) has described four possible types of mechanisms for b-HFO formation:

(1) heterogeneous nucleation on the bacterial cell wall, the so-called template effect,
(2) homogeneous nucleation and growth in the near-cell chemical environment,
(3) enhanced precipitation from ancillary organic compounds from the bacteria, and
(4) catalytically enhanced nucleation and growth where the bacterial surface functional groups act as catalytic agents.

The first of these is the organic equivalent of epitaxial growth and includes surface nucleation barrier reduction by contact. It is presently the preferred picture in the literature but has not been substantiated by relevant observations.

Thibault (2001) has made comparisons of synthetic a-HFO and b-HFO samples made under identical chemical conditions but in the absence or presence of washed bacteria (*Bacillus subtilis* or *Bacillus licheniformis*), respectively. He combined TEM, Mössbauer spectroscopy, X-ray diffraction, and constant field magnetometry measurements to conclude that there were significant differences between a-HFO and b-HFO. The main differences were:

(1) smaller particle sizes and a less dense colloidal network in b-HFO, as inferred from TEM pictures,
(2) much lower SP blocking temperatures or SF ordering temperatures in b-HFO, as seen in both magnetometry and Mössbauer measurements,
(3) significantly different quadrupole splitting distributions extracted from RT Mössbauer spectra, showing different distributions of local distortion environments, and
(4) the presence of particle-complexed Fe^{2+} in b-HFO samples only, with the amount having a systematic variation with synthesis pH.

Rancourt et al. (1999a, 2001a) have made the first spectroscopic measurements of surface complexed Fe on bacterial cell walls. They find that sorbed-Fe (s-Fe) is easily resolved from any b-HFO present by cryogenic Mössbauer measurements since the s-Fe does not order magnetically down to 4.2 K where all types of HFO always do. Evidence was found that the phosphate functional groups could form tridentate coordinations that ligand field stabilize the Fe^{2+} cation. This may be part of the mechanism that produces Fe^{2+} in b-HFO samples.

Indeed, P is known to complex strongly to HFO, as does As. In a recent study of natural As-rich and Si and C-bearing a-HFO samples, Rancourt et al. (2001b) showed many interesting phenomena. Si and C were found to compete with As (and presumably P) for surface complexation to the HFO, leading to banding of As-rich (Si and C-poor, yellowish) and As-poor (Si and C-rich, reddish) HFO deposits. The As was found to cause large Fe local environment distortions, directly measured by Mössbauer spectroscopy, that in turn correlated to the colors and related visible band edge positions, thereby directly illustrating the color mechanism involving Fe^{3+} valence orbital transitions. Dramatic differences between the magnetic behaviors of the As-rich and As-poor samples and synthetic a-HFO samples were seen. These were discussed in terms of particle sizes and supermoment formation mechanisms and illustrate the remarkable sensitivity of magnetic properties to the characteristics of surface complexed HFO nanoparticles.

Hydroxyhematite, nanohematite, and the Morin transition

Most authors who have studied small particle hematite have, as a first approximation, considered their samples to be small pieces of the ideal bulk material α-Fe_2O_3, where the changes in physical properties are predominantly controlled by particle size (e.g., Vandenberghe et al. 2000). In contrast, Dang et al. (1998) have stressed that even bulk hematite has a complex crystal chemistry involving the known OH/vacancy substitutional mechanism,

$$2\,Fe^{3+} + 3\,O^{2-} + 3\,H\uparrow \;= Fe^{3+} + \square + 3\,OH^{-} + Fe\downarrow\,,$$

and at least one other crystal chemical mechanism that significantly affects the lattice parameters and that the resulting compositions of nanoparticles will be highly correlated to particle size for a given set of synthesis circumstances. They found that the chemistry of hematite particles predominantly determined the lattice parameters and that the lattice parameters, in turn, predominantly determined the Morin transition. A plot of c versus a for many samples showed a well defined OH/vacancy line and many points above this line that could be made to join the OH/vacancy line by annealing in air to 200°C. Annealing to progressively higher temperatures then caused evolution along the OH/vacancy line towards the stoichiometric hematite end-point. The position on this c-a plot completely determined both the existence of a Morin transition (down to 4.2 K) and the value of the Morin transition temperature for particles with d ~ 30 nm or larger. Continued unpublished work is aimed at: understanding the crystal chemistry of samples having a-c values above the OH/vacancy line, exploring particle size effects in smaller hematite particles, and exploring the effects of different solvating anions used in the hydrothermal syntheses. Dang et al. (1998) also showed that the baseline magnetic susceptibility in the AF state (below the Morin transition temperature) decreased with increasing annealing temperature, in a way that is consistent with a vacancy mechanism for supermoment formation.

Mineral magnetism of loess/paleosol deposits

Loess deposits are wind blown deposits that cover approximately 10% of the terrestrial landmass. Chinese loess deposits may constitute the best land-based records of paleoclimates for the last several Ma (Kukla 1987, Kukla and An 1989, Liu 1988, 1991; Ding et al. 2001, Zhang et al. 2001). Alternating layers of loess and paleosol record different climates, relatively dry and cold versus wet and warm, respectively, where the main mineralogical differences are due to the varying degrees of pedogenesis and the associated authigenic and biogenic materials that have undergone various degrees of diagenesis since deposition. Magnetic measurements have been used both to provide dating from magnetic reversals (Heller and Evans 1995) and to provide sensitive and high resolution depth profiles that have been correlated to isotopic ratios from deep sea sediments (Evans et al. 1996).

Dang et al. (1999) performed a mineral magnetic and mineralogical study of a loess/paleosol couplet from the Chinese loess plateau in an effort to uncover which climate feature(s) the magnetic signal primarily records, by attempting to determine the magnetic mineralogy. They found the paleosol to contain more hydroxyhematite and maghemite that appeared to have formed at the expense of chlorite and layer silicate phases. Surprisingly, only the loess contained stoichiometric magnetite that displayed a Verwey transition at 119 K (Table 2), although its RT susceptibility was much smaller than that of the paleosol. The latter observation is particularly interesting when compared to observations of Siberian loess/paleosol deposits, that were deposited in arctic conditions. The Siberian susceptibility profiles also match deep sea isotopic ratio profiles but with a reversed sign, compared to the Chinese profiles (Chlachula et al. 1997, 1998). That is, Chinese paleosol units have larger susceptibilities than Chinese loess units whereas Siberian paleosol units have smaller susceptibilities than Siberian loess units.

Sabourin et al. (2001) have undertaken a detailed mineral magnetic and mineralogical study of a Siberian profile, in comparison to Chinese loess. They find that Siberian paleosols have undergone far less pedogenic transformation than Chinese paleosols, as expected, and that the first step relevant to magnetic measurements is the oxidation and transformation of the petrogenic stoichiometric magnetite. It appears that in Chinese paleosols the formation of pedogenic magnetic oxides far outweighs the loss of petrogenic magnetite whereas in Siberian paleosols the loss of petrogenic magnetite is not compensated by sufficient formation of pedogenic magnetic oxides. In both Chinese and Siberian paleosols there are significant fractions of SP magnetic particles that need to be characterized more completely.

Synthetic and meteoritic nanophase Fe-Ni and Earth's geodynamo

The iron meteorites are mostly composed of the Fe-Ni alloy minerals kamacite, taenite, and tetrataenite, and are believed to be remnants of the cores of parent bodies. Grains of Fe-Ni alloys are also common in most other types of meteorites. Because of exceedingly low cooling rates from the melt (typically ~1 K/Ma), the recovered materials are believed to be close to equilibrium structures and are therefore of interest to metallurgists, especially given the importance of industrial Fe-Ni alloys such as Invar ($Fe_{65}Ni_{35}$), Elinvar, and Permalloy.

Rancourt and Scorzelli (1995) recently proposed that the ubiquitous "PM taenite phase" in meteorites was a LM taenite phase, distinct from its HM taenite counterpart and analogous to the LM phase that had been theoretically predicted to occur but never observed or recognized as such. The LM phase was proposed to have low atomic moment magnitudes ($\mu_{Fe} \sim 0.5\ \mu_B$) and AF exchange interactions but the same FCC crystal structure as the HM taenite phase that has $\mu_{Fe} \sim 2.8\ \mu_B$ and predominantly FM interactions, as predicted by several ESCs. This would be the first example where distinct mineral phases can have the same compositions and crystal structures but different electronic structures and associated magnetic, material, and chemical properties.

Rancourt and Scorzelli suggested the mineral name antitaenite for the LM phase and proposed that it may be stabilized by an epitaxial interaction with tetrataenite, since it is always found in microstructural association with the latter phase and with an indistinguishable lattice parameter.

Rancourt et al. (1999b) later provided an experimental proof that antitaenite has an electronic structure that is distinct from that of any known natural or synthetic HM Fe-Ni alloy, in the form of direct local electronic density evaluations from ^{57}Fe Mössbauer spectroscopy. High resolution scanning electron microscopy of an unoxidized piece of the Santa Catharina ataxite showed a nanoscale microstructure consisting of 26-33 nm islands of tetrataenite in a honeycomb matrix phase of antitaenite having wall thicknesses of ~5 nm or less, consistent with the idea of epitaxial stabilization. It was also found that synthetic LM phase material could be produced by mechanical alloying of Fe and Ni, a method known to produce nanophase materials having nanoscale crystallites and nanoscale inter-grain boundaries.

Recently, Lagarec et al. (2001) located the composition-controlled equilibrium LM/HM transition of the synthetic FCC Fe-Ni system for the first time. They also used ESCs to show that the measured changes in local electronic density were precisely of the same magnitudes and signs as those predicted to occur on crossing the LM/HM transition. This work implies that, to the extent that FCC Fe-Ni can be stabilized with respect to the martensitic transition to the BCC structure, the electronic structure that has lowest energy on the bulk FCC structure above ~70 at % Fe is the LM structure. Lagarec et al. also showed that an FCC alloy having a LM ground state (including high temperature, FCC, γ-Fe and possibly high pressure, hexagonal close packed, ε-Fe) undergoes an entropic stabilization and gradual transition to a HM state as temperature is increased. Concomitantly, the dominant magnetic exchange interaction becomes less AF, crosses zero, and becomes FM. The latter composition, pressure, and temperature effects on the LM/HM nature of close packed Fe-Ni phases have important repercussions on the geodynamo, since the magnetic nature of Earth's solid core potentially plays an important role (Gilder and Glen 1998, Saxena and Littlewood 2001).

NEW DIRECTIONS: ENVIRONMENTAL MODELLING

What can magnetism do better than other methods? I have stressed that the PM cations (Table 1) are also the ones that can have several different valence states, thereby giving them prominent roles in biogeochemical reactions. Mineral magnetism, therefore, is a mineralogy focussed on the key reactive players in the environmental cycling of nutrients and toxic substances, most of which are nanoparticles or molecular clusters or PM active center metabolites.

Mineral magnetism, the extraction of crystal chemical and nano- and microstructural information from magnetic measurements, is presently underdeveloped. The low temperature magnetic properties of minerals are mostly unexplored yet the potential for widespread applications is high because of the sensitivity of magnetic properties to material properties, including subtle variations that cannot easily be detected by other methods. ESCs and other calculation methods are presently advanced enough for most of the connections between magnetism and underlying structure and chemistry to be made unambiguously. The real challenge is in assembling interdisciplinary teams that combine several measurement methods, appropriate theoretical and calculation methods, laboratory synthesis of realistic natural analogues, and problem-focussed field work.

Ultimately, in the environmental context, all mineral characterizations, including mineral magnetism, must serve to understand environmental systems, on all length scales from surface reactions, to near cell environments, to weathering depths, to early diagenetic scales, to lake and catchment areas, to continental, ocean, and global scales. Phenomena on all relevant time scales must also be included in realistic models, from local reaction rates, to transport times on all length scales, to reservoir residence times for reservoirs on all length scales, and to paleoclimatic driving times. In constructing environmental models with true predictive and explanatory capabilities, one must strive to work all the way down to the molecular level where nanoparticle properties determine the relevant dissolution and precipitation rates, sorption affinities and capacities, bioavailabilities to different organisms, settling and transport rates, etc. Environmental mineralogists must move from static crystal chemical descriptions to include evaluations of the relevant reactions that geochemists and others attempt to use in their models.

In my own present work, as a member of the Lake Sediment Structure and Evolution (LSSE) group and the Geological Survey of Canada Metals in the Environment (GSC-MITE) Phase-II project, my group provides detailed mineralogical characterizations of selected lacustrine sediment profiles as one input to developing realistic reaction transport models (RTMs) of the evolution of the aquatic sediment profiles. A first step in writing down the relevant reactions is to simply identify and quantify the solid phases, especially the reactive ones. The LSSE/GSC-MITE collaboration also involves a complete suite of biogeochemical characterizations including porewater geochemistry, bacterial enumerations, stable isotope methods, radioactive isotope dating methods, diatom and pollen enumerations, field measurements, etc. We find that one bottleneck in establishing the RTMs is a lack of characterizations of the biogeochemical reactions themselves involving the key solid phases. One problem is that the synthetic model materials used to evaluate reaction parameters in published studies are often not sufficiently characterized and are often not sufficiently realistic.

ACKNOWLEDGMENTS

I thank Marie Wang for collecting, organizing, and entering the references. I thank my students and collaborators for helpful discussions. Financial support from the Natural Sciences and Engineering Research Council of Canada is gratefully acknowledged, as is travel support from the MSA sponsors.

REFERENCES

Acremann Y, Back CH, Buess M, Portmann O, Vaterlaus A, Pescia D, Melchior H (2000) Imaging precessional motion of the magnetization vector. Science 290:492-495

Aharoni A (1969) Effect of a magnetic field on the superparamagnetic relaxation time. Phys Rev 177: 793-796

Aharoni A (1992) Susceptibility resonance and magnetic viscosity. Phys Rev B 46:5434-5441

Allen PD, St-Pierre TG, Street R (1998) Magnetic interactions in native horse spleen ferritin below the superparamagnetic blocking temperature. J Magnet Magnetic Mater 177-181:1459-1460

Altbir D, Vargas P, D'Albuquerque e Castro J, Raff U (1998) Dipolar interaction and magnetic ordering in granular metallic materials. Phys Rev B 57:13604-13609

American Society for Metals (1986) Materials Characterization—ASM Handbook. American Society for Metals International

Amin N, Arajs S (1987) Morin temperature of annealed submicronic α-Fe_2O_3 particles. Phys Rev B 35:4810-1811

Aragon R, Gehring PM, Shapiro SM (1993) Stoichiometry, percolation, and Verwey ordering in magnetite. Phys Rev Lett 70:1635-1638

Ashcroft NW, Mermin ND (1976) Solid State Physics. Saunders College, Philadelphia, Pennsylvania

Awschalom DD, McCord MA, Grinstein G (1990) Observation of macroscopic spin phenomena in nanometer-scale magnets. Phys Rev Lett 65:783-882

Awschalom DD, Smyth JF, Grinstein G, DiVincenzo DP, Loss D (1992) Macroscopic quantum tunneling in magnetic proteins. Phys Rev Lett 68:3092-3095

Backus G, Parker R, Constable C (1996) Foundations of magnetism. Cambridge University Press, Cambridge, UK

Bacri J-C, Boué F, Cabuil V, Perzynski R (1993) Ionic ferrofluids: Intraparticle and interparticle correlations from small-angle neutron scattering. Colloids Surf A: Physiochem Engin Aspects 80: 11-18

Balcells L, Iglesias O, Labarta A (1997) Normalization factors for magnetic relaxation of small-particle systems in a nonzero magnetic field. Phys Rev B 55:8940-8944

Ballet O, Coey JMD (1982) Magnetic properties of sheet silicates; 2:1 layer minerals. Phys Chem Minerals 8:218-229

Banerjee SK (1991) Magnetic properties of Fe-Ti oxides. Rev Mineral 25:107-128

Banfield JF, Nealson KH (eds) (1997) Geomicrobiology: Interactions Between Microbes and Minerals. Rev Mineral 35, 448 p

Barbara B, Chudnovsky EM (1990) Macroscopic quantum tunneling in antiferromagnets. Phys Lett A 145:205-208

Barbara B, Sampaio LC, Marchand A, Kubo O, Takeuchi H (1994) Two-variables scaling of the magnetic viscosity in Ba-ferrite nano-particles. J Magnet Magnetic Mater 136:183-188

Barra A-L, Debrunner P, Gatteschi D, Schulz ChE, Sessoli R (1996) Superparamagnetic-like behavior in an octanuclear iron cluster. EuroPhys Lett 35:133-138

Basso V, Beatrice C, LoBue M, Tiberto P, Bertotti G (2000) Connection between hysteresis and thermal relaxation in magnetic materials. Phys Rev B 61:1278-1285

Bauminger ER, Nowik I (1989) Magnetism in plant and mammalian ferritin. Hyperfine Interactions 50:484-498

Beausoleil N, Lavallée P, Yelon A, Ballet O, Coey JMD (1983) Magnetic properties of biotite. J Appl Phys 54:906-915

Benn K (1999) Applications of magnetic anisotropies to fabric studies of rocks and sediments. Tectonophysics 307:7-10

Berger R, Bissey J-C, Kliava J, Soulard B (1997) Superparamagnetic resonance of ferric ions in devitrified borate glass. J Magnet Magnetic Mater 167:129-135

Berger R, Kliava J, Bissey J-C, Baïetto V (1998) Superparamagnetic resonance of annealed iron-containing borate glass. J Phys: Condensed Matter 10:8559-8572

Berkowitz AE, Kodama RH, Makhlouf SA, Parker FT, Spada FE, McNiff EJ Jr, Foner S (1999) Anomalous properties of magnetic nanoparticles. J Magnet Magnetic Mater 196/197:591-594

Bessais L, Ben Jaffel L, Dormann JL (1992) Relaxation time of fine magnetic particles in uniaxial symmetry. Phys Rev B 45:7805-7815

Binder K, Heermann DW (1992) Monte Carlo Simulation in Statistical Physics. Springer-Verlag, Berlin

Binder K, Young AP (1986) Spin glasses: Experimental facts, theoretical concepts and open questions. Rev Mod Phys 58:801-976

Bitoh T, Ohba K, Takamatsu M, Shirane T, Chikazawa S (1996) Comparative study of linear and nonlinear susceptibilities of fine-particle and spin-glass systems: Quantitative analysis based on the superparamagnetic blocking model. J Magnet Magnetic Mater 154:59-65

Blaes N, Fischer H, Gonser U (1985) Analytical expression for the Mossbauer line shape of ^{57}Fe in the presence of mixed hyperfine interactions. Nuclear Instr Methods Phys Res B9:201-208

Blaes N, Preston RS, Gonser U (1985) Mössbauer spectra for nuclear motion correlated with EFG reorientation. *In* Applications of the Mössbauer Effect—ICAME-83, Alma-Ata, p 1491-1496

Blaes R, Guillin J, Bominaar EL, Grodzicki M, Marathe VR, Trautwein AX (1987) Spin-polarized SCC-Xα calculations for electronic- and magnetic-structure properties of 2Fe-2s ferrodoxin models. J Phys B 20:5627-5637

Bocquet S (1996) Superparamagnetism and the Mossbauer spectrum of goethite: a comment on a recent proposal by Coey et al. J Phys: Condensed Matter 8:111-113

Bødker F, Hansen MF, Bender Koch C, Lefmann K, Mørup S (2000) Magnetic properties of hematite nanoparticles. Phys Rev B 61:6826-6838

Bodker F, Morup S (2000) Size dependence of the properties of hematite nanoparticles. EuroPhys Lett 52:217-223

Bonet E, Wernsdorfer W, Barbara B, Benoit A, Mailly D, Thiaville A (1999) Three-dimensional magnetization reversal measurements in nanoparticles. Phys Rev Lett 83:4188-4191

Bonnemain B (1998) Superparamagnetic agents in magnetic resonance imaging: Physicochemical characteristics and clinical applications. A review. J Drug Targeting 6:167-174

Bradley JP, McSween Jr HY, Harvey RP (1998) Epitaxial growth of nanophase magnetite in Martian meteorite Allan Hills 84001: Implications for biogenic mineralization. Meteoritics Planet Sci 33: 765-773

Braun H-B (1993) Thermally activated magnetization reversal in elongated ferromagnetic particles. Phys Rev Lett 71:3557-3560

Broomberg J, Gélinas S, Finch JA, Xu Z (1999) Review of magnetic carrier technologies for metal ion removal. Magnetic and Electrical Separation 9:169-188

Brown JrWF (1963) Thermal fluctuations of a single-domain particle. Phys Rev 130:1677-1686

Brown JrWF (1979) Thermal fluctuations of fine ferromagnetic particles. I E E E Trans Magnetics 15: 1196-1208

Broz JS, Braun HB, Brodbeck O, Baltensperger W, Helman JS (1990) Nucleation of magnetization reversal via creation of pairs of Bloch walls. Phys Rev Lett 65:787-789

Brug JA, Anthony TC, Nickel JH (1996) Magnetic recording head materials. Mater Res Soc Bull, September 1996:23-27

Burton BP (1991) The interplay of chemical and magnetic ordering. Rev Mineral 25:303-321

Cabuil V (2000) Phase behavior of magnetic nanoparticles dispersions in bulk and confined geometries. Current Opinion Colloid Interface Sci 5:44-48

Cahn RW, Lifshin E (1993) Concise Encyclopedia of Materials Characterization. Pergamon Press, Oxford

Campbell WH (2000) Earth Magnetism. Academic Press, New York

Carlin RL (1986) Magnetochemistry. Springer-Verlag, Berlin

Chalupa J (1977) The susceptibilities of spin glasses. Solid State Commun 22:315-317

Chamberlin RV (2000) Mean-field cluster model for the critical behavior of ferromagnets. Nature 408: 337-339

Chamberlin RV, Haines DN (1990) Percolation model for relaxation in random systems. Phys Rev Lett 65:2197-2200

Chang C-R, Yang J-S, Klik I (1997) Thermally activated magnetization reversal through multichannels. J Appl Phys 81:5750-5752

Chang S-BR, Kirschvink JL (1985) Possible biogenic magnetite fossils from the late miocene potamida clays of Crete. *In* Kirschvink JL, Jones DS, MacFadden BJ (eds) Magnetite Miomineralization and Magnetoreception in Organisms—A New Biomagnetism. Plenum Press, New York, p 647-669

Chantrell RW, Walmsley N, Gore J, Maylin M (2000) Calculations of the susceptibility of interacting superparamagnetic particles. Phys Rev B 63:24410-1–24410-14

Chantrell RW, Walmsley NS, Gore J, Maylin M (1999) Theoretical studies of the field-cooled and zero-field cooled magnetization of interacting fine particles. J Appl Phys 85:4340-4342

Chaput F, Boilot J-P, Canva M, Brun A, Perzynski R, Zins D (1993) Permanent birefringence of ferrofluid particles trapped in a silica matrix. J Non-Crystal Solids 160:177-179

Chlachula J, Evans ME, Rutter NW (1998) A magnetic investigation of a late quaternary loess/palaelosol record in Siberia. Geophys J Int'l 132:128-132

Chlachula J, Rutter NW, Evans ME (1997) A late quaternary loess—Paleosol record at Kurtak, southern Siberia. Canadian J Earth Sci 34:679-686

Chowdhury D (1986) Spin Glasses and Other Frustrated Systems. Princeton University Press, Princeton, NJ

Cianchi L, Mancini M, Spina G, Tang H (1992) Mossbauer spectra of ferrihydrite: Superferromagnetic interactions and anisotropy local energy. J Phys—Condensed Matter 4:2073-2077

Clark DA (1983) Comments on magnetic petrophysics. Bull Aust Soc Explor Geophys 14:49-62

Coey JMD (1978) Amorphous magnetic order. J Appl Phys 49:1646-1652

Coey JMD, Ghose S (1988) Magnetic phase transitions in silicate minerals. Adv Phys Geochem 17: 162-184

Coey JMD, Readman PW (1973a) New spin structure in an amorphous ferric gel. Nature 246:476-478

Coey JMD, Readman PW (1973b) Characterization and magnetic properties of natural ferric gel. Earth Planet Sci Lett 21:45-51

Coffey WT, Cregg PJ, Crothers DSF, Waldron JT, Wickstead AW (1994a) Simple approximate formulae for the magnetic relaxation time of single domain ferromagnetic particles with unaxial anisotropy. J Magnet Magnetic Mater 131:L301-L303

Coffey WT, Crothers DSF (1996) Comparison of methods for the calculation of superparamagnetic relaxation times. Phys Rev E 54:4768-4774

Coffey WT, Crothers DSF, Dormann JL, Geoghegan LJ, Kalmykov YP, Waldron JT, Wickstead AW (1995a) Effect of an oblique magnetic field on the superparamagnetic relaxation time. Phys Rev B 52:15951-15965

Coffey WT, Crothers DSF, Dormann JL, Geoghegan LJ, Kalmykov YP, Waldron JT, Wickstead AW (1995b) The effect of an oblique magnetic field on the superparamagnetic relaxation time. J Magnet Magnetic Mater 145:L263-L267

Coffey WT, Crothers DSF, Dormann JL, Geoghegan LJ, Kennedy EC, Wernsdorfer W (1998a) Range of validity of Kramers escape rates for non-axially symmetric problems in superparamagnetic relaxation. J Phys: Condensed Matter 10:9093-9109

Coffey WT, Crothers DSF, Dormann JL, Kalmykov YP, Kennedy EC, Wernsdorfer W (1998b) Thermally activated relaxation time of a single domain ferromagnetic particle subjected to a uniform field at an oblique angle to the easy axis: Comparison with experimental observations. Phys Rev Lett 80:5655-5658

Coffey WT, Crothers DSF, Kalmykov YP, Massawe ES, Waldron JT (1993) Exact analytic formulae for the correlation times for single domain ferromagnetic particles. J Magnet Magnetic Mater 127: L254-L260

Coffey WT, Crothers DSF, Kalmykov YP, Massawe ES, Waldron JT (1994b) Exact analytic formula for the correlation time of a single-domain ferromagnetic particle. Phys Rev E 49:1869-1882

Coffey WT, Crothers DSF, Kalmykov YP, Waldron JT (1995c) Constant-magnetic-field effect in Néel relaxation of single-domain ferromagnetic particles. Phys Rev B 51:14947-15956

Coffey WT, Kalmykov YP, Massawe ES (1994c) The effective eigenvalue method and its application to stochastic problems in conjunction with the nonlinear Langevin equation. Adv Chem Phys 85:667-791

Collinson DW (1983) Methods in rock magnetism and paleomagnetism. Chapman & Hall, London

Comello V (1998) Magnetic storage research aiming at high areal densities. R&D Magazine, December 1998, p 14-19

Continentino M, Malozemoff AP (1986) Dynamical susceptibility of spin glasses in the fractal cluster model. Phys Rev B 34:471-474

Cornell RM, Schwertmann U (1996) The Iron Oxides—Structure, properties, reactions, occurrence and uses. VCH, Weinheim

Cowburn RP, Koltsov DK, Adeyeye AO, Welland ME (2000) Sensing magnetic fields using superparamagnetic nanomagnets. J Appl Phys 87:7082-7084

Creer KM, Tucholka P, Barton CE (1983) Geomagnetism of Baked Clays and Recent Sediments. Elsevier Science, Amsterdam

Cregg PJ, Crothers DSF, Wickstead AW (1994) An approximate formula for the relaxation time of a single domain ferromagnetic particle with uniaxial anisotropy and collinear field. J Appl Phys 76:4900-4902

Dang MZ, Rancourt DG (1996) Simultaneous magnetic and chemical order-disorder phenomena in Fe_3Ni, FeNi, and $FeNi_3$. Phys Rev B 53:2291-2301

Dang M-Z, Rancourt DG, Dutrizac JE, Lamarche G, Provencher R (1998) Interplay of surface conditions, particle size, stoichiometry, cell parameters, and magnetism in synthetic hematite-like materials. Hyperfine Interactions 117:271-319

Dang M-Z, Rancourt DG, Lamarche G, Evans ME (1999) Mineralogical analysis of a loess/paleosol couplet from the Chinese loess plateau. *In* Kodama H, Mermut AR, Torrance JK (eds) Clays for Our Future. Proc 11th Int'l Clay Conf, Ottawa, Canada, p 309-315

de Biasi RS, Devezas TC (1978) Anisotropy field of small magnetic particles as measured by resonance. J Appl Phys 49:2466-2469

de Jongh LJ, Miedema AR (1974) Experiments on simple magnetic model systems. Adv Phys 23:1-260

Deutschlander ME, Borland SC, Phillips JB (1999) Extraocular magnetic compass in newts. Nature 400:324-325

Dickson DPE, Reid NMK, Hunt C, Williams HD, El-Hilo M, O'Grady K (1993) Determination of f_0 for fine magnetic particles. J Magnet Magnetic Mater 125:345-350

Dimitrov DA, Wysin GM (1996) Magnetic properties of superparamagnetic particles by a Monte Carlo method. Phys Rev B 54:9237-9341

Ding ZL, Sun JM, Yang SL, Liu TS (2001) Geochemistry of the Pliocene red clay formation in the Chinese Loess Plateau and implications for its origin, source provenance and paleoclimate change. Geochim Cosmochem Acta 65:901-913

Dixon JB, Weed SB (1989) Minerals in Soil Environments. Soil Science Society of America, Madison, Wisconsin

Djurberg C, Svedlindh P, Nordblad P, Hansen MF, Bodker F, Morup S (1997) Dynamics of an interacting particle system: Evidence of critical slowing down. Phys Rev Lett 79:5154-5157

Domany E (1999) Superparamagnetic clustering of data—The definitive solution of an ill-posed problem. Physica A 263:158-169

Domany E, Blatt M, Gdalyahu Y, Weinshall D (1999) Superparamagnetic clustering of data: Application to computer vision. Computer Phys Commun 121-122:5-12

Dormann JL (1981) La phénomène de superparamagnétisme. Rev Phys Appl 16:275-301

Dormann JL, Bessais L, Fiorani D (1988) A dynamic study of small interacting particles: Superparamagnetic model and spin-glass laws. J Phys C: Solid State Phys 21:2015-2034

Dormann JL, Cherkaoui R, Spinu L, Noguès M, Lucari F, D'Orazio F, Fiorani D, Garcia A, Tronc E, Jolivet JP (1998) From pure superparamagnetic regime to glass collective state of magnetic moments in γ-Fe_2O_3 nanoparticle assemblies. J Magnet Magnetic Mater 187:L139-L144

Dormann JL, D'Orazio F, Lucari F, Tronc E, Prené P, Jolivet JP, Fiorani D, Cherkaoui R, Noguès M (1996) Thermal variation of the relaxation time of the magnetic moment of γ-Fe_2O_3 nanoparticles with interparticle interactions of various strengths. Phys Rev B 53:14291-14297

Dormann JL, Fiorani D (1995) Nanophase magnetic materials: Size and interaction effects on static and dynamical properties of fine particles (invited paper). J Magnet Magnetic Mater 140-144:415-418

Dormann JL, Fiorani D, El Yamani M (1987) Field dependence of the blocking temperature in the superparamagnetic model: $H^{2/3}$ coincidence. Phys Lett A 120:95-99

Dormann JL, Fiorani D, Tronc E (1997) Magnetic relaxation in fine-particle systems. *In* Prirogine I, Rice SA (eds) Adv Chem Phys, volume XCVIII. John Wiley & Sons, New York, p 283-494

Dormann JL, Spinu L, Tronc E, Jolivet JP, Lucari F, D'Orazio F, Fiorani D (1998) Effect of interparticle interactions on the dynamical properties of γ-Fe_2O_3 nanoparticles. J Magnet Magnetic Mater 183:L255-L260

Dubowik J, Baszynski J (1968) FMR study of coherent fine magnesioferrite particles in MgO-line shape behavior. J Magnet Magnetic Mater 59:161-168

Dunlop DJ (1990) Developments in rock magnetism. Rep Prog Phys 53:707-792

Dunlop DJ, Özdemir Ö (1997) Rock magnetism. Fundamentals and frontiers. Cambridge University Press, Cambridge, UK

Dzyaloshinsky I (1958) A thermodynamic theory of "weak" ferromagnetism of antiferromagnetics. J Phys Chem Solids 4:241-255

El-Hilo M, O'Grady K, Chantrell RW (1992a) Susceptibility phenomena in a fine particle system. I. Concentration-dependence of the peak. J Magnet Magnetic Mater 114:295-306

El-Hilo M, O'Grady K, Chantrell RW (1992b) The ordering temperature in fine particle systems. J Magnet Magnetic Mater 117:21-28

Essam JW (1980) Percolation theory. Rep Prog Phys 43:833-912

Evans ME, Ding Z, Rutter NW (1996) A high resolution magnetic susceptibility study of a loess/paleosol couplet at Baoji, China. Studia Geophys Geod 40:225-233

Ezzir A, Dormann JL, Hachkachi H, Nogues M, Godinho M, Tronc E, Jolivet JP (1999) Superparamagnetic susceptibility of a nanoparticle assembly: Application of the Onsager model. J Magnet Magnetic Mater 196-197:37-39

Ferchmin AR, Kobe S (1983) Amorphous Magnetism and Metallic Magnetic Materials—Digest. North-Holland Publishing Company, New York

Ferré R, Barbara B, Fruchart D, Wolfers P (1995) Dipolar interacting small particles: Effects of concentration and anisotropy. J Magnet Magnetic Mater 140-144:385-386

Ferris FG (1997) Formation of authigenic minerals by bacteria. *In* McIntosh JM, Groat LA (eds) Biological-Mineralogical Interactions. Mineral Assoc Canada Short Course Ser 25:187-208

Fiorani D, Dormann JL, Cherkaoui R, Tronc E, Lucari F, D'Orazio F, Spinu L, Nogues M, Garcia A, Testa AM (1999) Collective magnetic state in nanoparticles systems. J Magnet Magnetic Mater 196/197:143-147

Fiorani D, Tholence J, Dormann JL (1986) Magnetic properties of small ferromagnetic particles (Fe-Al_2O_3 granular thin films): Comparison with spin glass properties. J Phys C.: Solid State Phys 19:5495-5507

Fischer KH, Hertz JA (1991) Spin Glasses. Cambridge University Press, Cambridge, UK

Fisher ME, Privman V (1985) First-order transitions breaking O(n) symmetry: Finite-size scaling. Phys Rev B 32:447-464

Fortin D, Ferris FG, Beveridge TJ (1997) Surface-mediated mineral development by bacteria. Rev Mineral 35:161-177

Friedman JR, Sarachik MP, Tejada J, Ziolo R (1996) Macroscopic measurement of resonant magnetization tunneling in high-spin molecules. Phys Rev Lett 76:3830-3833

Friedman JR, Voskoboynik U, Sarachik MP (1997) Anomalous magnetic relaxation in ferritin. Phys Rev B 56:10793-10796

Frost BR (1991) Magnetic petrology: factors that control the occurrence of magnetite in crustal rocks. Rev Mineral 25:489-509

Garanin DA (1996) Integral relaxation time of single-domain ferromagnetic particles. Phys Rev E 54: 3250-3256

Garanin DA (1997) Quantum thermoactivation of nanoscale magnets. Phys Rev E 55:2569-2572

Garanin DA (1999) New integral relaxation time for thermal activation of magnetic particles. EuroPhys Lett 48:486-490

Garanin DA, Kladko K, Fulde P (2000) Quasiclassical hamiltonians for large-spin systems. EuroPhys Lett B14:293-300

García-Palacios JL, Lázaro FJ (1997) Anisotropy effects on the nonlinear magnetic susceptibilities of superparamagnetic particles. Phys Rev B 55:1006-1010

García-Palacios JL, Svedlindh P (2000) Large nonlinear dynamical response of superparamagnets: Interplay between precession and thermoactivation in the stochastic Landau-Lifshitz equation. Phys Rev Lett 85:3724-3727

García-Palacios JL, Svedlindh P (2001) Derivation of the basic system equations governing superparamagnetic relaxation by the use of the adjoint Fokker-Planck operator. Phys Rev B 63:172417-1-172417/4

Gatteschi D, Sessoli R (1996) Origin of superparamagnetic-like behavior in large molecular clusters. *In* Molecule-based Magnetic Materials. American Chemical Society, p 157-169

Gayraud J, Robin E, Rocchia R, Froget L (1996) Formation conditions of oxidized Ni-rich spinel and their relevance to the K/T boundary event. Geol Soc Am 307:425-443

Gazeau F, Bacri JC, Gendron F, Perzynski R, Raikher YuL, Stepanov VI, Dubois E (1998) Magnetic resonance of ferrite nanoparticles: Evidence of surface effects. J Magnet Magnetic Mater 186:175-187

Gazeau F, Dubois E, Hennion M, Perzynski R, Raikher Yu (1997) Quasi-elastic neutron scattering on γ-Fe_2O_3 nanoparticles. EuroPhys Lett 40:575-580

Gee JS, Cande SC, Hildebrand JA, Donnelly K, Parker RL (2000) Geomagnetic intensity variations over the past 780 kyr obtained from near-seafloor magnetic anomalies. Nature 408:827-832

Gider S, Awschalom DD, Douglas T, Mann S, Chaparala M (1995) Classical and quantum magnetic phenomena in natural and artificial ferritin proteins. Science 268:77-80

Gider S, Awschalom DD, Douglas T, Wong K, Mann S, Cain G (1996) Classical and quantum magnetism in synthetic ferritin proteins. J Appl Phys 79:5324-5326

Gilder S, Glen J (1998) Magnetic properties of hexagonal closed-packed iron deduced from direct observations in a diamond anvil cell. Science 279:72-74

Gilles C, Bonville P, Wong KKW, Mann S (2000) Non-Langevin behavior of the uncompensated magnetization in nanoparticles of artificial ferritin. EuroPhys J B17:417-427

Gittleman JI, Abeles B, Bozowski S (1974) Superparamagnetism and relaxation effects in granular Ni-SiO_2 and Ni-Al_2O_3 films. Phys Rev B 9:3891-3897

Golden DC, Ming DW, Schwandt CS, Lauer Jr. HV, Socki RA, Morris RV, Lofgren GE, McKay GA (2001) A simple inorganic process for formation of carbonates, magnetite, and sulfides in Martian meterorite ALH84001. Am Mineral 86:370-375

Goodman AA, Whittet DCB (1955) A point in favor of the superparamagnetic grain hypothesis. Astrophys J 455:L181-L184

Griscom DL, Friebele EJ, Shinn DB (1979) Ferromagnetic resonance of spherical particles of α-iron precipitated in fused silica. J Appl Phys 50:2402-2404

Grodzicki M (1980) A self-consistent charge Xα method I. Theory. J Phys B 13:2683-2690

Grossmann B, Rancourt DG (1996) Simulation of magneto-volume effects in ferromagnets by a combined molecular dynamics and Monte-Carlo approach. Phys Rev B 54:12294-12301

Grun R (2000) Electron spin resonance dating. *In* Ciliberto E, Spoto G (eds) Modern Analytical Methods in Art and Archaeology. John Wiley & Sons, New York, p 641-679

Grüttner C, Teller J (1999) New types of silica-fortified magnetic nanoparticles as tools for molecular biology applications. J Magnet Magnetic Mater 194:8-15

Gunther L, Mohie-Eldin M-EY (1994) Motional narrowing and magnetostrictive broadening of the Mossbauer spectrum due to superparamagnetism. J Magnet Magnetic Mater 129:334-338

Hansen MF, Bender Koch C, Mørup S (2000) Magnetic dynamics of weakly and strongly interacting hematite nanoparticles. Phys Rev B 62:1124-1135

Hansen MF, Bødker F, Mørup S, Lefmann K, Clausen KN, Lindgård P-A (1997) Dynamics of magnetic nanoparticles studied by neutron scattering. Phys Rev B 79:4910-4913

Hansen MF, Mørup S (1998) Models for the dynamics of interacting magnetic nanoparticles. J Magnet Magnetic Mater 184:262-274

Hanzlik M, Heunemann C, Holtkamp-Rötzler E, Winklhofer M, Petersen N, Fleissner G (2000) Superparamagnetic magnetite in the upper beak tissue of homing pigeons. BioMetals 13:325-331

Hartmann-Boutron F, Politi P, Villain J (1996) Tunneling and magnetic relaxation in mesoscopic molecules. Int'l J Modern Phys B 10:2577-2639

Hawkins C, Williams JM, Hudson AJ, Andrews SC, Treffry A (1994) Mossbauer studies of the ultrafine antiferromagnetic cores of ferritin. Hyperfine Interactions 91:827-833

Hawthorne FC (ed) (1988) Spectroscopic Methods in Mineralogy and Geology. Rev Mineral 18, 698 p

Hayashi M, Susa M, Nagata K (1997) Magnetic interaction between magnetite particles dispersed in calciumsilicate glasses. J Magnet Magnetic Mater 171:170-178

Heller F, Evans ME (1995) Loess magnetism. Rev Geophys 33:211-240

Himpsel FJ, Ortega JE, Mankey GJ, Willis RF (1998) Magnetic nanostructures. Adv Phys 47:511-597

Hohenberg PC (1967) Existence of long-range order in one and two dimensions. Phys Rev 158:383-386
Honda H, Kawabe A, Shinkai M, Kobayashi T (1998) Development of Chitosan-conjugated magnetite for magnetic cell separation. J Fermentation Bioengin 86:191-196
Hopkins PF, Moreland J, Malhotra SS, Liou SH (1996) Superparamagnetic magnetic force microscopy tips. J Appl Phys 79:6448-6450
Ibrahim MM, Darwish S, Seehra MS (1995) Nonlinear temperature variation of magnetic viscosity in nanoscale FeOOH particles. Phys Rev B 51:2955-2959
Ibrahim MM, Edwards G, Seehra MS, Ganguly B, Huffman GP (1994) Magnetism and spin dynamics of nanoscale FeOOH particles. J Appl Phys 75:5873-5875
Ibrahim MM, Zhao J, Seehra MS (1992) Determination of particle size distribution in an Fe_2O_3-based catalyst using magnetometry and X-ray diffraction. J Mater Res 7:1856-1860
Ignatchenko VA, Mironov YE (1991) Magnetic structures with a finite number of domain walls. J Magnet Magnetic Mater 94:170-178
Jacobs JA (1991) Geomagnetism. Academic Press, London
Jambor JL, Dutrizac JE (1998) Occurrence and constitution of natural and synthetic ferrihydrite, a widespread iron oxyhydroxide. Chem Rev 98:2549-2585
Jiles DC, Atherton DL (1984) Theory of ferromagnetic hysteresis (invited). J Appl Phys 55:2115-2120
Jones DH, Srivastava KKP (1989) A re-examination of models of superparamagnetic relaxation. J Magnet Magnetic Mater 78:320-328
Jönsson P, Jonsson T, García-Palacios JL, Svedlindh P (2000) Nonlinear dynamic susceptibilities of interacting and noninteracting magnetic nanoparticles. J Magnet Magnetic Mater 222:219-226
Jonsson T, Mattsson J, Djurberg C, Khan FA, Nordblad P, Svedlindh P (1995b) Aging in a magnetic particle system. Phys Rev Lett 75:4138-4141
Jonsson T, Mattsson J, Nordblad P, Svedlindh P (1997) Energy barrier distribution of a noninteracting nano-sized magnetic particle system. J Magnet Magnetic Mater 168:269-277
Jonsson T, Nordblad P, Svedlindh P (1998) Dynamic study of dipole-dipole interaction effects in a magnetic nanoparticle system. Phys Rev B 57:497-504
Jonsson T, Svedlindh P, Hansen MF (1998) Static scaling on an interacting magnetic nanoparticle system. Phys Rev Lett 81:3976-3979
Jonsson T, Svedlindh P, Nordblad P (1995a) AC susceptibility and magnetic relaxation studies on frozen ferrofluids. Evidence for magnetic dipole-dipole interactions. J Magnet Magnetic Mater 140-144: 401-402
Jung CW, Rogers JM, Groman EV (1999) Lymphatic mapping and sentinel node location with magnetite nanoparticles. J Magnet Magnetic Mater 194:210-216
Kahn O (1999) The magnetic turnabout. Nature 399:21-23
Kalmykov YP (2000) Longitudinal dynamic susceptibility and relaxation time of superparamagnetic particles with cubic anisotropy: Effect of a biasing magnetic field. Phys Rev B 61:6205-6212
Kalmykov YP, Titov SV (1999) Calculating coefficients for a system of moment equations used to describe the magnetization kinetics of a superparamagnetic particle in a fluctuating field. Phys Solid State 41:1854-1861
Kalmykov YP, Titov SV (2000) Derivation of matrix elements for the system of moment equations governing the kinetics of superparamagnetic particles. J Magnet Magnetic Mater 210:233-243
Kechrakos D, Trohidou KN (1998a) Effects of dipolar interactions on the magnetic properties of granular solids. J Magnet Magnetic Mater 177-181:943-944
Kechrakos D, Trohidou KN (1998b) Magnetic properties of dipolar interacting single-domain particles. Phys Rev B 58:12169-12177
Kirschvink JL (1989) Magnetite biomineralization and geomagnetic sensitivity in higher animals: An update and recommendations for future study. Bioelectromagnetics 10:239-259
Kirschvink JL, Jones DS, MacFadden BJ (1985) Magnetite Biomineralization and Magnetoreception in Organisms—A New Biomagnetism. Plenum Press, New York
Kirschvink JL, Kirschvink-Kobayashi A, Woodford BJ (1992) Magnetite biomineralization in the human brain. Proc Natl Acad Sci 89:7683-7687
Kleemann W, Petracic O, Binek Ch, Nakazei GN, Pogorelov YuG, Sousa JB, Cardoso S, Freitas PP (2001) Interacting ferromagnetic nanoparticles in discontinuous $Co_{80}Fe_{20}/Al_2O_3$ multilayers: From superspin glass to reentrant superferromagnetism. Phys Rev B 63:134423-1-134423/5
Klein C, Hurlbut CS Jr (1999) Manual of Mineralogy. John Wiley & Sons, New York
Kliava J, Berger R (1999) Size and shape distribution of magnetic nanoparticles in disordered systems: Computer simulations of superparamagnetic resonance spectra. J Magnet Magnetic Mater 205:328-342
Kneller E (1968) Magnetic-interaction effects in fine-particle assemblies and in thin films. J Appl Phys 39:945-955

Kneller E, Puschert W (1966) Pair interaction models for fine particle assemblies. IEEE Trans Magnetics MAG-2:250

Kodama RH (1999) Magnetic nanoparticles. J Magnet Magnetic Mater 200:359-372

Kodama RH, Berkowitz AE (1999) Atomic-scale magnetic modeling of oxide nanoparticles. Phys Rev B 59:6321-6336

Kröckel M, Grodzicki M, Papaefthymiou V, Trautwein AX, Kostikas A (1996) Tuning of electron delocalization in polynuclear mixed-valence clusters by super-exchange and double exchange. J Biol Inorg Chem 1:173-176

Krupicka S, Novak P (1982) Oxide Spinels. *In* Wohlfarth EP (ed) Ferromagnetic Materials. North-Holland, Amsterdam, p 189-304

Kryder MH (1996) Ultrahigh-density recording technologies. Mater Res Bull, September 1996, 17-22

Kukla G (1987) Loess stratigraphy in Central China. Quaternary Sci Rev 6:191-219

Kukla G, An Z (1989) Loess stratigraphy in Central China. Palaeogeogr Palaeoclimatol Palaoecol 72: 203-225

Kundig W, Bommel H, Constabaris G, Lindquist RH (1966) Some properties of supported small α-Fe_2O_3 particles determined with the Mossbauer effect. Phys Rev 142:327-333

Kyte FT (1998) A meteorite from the Cretaceous/Tertiary boundary. Nature 396:237-239

Kyte FT, Bohor BF (1995) Nickel-rich magnesiowüstite in Cretaceous/Tertiary boundary spherules crystallized from ultramafic, refractory silicate liquids. Geochim Cosmochem Acta 59:4967-4974

Kyte FT, Bostwick JA (1995) Magnesioferrite spinel in Cretaceous/Tertiary boundary sediments of the Pacific basin: Remnants of hot, early ejecta from the Chicxulub impact? Earth Planet Sci Lett 132:113-127

Lagarec K, Rancourt DG (2000) Fe_3Ni-type chemical order in $Fe_{65}Ni_{35}$ films grown by evaporation: Implications regarding the Invar problem. Phys Rev B 62:978-985

Lagarec K, Rancourt DG, Bose SK, Sanyal B, Dunlap RA (2001) Observation of a composition-controlled high-moment/low-moment transition in the face centered cubic Fe-Ni system: Invar effect is an expansion, not a contraction. J Magnet Magnetic Mater (in press)

Lalonde AE, Rancourt DG, Ping JY (1998) Accuracy of ferric/ferrous determinations in micas: A comparison of Mössbauer spectroscopy and the Pratt and Wilson wet-chemical methods. Hyperfine Interactions 117:175-204

Lear PR (1987) The role of iron in nontronite and ferrihydrite. PhD Dissertation. University of Illinois, Champagne-Urbana

Lifshin E (1992) Characterization of Materials, Part I. VCH, Weinheim

Lifshin E (1994) Characterization of Materials, Part II. VCH, Weinheim

Ligenza S, Dokukin EB, Nikitenko YuV (1995) Neutron depolarization studies of magnetization process in superparamagnetic cluster structures. J Magnet Magnetic Mater 147:37-44

Lin D, Nunes AC, Majkrzak CF, Berkowitz AE (1995) Polarized neutron study of the magnetization density distribution within a $CoFe_2O_4$ colloidal particle II. J Magnet Magnetic Mater 145:343-348

Liou SH, Malhotra SS, Moreland J, Hopkins PF (1997) High resolution imaging of thin-film recording heads by superparamagnetic magnetic force microscopy tips. Appl Phys Lett 70:135-137

Liu C, Zou B, Rondinone AJ, Zhang ZJ (2000) Reverse micelle synthesis and characterization of superparamagnetic $MnFe_2O_4$ spinel ferrite nanocrystallites. J Phys Chem B 104:1141-1145

Liu T (1988) Loess in China. Springer-Verlag, Berlin

Liu T, Liu T (1991) Loess, Environment and Global Change. Science Press Beijing, China

Lodder JC (1995) Magnetic microstructures of perpendicular magnetic-recording media. Mater Res Soc Bull, October 1995:59-63

López A, Lázaro FJ, García-Palacios JL, Larrea A, Pankhurst QA, Martínez C, Corma A (1997) Superparamagnetic particles in ZSM-5-type ferrisilicates. J Mater Res 12:1519-1529

Lowes FJ (1989) Geomagnetism and Paleomagnetism. Kluwer Academic Publishers, Boston

Lucínski T, Elefant D, Reiss G, Verges P (1966) The concept of the existence of interfacial superparamagnetic entities in Fe/Cr multilayers. J Magnet Magnetic Mater 162:29-37

Luo W, Nagel SR, Rosenbaum TF, Rosensweig RE (1991) Dipole interactions with random anisotropy in a frozen ferrofluid. Phys Rev Lett 67:2721-2724

Madsen MB, Morup S, Koch CJW (1986) Magnetic properties of ferrihydrite. Hyperfine Interactions 27:329-332

Maher BA, Thompson R (1999) Quaternary climates, environments and magnetism. Cambridge University Press, Cambridge, UK

Makhlouf SA, Parker FT, Berkowitz AE (1997b) Magnetic hysteresis anomalies in ferritin. Phys Rev B 55:R14717-R14720

Makhlouf SA, Parker FT, Spada FE, Berkowitz AE (1997a) Magnetic anomalies in NiO nanoparticles. J Appl Phys 81:5561-5563

Malaescu I, Marin CN (2000) Deviation from the superparamagnetic behavior of fine-particle systems. J Magnet Magnetic Mater 218:91-96
Mamiya H, Nakatani L, Furubayashi T (1998) Blocking and freezing of magnetic moments for iron nitride fine particle systems. Phys Rev Lett 80:177-180
Marin ML, Ortuno M, Hernandez A, Abellan J (1990) Percolative treatment of the Verwey transition in cobalt-iron and nickel-iron ferrites. Physica Status Solidi (B) 157:275-280
McElhinny MW, McFadden PL (1999) Paleomagnetism. Academic Press, New York
McVitie S, Chapman JN (1995) Coherent Lorentz imaging of soft thin-film magnetic materials. Mater Res Soc Bull, October 1995:55-58
Menear S, Bradbury A, Chantrell RW (1984) A model of the properties of colloidal dispersions of weakly interacting fine ferromagnetic particles. J Magnet Magnetic Mater 43:166-176
Mermin ND, Wagner H (1966) Absence of ferromagnetism or antiferromagnetism in one-or-two-dimensional isotropic Heisenberg models. Phys Rev Lett 17:1133-1136
Merrill RT, McElhinny MW, McFadden PL (1998) The Magnetic Field of the Earth. Academic Press, London
Miltat J, Thiaville A (2000) Magnets fast and small. Science 290:466-467
Mira J, López-Pérez JA, López-Quintela MA, Rivas J (1997) Magnetic iron oxide nanoparticles synthesized via microemulsions. Mater Sci Forum 235-238:297-302
Moriya T (1960) Anisotropic superexchange interaction and weak ferromagnetism. Phys Rev 120:91-98
Morris RV, Golden DC, Ming DW, Shelfer TD, Jorgensen LC, Bell III JF, Graff TG, Mertzman SA (2001) Phyllosilicate-poor palagonitic dust from Mauna Kea Volcano (Hawaii): A mineralogical analogue for magnetic Martian dust? J Geophys Res 106:5057-5083
Morris RV, Golden DG, Shelfer TD, Lauer Jr. HV (1998) Lepidocrocite to maghemite to hematite: A pathway to magnetic and hematitic martian soil. Meteoritics Planet Sci 33:743-751
Morup S (1994a) Superparamagnetism and spin glass ordering in magnetic nanocomposites. EuroPhys Lett 28:671-676
Morup S (1994b) Superferromagnetic nanostructures. Hyperfine Interactions 90:171-185
Morup S, Bodker F, Hendriksen PV, Linderoth S (1995) Spin-glass-like ordering of the magnetic moments of interacting nanosized maghemite particles. Phys Rev B 52:287-294
Morup S, Madsen MB, Franck J, Villadsen J, Koch CJW (1983) A new interpretation of Mossbauer spectra of microcrystalline goethite: "super-ferromagnetism" or "super-spin-glass" behavior? J Magnet Magnetic Mater 40:163-174
Morup S, Topsoe H, Lipka J (1976) Modified theory for Mossbauer spectra of superparamagnetic particles: Application to Fe_3O_4. J Physique, Colloque C6, Suppl n° 12:C6-287–C6-291
Morup S, Tronc E (1994) Superparamagnetic relaxation of weakly interacting particles. Phys Rev Lett 72:3278-3281
Moskowitz BM, Frankel RB, Walton SA, Dickson DPE, Wong KKW, Douglas T, Mann S (1997) Determination of the preexponential frequency factor for superparamagnetic maghemite particles in magneto-ferritin. J Geophys Res 102:22671-22680
Mullins CE, Tite MS (1973) Magnetic viscosity, quadrature susceptibility, and frequency dependence of susceptibility in single-domain assemblies of magnetite and maghemite. J Geophys Res 78:804-809
Murad E, Bowen LH, Long GL, Quin TG (1988) The influence of crystallinity on magnetic ordering in natural ferrihydrites. Clay Minerals 23:161-173
Nagata T (1961) Rock Magnetism. Maruzen, Tokyo
Néel L (1949a) Théorie du trainage magnétique des ferromagnétiques en grains fins avec applications aux terres cuites. Annal Géophys 5:99-136
Néel L (1949b) Influence des fluctuations thermiques sur l'aimantation de grains ferromagnétiques très fins. Comptes-rendus des séances de l'Académie des sciences 228:664-666
Néel L (1953) L'anisotropie superficielle des substances ferromagnétiques. Comptes-rendus des séances de l'Académie des sciences 237:1468-1470
Néel L (1954) Anisotropie magnétique superficielle et surstructures d'orientation. Journal de physique et le radium 15:225-239
Néel L (1955) Some theoretical aspects of rock magnetism. Adv Phys 4:191-243
Néel L (1961a) Superparamagnétisme des grains très fins antiferromagnétiques. Comptes-rendus des séances de l'Académie des sciences 252:4075-4080
Néel L (1961b) Superantiferromagnétisme dans les grains fins. Comptes-rendus des séances de l'Académie des sciences 253:203-208
Néel L (1961c) Superposition de l'antiferromagnétisme et du superparamagnétisme dans un grain très fin. Comptes-rendus des séances de l'Académie des sciences 253:9-12
Néel L (1961d) Sur le calcul de la susceptibilité additionnelle superantiferromagnétique des grains fins et sa variation thermique. Comptes-rendus des séances de l'Académie des sciences 253:1286-1291

Nordström L, Singh DJ (1996) Noncollinear intra-atomic magnetism. Phys Rev Lett 76:4420-4423
Nowak U, Chantrell RW, Kennedy EC (2000) Monte Carlo simulation with time step quantification in terms of Langevin dynamics. Phys Rev Lett 84:163-166
Nowak U, Hinzke D (1999) Magnetization switching in small ferromagnetic particles: Nucleation and coherent rotation. J Appl Phys 85:4337-4339
Nuth III JA, Wilkinson GM (1995) Magnetically enhanced coagulation of very small iron grains: A correction of the enhancement factor due to dipole-dipole interactions. Icarus 117:431-434
O'Reilly W (1984) Rock and Mineral Magnetism. Blackie, Glasgow and London
Oka T, Grun R, Tani A, Yamanaka C, Ikeya M, Huang HP (1997) ESR microscopy of fossil teeth. Radiation Measurements 27:331-337
Omari R, Prejean JJ, Souletie J (1983) Critical measurements in the spin glass CuMn. J Physique 44: 1069-1083
Onodera S, Kondo H, Kawana T (1996) Materials for magnetic-tape media. Mater Res Soc Bull, September 1996:35-41
Opdyke ND, Channell JET (1996) Magnetic Stratigraphy. Academic Press, New York
Pankhurst QA, Pollard RJ (1992) Structural and magnetic properties of ferrihydrite. Clays Clay Minerals 40:268-272
Pankhurst QA, Pollard RJ (1993) Fine-particle magnetic oxides. J Phys: Condensed Matter 5:8487-8508
Parker FT, Foster MW, Margulies DT, Berkowitz AE (1993) Spin canting, surface magnetization, and finite-size effects in γ-Fe_2O_3 particles. Phys Rev B 47:7885-7891
Parkinson WD (1983) Introduction to geomagnetism. Scottish Academic Press, Edinburgh
Pastor GM, Dorantes-Davila J, Pick S, Dreyssé H (1995) Magnetic anisotropy of 3d transition-metal clusters. Phys Rev Lett 75:326-329
Pathria RK (1988) Statistical Mechanics. Pergamon Press, Exeter, UK
Petersen N, von Dobeneck T, Vali H (1986) Fossil bacterial magnetite in deep-sea sediments form the South Atlantic Ocean. Nature 320:611-615
Pfannes H-D (1997) Simple theory of superparamagnetism and spin-tunneling in Mossbauer spectroscopy. Hyperfine Interactions 110:127-134
Pfannes H-D, Mijovilovich A, Magalhães-Paniago R, Paniago R (2000) Spin-lattice-relaxation-like model for superparamagnetic particles under an external magnetic field. Phys Rev B 62:3372-3380
Pieters CM, Taylor LA, Noble SK, Keller LP, Hapke B, Morris RV, Allen CC, McKay DS, Wentworth S (2000) Space weathering on airless bodies: Resolving a mystery with lunar samples. Meteoritics Planet Sci 35:1101-1107
Pietzsch O, Kubetzka A, Bode M, Wiesendanger R (2001) Observation of magnetic hysteresis at the nanometer scale by spin-polarized scanning tunneling spectroscopy. Science 292:2053-2056
Pindor AJ, Staunton J, Stocks GM, Winter H (1983) Disordered local moment state of magnetic transition metals: A self-consistent KKR CPA calculation. J Phys F: Metal Phys 13:979-989
Politi P, Rettori A, Hartmann-Boutron F, Villain J (1995) Tunneling in mesoscopic magnetic molecules. Phys Rev Lett 75:537-540
Pollard RJ, Cardile CM, Lewis DG, Brown LG (1992) Characterization of FeOOH polymorphs and ferrihydrite using low-temperature, applied-field, Mössbauer spectroscopy. Clay Minerals 27:57-71
Popkov AF, Savchenko LL, Vorotnikova NV (1999) Thermally activated transformation of magnetization-reversal modes in ultrathin nanoparticles. JETP Lett 69:596-602
Popplewell J, Sakhnini L (1995) The dependence of the physical and magnetic properties of magnetic fluids on particle size. J Magnet Magnetic Mater 149:72-78
Preisach F (1935) Über die magnetische Nachwirkung. Z Physik 94:277-302
Prozorov R, Yeshurun Y, Prozorov T, Gedanken A (1999) Magnetic irreversibility and relaxation in assembly of ferromagnetic nanoparticles. Phys Rev B 59:6956-6965
Pshenichnikov AF, Mekhonoshin VV (2000) Equilibrium magnetization and microstructure of the system of superparamagnetic interacting particles: Numerical simulation. J Magnet Magnetic Mater 213: 357-369
Qiu ZQ, Bader SD (1995) Surface magnetism and Kerr spectroscopy. Mater Res Soc Bull, October 1995:34-37
Quin TG, Long GL, Benson CG, Mann S, Williams RJP (1988) Influence of silicon and phosphorus on structural and magnetic properties of synthetic goethite and related oxides. Clays Clay Minerals 36:165-175
Raikher YL, Shliomis MI (1975) Theory of dispersion of the magnetic susceptibility of fine ferromagnetic particles. Soviet Physics JETP 40:526-532
Raikher YL, Shliomis MI (1994) The effective field method in the orientational kinetics of magnetic fluids and liquid crystals. Adv Chem Phys 87:595-751

Raikher YL, Stepanov VI (1995) Intrisic magnetic resonance in superparamagnetic systems. Phys Rev B 51:16428-16431
Raikher YL, Stepanov VI (1995a) Stochastic resonance and phase shifts in superparamagnetic particles. Phys Rev B 52:3493-3498
Raikher YL, Stepanov VI (1995b) Magnetic resonances in ferrofluids: temperature effects. J Magnet Magnetic Mater 149:34-37
Raikher YL, Stepanov VI (1997) Linear and cubic dynamic susceptibilities of superparamagnetic fine particles. Phys Rev B 55:15005-15017
Rancourt DG (1985) New theory for magnetic GICs: Superferromagnetism in two dimensions. J Magnet Magnetic Mater 51:133-140
Rancourt DG (1986) Low temperature behavior of Ising magnetic chains; decorated solitons, locally enhanced exchange and diffusive propagation. Solid State Commun 58:433-440
Rancourt DG (1987) Magnetic phenomena in layered and intercalated compounds. *In* Legrand AP, Flandrois S (eds) Chemical Physics of Intercalation. Plenum, New York, p 79-103
Rancourt DG (1988) Pervasiveness of cluster excitations as seen in the Mossbauer spectra of magnetic materials. Hyperfine Interactions 40:183-194
Rancourt DG (1994a) Mossbauer spectroscopy of minerals. I. Inadequacy of lorentzian-line doublets in fitting spectra arising from quadrupole splitting distributions. Phys Chem Minerals 21:244-249
Rancourt DG (1994b) Mossbauer spectroscopy of minerals. II. Problem of resolving *cis* and *trans* octahedral Fe^{2+} sites. Phys Chem Minerals 21:250-257
Rancourt DG (1998) Mossbauer spectroscopy in clay science. Hyperfine Interactions 117:3-38
Rancourt DG, Dang MZ (1996) Relation between anomalous magneto-volume behavior and magnetic frustration in Invar alloys. Phys Rev B 54:12225-12231
Rancourt DG, Dang M-Z, Lalonde AE (1992) Mossbauer spectroscopy of tetrahedral Fe^{3+} in trioctahedral micas. Am Mineral 77:34-43
Rancourt DG, Daniels JM (1984) Influence of unequal magnetization direction probabilities on the Mossbauer spectra of superparamagnetic particles. Phys Rev B 29:2410-2414
Rancourt DG, Daniels JM, Nazar LF, Ozin GA (1983) The superparamagnetism of very small particles supported by zeolite-Y. Hyperfine Interactions 15/16:653-656
Rancourt DG, Dube M, Heron PRL (1993) General method for applying mean field theory to disordered magnetic alloys. J Magnet Magnetic Mater 125:39-48
Rancourt DG, Ferris FG, Fortin D (1999a) Sorbed iron on the cell wall of Bacillus subtilis characterized by Mössbauer spectroscopy: *In* Evidence for Bioreduction. Ferris FG (ed) Abstracts: Int'l Symp Environ Biogeochem, Hunstville, Canada. XIV:27
Rancourt DG, Flandrois S, Biensan P, Lamarche G (1990a) Magnetism of a graphite bi-intercalation compound with two types of ferromagnetic layers: Double hysteretic transition in $CrCl_3$-$NiCl_2$-C. Canadian J Phys 68:1435-1439
Rancourt DG, Fortin D, Pichler T, Thibault P-J, Lamarche G, Morris RV, Mercier PHJ (2001b) Mineralogy of a natural As-rich hydrous ferric oxide coprecipitate formed by mixing of hydrothermal fluid and seawater: Implications regarding surface complexation and color banding in ferrihydrite deposits. Am Mineral 86:1-18 (in press)
Rancourt DG, Julian SR, Daniels JM (1985) Mossbauer characterization of very small superparamagnetic particles: Application to intra-zeolitic α-Fe_2O_3 particles. J Magnet Magnetic Mater 49:305-316
Rancourt DG, Lagarec K, Densmore A, Dunlap RA, Goldstein JI, Reisner RJ, Scorzelli RB (1999b) Experimental proof of the distinct electronic structure of a new meteoritic Fe-Ni alloy phase. J Magnet Magnetic Mater 191:L255-L260
Rancourt DG, Lamarche G, Tume P, Lalonde AE, Biensan P, Flandrois S (1990b) Dipole-dipole interactions as the source of spin-glass behavior in exchangewise two-dimensional ferromagnetic layer compounds. Canadian J Phys 68:1134-1137
Rancourt DG, Meschi C, Flandrois S (1986) S = 1/2 antiferromagnetic finite chains effectively isolated by frustration: $CuCl_2$ graphite intercalation compounds. Phys Rev B 33:347-255
Rancourt DG, Ping JY (1991) Voigt-based methods for arbitrary-shape static hyperfine parameter distributions in Mössbauer spectroscopy. Nuclear Instruments and Methods in Physics Research B (NIMB) 58:85-97
Rancourt DG, Ping JY, Berman RG (1994) Mossbauer spectroscopy of minerals. III. Octahedral-site Fe^{2+} quadrupole splitting distributions in the phlogopite-annite series. Phys Chem Minerals 21:258-267
Rancourt DG, Scorzelli RB (1995) Low-spin γ-Fe-Ni(γ_{LS}) proposed as a new mineral in Fe-Ni-bearing meteorites: Epitaxial intergrowth of γ_{LS} and tetrataenite as a possible equilibrium state at ~20-40 at % Ni. J Magnet Magnetic Mater 150:30-36

Rancourt DG, Thibault P-J, Ferris FG (2001a) Resolution and quantification of Fe sorbed to bacterial cell walls, biogenic ferrihydrite, and abiotic ferrihydrite by cryogenic ^{57}Fe Mössbauer spectroscopy. Proc of ICOBTE-2001, 6th Int'l Conf Biogeochemistry of Trace Elements. Extended abstr GO448, p 360

Ribas R, Labarta A (1996a) Magnetic relaxation of a one-dimensional model for small particle systems with dipolar interaction: Monte Carlo simulation. J Appl Phys 80:5192-5199

Ribas R, Labarta A (1996b) Monte Carlo simulation of magnetic relaxation in small-particle systems with dipolar interactions. J Magnet Magnetic Mater 157-158:351-352

Ricci TF, Scherer C (1997) Linear response and stochastic resonance of superparamagnets. J Statist Phys 86:803-819

Richter C, van der Pluijm BA (1994) Separation of paramagnetic and ferrimagnetic susceptibilities using low temperature magnetic susceptibilities and comparison with high field methods. Phys Earth Planet Inter 82:113-123

Rikitake T, Honkura Y (1986) Solid earth geomagnetism. D Reidel Pub Co, Dordrecht, The Netherlands

Roberts AP, Cui Y, Verosub KL (1995) Wasp-waisted hysteresis loops: Mineral magnetic characteristics and discrimination of components in mixed magnetic systems. J Geophys Res 100:17909-17924

Robin E (1996) Le verdict du spinelle: les derniers vestiges de la météorite elle-même ont été retrouvés. La Recherche 293:58-60

Roch A, Muller RN (1999) Theory of proton relaxation induced by superparamagnetic particles. J Chem Phys 110:5403-5411

Rochette P, Fillion G (1988) Identification of multicomponent anisotropies in rocks using various field and temperature values in a cryogenic magnetometer. Phys Earth Planet Interiors 51:379-386

Ronov AB, Yaroshevsky AA (1969) The Earth's Crust and Upper Mantle. American Geophysical Union, Washington, DC

Rose K, Gurewitz E, Fox GC (1990) Statistical mechanics and phase transitions in clustering. Phys Rev Lett 65:945-948

Rosensweig RE (1985) Ferrohydrodynamics. Cambridge University Press, Cambridge, UK

Rubens SM (1979) William Fuller Brown, Jr. I E E E Trans Magnetics 15:1192-1195

Sabourin N, Rancourt DG, Dang M-Z, Evans ME (2001) Mineral magnetic, and mineralogical, and geochemical study of a Siberian loess/paleosol sequence. (manuscript in preparation)

Sadykov EK, Isavnin AG (1996) Theory of dynamic magnetic susceptibility in uniaxial superparamagnetic particles. Phys Solid State 38:1160-1164

Safarik I, Safarikova M (1997) Copper phthalocyanine dye immobilized on magnetite particles: An efficient adsorbent for rapid removal of polycyclic aromatic compounds from water solutions and suspensions. Separation Sci Technol 32:2385-2392

Sappey R, Vincent E, Hadacek N, Chaput F, Boilot JP, Zins D (1997a) Nonmonotonic field dependence of the zero-field cooled magnetization peak in some systems of magnetic nanoparticles. Phys Rev B 56:14551-14559

Sappey R, Vincent E, Ocio M, Hammann J, Chaput F, Boilot JP, Zins D (1997b) A new experimental procedure for characterizing quantum effects in small magnetic particle systems. EuroPhys Lett 37:639-644

Saxena SS, Littlewood PB (2001) Iron cast in exotic role. Nature 412:290-291

Scheinfein MR, Schmidt KE, Heim KR, Hembree GG (1996) Magnetic order in two-dimensional arrays of nanometer-sized superparamagnets. Phys Rev Lett 76:1541-1544

Schmidt H, Ram RJ (2001) Coherent magnetization reversal of nanoparticles with crystal and shape anisotropy. J Appl Phys 89:507-513

Seehra MS, Babu VS, Manivannan A (2000) Neutron scattering and magnetic studies of ferrihydrite nanoparticles. Phys Rev B 61:3513-3518

Sestier C, Da-Silva MF, Sabolovic D, Roger J, Pons JN (1998) Surface modification of superparamagnetic nanoparticles (ferrofluid) studied with particle electrophoresis: Application to the specific targeting of cells. Europhoresis 19:1220-1226

Sestier C, Sabolovic D (1998) Particle electrophoresis of micrometic-sized superparamagnetic particles designed for magnetic purification of cells. Europhoresis 19:2485-2490

Sharma R, Saini S, Ros PR, Hahn PF, Small WC, de Lange EE, Stillman AE, Edelman RR, Runge VM, Outwater EK, Morris M, Lucas M (1999) Safety profile of ultrasmall superparamagnetic iron oxide ferumoxtran-10: Phase II clinical trial data. J Magnet Reson Imaging 9:291-294

Sharma VK, Baiker A (1981) Superparamagnetic effects in the ferromagnetic resonance of silica supported nickel particles. J Chem Phys 75:5596-5601

Sharma VK, Waldner F (1977) Superparamagnetic and ferrimagnetic resonance of ultrafine Fe_3O_4 particles in ferrofluids. J Appl Phys 48:4298-4302

Sibilia JP (1996) A Guide to Materials Characterization and Chemical Analysis. VCH Publishers, Weinheim, FRG

Simkiss K, Taylor MG (2001) Trace element speciation at cell membanes: aqueous, solid and lipid phase effects. J Environ Monitoring 3:15-21

Smith NV, Padmore HA (1995) X-ray magnetic circular dichroism spectroscopy and microscopy. Mater Res Soc Bull October:41-44

Soltis FG (2001) Magnetic storage: How much longer can it keep going? News/400, February 2001:25-29

Song T, Roshko RM (2000) Preisach model for systems of interacting superparamagnetic particles. IEEE Trans Magnetics 36:223-230

Spinu L, Stancu A (1998) Modelling magnetic relaxation phenomena in fine particles systems with a Preisach-Néel model. J Magnet Magnetic Mater 189:106-114

Srivastava KKP, Jones DH (1988) Toward a microscope description of superparamagnetism. Hyperfine Interactions 42:1047-1050

St-Pierre TG, Chan P, Bauchspiess KR, Webb J, Betteridge S, Walton S, Dickson DPE (1996) Synthesis, structure and magnetic properties of ferritin cores with varying composition and degrees of structural order: Models of iron oxide deposits in iron-overload diseases. Coordination Chem Rev 151:125-143

St-Pierre TG, Jones DH, Dickson DPE (1987) The behavior of superparamagnetic small particles in applied magnetic fields: A Mossbauer spectroscopy study of ferritin and haemosiderin. J Magnet Magnetic Mater 69:276-284

St-Pierre TG, Webb J, Mann S (1989) Ferritin and hemosiderin: Structural and magnetic studies of the iron core. *In* Mann S, Webb J, Williams RJP (eds) Biomineralization—Chemical and Biochemical Perspectives. VCH Publishers, Weinheim, p 295-344

Stacey FD, Banerjee SK (1974) Physical principles of rock magnetism. *In* Developments in Solid Earth Geophysics. Elsevier, Amsterdam

Stacey GD (1963) The physical theory of rock magnetism. Adv Phys 12:45-133

Stamp PCE (1991) Quantum dynamics and tunneling of domain walls in ferromagnetic insulators. Phys Rev Lett 66:2802-2805

Stancu A, Spinu L (1998) Temperature- and time-dependent Preisach model for a Stoner-Wohlfarth particle system. IEEE Trans Magnetics 34:3867-3875

Stein DL (1992) Spin glasses and biology. World Scientific, Singapore

Steiner M, Villain J, Windsor G (1976) Theoretical and experimental studies on one-dimensional magnetic systems. Adv Phys 25:87-209

Stokroos I, Litinetsky L, van der Want JJL, Ishay JS (2001) Keystone-like crystals in cells of hornet combs. Nature 411:654

Stoner EC, Wohlfarth EP (1948) A mechanism of magnetic hysteresis in heterogeneous alloys. Philos Trans Royal Soc London 240:599-644

Street R, Woolley JC (1949) A study of magnetic viscosity. Proc Phys Soc A 62:562-572

Suber L, Fiorani D, Imperatori P, Foglia S, Montone A, Zysler R (1999) Effects of thermal treatments on structural and magnetic properties of acicular α-Fe_2O_3 nanoparticles. NanoStructured Mater 11: 797-803

Suber L, Santiago AG, Fiorani D, Imperatori P, Testa AM, Angiolini M, Montone A, Dormann JL (1998) Structural and magnetic properties of α-Fe_2O_3 nanoparticles. Appl Organometal Chem 12:347-351

Sutton AP (1993) Electronic Structure of Materials. Clarendon Press, Oxford, UK

Suzuki T (1996) Magneto-optic recording materials. Mater Res Soc Bull, September 1996:42-47

Svedlindh P, Jonsson T, García-Palacios JL (1997) Intra-potential-well contribution to the AC susceptibility of a noninteracting nano-sized magnetic particle system. J Magnet Magnetic Mater 169:323-334

Szabo A, Ostlund NS (1996) Modern quantum chemistry. Introduction to advanced electronic structure theory. Dover, New York

Tarduno JA (1995) Superparamagnetism and reduction diagenesis in pelagic sediments: Enhancement or depletion? Geophys Res Lett 22:1337-1340

Tarling DH, Hrouda P (1993) The Magnetic Anisotropy of Rocks. Chapman and Hall, New York

Tauxe L, Mullender TAT, Pick T (1996) Potbellies, wasp-waists, and superparamagnetism in magnetic hysteresis. J Geophys Res 101:571-583

Tejada J, Zhang XX (1994) On magnetic relaxation in antiferromagnetic horse-spleen ferritin proteins. J Phys: Condensed Matter 6:263-266

Tejada J, Zhang XX, del Barco E, Hernández JM, Chudnovsky EM (1997) Macroscopic resonant tunneling of magnetization in ferritin. Phys Rev Lett 79:1754-1757

Tejada J, Ziolo RF, Zhang XX (1996) Quantum tunneling of magnetization in nanostructured materials. Chem Mater 8:1784-1792

Thibault P-J (2001) Charactérization de la ferrihydrite authigénique synthétisée sous différentes conditions. MsC Thesis, Department of Earth Sciences, University of Ottawa (submitted)

Thomas L, Lionti F, Ballou R, Gatteschi D, Sessoli R, Barbara B (1996) Macroscopic quantum tunnelling of magnetization in a single crystal of nanomagnets. Nature 383:145-147

Thompson R, Oldfield F (1986) Environmental Magnetism. Allen & Unwin, St. Leonards, UK

Tiefenauer LX, Kühne G, Andres RY (1993) Antibody-magnetite nanoparticles: In vitro characterization of a potential tumor-specific contrast agent for magnetic resonance imaging. Bioconjugate Chem 4:347-352

Upadhyay RV, Sutariya GM, Mehta RV (1993) Particle size distribution of a laboratory-synthesized magnetic fluid. J Magnet Magnetic Mater 123:262-266

Uyeda C (1993) Diamagnetic anisotropies of oxide minerals. Phys Chem Minerals 20:77-81

Vacquier V (1972) Geomagnetism in Marine Geology. Elsevier Science, Amsterdam

Van Der Giessen AA (1967) Magnetic properties of ultra-fine iron (3) oxide-hydrate particles prepared from iron (3) oxide-hydrate gels. J Phys Chem Solids 28:343-346

van der Kraan AM (1971) Superparamagnetism of small alpha-Fe_2O_3 crystallites studied by means of the Mossbauer effect. J Physique 32:C1-1034-C1-1036

Van der Woude F, Dekker AJ (1965a) Interpretation of Mössbauer spectra of paramagnetic materials in a magnetic field. Solid State Commun 3:319-321

Van der Woude F, Dekker AJ (1965b) The relation between magnetic properties and the shape of Mossbauer spectra. Physica Status Solidi 9:775-786

van Lierop J, Ryan DH (2001) Mossbauer spectra of single-domain fine particle systems described using a multiple-level relaxation model for superparamagnets. Phys Rev B 63:64406–1-64406-8

Van Ruitenbeek JM, Van Leeuwen DA, De Jongh LJ (1994) Magnetic properties of metal cluster compounds. *In* De Jongh LJ (ed) Physics and Chemistry of Metal Cluster Compounds. Kluwer Academic Publishers, Dordrecht, The Netherlands, p 277-306

Van Vleck JH (1959) The Theory of Electric and Magnetic Susceptibilities. Oxford University Press, London

Vandenberghe RE, Barrero CA, da Costa GM, Van San E, De Grave E (2000) Mossbauer characterization of iron oxides and (oxy)hydroxides: The present state of the art. Hyperfine Interactions 126:247-259

Vasquez-Mansilla M, Zysler RD, Arciprete C, Dimitrijewits MI, Saragovi C, Greneche JM (1999) Magnetic interaction evidence in α-Fe_2O_3 nanoparticles by magnetization and Mössbauer measurements. J Magnet Magnetic Mater 204:29-35

Vaz C, Godinho M, Dormann JL, Noguès M, Ezzir A, Tronc E, Jolivet JP (1997) Superparamagnetic regime in γ-Fe_2O_3 nanoparticle systems; effect of the applied magnetic field. Mater Sci Forum 325-238:813-817

Victora RH (1989) Predicted time dependence of the switching field for magnetic materials. Phys Rev Lett 63:457-460

Vincent E, Yuan Y, Hammann J, Hurdequint H, Guevara F (1996) Glassy dynamics of nanometric magnetic particles. J Magnet Magnetic Mater 161:209-219

Wang J-S, Swendsen RH (1990) Cluster Monte Carlo algorithms. Physica A 167:565-579

Warren LA, Ferris FG (1998) Continuum between sorption and precipitation of Fe(3) on microbial surfaces. Environ Sci Technol 32:2331-2337

Watson JHP, Cressey BA, Roberts AP, Ellwood DC, Charnock JM, Soper AK (2000) Structural and magnetic studies on heavy-metal-adsorbing iron sulphide nanoparticles produced by sulphate-reducing bacteria. J Magnet Magnetic Mater 214:13-30

Weaver JC, Vaughan TE, Astumian RD (2000) Biological sensing of small field differences by magnetically sensitive chemical reactions. Nature 405:707-709

Weiss PR (1948) The application of the Bethe-Peierls method to ferromagnetism. Phys Rev 74:1493-1504

Wernsdorfer W, Orozco EB, Hasselbach K, Benoit A, Barbara B, Demoncy N, Loiseau A, Pascard H, Mailly D (1997a) Experimental evidence of the Néel-Brown model of magnetization reversal. Phys Rev Lett 78:1791-1794

Wernsdorfer W, Orozco EB, Hasselbach K, Benoit A, Mailly D, Kubo O, Nakano H, Barbara B (1997b) Macroscopic quantum tunneling of magnetization of single ferrimagnetic nanoparticles of barium ferrite. Phys Rev Lett 79:4014-4017

Wickman HH, Klein MP, Shirley DA (1966) Paramagnetic hyperfine structure and relaxation effects in Mossbauer spectra: Fe^{57} in ferrichrome. Phys Rev 152:345-357

Wiltschko R, Wiltschko W (1995) Magnetic Orientation in Animals. Springer-Verlag, Berlin

Wiser N (1996) Phenomenological theory of the giant magnetoresistance of superparamagnetic particles. J Magnet Magnetic Mater 159:119-124

Wohlfarth EP (1964) A review of the problem of fine-particle interactions with special reference to magnetic recording. J Appl Phys 35:783-790

Worm H-U (1998) On the superparamagnetic-stable single domain transition for magnetite, and frequency dependence of susceptibility. Geophys J Int'l 133:201-206

Xu J, Hickey BJ, Howson MA, Greig D, Cochrane R, Mahon S, Achilleos C, Wiser N (1997) Giant magnetoresistance in AuFe alloys: Evidence for the progressive unblocking of superparamagnetic particles. Phys Rev B 56:14602-14606

Yeh T-C, Zhang W, Ildstad ST, Ho C (1993) Intracellular labeling of T-cells with superparamagnetic contrast agents. MRM (Magnetic Resonance in Medicine) 30:617-625

Yelon A, Movaghar B (1990) Microscopic explanation of the compensation (Meyer-Neldel) rule. Phys Rev Lett 65:618-620

Zarutskaya T, Shapiro M (2000) Capture of nanoparticles by magnetic filters. J Aerosol Sci. 31:907-921

Zhang P, Molnar P, Downs WR (2001) Increased sedimentation rates and grain sizes 2-4 Myr ago due to the influence of climate change on erosion rates. Nature 410:891-897

Zhao J, Huggins FE, Feng Z, Huffman GP (1996) Surface-induced superparamagnetic relaxation in nanoscale ferrihydrite particles. Phys Rev B 54:3403-3407

Zhu J-G (1995) Micromagnetic modeling: Theory and applications in magnetic thin films. Mater Res Soc Bull, October 1995:49-54

Zysler RD, Fiorani D, Testa AM (2001) Investigation of magnetic properties of interacting Fe_2O_3 nanoparticles. J Magnet Magnetic Mater 224:5-11

8 Atmospheric Nanoparticles

Cort Anastasio

Atmospheric Science Program
Department of Land, Air & Water Resources
University of California-Davis
One Shields Avenue
Davis, California 95616

Scot T. Martin

Division of Engineering and Applied Sciences
29 Oxford Street, Pierce Hall
Harvard University
Cambridge, Massachusetts 02138

INTRODUCTION

Atmospheric nanoparticles are of growing interest to many investigators for two main reasons. First, nanoparticles are important precursors for the formation of larger particles, which are known to strongly influence global climate, atmospheric chemistry, visibility, and the regional and global transport of pollutants and biological nutrients. Second, atmospheric nanoparticles may play critical roles in the deleterious human health effects associated with air pollution. In addition to these two well-recognized roles, nanoparticles may also significantly influence the chemistry of the atmosphere. Because their composition and reactivity can be quite different from larger particles, the presence of nanoparticles may open novel chemical transformation pathways in the atmosphere. There is possibly also an important role for nanostructures within larger particles, which often contain nanoscale features such as mineral grain agglomerates, soot spherules, or layer coatings of sulfates and nitrates. These complex morphological features likely influence a number of properties. For example, nanostructures probably affect water uptake via capillary condensation and nanoscale aqueous surface films may provide a medium for heterogeneous chemistry. In addition, nanoscale active sites on surfaces may influence particle phase transitions through heterogeneous nucleation. However, with a few notable exceptions, the potential roles and implications in atmospheric chemistry for nanoparticles and nanostructures have not been quantitatively examined.

The goal of this chapter is to provide a survey of what is currently known about nanoparticles in the atmosphere, especially in terms of their formation and growth, number concentrations and chemical composition, and chemical, physical, and mechanical properties. Although combustion sources are important sources of nanoparticles, we do not discuss the special conditions (e.g., in terms of particle dynamics) encountered in combustion plumes. This chapter complements two recent journal issues, one focused on nano- and ultrafine particles in the atmosphere (*Philosphical Transactions of the Royal Society of London* A, vol. 358, no. 1775, 2000) and the other containing papers related to nanoparticles in technology and in the atmosphere (*Journal of Aerosol Science*, vol. 29, no. 5-6, 1998). Given the mineralogy context of the volume in which this chapter appears, we also provide a general introduction to the occurrence and physicochemical properties of atmospheric particles.

At the outset of this chapter, one point of distinction is necessary regarding the term "nano" in the context of atmospheric particles. In contemporary scientific usage in many fields such as materials science, chemistry, and physics, nanoscale is understood to mean

1529-6466/00/0044-0008$10.00

at a lengthscale where properties diverge from the bulk. In this chapter, we also adopt this definition on occasion. However, there is a second usage employed by aerosol scientists, who in recent years have pushed the detection limits of their instrumentation to particles as small as 3 nm. In this framework, a nanoparticle is defined solely by size. A common contemporary understanding is that a nanoparticle has a diameter under 50 nm, and we will generally use this definition in this chapter. The exact usage in each case should be clear from its context.

As this review will show, much must still be elucidated in the field of atmospheric nanoparticles and many important unknowns remain. However, a recent surge of research activities in this area should greatly contribute to our knowledge in the coming years.

BACKGROUND CONCEPTS

This section provides a general overview on the physicochemical properties of atmospheric aerosols. An aerosol is defined as a suspension of solid or liquid particles in a gas, though sometimes the term is used in the vernacular to refer solely to the suspended particles or particulate matter. Excellent references on atmospheric aerosols include Finlayson-Pitts and Pitts (2000), Seinfeld and Pandis (1998), Kreidenweis et al. (1999), and Friedlander (2000).

Size distributions of atmospheric particles

Atmospheric particles have spherical equivalent diameters (D_p) ranging from 1 nm to 100 μm. Plots of particle number concentration (as well as surface area and volume) as a function of particle size usually show that an atmospheric aerosol is composed of three or more modes, as illustrated in Figure 1. By convention, particles are classified into three approximate categories according to their size: Aitken (or transient) nuclei mode ($D_p < 0.1$ μm), accumulation mode ($0.1 < D_p < 2.5$ μm), and coarse mode ($D_p > 2.5$ μm) (Seinfeld and Pandis 1998). Particles smaller than 2.5 μm are generally classified as fine. The terms $PM_{2.5}$ and PM_{10} refer to particulate matter with aerodynamic equivalent diameters under 2.5 and 10 μm, respectively. These terms are often used to describe the total mass of particles with diameters smaller than the cutoff size.

In addition to the three modes described above, recent measurements have shown that there is often a distinct particle mode under 10-nm diameter (Fig. 2). There is no current agreement for the name of particles in this mode, which are interchangeably called ultrafine particles, nanoparticles, or nucleation mode particles. There are also alternative definitions for these terms, which can be a source of confusion. For example, the term ultrafine particles is sometimes employed to refer solely to particles with D_p = 3 -10 nm (e.g., in nucleation studies) or to all particles with $D_p < 100$ nm (e.g., in health and emission studies). Similarly, the term nanoparticles is sometimes employed as a description for all particles of $D_p < 50$ nm (regardless of mode), sometimes for particles of 10-nm diameter or less, and occasionally for any particle with $D_p < 1$ μm. In this review we use the common current definitions of ultrafine particles as those with $D_p < 100$ nm and nanoparticles as those with $D_p < 50$ nm.

The different size modes reflect differences in particle sources, transformations, and sinks (Finlayson-Pitts and Pitts 2000). For example, coarse particles are generated by mechanical processes such as wind erosion of soil, wave action in the oceans, and abrasion of plant material. In contrast, many of the fine particles in the atmosphere are produced from either primary emissions from combustion sources or via atmospheric gas-to-particle conversions (i.e., new particle formation). The relative and absolute sizes of particle modes, as well as the number of modes, can vary greatly in different locations and at different times. In addition, the chemical composition of particles within one size

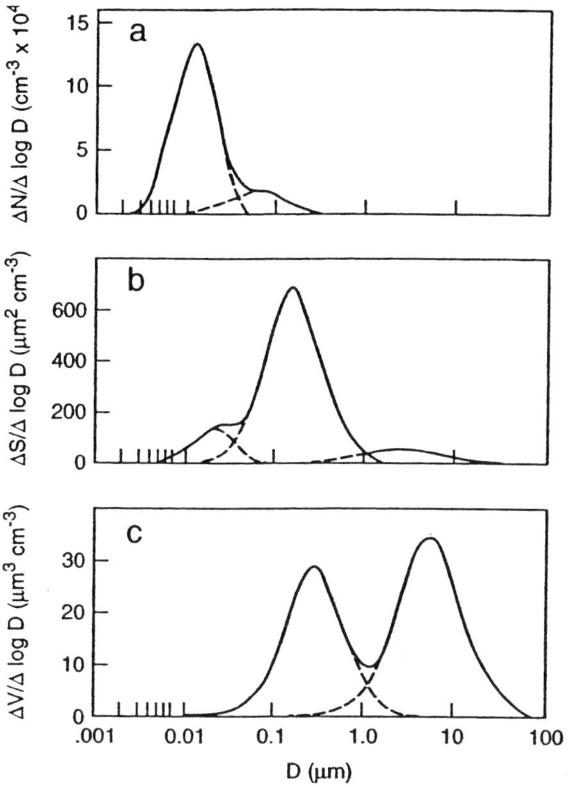

Figure 1. Number (ΔN/Δlog D), surface area (ΔS/Δlog D), and volume (ΔV/Δlog D) distributions for a typical urban aerosol. The solid lines are the size distributions, while the dashed lines show the tails between intersecting modes. The total number concentration, surface area, and volume equal the areas under the curves of each mode. From Finlayson-Pitts and Pitts (2000). Used by permission of Academic Press.

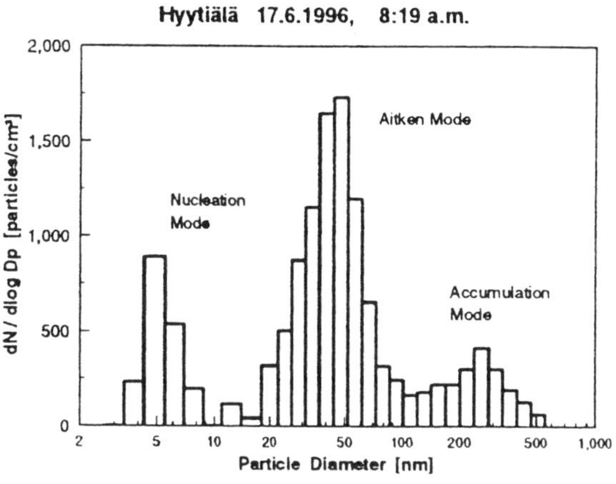

Figure 2. Trimodal structure of the submicron particle number size distribution observed at a boreal forest in Hyytiälä, Finland on June 17, 1996, 08:09-08:19. The total particle number concentration of the submicron aerosol is 1011 particles cm^{-3}. From Mäkelä et al. (1997). Used by permission of the American Geophysical Union.

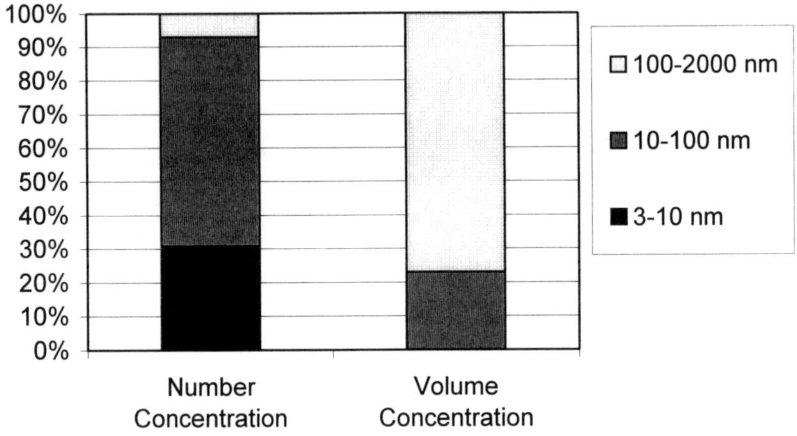

Figure 3. Percent of total ($D_p \leq 2$ μm) particle number and volume concentrations in three size ranges in Atlanta, Georgia. From Woo et al. (2001). Used by permission of Taylor & Francis, Inc.

mode often differs considerably from those in other size modes. Even within one size mode, there is often a wide variety of particle compositions, i.e., the aerosol is externally mixed (Noble and Prather 1996). The opposite situation, where all particles have the same composition, is called an internally mixed aerosol.

As shown in Figure 1, within an atmospheric aerosol the smallest particles usually dominate the total number of particles, while the accumulation and coarse modes often determine the total surface area and volume (i.e., mass), respectively. For example, Figure 3 shows results from a study in Atlanta where nanoparticles (D_p = 3-10 nm) and nano- and ultrafine particles (D_p = 10-100 nm) contributed approximately 30 and 60%, respectively, to the total particle number concentration ($D_p \leq 2$ μm). However, in terms of particle mass, the accumulation mode particles were dominant, and nanoparticles with $D_p < 10$ nm contributed insignificantly.

Sources and sinks of atmospheric particles

Atmospheric particles are classified as primary when they are emitted directly into the atmosphere and as secondary when they form in the atmosphere from reactions of atmospheric gases. The dominant particle sources by mass are mineral dust from wind-blown soils and sea-salt particles from wave breaking (Table 1). These primary sources produce high numbers of coarse particles that dominate the mass distribution of the aerosol, but measurements during the last decade show that these sources can also produce appreciable numbers of nanoparticles. The dominant sources of fine particles include anthropogenic primary emissions from combustion processes and secondary aerosol formation from the oxidation of species such as gaseous SO_2 or organic compounds (Table 1). The estimates in Table 1 show that anthropogenic emissions of particles and their gaseous precursors have more than doubled the flux of fine particles into the atmosphere. This fact suggests that the burden of nanoparticles in the atmosphere has also greatly increased because of anthropogenic emissions.

There are three major sinks that act to remove particles from the atmosphere: diffusion (Brownian motion), wet deposition, and gravitational settling. The relative importance of each mechanism depends primarily upon particle size (Seinfeld and Pandis 1998; Kreidenweis et al. 1999; Friedlander 2000). As shown in Figure 4, diffusion is the

Table 1. Global emission estimates for major aerosol types in the 1980s. From Seinfeld and Pandis (1998). Used by permission of Wiley-Interscience.

Source	*Estimated flux* (Tg yr $^{-1}$) *Low*	*High*	*Best*	*Particle size category*[a]
NATURAL				
Primary				
Soil dust (mineral aerosol)	1000	3000	1500	Mainly coarse[b]
Sea salt	1000	10000	1300	Mianly coarse[b]
Volcanic dust	4	10000	30	Coarse
Biological debris	26	80	50	Coarse
Secondary				
Sulfates from biogenic gases	80	150	130	Fine
Sulfates from volcanic SO_2	5	60	20	Fine
Organics from biogenic gases	40	200	60	Fine
Nitrates from NO_x	15	50	30	Fine and coarse
Total natural	2200	23500	3100	
ANTHROPOGENIC				
Primary				
Industrial dust, etc. (except soot)	40	130	100	Fine and Coarse
Soot	5	20	10	Mainly Coarse
Secondary				
Sulfates from SO_2	170	250	190	Fine
Biomass burning	60	150	90	Fine
Nitrates from NO_x	25	65	50	Mainly Coarse
Organics from anthropogenic gases	5	25	10	Fine
Total anthropogenic	300	650	450	
TOTAL	2500	24000	3600	

[a] Coarse and fine size categories refer to mean D_p above and below 1 μm, respectively.
[b] By mass, coarse particles dominate. By number the fine mode may dominate.
Note that sulfates and nitrates are assumed to occur as ammonium salts.
Flux units: 10^{12} g yr $^{-1}$ (dry mass).

dominant removal mechanism for nanoparticles (D_p < 50 nm) because of their small sizes. Nanoparticles can be removed by diffusing to the Earth's surface (dry deposition), diffusing and colliding with larger particles (intermodal coagulation), or by growing out of the nanoparticle size range (through condensation of gases). Nanoparticle residence times thus depend both upon diameter (which affects the diffusion rate) and upon the atmospheric aerosol surface area. Under typical conditions, nanoparticles are believed to have residence times ranging from minutes to a few days. Coarse particles also have atmospheric residence times that vary from minutes to days, but due to their relatively large sizes, gravitational settling is their dominant removal mechanism. Intermediate between these sizes are the accumulation mode particles. These particles are removed efficiently neither by diffusion nor by settling, and thus they tend to have the longest atmospheric lifetimes (typically days to weeks). The dominant sink for accumulation

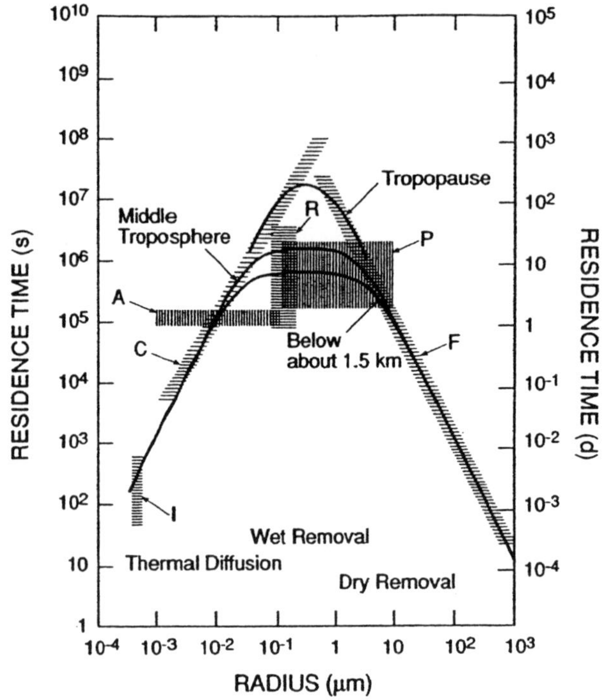

Figure 4. Residence time of particles in seconds (left axis) and days (right axis) as a function of particle radius. The shaded areas represent estimates of the lifetimes made as follows: I, molecular or ionic clusters; C, coagulation of particles; P, removal by precipitation; F, gravitational settling; A, derived from spatial distribution of Aitken particles; R, derived from the distribution of small radioactive particles. From Kreidenweis et al. (1999) in *Atmospheric Chemistry and Global Change* by Brasseur et al. © 1999 by Oxford University Press, Inc. Used by permission.

mode particles is generally removal by precipitation, either as rain-out or wash-out (Finlayson-Pitts and Pitts 2000).

Visibility reduction

One of the most widely recognized aspects of atmospheric pollution is reduced visibility, which is due primarily to light extinction by particles. A common measure of visibility is visual range (V_R), which is defined as the maximum distance at which a black object can be seen against the horizon, as follows:

$$V_R = \frac{\ln(C/C_0)}{b_{ext}} \tag{1}$$

where C/C_0 is the contrast that can be discerned by an observer (typically 0.02-0.05) and b_{ext} is the extinction coefficient (units of length^{-1}) (Pilinis 1989; Finlayson-Pitts and Pitts 2000). Thus regions with greater extinction coefficients have lower visual ranges and visibility. Four components contribute to b_{ext}: scattering of light by particles (b_{sp}), absorption of light by particles (b_{ap}), scattering by gases (b_{sg}), and absorption by gases (b_{ag}).

In areas with reduced visibility, scattering of light by particles is usually the dominant extinction term. Typical values of b_{sp} are 10^{-5} to 10^{-3} m^{-1} (Mathai 1990; Sloane et al. 1991; Eldering et al. 1994; Finlayson-Pitts and Pitts 2000; Pryor and Barthelmie 2000). Light scattering is conveniently classified into three regimes based on the size parameter, α, expressed as $\alpha = 2D_p/\lambda$ where λ is the wavelength of light. Rayleigh scattering occurs for small α, Mie scattering for $\alpha \sim 1$, and geometric scattering occurs for large α (Friedlander 2000; Finlayson-Pitts and Pitts 2000). For visible wavelengths, which are relevant for visibility reductions as well as the direct climate effects discussed

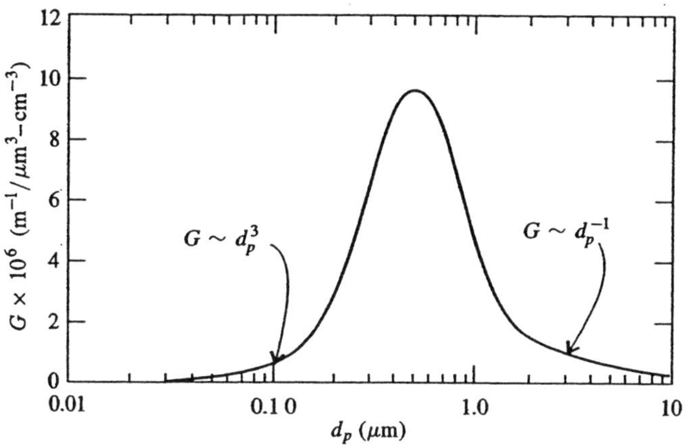

Figure 5. Light scattering per unit volume of aerosol material (G) as a function of particle diameter (d_p), integrated over the wavelengths 360-680 nm for a refractive index of 1.5. The curve is independent of the particle size distribution. From Friedlander (2000) © by Oxford University Press, Inc. Used by permission.

below, Mie scattering is typically the most important mechanism for scattering by atmospheric aerosols. In this regime accumulation mode particles are the most efficient at scattering visible light (Friedlander 2000), as shown in Figure 5. Because of this, and because accumulation mode particles typically dominate fine particle mass, values of b_{ext} for atmospheric aerosols often correlate well with the mass of the fine particle fraction (Finlayson-Pitts and Pitts 2000). Fine particulate components such as sulfate, nitrate, and organic carbon often contribute significantly to b_{sp} (Pilinis 1989; Eldering et al. 1994; Pryor and Barthelmie 2000). In contrast, ultrafine and nanoparticles make small contributions to the scattering of visible light (cf. Fig. 5) (Friedlander 2000).

Absorption of light by particles (b_{ap}) is sometimes an important contribution to the value of b_{ext}. In urban areas, b_{ap} can account for approximately 13-42% of the total extinction coefficient (Finlayson-Pitts and Pitts 2000). The chemical components that contribute most substantially to b_{ap} are soot, organic carbon, and soil dust (Finlayson-Pitts and Pitts 2000). It is unclear whether nano- and ultrafine particles contribute significantly to light absorption by particles. On the one hand, these particle sizes account for an appreciable number fraction of the soot emitted from combustion sources such as diesel engines. On the other hand, unless the particles are highly absorbing, light absorption scales with particle mass, which for atmospheric aerosol contains an insignificant contribution from nanoparticles.

Radiative forcing and climate change

Over the course of the past century, human activities have significantly changed the composition of the atmosphere, which in turn has apparently altered the Earth's climate (IPCC 2001). For example, anthropogenic emissions of long-lived greenhouse gases, such as CO_2, have made the atmosphere less transparent to longwave radiation. Consequently, the atmosphere more efficiently traps longwave radiation emitted by the Earth, which is expected to lead to increased surface temperatures (Crowley 2000; IPCC 2001). However, it is unclear to what extent this increase in temperature might be damped, or possibly enhanced, by climate feedbacks involving clouds, desertification, or changes in ocean circulation. Overall, anthropogenic increases in directly emitted greenhouse gases since pre-industrial times have caused a global-mean radiative forcing

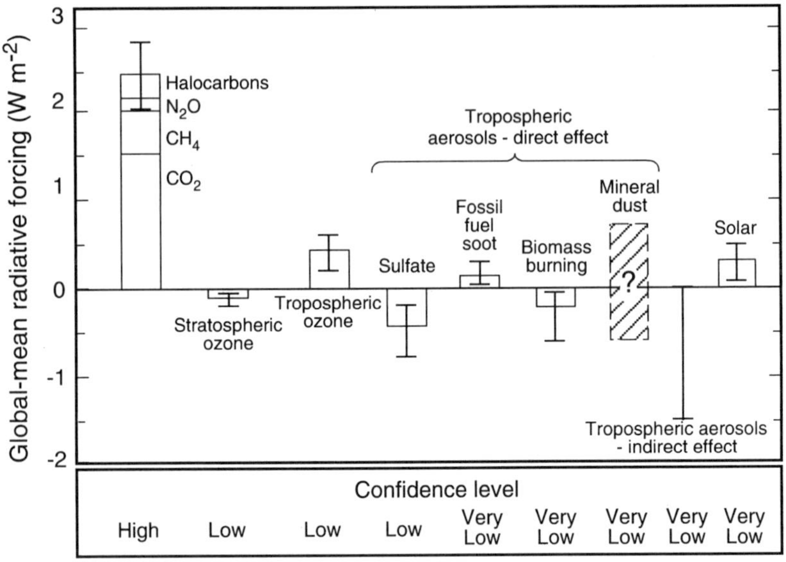

Figure 6. Estimates of globally averaged radiative forcings at the tropopause as a result of changes in greenhouse gases, aerosol particles, and solar activity from pre-industrial times to the present. From Buseck et al. (2000). Used by permission of the editor of *International Geology Review*.

of ca. +2.5 W m^{-2} (Fig. 6). According to most current models, this greenhouse gas forcing is primarily responsible for the globally averaged 0.6°C increase in temperature that has occurred over the past century (Barnett et al. 1999; Crowley 2000; IPCC 2001).

Based on our current understanding, this greenhouse warming would have been larger if not for the effects of atmospheric particles (IPCC 2001). Anthropogenic increases in atmospheric particle loadings have provided a net cooling effect as a result of both direct (i.e., backscatter of incoming solar radiation to space) and indirect effects (i.e., by their effects on clouds) (Boer et al. 2000a,b; Schwartz and Buseck 2000; IPCC 2001). One of the most investigated direct effects is the scattering of incoming solar radiation by sulfate particles. This scattering is believed to cause a negative radiative forcing of approximately -0.5 W m^{-2}, though the exact value is uncertain (Fig. 6) (Penner et al. 1998; Harvey 2000; Kiehl et al. 2000). Anthropogenic sulfate particles are derived primarily from fossil fuel combustion, which releases SO_2 that is subsequently oxidized in the atmosphere to H_2SO_4 (Finlayson-Pitts and Pitts 2000). The combustion of fossil fuels, especially coal, has approximately doubled the mass of sulfate particles in the atmosphere (Table 1).

Other types of particles also contribute significantly to changes in radiative forcing (Fig. 6). Carbonaceous aerosols from biomass burning are thought to cause a net negative forcing due to scattering from organic components. In contrast, soot from fossil fuel combustion is believed to have a net positive forcing because of light absorption (Penner et al. 1998). Recent work by Jacobson (2001) estimates the forcing from soot particles to be +0.55 W m^{-2}. This value is much larger than previously thought (e.g., Fig. 6) and arises from the internal mixing of soot with other particle types such as sulfate. It has been suggested that this atmospheric warming by soot particles could reduce cloud cover (Ackerman et al. 2000). However, modeling by Lohmann and Feichter (2001) indicates that on a global scale this "semi-direct" effect is small compared to the indirect effects of aerosols.

Another important particle type is mineral dust. The atmospheric loading of mineral dust is increasing due to land-use changes accompanying anthropogenic activities. These particles may have significant climatic effects, though the sign and the magnitude of this forcing are not known (Fig. 6) (Buseck and Pósfai 1999; Buseck et al. 2000). This uncertainty arises because mineral dusts both scatter incoming solar radiation (a negative forcing) and absorb solar and longwave radiation (a positive forcing) (Miller and Tegen 1998; Buseck and Pósfai 1999).

In addition to their direct effects, particles also affect climate indirectly by modifying the optical properties, frequencies, and lifetimes of clouds (IPCC 2001). This indirect effect is estimated to have a very large negative forcing, though the magnitude of the effect is highly uncertain (Fig. 6). The indirect effect is a result of anthropogenic increases in the number of cloud condensation nuclei (CCN), which are hygroscopic particles that serve as centers for water condensation and thus the formation of cloud droplets (Finlayson-Pitts and Pitts 2000). For a fixed cloud liquid water content, an increase in the number of CCN yields smaller cloud drops. This causes an increase in the albedo (reflectivity) of the cloud and thereby enhances its efficiency for scattering incoming solar radiation (Twomey 1977; Charlson et al. 1992). In addition, clouds with smaller drops precipitate less frequently and thus have longer lifetimes, which amplifies the climatic impact (Albrecht 1989; Borys et al. 2000; Rosenfeld 2000). Because nanoparticles are important precursors for CCN, there is much current research activity aimed at understanding how nanoparticles form and grow into CCN.

Health effects

Epidemiologic analyses link ambient atmospheric particles with acute and chronic adverse health effects, including respiratory disease, reduced lung function, cardiovascular effects, and mortality (Pope et al. 1995a; Pope 2000; Schlesinger 2000). The associations found in these analyses are particularly strong for susceptible populations such as the elderly, children with asthma, and others with pre-existing respiratory problems (Utell and Frampton 2000). There is no consensus on what types or sizes of atmospheric particles are most responsible for these effects. While a number of epidemiological studies find that fine particle mass (i.e., $PM_{2.5}$) is strongly associated with adverse health effects (Dockery et al. 1993; Pope et al. 1995b; Brunekreef 2000; Pope 2000), other studies indicate coarse particles (e.g., PM_{10}) are more strongly associated (Pope et al. 1995a; Pekkanen et al. 1997; Brunekreef 2000; Loomis 2000). In addition, a few epidemiological studies have shown that impairment in pulmonary function is best correlated to the number of ultrafine ($D_p < 100$ nm) particles (Peters et al. 1997; Hauser et al. 2001).

There are numerous hypotheses—but few definitive results—as to what physicochemical characteristics of atmospheric particles are responsible for adverse health effects (Samet 2000; Schlesinger 2000). Hypotheses include general properties such as mass, surface area, or size, as well as more specific chemical properties such as acidity or elevated concentrations of transition metals (Dreher et al. 1997; Samet 2000). For example, it has been suggested that particulate iron is toxic due to its ability to generate the strongly oxidizing hydroxyl radical through the Fenton reaction (Ghio et al. 1996; Smith and Aust 1997; Donaldson et al. 1998; van Maanen et al. 1999):

$$\mathrm{Fe(II)} + \mathrm{HOOH} \longrightarrow \mathrm{Fe(III)} + {}^{\bullet}\mathrm{OH} + {}^{-}\mathrm{OH} \quad (2)$$

In biological systems the hydroxyl radical can lead to damage such as lipid peroxidation and DNA breakage (van Maanen et al. 1999). In support of this hypothesis, iron is often the most abundant transition metal in atmospheric particles, and there are numerous reports that particulate Fe(III) can be reduced to Fe(II) as a result of atmospheric reactions (Faust 1994).

Animal studies provide additional evidence that ultrafine or nanoparticles might be at least partially responsible for the health effects associated with atmospheric particles. For example, for the same mass of particles instilled into rat lungs, TiO_2 nanoparticles (D_p = 20 nm) yield a greater inflammatory response compared to that from fine TiO_2 particles (250 nm) (Oberdörster 2001). However, when the doses are normalized to particle surface area, the inflammatory response is the same, suggesting that surface reactions might have been responsible for the observed effects. These results also indicate that there is no special reactivity for the TiO_2 nanoparticles compared to the larger particles. In contrast, Donaldson et al. (2000) suggest that, in terms of adverse health effects, nanoparticle surfaces might be more active than the surfaces of larger particles. A number of other studies show that inhaled or instilled nanoparticles give a greater inflammatory response and cause more oxidative stress than fine particles, though the confounding role of surface area is not evaluated (Donaldson et al. 1998, 2000, 2001). Not all studies, however, show that nanoparticles cause measurable responses. For example, in one study rats and mice exposed to nanoparticles of black carbon ($D_p \sim 40$ nm; 10^5-10^6 cm^{-3}) show no signs of lung injury, while those animals exposed to greater masses of fine particles ($D_p \sim 400$ nm; 10^4-10^5 cm^{-3}) show some adverse effects (Arts et al. 2000).

In addition to the greater surface area discussed above, a number of other hypotheses have been presented as potential explanations for the possibly greater toxicity associated with nano- and ultrafine particles. First, these particles could have increased toxicity because of their high number concentrations in the atmosphere and significant deposition in the lungs. Hughes et al. (1998) have estimated that ca. 10^{11} ultrafine particles are deposited daily in the lungs of a typical person in the Los Angeles area during winter. Ultrafines are efficiently deposited in human lungs (e.g., 40-50% deposition fraction for inhaled 40-nm particles (Jaques and Kim 2000)), and a large fraction of this deposition occurs in the alveoli (i.e., the deepest reaches of the lungs) (Oberdörster 2001). By contrast, accumulation mode particles typically have lower number concentrations, a lower overall lung deposition rate, and a smaller fraction deposited in the alveoli (Oberdörster 2001). Second, there is some evidence that nanoparticles inhibit or overwhelm phagocytosis, perhaps because of their great numbers, and thus decrease the ability of alveolar macrophages to clear out foreign particles (Donaldson et al. 2001). This causes an increased contact time between particles and lung epithelial cells (Donaldson et al. 1998). In conjunction with the small size of nanoparticles, this might allow these particles to cross the alveolar epithelial barrier and enter interstitial spaces (Donaldson et al. 2001). Subsequent transport to other organs could then be possible (Oberdörster 2001). Finally, at least one study indicates that nanoparticles act synergistically or additively with ozone to cause adverse health effects, especially in older rats and mice pre-treated with endotoxin to simulate the compromised lungs of susceptible groups (Elder et al. 2000). This result suggests that health effects from particles are strongly tied with co-exposure to other pollutants (Samet 2000; Oberdörster 2001).

The evidence of adverse health effects from respirable particles, and perhaps nanoparticles in particular, raises the question of what regulatory strategy would be most effective to protect human health. In response to the epidemiological evidence linking fine particles with adverse health effects, the EPA in 1997 proposed a new mass-based $PM_{2.5}$ standard for ambient air quality (Seinfeld and Pandis 1998). However, in light of the limited evidence that nanoparticles in particular, and perhaps particle number in general, are more important components of particle toxicity, there is concern that the $PM_{2.5}$ standard is not the most effective or economic regulatory approach to protect human health. This is because most of the $PM_{2.5}$ mass results from the larger accumulation mode particles, but most of the aerosol particle number generally results from the nano- and ultrafine particles. The result is that particle number and mass are expected to generally be unrelated (Fig. 7) (Keywood et al. 1999; Woo et al. 2001). Thus control of fine particle mass is likely to have little effect upon nanoparticle number concentrations in the atmosphere. Similarly, a mass-based standard is unlikely to be effective or economic if specific chemical species within atmospheric particles are responsible for the observed adverse health effects.

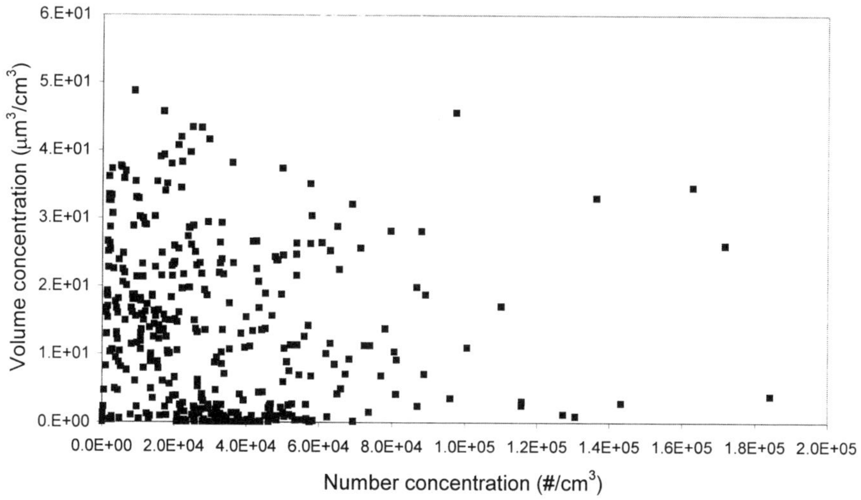

Figure 7. Relationship between volume and number concentrations for particles with D_p = 3 nm to 2 μm in Atlanta, Georgia, during August 1998 through August 1999. From Woo et al. (2001). Used by permission of Taylor & Francis, Inc.

Chemical reactions of atmospheric particles

Reactions involving atmospheric particles have significant effects upon the composition of the atmosphere (Andreae and Crutzen 1997). Perhaps the most well known example is the particle-catalyzed springtime destruction of stratospheric ozone over the Antarctic. Particle reactions in the troposphere are also significant and affect concentrations of important gaseous pollutants such as ozone, nitric acid, and sulfur dioxide. These reactions often alter the composition of the particles as well, which can alter the ability of particles to act as cloud condensation nuclei, change their efficiency for scattering and absorbing light, and perhaps influence their health effects.

Three examples of chemistry occurring on or in atmospheric particles are provided below. These examples demonstrate how particulate-based chemical reactions can alter the composition both of particles and of the gas phase. However, the examples are by no means comprehensive of all chemical processes that occur in the troposphere. The particles involved in these examples are generally accumulation or coarse mode, though in some cases similar reactions on nanoparticles of the same material might also occur. The chemistry of nanoparticles is discussed separately later in this chapter.

Mineral dusts. The northern African and Gobi deserts are large sources of mineral dusts (Pye 1987; Charlson and Heintzenberg 1994; Prospero 1999a), which are regularly carried by atmospheric circulation patterns to the eastern USA and Brazil as well as the Pacific Ocean (Swap et al. 1992; Leinen et al 1994; Prospero 1999a). The dusts are monitored both by analysis of collected aerosol particles (Prospero 1999b) and by satellite observations (Husar et al. 1997; Herman et al. 1997; Chiapello et al. 1999) (Fig. 8). In addition to these global sources, there are also regional emissions of mineral dusts. In total, nearly one-third of the Earth's land surface is arid and hence a possible dust source (Tegen and Fung 1994), and land use changes associated with human activities are expected to increase the global burden of mineral dusts (Tegen and Fung 1995; Tegen et al. 1996). In addition to the chemistry described below, mineral dusts are also important as an essential component of open ocean fertility, primarily as a source of iron (Duce et al. 1980; Martin and Gordon 1998; Boyd et al. 2000).

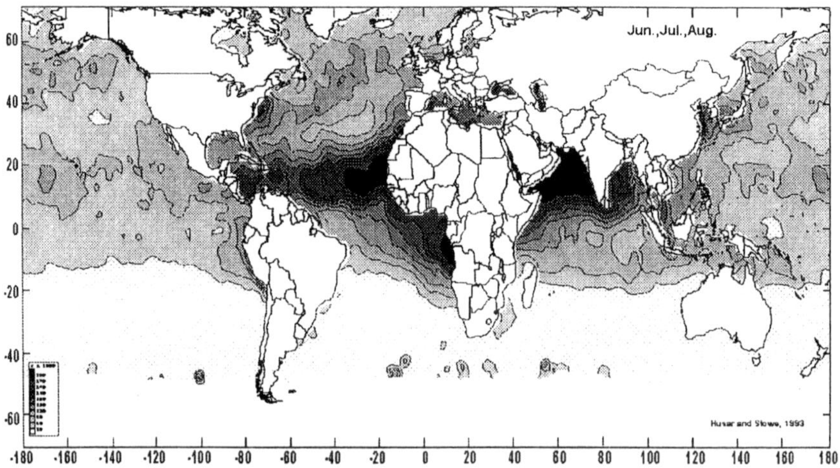

Figure 8. Radiatively equivalent aerosol optical thickness (EAOT × 1000) over the oceans derived from NOAA AVHRR satellites for summer. Darker shades indicate greater values of EAOT. Adapted from Husar et al. (1997). Used by permission of the American Geophysical Union.

The mode of the number size distributions of mineral dusts outside of source regions (i.e., after long-range transport) is approximately 100 nm, though the mass mode typically lies between diameters of 2 to 5 μm (D'Almedia and Schütz 1983). The small particles usually arise from submicron clay particles that are initially attached to larger mineral grains but break free during the abrasive events of a dust storm. Analysis of mineral dust indicates the particles are largely composed of silicates (clay minerals, feldspars, and quartz) and occasionally carbonates and sulfates (Schütz and Sebert 1987; Schütz 1989, 1997; Merrill et al. 1994; Claquin et al. 1999). Dominant clays are illite and kaolinite with contributions by smectite (montmorillonite) and chlorite. Iron and aluminum oxides and hydroxides also contribute, especially as surface coatings, which then dominate the surface chemistry of agglomerated mineral particles (Dixon and Weed 1989). On average, dust composition reflects that of the Earth's crust, though northern Saharan dust contains an unusually high proportion of calcite and gypsum. Submicron dust arises from heavily weathered soil components.

During their residence time in the atmosphere, mineral dusts become coated by sulfates, nitrates, and other species (Dentener et al. 1996; Buseck and Pósfai 1999; Zhang and Carmichael 1999; Song and Carmichael 1999; Buseck et al. 2000). These coatings are formed through chemical reactions such as the oxidation of SO_2 and NO_2 at the gas-solid interface, as well as by condensation of sulfuric and nitric acids. Once coated, the hygroscopic dusts act as cloud condensation nuclei and further oxidation reactions can take place in the aqueous medium (Wurzler et al. 2000). Subsequent evaporation of the cloud droplet yields a coated particle.

Recent work has investigated whether surfaces of mineral dusts may provide stoichiometric or catalytic reaction centers that are significant enough to perturb important gas-phase cycles (Dentener et al. 1996; Zhang and Carmichael 1999; Song and Carmichael 1999; de Reus et al. 2000a). For example, the kinetics of the oxidation of SO_2 and NO_2 in the atmosphere may be enhanced by mineral surface reactions. Compared to the gas-phase oxidation of SO_2 and subsequent formation of new submicron sulfate particles, the scavenging of SO_2 by mineral surfaces reduces the climate cooling effect of sulfate aerosol by reducing particle number. In addition, interactions of O_3, HNO_3, and $HO_2^{\bullet}$-radicals with mineral surfaces may perturb the photooxidant cycle downwind of dust sources and lead to reduced levels of tropospheric ozone. A number of studies have examined the uptake of HNO_3 and O_3 on mineral dusts or surrogate surfaces (Fenter et al. 1995; Dentener et al. 1996; Goodman et al. 2000; Hanisch and Crowley 2001; Underwood et al 2001). In some cases values range widely, as discussed by Underwood et al. (2000) and Hanisch and Crowley (2001). The uptake coefficients for NO_2 reaction on alumina and other crustal constituents are regarded as too low to have an impact on gas-phase NO_2, HNO_3, or O_3 (Borensen et al. 2000; Underwood et al. 2001), though coating of mineral surfaces may still be significant. The uptake coefficients for low molecular weight organic molecules such as acetaldehyde, acetone, and propionaldehyde have been measured on SiO_2, Al_2O_3, Fe_2O_3, TiO_2, and CaO. Model results with these uptake values suggest that heterogeneous loss of these organics to mineral dusts is comparable to their loss by direct photolysis or through reaction with hydroxyl radical ($^{\bullet}OH$) in the middle to upper troposphere (Li et al. 2001). In contrast to nitrate, a critical unknown is the rate of SO_2 uptake (and oxidation) on mineral surfaces. More laboratory work is necessary to reduce these uncertainties; current modeling studies must employ assumed values.

Polar stratospheric clouds. During winter over the polar regions of the Earth, temperatures are cold enough that polar stratospheric cloud (PSC) particles composed of sulfuric and nitric acids and water are present (Peter 1997, 1999; Zondlo et al. 2000). The PSC particles form on background H_2SO_4 particles, which have diameters of approximately 50 nm, and grow by condensation of gaseous H_2O and HNO_3 to form particles with diameters of up to several micrometers. Particles as large as 25 μm have recently been reported (Fahey et al. 2001). The chemical constituents of PSC particles may be arranged in several possible phases, including aqueous, ice, and hydrates of sulfuric and nitric acid (Kolb et al. 1995; Martin 2000). Common hydrates include $H_2SO_4{\cdot}4H_2O$ (sulfuric acid tetrahydrate, SAT), $HNO_3{\cdot}2H_2O$ (nitric acid dihydrate, NAD), and $HNO_3{\cdot}3H_2O$ (nitric acid trihydrate, NAT).

A critical step in the annual depletion of polar ozone is the activation of chlorine on the surface of PSC particles, i.e., the conversion of non-ozone-destroying chlorine forms such as $ClONO_2$ into reactive, ozone-destroying chlorine compounds (Abbatt and Molina 1993; Anderson 1995). Critical heterogeneous reactions, for example, are:

$$ClONO_2 + H_2O \xrightarrow{PSC} HOCl + HNO_3 \quad (3)$$

$$ClONO_2 + HCl \xrightarrow{PSC} Cl_2 + HNO_3 \quad (4)$$

HOCl and Cl_2 both rapidly photolyze during polar spring to yield Cl radicals which catalytically destroy O_3. In addition, the HNO_3-containing PSC particles sediment out of the stratosphere, thus removing HNO_3 and serving as a sink for NO_x (i.e., NO and NO_2). This denitrification contributes to sustaining Cl as an active radical because it slows several important chlorine deactivation pathways such as:

$$ClO + NO_2 \xrightarrow{PSC} ClONO_2 \quad (5)$$

One critical research question relates to the kinetics of PSC formation (e.g., nucleation rates of ice and the various acid hydrates) from supercooled aqueous droplets of sulfuric and nitric acids (Tolbert and Toon 2001). Nucleation rates measured in the laboratory appear too slow in many cases to explain the occurrence of PSC particles observed in the polar stratosphere. The obvious hypothesis, that the particles contain heterogeneous nuclei, has not found support because of the low occurrence of nuclei in the stratosphere and the low activity of candidate materials tested in the laboratory. An active area of research remains investigating mechanisms that could explain the formation of PSCs.

Sea-salt particles. Particles formed from sea spray initially have a composition that is similar to bulk seawater but with enhanced levels of organic compounds. The inorganic species in sea-salt particles are chiefly Na^+ and Cl^- with contributions from SO_4^{2-}, Ca^{2+}, Mg^{2+}, and K^+ (Holland 1978). The organics include surfactants and other compounds that are enriched in the surface layer of the ocean. It has been suggested that these compounds form organic layers on the surfaces of sea-salt particles that might impede the transport of gases to and from the largely inorganic core (Ellison et al. 1999). Although most sea-salt particle mass occurs in the coarse mode, there are also significant numbers of sea-salt particles at sizes below 100 nm (Gong et al 1997). For example, Berg et al. (1998) report the occurrence of sea-salt particles in the smallest size mode (35 nm) of their measurements.

Atmospheric processing alters the chemical composition of sea-salt particles, for example by increasing the amounts of SO_4^{2-} and NO_3^- and depleting Cl^- (Duce, 1969; Finlayson-Pitts et al. 1989; Chameides and Stelson 1992; McInnes et al. 1994). One pathway in this processing is:

$$NaCl(s,aq) + HNO_3(g) \longrightarrow NaNO_3(s,aq) + HCl(g) \quad (6)$$

An analogous reaction occurs with sulfuric acid. H_2SO_4 also forms within deliquesced sea-salt particles by the oxidation of aqueous SO_2 species by ozone (Sievering et al. 1992). Figure 9 provides an example of a 100-nm atmospheric sea-salt particle enriched in sulfate due to atmospheric processing. In addition to this chemistry, sea-salt particles are also important sources of reactive halogen species (e.g., Br_2 or BrCl) in the marine boundary layer (Pszenny et al. 1993; De Haan et al. 1999; Foster et al. 2001). These reactive halogens evaporate from sea-salt particles and undergo direct photolysis to yield gas-phase bromine and chlorine radicals, which in turn affect the chemistry of tropospheric ozone, dimethyl sulfide, and hydrocarbons (Vogt et al. 1996; Keene et al. 1998).

Many characteristics of sea-salt particles, such as chemical reactivity and light scattering properties, depend on the particle phase. At sufficiently low relative humidity

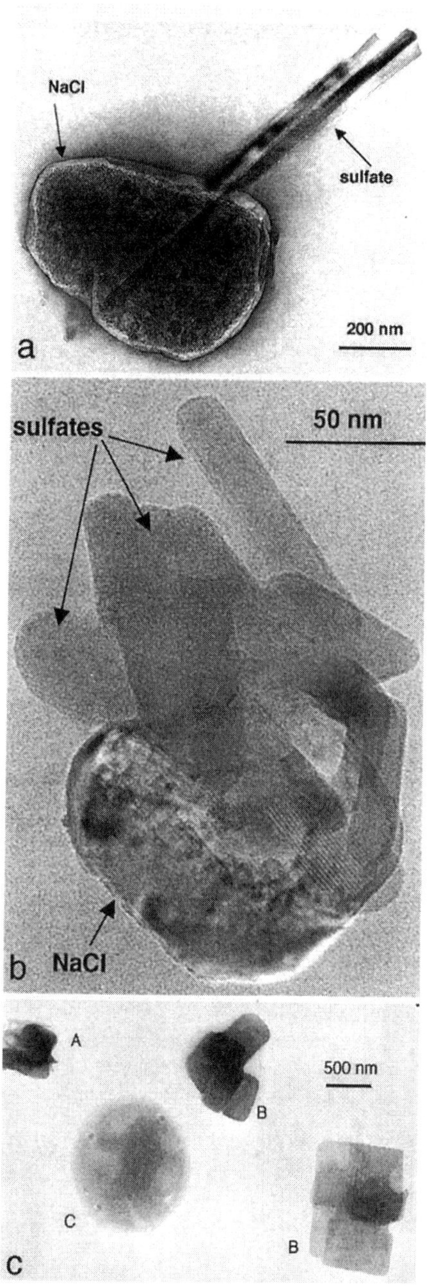

Figure 9. TEM images of sea salt. (a,b) Subhedral halite (NaCl) and euhedral sulfate crystals. The particle in (b) belongs to the smallest sea-salt particles that occur in the ACE-1 samples (Southern Ocean, Cape Grim, ACE-1). (c) Halite particles in various stages of conversion to sulfate and nitrate. Grain A is partly converted, whereas C has been completely converted to nitrate and grains B to sulfates. (Azores, North Atlantic, ASTEX/MAGE.) Images by Mihaly Pósfai. Figure adapted from Buseck and Pósfai (1999). Used by permission of the editor of *Proceedings of the National Academy of Sciences, USA*.

(RH), crystalline phases precipitate from the aqueous droplets (Martin 2000). Solid formation during the evaporation of relatively large volumes of seawater begins with $CaSO_4 \cdot 2H_2O$, which is followed by NaCl and finally $Na_2Ca(SO_4)_2$ at lower RH (McCaffrey et al. 1987; Marion and Farren 1999). However, at the smaller volumes characteristic of atmospheric particles, significant supersaturation is possible and even likely according to laboratory studies (Tang et al. 1997). A critical uncertainty in the laboratory work to date, especially for aerosols, is that ambiguity remains as to which phases nucleate rapidly and thus crystallize in atmospheric sea-salt particles and in what relative order they do so. It is not known, for example, if $CaSO_4 \cdot 2H_2O$ is the first salt to crystallize. It is important to emphasize that laboratory studies on NaCl surfaces indicate that several monolayers of water coat the surface even at low RH (Barraclough and Hall 1974; Peters and Ewing 1997a; Hemminger 1999). Thus chemical transformations of gas-phase species at the surfaces of crystalline sea-salt particles usually have reactivity and pathways similar to an aqueous environment (Finlayson-Pitts and Hemminger 2000). However, the optical properties of these particles are described well by consideration of the crystalline core, which is much more massive than the aqueous surface coating.

NUCLEATION

The formation of nanoparticles from gaseous precursors is the dominant mechanism for new particle formation in the atmosphere. Secondary mechanical effects, such as shattering of ice or rain particles or abrasion of mineral dust, also lead to increases in particle number density but at sizes larger than the nanoparticle range. As illustrated in Figure 10, the first step in nanoparticle formation is generally the oxidation of precursor gases such as SO_2 or

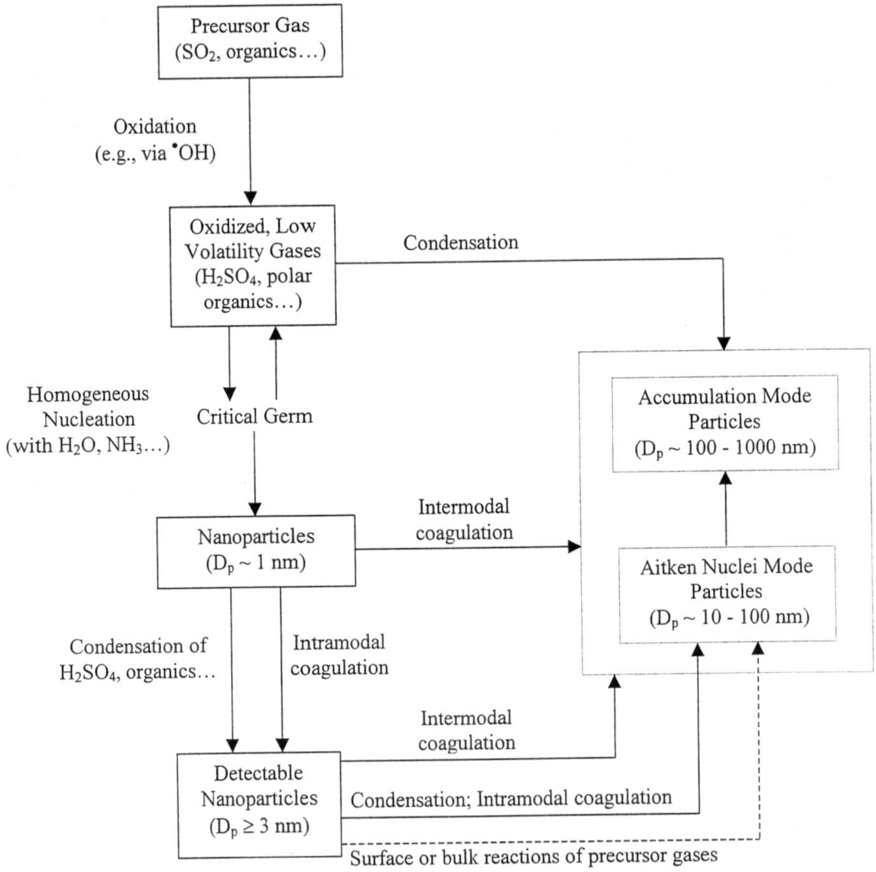

Figure 10. Schematic diagram of nanoparticle formation and growth.

organic compounds to form low-volatility products such as H_2SO_4 or polar organics. At sufficient supersaturations, these products condense to form a critical germ (also called a critical nucleus), which is a polymeric unit just sufficient in size to begin growth into a larger particle. Growth of the critical germ by condensation of low-volatility gases yields a nanoparticle (diameter of ~1 nm). These fresh nanoparticles can eventually join the accumulation mode via further growth, e.g., through continued condensation of vapors.

Nucleation is the science investigating the kinetics and thermodynamics of the formation of a new phase of a material at a size just sufficient to be stable. In addition to their role in new particle formation, nucleation processes are also critical to an accurate understanding of a number of other atmospheric events, including cloud droplet activation on CCN, ice formation, and the deliquescence/efflorescence of particles. In this section we focus on the nucleation of new particles through homogeneous nucleation, i.e., from gaseous precursors. The theoretical treatment of new particle nucleation, as well as field and laboratory measurements of nanoparticle formation, are addressed.

Theoretical treatment of critical germ formation

The process of new particle formation is a first-order transition from a disordered to an ordered atomic arrangement (Doremus, 1985; Oxtoby 1992; Laaksonen et al. 1995; Martin 2000). Gas-to-liquid and liquid-to-crystalline conversions both proceed via first-order transitions, which require activated nucleation. Although the overall transformation is favored by a negative change in the Gibbs free energy, each microscopic step along the route is not. Thermal fluctuations in the medium must assist in overcoming the positive free energy change in a microscopic region. As an example, liquid water has a driving force to freeze below 273 K, but in fact submicron droplets readily supercool to 233 K. Liquid water monomers are dynamically associating in larger polymeric networks resembling ice. Most of these networks dissipate, however, without forming ice because the surface free energy of small networks exceeds the volume free energy. At this microscopic step in the freezing process (i.e., the *n*-mer or cluster level), the Gibbs free energy change is positive and unfavorable. However, the liquid is incessantly exploring possible configurations of its spatial and energy coordinates through thermal fluctuations, and at some time point an *n*-mer forms that is large enough such that the negative volume free energy just offsets the positive surface free energy. This large *n*-mer, called a critical germ, then grows freely into a large crystal. These ideas have been developed quantitatively by classical nucleation theory and more recently by density functional theory and cluster kinetic theory (Laaksonen 1995; Kusaka et al. 1998a,b, 1999; Bowles et al. 2000).

Classical homogeneous nucleation theory is widely employed to describe the formation of new particles in the atmosphere (Pruppacher and Klett 1997; Seinfeld and Pandis 1998). The free energy of germ formation, ΔG_{germ}, in classical homogeneous nucleation theory is described as follows:

$$\Delta G_{germ} = \frac{16\pi v^2 \sigma^3_{germ}}{3(kT \ln S)^2} \tag{7}$$

where v is the molecular volume, σ_{germ} is the surface tension of the germ in the medium, k is the Boltzmann constant, T is temperature, and S is the saturation ratio of a metastable phase with respect to a stable phase. The third-order dependence on σ_{germ} makes this term a critical factor in calculations. The volume nucleation rate, J, is as follows:

$$J = n\frac{kT}{h}\exp\left(-\frac{\Delta G_{germ}}{kT}\right) \tag{8}$$

where n is the molecular concentration in the liquid phase and h is the Planck constant. The probability, P, of a chemical system forming a critical nucleus after time t is:

$$P(t) = 1 - \exp(-JVt) \tag{9}$$

where V is the system volume.

Nucleation of an ordered phase is often favored when there is an available surface of foreign matter. The free energy of germ formation is then reduced by a favorable interaction of the critical germ with the foreign surface. Classical heterogeneous nucleation theory provides the surface nucleation rate as follows:

$$j = n_s\frac{kT}{h}\exp\left(-f(m)\frac{\Delta G_{germ}}{kT}\right) \tag{10}$$

where n_s is the number of monomer units per unit surface area. An important assumption inherent in this theory is an undifferentiated surface (i.e., a large terrace structure). $f(m)$ describes the favorability of the interaction between the germ and the foreign surface, as follows:

$$f(m) = \frac{(2+m)(1-m)^2}{4} \tag{11}$$

The term m follows from $m = \cos\theta$ where θ is the contact angle, as given by the surface tension relationships of Young's equation. Because $-1 \leq m \leq 1$, it follows that $1 \geq f \geq 0$. The probability of a chemical system forming a critical nucleus is:

$$P(t) = 1 - \exp(-jAt) \tag{12}$$

where A is the area of the foreign surface.

Chemical composition of critical germ

The chemical composition of a critical germ often differs significantly from both the constitution of the supersaturated mother liquor and the equilibrium composition of the stable phase. For example, for $J = 1$ cm^{-3} s^{-1} at 298 K, 2.2×10^9 molecules H_2SO_4 cm^{-3} and 4.2×10^{17} molecules H_2O cm^{-3} (50% RH) in the gas phase are calculated to yield a critical germ containing 39 molecules and a H_2SO_4 mole fraction of 0.205, in contrast to the gas-phase H_2SO_4 mole fraction of 5.0×10^{-9} (Jaecker-Voirol and Mirabel 1988). The chemical composition of the germ depends on the details of the energy surface, as specified by $\Delta G(n_1, n_2, \ldots, n_i)$ for an i-component system (e.g., $i = 2$ for H_2SO_4/H_2O) where n_i denotes the number of molecules in a candidate germ. Experimental studies on cluster formation are useful in elucidating the shape of the energy surface (Jaecker-Voirol et al. 1987; Mirabel and Ponche 1991; MacTaylor and Castleman 2000). Movement across the surface from monomers to a macroscopic particle requires surmounting an activation barrier denoted by $\Delta G_{germ}\left(n_1^*, n_2^*, \ldots, n_i^*\right)$, where n_i^* is the number of molecules of component i in the critical germ. The height of the barrier is specified by the condition:

$$\left(\frac{\Delta G_{germ}}{n_1^*}\right)_{n_i;i\neq 1} = \left(\frac{\Delta G_{germ}}{n_2^*}\right)_{n_i;i\neq 2} = \ldots = \left(\frac{\Delta G_{germ}}{n_i^*}\right)_{n_i;i\neq i} = 0 \tag{13}$$

Experimental data of the dependence of J on concentration can yield an estimate of n_i^*, especially for the conversion of vapors to condensed phases. The essential relation is provided in the nucleation theorem, as follows (Oxtoby and Kashchiev 1994):

$$\frac{\partial \Delta G_{germ}}{\partial \mu_{o,i}} = -\Delta n_i^* \tag{14}$$

This equation states that the change in the free energy of the critical germ with the chemical potential $\mu_{o,i}$ per molecule of species i in the *original* phase (i.e., the mother liquor) equals the negative of the excess number Δn_i^* of molecules of type i in the nucleus over that present in the same volume of original space. The nucleation theorem is independent of the model and of the transition: it holds true for classical nucleation theory, density functional theory, or cluster kinetic analysis and for gas-to-liquid or liquid-to-solid conversions.

In the particular case of gas-to-condensed-phase transitions, Equations (8) and (14) can be developed together to yield the following accurate approximation for isothermal binary nucleation (e.g., for H_2SO_4/H_2O):

$$\Delta n_i^* = \frac{\partial (kT \ln J)}{\partial \mu_{o,i}} - \delta \tag{15}$$

where δ is a small correction factor (typically between 0 and 1) related to the dependence of the preexponential factor on chemical potentials (i.e., usually small). The chemical potential relates to chemical composition by $\mu_{o,i} = \mu^0_{o,i} + kT \ln f_i x_{o,i}$ where f_i is the activity coefficient and $x_{o,i}$ is the mole fraction composition. When the activity coefficient is unity or otherwise ignored, Equation (15) yields:

$$\partial(\ln J) = \left(\Delta n_i^* + \delta\right)\partial\left(\ln x_{o,i}\right) \tag{16}$$

Under the assumption that Δn_i^* is independent of $x_{o,i}$, the slope of a log-log plot of the experimental dependence of J on the composition of component i in the mother liquor yields the composition of the germ component i. It should be emphasized that $\Delta n_i^* = f(x_{o,1}, x_{o,2}, \ldots, x_{o,i})$, which is factorable as $\Delta n_i^* = N_i\, f(x_{o,1}, x_{o,2}, \ldots, x_{o,i-1})$ under the assumption of independence in $x_{o,i}$. The point is that Δn_i^* has a specific value N_i when $x_{o,1}, x_{o,2}, \ldots, x_{o,i-1}$ are held constant. When these values vary, N_i also varies.

Employing Equation (16), Ball et al. (1999) analyze their data of particle formation rates from vapors of H_2SO_4/H_2O to determine that there are 13 H_2SO_4 monomers in a critical germ at 2 to 12×10^9 molecules H_2SO_4 cm^{-3} and 2.3% RH at 295 K. As RH increases to 15.3%, the number of H_2SO_4 monomers drops to 7. While the H_2SO_4 monomers drop from 13 to 7, the H_2O monomers in the critical germ increase from 4 to 6, and the acid mole fraction in the critical germ shifts from 0.76 to 0.53. In breakthrough work, Eisele and Hanson (2000) succeed in directly measuring H_2SO_4/H_2O clusters containing 3 to 8 H_2SO_4 molecules by mass spectrometric techniques. The measurements are under conditions where new particle formation is not rapid, i.e., the dynamic cluster distribution is being observed. Kulmala et al. (1998a) provide a parameterization, which is convenient for inclusion in computer codes, of the mole fraction composition of the critical germ and the nucleation rate of new particles from H_2SO_4/H_2O gases.

In the Ball et al. (1999) work mentioned above, addition of NH_3 at levels of tens of parts per trillion by volume (pptv) increases the particle nucleation rate and at 15% RH reduces the number of H_2SO_4 monomers in the critical germ from 8 to 5. It would be desirable in future studies to extend this experimental work to higher RH values because the strongly nonlinear dependencies make extrapolations very uncertain, although modeling work by Korhonen et al. (1999) suggests the nucleation rate from $H_2SO_4/H_2O/NH_3$ could be approximately independent of RH under atmospheric conditions. According to this model description (Korhonen et al. 1999), a critical germ contains 8 H_2SO_4, 4 NH_3, and 6 H_2O molecules at 298 K, 52.3% RH, 5 pptv NH_3, and 10^8 molecules H_2SO_4 cm^{-3}. In this model the presence of NH_3 increases the nucleation rate by several orders of magnitude.

Equations (15) and (16) develop Equation (14) for gas-to-condensed-phase transitions. It would also be desirable to develop Equation (14) for liquid-to-crystalline nucleation to assign physical meaning to Δn_i^*. However, development of Equation (14) is difficult both because δ can be a large correction factor for liquid/solid interfaces and because the molecular density differences between liquids and solids are smaller than between gases and condensed phases (Kashchiev 1982).

New particle formation in the atmosphere

Understanding new particle formation has been a focus of field, laboratory, and modeling efforts in recent years. This research has been driven by observations that rates of new particle production in field studies often exceed modeled rates that are based upon laboratory studies of nucleation (Covert et al. 1992; Weber et al. 1996, 1997, 1998a, 1999; Clarke et al. 1998; Kulmala et al. 1998b).

Nanoparticle formation in the atmosphere is initiated by the process of homogenous nucleation from gas-phase precursors (Eqn. 8). As shown in Figure 10, the first step in this process is the conversion of gases to lower volatility products, via reactions with species such as hydroxyl radical ($^{\bullet}OH$). When the resulting partial pressures of low volatility vapors exceed saturation, vapor condensation to form new particles with diameters of approximately 1 nm is thermodynamically favored (Seinfeld and Pandis 1998; Friedlander 2000). However, new particle formation is in competition with scavenging of the low volatility gas by condensation to pre-existing particles (Kulmala et al. 1995; Weber et al. 1997; de Reus et al. 1998; Pirjola and Kulmala 1998; Pirjola et al. 1999; Kerminen et al. 2000; Clement et al. 2001). Thus new particle formation is not constant in time but instead occurs in bursts under conditions where there is little pre-existing aerosol but sufficient concentrations of (generally) photochemically produced low volatility vapors, as illustrated in Figure 11.

Sulfuric acid. New particle formation involving H_2SO_4 has been investigated most intensively because there is strong evidence that sulfuric acid can be an important particle precursor. Gaseous sulfuric acid is formed via the oxidation of sulfur dioxide (SO_2) by hydroxyl radical in the following multistep reaction:

$$SO_2 + {}^{\bullet}OH + O_2 + H_2O \longrightarrow H_2SO_4 + HO_2{}^{\bullet} \tag{17}$$

where the rate-limiting elementary step is the attack of $^{\bullet}OH$ on SO_2 (Finlayson-Pitts and Pitts 2000). Since hydroxyl radical is primarily formed via reactions that require sunlight,

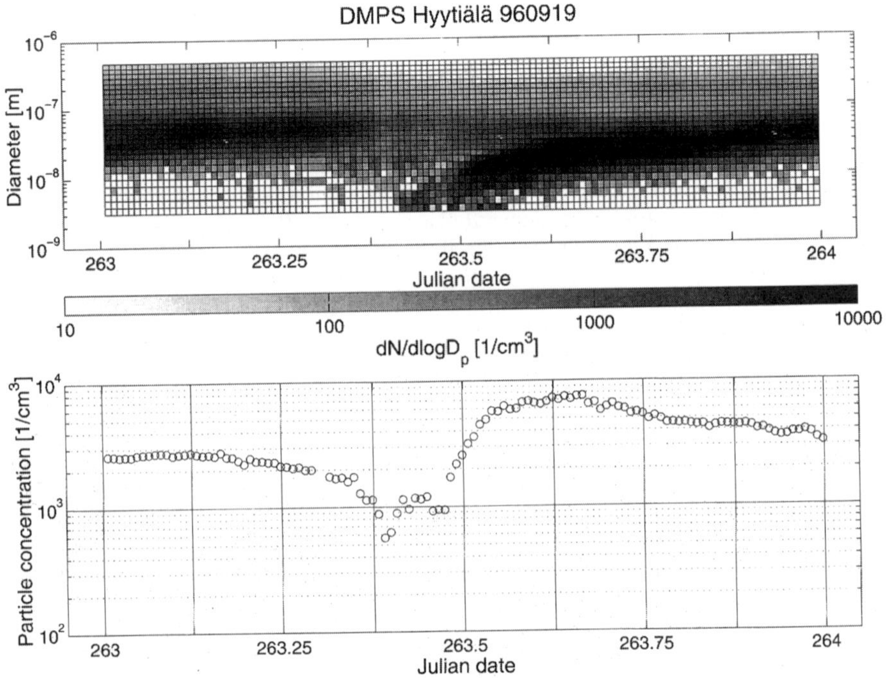

Figure 11. Particle size distributions (top panel) and particle number concentrations (bottom panel) at Hyytiälä, Finland as a function of time of day (Julian day 263.5 = noon on September 19, 1996). Note the burst of nanoparticle nucleation occurring near noon and its subsequent growth. From Clement et al. (2001). Used by permission of the editor of *The Journal of Aerosol Science*.

the formation of gaseous H_2SO_4 (and many other secondary low-volatility gases) has a strong diurnal dependence. Correspondingly, new particle formation also shows a diurnal dependence (e.g., Fig. 11).

The photochemical dependence for the production of $^{\bullet}OH$, H_2SO_4, and new particles is apparent in Figure 12 for data from Idaho Hill, Colorado, which is a remote continental site (Weber et al. 1997). As seen in the top panel (a), $^{\bullet}OH$ formation begins at sunrise, and concentrations correlate well with the UV irradiance. H_2SO_4 concentrations (c) also rise at sunrise and correlate with UV irradiance, which follows from Equation (17). On the other hand, SO_2 mixing ratios (b) show no diurnal trends because SO_2 has a long lifetime (days) with respect to reaction with $^{\bullet}OH$. The bottom panel in this sequence (d) reveals that the formation of 3- to 4-nm diameter particles is associated with the photochemical cycle and that the greatest particle number concentrations occur near solar noon. However, the appearance of these particles does not begin until approximately an hour after sunrise. This induction period is attributed to the time required for the initial particle nuclei ($D_p \sim 1$ nm) to grow to measurable size (3 nm) (Weber et al. 1997).

In the Idaho Hill study described above, measured concentrations of gaseous H_2SO_4 and H_2O are much lower than the levels required for rapid new particle formation based on classical nucleation theory (Weber et al. 1997). Similar results are reported in several other studies (Wiedensohler et al. 1997; Pirjola et al. 1998, 2000; O'Dowd et al. 1999; Weber et al. 1999). These measurements show that homogeneous classical nucleation theory for H_2SO_4/H_2O generally underpredicts measured rates of new particle formation in the boundary layer. In contrast, in studies of high-altitude regions near cloud venting, observed rates of nucleation are similar to those predicted from binary nucleation theory (Clarke et al. 1999; Weber et al. 1999). In addition, a study of polluted air masses with high SO_2 concentrations in the Finnish Arctic found that application of classical nucleation theory to H_2SO_4 and H_2O can explain measured new particle formation rates reasonably well (Pirjola et al. 1998). Overall, these results suggest that binary nucleation of H_2SO_4 and H_2O is sometimes responsible for new particle formation in the atmosphere, but that other mechanisms are often more important. However, even in these latter cases there is evidence that sulfuric acid plays a role in nucleation, such as the correlation of gaseous H_2SO_4 with nanoparticle number concentrations at Idaho Hill (Fig. 12). Despite this evidence, it is also possible that H_2SO_4 is not a nucleating agent in these studies, but instead that its concentration co-varies with another, as yet unknown, photochemically produced agent (Weber et al. 1999).

Numerous authors suggest that a plausible alternative to H_2SO_4/H_2O binary nucleation is ternary nucleation involving H_2SO_4, H_2O, and NH_3 (Coffman and Hegg 1995; Weber et al. 1997, 1999; Kim et al. 1998; Korhonen et al. 1999). Ammonia is a common gas in the troposphere with mixing ratios that are typically under 50 pptv in remote regions and on the order of ppbv (parts per billion by volume) in regions near sources such as animal operations (Finlayson-Pitts and Pitts 2000). The H_2SO_4 vapor pressure is much smaller over aqueous ammonium sulfate solutions, as compared to over H_2SO_4/H_2O solutions, which suggests that NH_3 stabilizes H_2SO_4-H_2O clusters (Marti et al. 1997a). In addition, as described earlier, several recent laboratory and modeling studies show that the presence of pptv levels of NH_3 greatly reduces the critical concentration of H_2SO_4 required to form new particles (Ball et al. 1999; Korhonen et al. 1999; Pirjola et al. 2000). Whether the ternary $H_2SO_4/H_2O/NH_3$ system correctly predicts nucleation rates in a wide range of atmospheric conditions remains to be tested.

Recent work by Kulmala et al. (2000) indicates that the $H_2SO_4/H_2O/NH_3$ ternary nucleation mechanism yields nanoparticle ($D_p > 3$ nm) concentrations that match field observations at a Finnish study site. In addition, the authors present the interesting

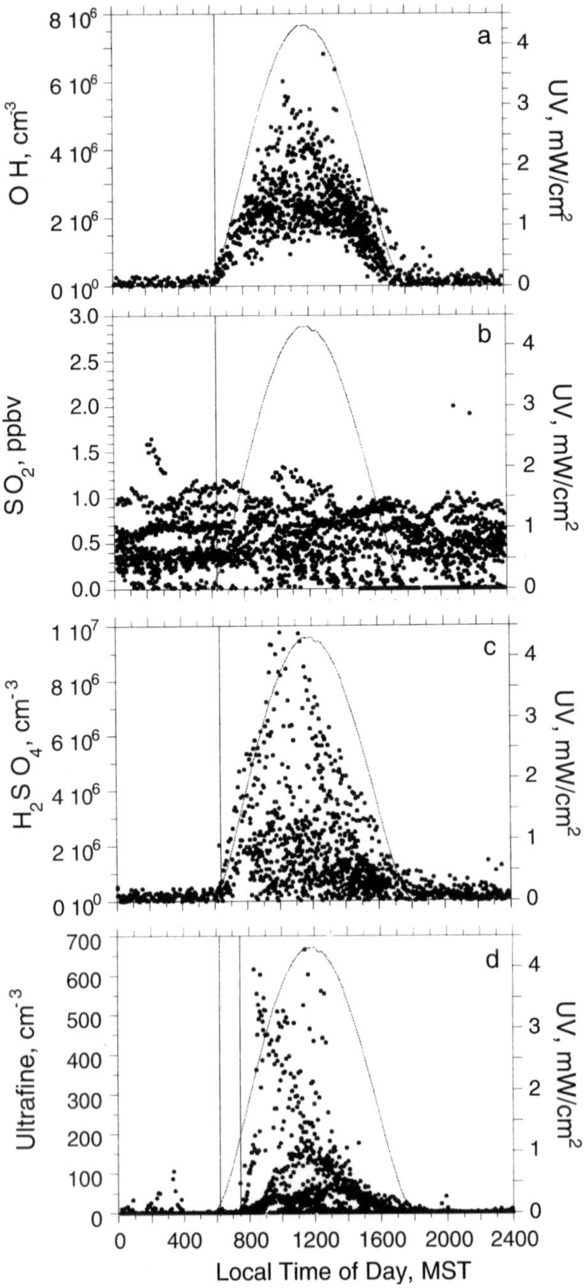

Figure 12. Concentrations of (a) $^{\bullet}OH$, (b) SO_2, (c) H_2SO_4, and (d) nanoparticles (D_p = 2.7-4 nm) measured at Idaho Hill, Colorado, from September 5 to 29, 1993, during periods of relatively clean (downslope) air flow. The solid curve on each plot (right axis) is the UV irradiance measured on a clear day (September 26) during the study. The vertical line at approximately 06:15 in each plot indicates sunrise. The second vertical line at approximately 07:30 in plot *d* indicates the beginning of measurable number concentrations of nanoparticles. From Weber et al. (1997). Used by permission of the American Geophysical Union.

hypothesis that 1- to 3-nm particles ("thermodynamic stable clusters") nucleate from $H_2SO_4/NH_3/H_2O$ continuously and are present in the atmosphere undetected. According to this hypothesis, these small particles are formed daily to yield peak number concentrations of approximately 10^5 cm^{-3}. However, their growth is slow due to the absence of condensable H_2SO_4. Their lifetime is also relatively short in areas with heavy loadings of larger particles due to intermodal coagulation. A measurement challenge (Eisele and Hanson 2000) remains to observe in situ these thermodynamically stable clusters and thus refute or support the hypothesis.

Organic compounds. Low volatility organic gases are also believed to be precursors for the formation of new particles. As in the case of SO_2 oxidation to H_2SO_4, the first step in the formation of an organic nanoparticle appears to be the oxidation of a precursor gas to a more polar, lower volatility gas (Fig. 10). While there are thousands of natural and anthropogenic organic gases present in the atmosphere (Graedel et al. 1986), only those that are emitted in significant quantities, react quickly in the atmosphere, and form very low volatility products are expected to be significant in particle nucleation. Examples of these potentially significant precursor gases include biogenic monoterpenes such as α- and β-pinene (Griffin et al. 1999; Kavouras et al. 1999) and, in urban areas, anthropogenic aromatics such as alkyl benzenes (Odum et al. 1997). Reactions of these organic gases with oxidants, such as $^{\bullet}OH$ and O_3, yield more oxygenated compounds, such as carboxylic acids and peroxides. These oxygenated species have lower volatilities than their parent compounds and thus increase the likelihood of forming new particles (e.g., Forstner et al. 1997; Barthelmie and Pryor 1999; Tobias and Ziemann 2000). The critical germs in these cases probably contain several different types of interacting organic species.

The Idaho Hill study discussed earlier also assesses the possible roles of biogenic terpenes and anthropogenic organic gases in nucleation (Weber et al. 1997; Marti et al. 1997b). The conclusions reached are that H_2SO_4 is generally a more important nucleating species than the measured organic gases but that there are possibly a few events where organics participate in nucleation. There is indirect evidence from other field studies that organic compounds at times lead to new particle formation in the boundary layer. For example, several studies find new particle formation in forests under conditions where organic species should be abundant (Mäkelä et al. 1997; Kulmala et al. 1998b; Becker et al. 1999; Kavouras et al. 1999). In addition, species such as carboxylic acids from terpene oxidation have been measured in accumulation mode or larger particles in forested regions (Kavouras et al. 1999; Yu et al. 1999). However, the most abundant secondary organic compounds in accumulation mode particles are likely to be the compounds condensed during particle growth, rather than any putatitive organics contributing to the critical germ (Kerminen et al. 2000). Thus while organic compounds are a major component of ultrafine particles in the atmosphere (Cass et al. 2000), much more work remains to identify and to evaluate the organic compounds that might serve as nucleating agents.

Primary emissions

In addition to the formation of new secondary nanoparticles, there are also primary emissions of nanoparticles to the atmosphere, most importantly from high temperature combustion. While there are no published emissions inventories specifically for nanoparticles, Cass et al. (2000) estimate that approximately 85% of the mass of primary ultrafine particles (D_p < 100 nm) in the Los Angeles area is emitted from combustion sources. On-road vehicles are estimated to account for approximately 40% of ultrafine particle mass. Similar results are reported for the United Kingdom, with a somewhat greater contribution (60% of total) from vehicles (Harrison et al. 2000). Observations in

Birmingham, where the number of nanoparticles decreases rapidly with distance from roads, also suggest that vehicles are an important source of primary nanoparticles (Shi et al. 1999).

Diesel engines are a particular research focus because they emit approximately 10-100 times more particles by mass than spark-ignition (gasoline) engines (Kittelson 1998). As shown in Figure 13, particulate matter from diesel engines generally contains both a nanoparticle mode (D_p < 50 nm) as well as an accumulation mode (D_p = 100-300 nm) (Kittelson 1998; Collings and Graskow 2000). The nanoparticle mode typically accounts for more than 90% of the total number of particles emitted from diesel engines, but less than 20% of the total particle mass (Kittelson 1998). The nanoparticle number size distributions recorded in laboratory measurements of vehicle emissions depend strongly upon the dilution process of the exhaust prior to analysis (Kittelson 1998; Shi and Harrison 1999; Collings and Graskow 2000). Even so, chase-car roadway measurements of particles reveal number size distributions similar to the laboratory measurements shown in Figure 13 (Collings and Graskow 2000). In an attempt to reduce diesel particle emissions in the United States, the EPA has enforced a series of increasingly strict mass-based standards (Kittelson 1998). It is unclear whether these regulations will impact the number of particles emitted from diesel engines. In fact, initial evidence suggests that new diesel engines designed to reduce emissions of particle mass might emit higher number concentrations of nanoparticles (Bagley et al. 1996; Kittelson 1998).

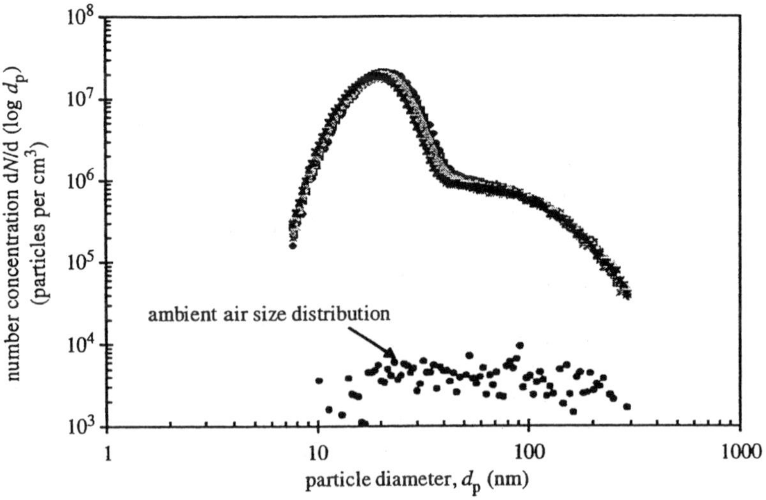

Figure 13. Particle number distributions for a diesel engine (2.5 L displacement, direct injection, 4-cylinder, 1500 rpm, 7.5 kW) run with 300 ppm sulfur content diesel fuel. Distributions are measured using a dilution tunnel with conditions simulating those expected in the environment (final dilution ratio of approximately 100:1). From Collings and Graskow (2000). Used by permission of the Royal Society.

Most of the nanoparticles emitted from diesel vehicles appear to nucleate from sulfuric acid and water during the cooling and dilution of the exhaust with ambient air (Kittelson 1998; Shi and Harrison 1999; Collings and Graskow 2000; Tobias et al. 2001). However, based on results from Shi and Harrison (1999), the measured rate of particle formation in diesel exhaust is much greater than the rate calculated from binary homogeneous nucleation of H_2SO_4 and H_2O. The implication is that other species are also participating. As the vehicle exhaust further cools, unburned fuel and lubricating oil condense upon the sulfuric acid nucleus, increasing its size and changing its composition (Burtscher et al. 1998; Kittelson 1998; Tobias et al. 2001). The organic composition of diesel nanoparticles varies with particle size, with a higher proportion of larger and less volatile organics present in the smaller particles (Tobias et al. 2001).

The accumulation mode of diesel particles consists primarily of elemental carbon in addition to significant amounts of organic carbon (including PAHs), sulfate, nitrate, ammonium, chloride, and sodium (Kittelson 1998; Schauer et al. 1999; Kleeman et al. 2000; Richter and Howard 2000). Many other species are also present in trace amounts, including Si, Fe, Ti, Zn, and Al (Schauer et al. 1999; Kleeman et al. 2000). Based on extrapolation of the size-resolved chemical composition measurements for particles with $D_p > 50$ nm (Kleeman et al. 2000), it appears that many of these same accumulation mode components are also present in diesel nanoparticles.

In addition to diesel vehicles, gasoline (spark ignition) vehicles also emit significant numbers of nanoparticles (Burtscher et al. 1998; Kittelson 1998; Kleeman et al. 2000). Although the total mass emission rate of particles from diesel engines is much greater than from gasoline engines, under highway cruise conditions diesel and gasoline engines have similar nanoparticle emission rates (Kittelson 1998). Other combustion sources of primary nanoparticles include the burning of coal and fuel oil (Huffman et al. 2000; Linak and Miller 2000; Senior et al. 2000; Fan and Zhang 2001; Zhuang and Biswas 2001), combustion of wood and other biomass (Rogge et al. 1998; Kleeman et al. 1999), aircraft (Brock et al. 2000; Kärcher et al. 2000), meat charbroiling (Rogge et al. 1991; Kleeman et al. 1999), and cigarettes (Rogge et al. 1994; Kleeman et al. 1999). The particles produced from these processes contain elemental carbon as well as a complex suite of organic compounds and numerous trace metals.

GROWTH

This section provides a conceptual framework and several examples of modeling and fieldwork on the growth of atmospheric nanoparticles. The growth of nanoparticles is an important source of Aitken mode and accumulation mode particles, including cloud condensation nuclei, especially in remote regions with few primary particle sources. For more quantitative descriptions of growth processes, as well as their parameterizations in models, see Kulmala (1993), Kulmala et al. (1993), Kerminen et al. (1997), Mattila et al. (1997), Vesala et al. (1997), Seinfeld and Pandis (1998), and Friedlander (2000).

The processes involved in nanoparticle growth are depicted in Figure 10. Phenomenological descriptions of early growth are presently limited because the formation of new nanoparticles, which have diameters of around 1 nm, cannot be detected by current state-of-the-art field-deployable instruments, which can measure particles with diameters of 3 nm and above. Model descriptions, however, indicate that growth is accomplished through one of three pathways: the condensation of low volatility gases such as sulfuric acid, coagulation with other newly formed nanoparticles (i.e., intramodal coagulation), or surface or bulk reactions that increase particle mass (i.e., reactive condensation) (Kerminen 1999). The growth of nanoparticles changes their chemical composition, especially at the surface, and therefore likely alters their health effects and chemical reactivity.

The rates and relative importance of condensation and intramodal coagulation depend upon the number concentration and size of nanoparticles, the partial pressures of condensable gases, and the accommodation coefficients of the gases onto the particles

(Friedlander 2000). While intramodal coagulation can be important under some conditions (e.g., in regions with very high nanoparticle number concentrations), condensation is the dominant growth mechanism under typical atmospheric conditions (Wexler et al. 1994; Kerminen et al. 1997). In addition, the growth of nanoparticles is influenced by the presence of larger particles in the aerosol due to collisions with these particles (i.e., intermodal coagulation) and a competition between the nanoparticles and larger particles in scavenging the condensable gases. All other factors being equal, the vapor pressures of chemical species over nanoparticles are greater than over larger particles due to the Kelvin effect. The Kelvin effect states that the vapor pressure over a curved surface, P, increases relative to that over a flat surface, P^0, as follows: $P = P^0\exp(4\sigma v/kTD_p)$, where σ is the surface tension and v is the condensed-phase molecular volume (Seinfeld and Pandis 1998). The overall effect of these factors is that insignificant growth, and possibly even evaporation, of nanoparticles occurs in the presence of high number concentrations of larger particles. The third growth mechanism, reactive condensation, is important for the growth of accumulation mode particles through reactions such as the aqueous oxidation of SO_2 by HOOH or O_3 (Finlayson-Pitts and Pitts 2000). However, these same reactions are calculated to be too slow to be significant for nanoparticle growth (Kerminen et al. 1997; Kerminen 1999).

In the remaining part of this section we focus on nonreactive condensation because this process usually contributes more to nanoparticle growth than does intramodal coagulation (except perhaps in urban environments). In most field studies, observed nanoparticle growth rates are much faster than can be explained by the condensation of H_2SO_4 and H_2O (Weber et al. 1997,1998a; O'Dowd et al. 1999; de Reus et al. 2000b). Models yield similar results, especially for conditions where particle formation occurs via $H_2SO_4/H_2O/NH_3$ ternary nucleation (Kerminen et al. 1997; Kulmala et al. 2000). In contrast to its effect on nucleation, the presence of NH_3 has little effect on growth rates in the H_2SO_4/H_2O system (Kerminen et al. 1997; O'Dowd et al. 1999). Model results reveal that the addition of HNO_3 or HCl to the ternary $H_2SO_4/H_2O/NH_3$ system greatly enhances growth rates under conditions found in continental regions, though not at the lower levels of gas-phase HNO_3 and HCl found in remote marine systems (Kerminen et al. 1997).

Several lines of evidence suggest that organic compounds play a significant role in nanoparticle growth. For example, terpenoic acids and other terpene oxidation products are present in accumulation mode particles in forested regions (Kavouras et al. 1999; Yu et al. 1999). In addition, Marti et al. (1997b) report that measured total particle surface areas and volumes ($D_p < 500$ nm) at a remote continental location roughly correlate with the estimated rates of formation of condensable products from terpene oxidation. Finally, measurements at two field sites where biogenic organic gases are likely prevalent (namely, a boreal forest in Finland during spring and summer (Kulmala et al. 1998b) and a coastal site during low tide (O'Dowd et al. 1999)) show that measured growth rates require unidentified condensable species with peak concentrations of greater than 10^7 molecules cm^{-3}.

These unidentified species might be organic compounds. Modeling by Kerminen and co-workers shows that condensation of organic gases leads to the growth of 5-nm nanoparticles to CCN size ($D_p > 50$ nm) in under 24 hours. The simulations employ realistic amounts of precursor organic gases and assume that at least some of the organic oxidation products have extremely low saturation vapor pressures on the order of 0.01 pptv or less (Kerminen 1999; Kerminen et al. 2000). These nonvolatile organics are required for the initial stage of growth where the nanoparticle radius is very small and thus the Kelvin effect is very large. As the particle grows, low volatility (rather than nonvolatile) organic gases are increasingly important as particle components, which is a combined result of a reduced Kelvin effect and the dependence of gas-particle partitioning of organics on particle mass (Kerminen et al. 2000). These results imply that the chemical composition of organic particles changes dramatically during growth and that the composition of the original nonvolatile nanoparticle nucleus is lost as a result of subsequent condensation of other products.

CHARACTERIZATION

Number concentrations

Observations of atmospheric nanoparticle number concentrations have increased greatly during the last five years. This advance has been made possible by new instruments and techniques that can measure particles with diameters as small as 3 nm. The new approaches include pulse height analysis of data from an ultrafine condensation particle counter (UCPC) (Stolzenburg and McMurry 1991; Wiedensohler et al. 1994; Weber et al. 1998b) and the combination of a nano-differential mobility analyzer (nano-DMA) (Chen and Pui 1997; Seto et al. 1997) with a UCPC.

There are large differences in nanoparticle number concentrations between remote and polluted regions: measured values in remote continental and marine areas are typically under 100 cm^{-3} (Weber et al. 1995, 1997; Covert et al. 1996) whereas number concentrations in urban areas are typically 10^4 cm^{-3} (Shi et al. 1999; McMurry et al. 2000). Higher concentrations (~10^5 cm^{-3}) occur near roads (Shi et al. 1999) and during bursts of new particle formation (McMurry et al. 2000; Woo et al. 2001). At a given location concentrations of the smallest measurable particles (D_p < 10 nm) are usually highly variable in time, occurring in bursts most commonly seen during periods of high solar radiation (e.g., Figs. 11 and 12). Figure 11 also shows that new nanoparticles (3 nm) grow into larger nanoparticles and nuclei mode particles over the course of several hours when conditions are favorable.

New particle formation in an urban site (Atlanta) is illustrated in Figure 14, which shows an afternoon burst of 3- to 10-nm particles during a period with high levels of SO_2. At its peak, this burst nearly quadruples the total number of particles present at the sampling site (Woo et al. 2001). Overall, number concentrations of the smallest nanoparticles (under 10 nm) are highest in Atlanta during midday, while concentrations of the larger nano- and ultrafine particles (10-100 nm) are typically higher during the morning and evening. These results suggest that vehicular emissions might be a significant source of the larger nanoparticles (Woo et al. 2001). An unexplained observation is that appreciable numbers of 3- to 10-nm particles are present in Atlanta even after sunset (Fig. 14), which is unlike the case for the remote Idaho Hill site (Fig. 12) or the Finnish boreal forest site (Fig. 11). These nighttime particles might be the result of vehicular emissions or other anthropogenic sources of primary nanoparticles in Atlanta.

Particle number distributions over the course of a year in Atlanta are shown in Figure 15. During the period August 1998 to April 1999, the number mode size of the aerosol number distribution typically occurs between 10- and 40-nm diameter. However, during the latter part of this study (April to August 1999), the smallest particles (those under 10 nm) dominate the particle number concentrations. During these months, the number of particles increases with decreasing particle size, suggesting that new particle formation events are common in this region (McMurry et al. 2000; Woo et al. 2001).

Chemical composition

Knowledge of the chemical composition of atmospheric nanoparticles is limited.

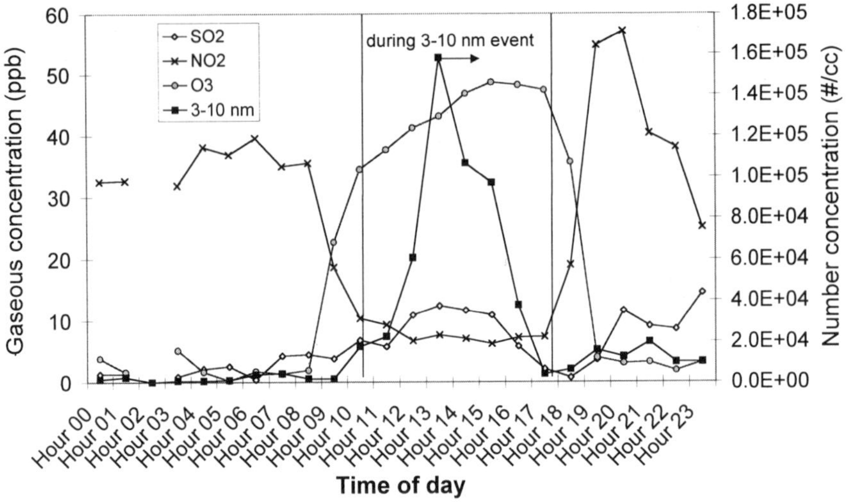

Figure 14. Nucleation burst of nanoparticles (D_p = 3-10 nm) and associated mixing ratios of gases in Atlanta, Georgia, on April 1, 1999. The vertical lines represent the period of enhanced number concentrations of 3-10 nm nanoparticles. Note, however, that the number concentrations of these particles are often still appreciable outside of this period, especially at night. From Woo et al. (2001). Used by permission of Taylor & Francis, Inc.

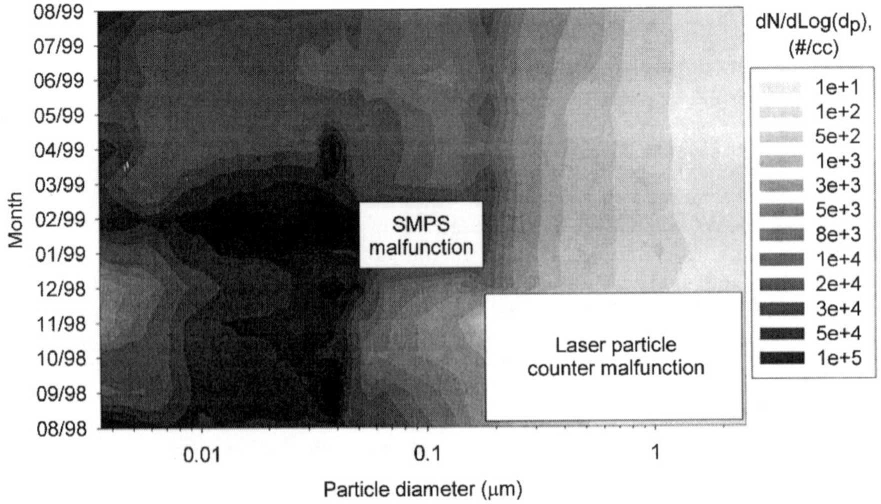

Figure 15. Contour plot of monthly average particle number distributions measured in Atlanta, Georgia. SMPS = scanning mobility particle sizer, used to measure numbers of particles with D_p = 20-250 nm. A laser particle counter is employed to measure particles with diameters of 0.1-2 μm. Nanoparticles with D_p = 3-50 nm are measured with a nano-DMA in conjunction with a UCPC. The white boxes without data indicate times of equipment failure. From Woo et al. (2001). Used by permission of Taylor & Francis, Inc.

Measurements are difficult both because these particles have very little mass and because their composition varies significantly from particle-to-particle, as well as temporally and spatially. As a first step in a description, the chemical composition of new particles should relate directly to their routes of origin and subsequent growth. For example, new particles that form from the condensation of H_2SO_4 should clearly contain significant amounts of sulfuric acid as well as any other species involved in the nucleation event (e.g., H_2O, NH_4^+, and perhaps organic species). Similarly, ammonium nitrate and low volatility organic species, when involved in nucleation, should also be important chemical components of new particles.

For fresh nanoparticles arising from primary emissions, it should be possible to roughly infer their composition based on studies that have characterized fine particles from the same emission source. For example, ultrafine and accumulation mode particles from medium duty diesel vehicles mainly contain elemental and organic carbon (Kleeman et al. 2000). Most individual molecules composing this particulate organic carbon have not been identified, but investigated compound classes include large *n*-alkanes, alkanoic acids, and polycyclic aromatic hydrocarbons (Schauer et al. 1999). Though not yet characterized, it is likely that nanoparticles from diesel engines contain similar types of compounds. Likewise, fresh nanoparticles from wood combustion should have compositions somewhat similar to larger wood smoke particles, which contain inorganic compounds as well as hundreds of organic species including levoglucosan, substituted methoxyphenols, and PAHs (Rogge et al. 1998; Simoneit et al. 1999; Nolte et al. 2001; Schauer et al. 2001). One caveat to this argument is that the relatively volatile components (e.g., many of the organics) present in the larger particles should be much less prevalent in nanoparticles from the same source (e.g., Kerminen et al. 2000; Tobias et al. 2001).

As discussed previously, growth and chemical reactions of nanoparticles lead to changes in their compositions. Thus the inferences in composition described above should hold only for fresh nanoparticles; correlations should weaken with atmospheric aging of the particles. It would be desirable to compare these expectations against actual field measurements of particle compositions. However, quantitative measurements of the chemical composition of ambient ultrafine particles are available only for the larger members of this class ($D_p \approx$ 50-100 nm). Available data, from urban areas in Southern California, indicate that organic compounds represent approximately half of the ultrafine particle mass. The remaining mass is contributed by trace metal oxides, elemental carbon, sulfate, nitrate, ammonium, sodium, and chloride (Cass et al. 2000).

Some examples of the composition of 50- to 100-nm particles from different locations in Southern California are given in Figure 16. These data indicate that the relative amounts of ultrafine particle components vary widely at different locations, a reflection of differences in particle sources and the condensation of low-volatility species. It is unknown what fraction of each chemical component in these samples was present at the point of primary emission (or initial formation) and what fraction resulted from atmospheric aging. Condensation likely contributes most significantly for chemical species that have strong vapor sources, such as organic carbon, sulfate, nitrate and ammonium. In contrast, condensation should contribute insignificantly for nonvolatile components such as metals. As shown in Table 2, Fe, Na, K, and Ti are the most abundant metals in the ultrafine particles measured, and there are also significant concentrations of Cr, Zn, and Ba. These metals are potentially significant because of their possible catalytic chemical activities and potential to contribute to the adverse human health effects associated with particles. Other metals of possible significance, such as Pt, Pd, and Rh, are not determined in the samples because of limitations of the analytical method (Hughes et al. 1998).

Future measurements should greatly increase our knowledge of the chemical composition of nanoparticles. The recent development of a novel laser desorption/ionization single-particle mass spectrometer (RSMS-II) (Carson et al. 1997; Ge et al. 1998; Phares

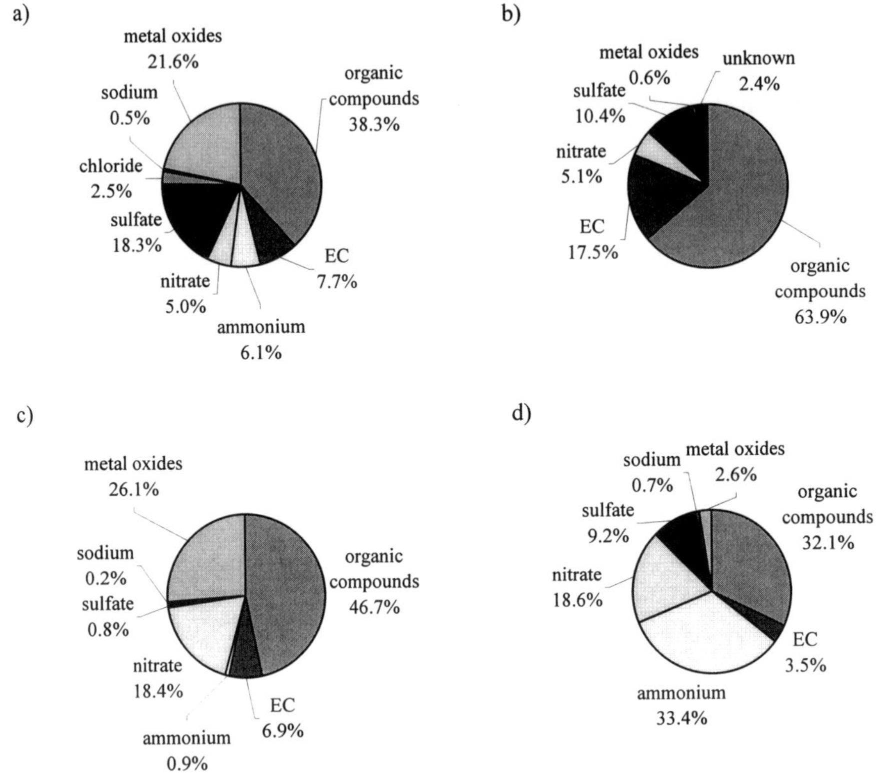

Figure 16. Chemical composition of atmospheric ultrafine particles (D_p = 56-100 nm) measured in four cities in Southern California: (a) Fullerton, September-October, 1996 (average mass concentration of 56-100 nm particles = 0.64 $\mu g\ m^{-3}$), (b) Diamond Bar, September-November, 1997 (0.55 $\mu g\ m^{-3}$), (c) Mira Loma, September-November 1997 (0.58 $\mu g\ m^{-3}$), and (d) Riverside, August-November, 1997 (0.91 $\mu g\ m^{-3}$). Note that elemental carbon is abbreviated as "EC". Adapted from Cass et al. (2000). Used by permission of the Royal Society.

et al. 2001a) enables real-time measurements of the chemical components of ambient nanoparticles (aerodynamic $D_p \geq 10$ nm). At this time, this instrument does not quantitatively determine chemical composition because the detection of particles, and the efficiency with which they are ablated by the laser, depend upon particle size and composition (Kane and Johnston 2000; Phares et al. 2001a). Even so, the instrument provides insight into the elements and compound classes present in ambient nanoparticles. As configured in the two studies described below, the instrument measures particles in 9 or 13 size bins ranging from $D_p > 1$ μm down to nanoparticles as small as 14 nm.

Results are discussed here for the RSMS-II deployed in Atlanta during August 1999 and in Houston during August-September 2000 (Rhoades et al. 2001; Phares et al. 2001b). In the Atlanta study, approximately 16,000 particles, including both nanoparticles and larger particles, are classified into nine major composition classes. Organic carbon is the most abundant class (representing 74% of all particles). Organic particles are even more prevalent among the measured nanoparticles, accounting for 85-90% of the total number of detected particles with diameters under 50 nm (Rhoades et al. 2001). This class of particles is dominated by C^+-ions and other carbon fragments, but it

Table 2.

Concentrations (ng m^{-3}-air) of trace metals in ultrafine particles ($D_p \approx$ 50-100 nm)

	Pasadena, CA [a] (Jan.–Feb., 1996)		*Central Los Angeles, CA* [b] (Aug.– Sept., 1997)	
Element	*Mean*	*Range*	*Mean*	*Range*
Group I and II metals				
Na	27.0	0.75 - 34.8	85	bdl - 249
K	bdl	bdl	88	bdl - 93
Cs	0.016	0.0081 - 0.028	0.100	bdl - 0.34
Ba	1.04	bdl - 2.8	19	bdl - 19
Transition metals				
Sc	0.0046	bdl - 0.0081	0.028	bdl - 0.054
Ti	7.65	4.07 - 10.2	43	bdl - 43
V	0.059	bdl - 0.14	bbl	--
Cr	7.32	bdl - 26.2	6.7	bdl - 15
Mn	0.74	bdl - 2.43	bbl	bbl - 0.056
Fe	67.5	bdl - 148.3	186	bdl - 470
Zn	3.68	bdl - 6.56	3.8	bbl - 10
Mo	0.072	bdl - 0.19	0.48	bdl - 0.68
Cd	0.061	bdl - 0.12	0.19	bbl - 0.49
Au	bdl	bdl	bdl	--
Hg	0.0052	0.00014 - 0.012	0.09	0.038 - 0.14
Lanthanides				
La	0.11	0.00082 - 0.51	0.021	bdl - 0.021
Ce	0.19	bdl - 0.82	1.2	bdl - 2.3
Sm	0.015	0.00030 - 0.73	0.012	bdl - 0.019
Eu	0.013	bdl - 0.023	0.20	bdl - 0.37
Yb	0.0028	bdl - 0.0059	0.26	bbl - 0.5
Lu	0.00038	bdl - 0.0013	0.014	0.0011 - 0.028
Actinides				
Th	0.0065	bdl - 0.018	bbl	—
U	0.0057	bdl - 0.019	bdl	—

bdl denotes 'below detection limit' bbl denotes 'below blank levels'

[a] Data from Hughes et al. (1998). Used by permission of the American Chemical Society.

[b] Data from Cass et al. (2000). Used by permission of the Royal Society.

also includes Si and K ions. These observations are consistent with a high temperature combustion source of primary particles. Given the high biogenic emissions in Atlanta during the summer (Chameides et al. 1988), gas-to-particle conversion of oxidation products from biogenic hydrocarbons also likely contributes to the organic aerosol. Other important particle classes of nanoparticles in Atlanta include potassium (a class with contributions from C, Ca, Na, Si and Al), calcium (with C, Fe, and Si), nitrate (with C, Si, and NH_4^+), and elemental carbon.

The compositions of nanoparticles in Houston differ markedly from those observed in Atlanta. In particular, the dominant composition in Houston is a Si-based particle rather than an organic type (Phares et al. 2001b). The Si particle type is also prevalent among the nanoparticles measured, accounting for approximately 70% and 60% of the number of detected particles with diameters of 35 and 50 nm, respectively. Because atmospheric Si is normally associated with coarse particles derived from crustal material (Finlayson-Pitts and Pitts 2000), the presence of nanoparticulate silica suggests an alternative source, possibly involving high temperature combustion (Wooldridge 1998; Wooldridge et al. 1999). The presence of Si in particles of 35-nm diameter, and perhaps smaller, also lends support to speculation that reactions on silica nanoparticles might occur in the troposphere. Other types of 35-nm particles prevalent in Houston include carbon (including several types of aliphatic amines), iron, mixed organic/mineral, potassium, and aluminum (Phares et al. 2001b). Sea-salt particles of 35-nm diameter are also present in Houston when the wind direction follows from the ocean, in agreement with previous observations of 35-nm diameter sea-salt nanoparticles in the Pacific Ocean (Berg et al. 1998).

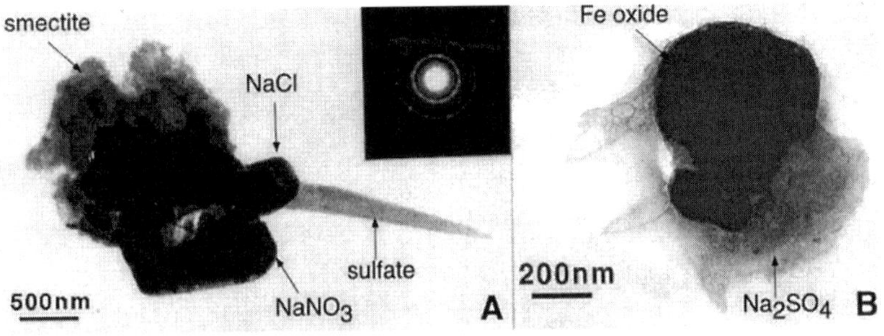

Figure 17. TEM images of internally mixed terrestrial and marine aerosol particles. (a) Terrestrial smectite with sea-salt particles. The selected-area electron diffraction pattern (upper right) confirms the identification of smectite (Azores, North Atlantic Ocean, ASTEX/MAGE). (b) Fe oxide emitted from coal burning, with Na_2SO_4 (Sagres, Portugal, ACE-2). Images by Mihaly Pósfai. Figure adapted from Buseck et al. (2000). Used by permission of the editor of *Proceedings of the National Academy of Sciences, USA*.

Morphology

The shape of atmospheric particles and the arrangement of chemically distinct components within a single particle have been investigated by sampling onto transmission electron microscopy (TEM) grids followed by imaging in the laboratory (Buseck and Pósfai 1999; Buseck et al. 2000). Figures 9 and 17 provide several examples of inorganic particles collected in different locations. A large inventory of images confirms what these figures suggest, i.e., that atmospheric particles in the large submicron to supermicron sizes are usually composed of grains of material in the nanoparticle regime. Nanoscale sulfate coatings, which are presumed to arise from SO_2 oxidation on mineral surfaces or from the condensation of H_2SO_4, are another common feature. Another particularly good example of morphology at the nanoscale is found in soot particles, which are commonly composed of ca. 50-nm spherical units agglomerated in chains to form larger particles. *In situ* mass spectrometric analysis of single atmospheric particles also confirms extensive internal mixing of chemical components (Noble and Prather, 2000). An understanding of atmospheric nanoparticles is thus incomplete without recognizing that many, if not most, larger atmospheric particles are agglomerated domains of nanophase materials.

Although environmental TEM instruments are now beginning to come on-line, a drawback of the TEM imaging completed to date is that they required operation at very low pressures. Thus the effect of volatile components on particle morphology is unknown; water is of particular importance. Deliquescent components (e.g., NaCl and $NaNO_3$) likely form an aqueous layer around insoluble minerals (e.g., smectite), at least at some of the higher RH values common in the atmosphere. At lower RH values, complex morphological shapes at the nanoscale increase hygroscopic response by reducing the chemical potentials of aqueous solutions in interstices between particles (Xie and Marlow 1997). Organic surfactant layers may be present in some particles (Gill et al. 1983; Ellison et al. 1999), and one property of this interfacial nanostructure is believed to be a decrease in the water evaporation rate (Xiong et al. 1998).

At sufficiently low RH values, crystallization occurs. The result is rarely nonporous particles of idealized shapes such as spheres or cubes (Charlesworth and Marshall 1960; Leong 1981). Laboratory studies on the morphological characteristics of evaporatively dried salt particles provide some insight into possible processes affecting the morphology of atmospheric particles. Cziczo et al. (1997) and Weiss and Ewing (1999) show from infrared studies that submicron $NaCl/H_2O$ aerosols dried by passage through a diffusion dryer retain nanoscale pockets of water to 0% RH. These studies, however, are unable to provide information on the overall particle shape. Even NaCl particles generated by high-temperature aerosol condensation methods fail to yield nonporous cubic particles. Matteson et al. (1972) report that the density of particles generated by such methods are different from the bulk materials. The differences arise from nanoscale structural features. Craig et al. (1952) and Krämer et al. (2000) report that chain agglomerates of NaCl particles produced by condensation rearrange to cubic particles upon exposure to sub-deliquescent RH values. Although evaporation in diffusion dryers has been modeled, these studies omit the effects of the detailed microphysical structuring of the particles that accompanies drying (Xiong and Kodas 1993). Ge et al. (1996) study multicomponent aerosol crystallization by rapid evaporative drying of particles containing $KCl/NaCl/H_2O$, $KCl/KI/H_2O$, $(NH_4)_2SO_4/NH_4NO_3/H_2O$ followed by collection of the mass spectra of the surface layers by single-particle mass spectrometry. They find that nanoscale surface layers are enriched in the minor component. The implication is that the particle morphology consists of some kind of layering with the initially crystallized phase towards the core of the particle and the minor components, which precipitate later, in the surface layers.

The detailed morphology (including overall particle shape, grain sizes and relative juxtapositions, and possible water inclusions) adopted by dried atmospheric particles probably depends on the temperature and the rate of evaporation. For these reasons, it cannot be certain that morphological features observed by TEM for particles dried by exposure to vacuum are the same as those adopted by particles dried by atmospheric processes. A critical need is the development of techniques capable of in situ characterization of particle morphology. Because the particles themselves are often no larger than 1 μm, the heterogeneity of their features occurs on the 10 to 100-nm scale.

PROPERTIES

Motion

On a large scale, particles (as well as gases) are moved through the atmosphere by advection and turbulence, i.e., horizontal and vertical winds (Wexler et al. 1994; Seinfeld and Pandis 1998). Simultaneous with these large-scale motions are the smaller-scale processes that can transport particles across surface boundary layers (e.g., at the Earth's surface) and thus remove them. As discussed earlier, diffusion is the dominant removal mechanism for small particles because of their high diffusion coefficients and low gravitational settling velocities. Because of their very small sizes, nanoparticles can slip between gas molecules, with the result that diffusion coefficients for nanoparticles are much larger than predicted based on continuum fluid mechanics (Seinfeld and Pandis 1998; Friedlander 2000). The diffusion coefficient (D, cm^2 s^{-1}) for particles in air is given by:

$$D = \left(\frac{kT}{3\pi\mu \mathrm{D_p}} \right) C_c \tag{18}$$

where μ is the gas viscosity and C_c is the Cunningham slip correction factor. The term in parentheses is the Stokes-Einstein relation, which is the diffusion coefficient for particles large enough such that the surrounding gas behaves as a continuous fluid (i.e., $C_c = 1$) (Seinfeld and Pandis 1998).

The Cunningham correction factor has been empirically determined to be (Seinfeld and Pandis 1998):

$$C_c = 1 + \frac{2\lambda}{\mathrm{D_p}} \left[1.257 + 0.4 \exp\left(-\frac{1.1\mathrm{D_p}}{2\lambda} \right) \right] \tag{19}$$

where λ is the mean free path. As shown in Figure 18, $C_c = 22$ for a 10-nm particle and is 216 for a 1-nm particle. This slip correction is small (under 1.08) for particles 2 µm and greater. Limit analysis of Equations 18 and 19 shows that the diffusion coefficient for nanoparticles goes as $D \propto 1/\mathrm{D_p^2}$.

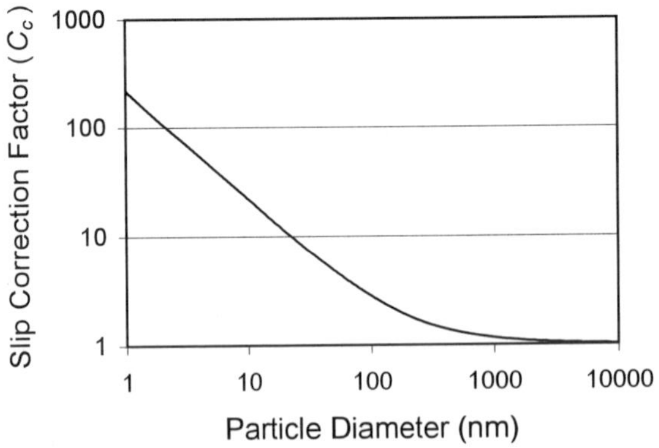

Figure 18. Cunningham slip correction factors (C_c) for spherical particles in air as calculated from Equation (19) at 298 K and 1 atm.

Although the terminal settling velocities for nanoparticles are extremely small and of little consequence, it is interesting to note that they are also much faster than predicted from continuum fluid dynamics. As with nanoparticle diffusion, this is a result of particle slip between gas molecules. The terminal settling velocity for particles ($\mathrm{D_p} \leq 20$ µm) in air is given by the following equation (Seinfeld and Pandis 1998):

$$v_t = \left(\frac{\mathrm{D_p^2} \rho_p g}{18\mu} \right) C_c \tag{20}$$

where ρ_p is the particle density and g is the gravitational acceleration (9.8 m s^{-2}). Thus the terminal settling velocities of nanoparticles are enhanced by the same factor (C_c) as are their diffusion coefficients.

Hygroscopic behavior

The phases of particles (e.g., aqueous or crystalline) affect their roles in atmospheric processes (Martin 2000). For example, light scattering is affected by both changes in refractive indices and particle sizes that accompany cycling between aqueous and crystalline states. Heterogeneous chemistry is also strongly influenced by the presence or absence of an aqueous phase. In atmospheric particles, salts comprise one important component in the hygroscopic response, though particulate organics are also increasingly recognized as playing a role (Pitchford and McMurry 1994; Saxena et al. 1995; Cruz and Pandis 2000).

In the lower troposphere, the dominant cycling of the inorganic component of particles is between aqueous and crystalline states as the relative humidity (RH) changes. At sufficiently high RH values, crystalline salts uptake water and deliquesce to form aqueous particles (Martin, 2000). With decreasing RH, crystallization (efflorescence) is often inhibited kinetically and high supersaturations are possible. For example, at 298 K deliquescence of $(NH_4)_2SO_4$ particles occurs at 80% RH but crystallization of the $(NH_4)_2SO_4/H_2O$ particles via homogeneous nucleation does not occur until 35% RH. Two ongoing research efforts focus on how this hysteresis cycle is altered by the presence of nanosized foreign surfaces such as mineral dusts inside the particle (Han and Martin 1999; Martin et al. 2001) and on how the deliquescence behavior is altered when the salt particles have a significant contribution from surface energies, i.e., nano-sized particles.

Effect of size on heterogeneous nucleation. Figure 17 shows an example of a mineral dust particle coated by hygroscopic salts. The deliquescence/efflorescence hysteresis cycle is reduced due to heterogeneous nucleation when foreign material, such

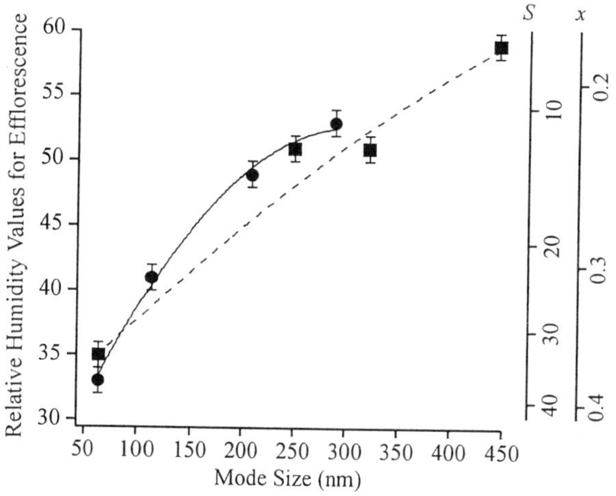

Figure 19. RH values observed for the efflorescence of ammonium sulfate (120 sec residence time) by heterogeneous nucleation as a function of mode diameter of the inclusions for corundum (●) and hematite (■). The lines show $\Phi = 0.50$ of the optimized fit to the active site model (Eqn. 24). Right axes show saturation ratios, *S*, of the aqueous phase with respect to crystalline ammonium sulfate and salt mole fractions, *x*, of the aqueous phase, as calculated from the model of Clegg et al. (1998) when assuming equilibrium between RH and water activity and omitting Kelvin effects. Adapted from Martin et al. (2001). Used by permission of the American Geophysical Union.

as mineral dusts, are present (Han and Martin 1999; Martin et al. 2001). Figure 19 shows that the RH value for $(NH_4)_2SO_4$ efflorescence increases with the size of submicron hematite and corundum inclusions (Martin et al. 2001). Rationalization of these data based upon changes in surface area is not possible within the framework of classical heterogeneous nucleation theory (Eqns. 10-12) when a reasonable range of physical parameters is employed. According to this theory, the dependence $d\mathrm{RH}_{\mathrm{eff}}/d\mathrm{D}_{\mathrm{p}}$ is nearly vertical. One critical assumption of classical theory is that the surface is everywhere the same, i.e., an infinitely extending terrace. An alternative theory, based on the occurrence of specific active sites, succeeds in explaining the size-dependence shown in Figure 19. According to the active-site theory (Fletcher 1969; Gorbunov and Kakutkina 1982), a surface is populated by nanoscale step edges, pits, kinks, and so on, of which each is possibly a heterogeneous nucleation center. This chapter follows the development by Han et al. (2001).

The central point of the active-site theory is that the areal density of nanoscale surface features is small and approaches the same scale as individual particle surface areas, especially in the regime of nanoparticles. The probability that a particle bears i active-sites is given by the Poisson distribution, as follows:

$$P_i = \exp(-4\pi R^2 n_0)\frac{(4\pi R^2 n_0)^i}{i!} \tag{21}$$

where R is the particle radius and n_0 is the number of active-sites per unit area (i.e., areal density). For each active-site, the probability distribution of the surface area, $P_s(A)$, is log-normal, as follows:

$$P_s(A) = B\exp\left\{-\gamma^2\left[\ln\left(\frac{A}{A_0}\right)\right]^2\right\} \tag{22}$$

where A is the active-site area, A_0 is the minimum active-site area, B constrains the integral $\int_{A_0}^{4\pi R^2} P_s(A)dA = 1$, and γ is the width of the distribution (Fletcher 1969). The probability, $P_g(A)$, of a critical embryo forming on an active-site of area A is given as follows:

$$P_g(A) = 1 - \exp[-A(j - j_0)\tau] \tag{23}$$

where j ($\mathrm{cm}^{-2}\,\mathrm{s}^{-1}$) is the heterogeneous nucleation rate over the active-site (see Eqn. 25), j_0 is the heterogeneous nucleation rate over a defect-free portion of the surface (i.e., a terrace as in Eqn. 10), and τ is the observation time. Based on Equations (21)-(23), Gorbunov and Kakutkina (1982) evaluated the fraction (i.e., probability), Φ, of particles undergoing a phase transition in an ensemble of monodisperse particles, as follows:

$$\Phi = 1 - \exp\left\{-4\pi R^2\left[j_0\tau + n_0\int_{A_0}^{4\pi R^2} P_s(A)P_g(A)dA\right]\right\} \tag{24}$$

For the infrared detection technique employed in Figure 19, the threshold sensitivity to the phase transition is $\Phi = 0.5$. The term j is calculated by a reduction in the free energy barrier due to an improved efficiency over the active site, as follows:

$$j = n\frac{kT}{h}\exp\left(-\frac{f(m)\Delta G_{germ} - A(m_1 - m)\sigma}{kT}\right) \tag{25}$$

where σ is the surface tension of the ammonium sulfate embryo/aqueous solution interface and m_1 describes the efficacy of heterogeneous nucleation at the active site (cf.

Eqn. 8). We assume perfect active sites (i.e., $m_1 = 1$) as done by Fletcher (1969). The active-site model embodied in Equation (24) rationalizes the data shown in Figure 19 for surfaces described as follows: {<0, 10.4 ± 0.4, 0.85 ± 0.15, 3 ± 1 × 10^{-15}} for α-Al_2O_3 and {0.04, 9.0 ± 0.2, 0.55 ± 0.05, 3 ± 1 × 10^{-15}} for α-Fe_2O_3 in the format {m, $\mathrm{Log}_{10}\, n_0$ (site/cm^2), γ, A_0 (cm^2)}.

The values for α-Al_2O_3 indicate that the maximum likelihood distribution for 100 particles of 10-nm diameter (Eqn. 21) is that 73 particles have no active sites, 23 have one site, 4 have two sites, and 1 has three or more sites. Each category of particle exhibits strikingly different efficiency as heterogeneous nuclei. Most notable is that those particles having no active sites yield nucleation rates no faster than homogeneous nucleation. In contrast, in a population of 1-μm particles, over 99% of the particles have 3150 ± 100 active sites. The result is a strong nonlinearity in the heterogeneous nucleation properties in scaling from nanoparticles to micron-sized particles.

Figure 19 empirically demonstrates the divergence of nanoparticle properties from bulk properties. At sufficiently large size (i.e., extrapolation beyond the micron), every surface contains a number of active sites and the hysteresis gap between deliquescence and efflorescence approaches a limiting value of zero. However, when the particles move into the nanoregime, the distribution statistics shown by the Poisson distribution in Equation (21) indicate that there are perceptible differences from one particle to the next in terms of how many active sites are on an individual particle. For the smallest particles (e.g., 10 nm), over half of the particles have no active sites at all and are very ineffective heterogeneous nuclei.

Effect of size on deliquescence relative humidities. For particle diameters below 1 μm, free energy is affected increasingly by the surface energy contribution. The result is a shift in the deliquescence relative humidity for small particle sizes (Hameri et al. 2000; Mirabel et al. 2000; LM Russell, unpublished results). Work in this area is new, and the experimental and theoretical results have not yet led to agreed-upon conclusions, i.e., whether decreasing particle size leads to higher or lower values of the deliquescence RH (DRH). A theoretical framework can be developed, as follows (Mirabel et al. 2000). The DRH value is the water activity at which the free energies are equal for water and salt molecules arranged as either vapor and crystalline salt or vapor and aqueous salt (Martin 2000). The respective free energy formulations are as follows:

$$G_{vapor/crystal} = N_{total}^{H_2O}\mu_{vapor}^{H_2O} + \left(N_{solute}\mu_{solute,bulk}^{crystal} + \sigma^{crystal/vapor}A^{crystal}\right)$$

$$G_{vapor/aqueous} = \left(\left(N_{total}^{H_2O} - N_{aqueous}^{H_2O}\right)\mu_{vapor}^{H_2O}\right) + \left(N_{solute}\mu_{solute,bulk}^{aqueous}(x) + N_{aqueous}^{H_2O}\mu_{H_2O,bulk}^{aqueous}(x) + \sigma^{aqueous/vapor}(x)A^{aqueous}\right) \quad (26)$$

The definitions of the symbols are: free energy (G), number of molecules (N), chemical potentials in the bulk (μ), surface area (A), surface tension (σ), and mole fraction composition (x). Equation (26) is developed by Mirabel et al. (2000) by assuming that particles are spherical (even when crystalline), that the solute is not volatile, that the vapor phase reservoir of water is infinite, and that all chemical potentials behave ideally with respect to x. Note that the first and fourth approximations are gross.

Figure 20 shows the solution to Equation (26) for several different assumed $\sigma^{crystal/vapor}$ values in the NaCl/H_2O system. The y-axis is the ratio of the DRH value anticipated from Equation (26) to the DRH value of a bulk system (DRH_{bulk} = 75% at 298 K) as a function of the number of solute molecules in the particle. In the nanosize regime, the DRH value is predicted to be shifted to much lower values in most cases. This result

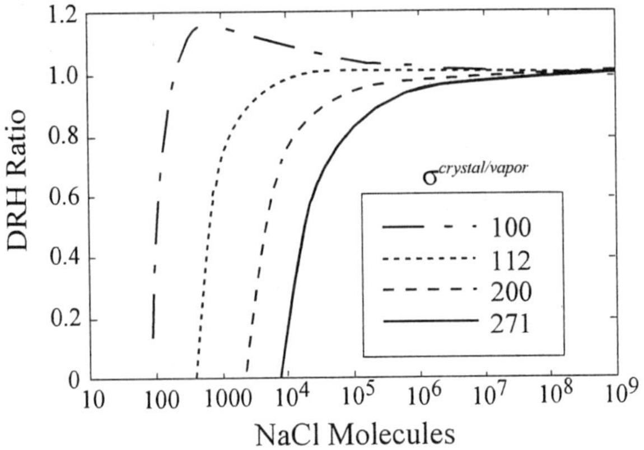

Figure 20. Plots of DRH relative to bulk crystal (viz. 75%) versus crystal size for various values of crystal/vapor surface tensions. Note the sensitivity of DRH to surface tension. Adapted from Mirabel et al. (2000). Used by permission of the American Chemical Society.

is driven primarily by the difference between $\sigma^{crystal/vapor}$ and $\sigma^{aqueous/vapor}$ (taken as 70 mJ m^{-2} independent of x in these calcuations). Because $\Delta\sigma > 0$ for normally behaving materials (i.e., surface tensions of solids against vapor are greater than liquids against vapor), there is a thermodynamic driving force to favor the aqueous state when sizes are small enough that the surface energy makes a significant contribution to the overall free energy of the particle. In a few cases (e.g., $\sigma^{crystal/vapor}$ = 100 mJ m^{-2}), a maximum is seen in Figure 20 where DRH of nanoparticles is greater than the bulk value. This subtle effect, which is explained in detail by Mirabel et al. (2000), arises because H_2O is volatile and NaCl is not volatile. In this case, the mole fraction composition of the smallest aqueous droplets are enriched in NaCl because the Kelvin effect drives the evaporation of water. The net effect on the free energies is that DRH increases as compared to bulk values.

Because NaCl is known to adsorb multilayers of water prior to the DRH value, the physical description of the chemical system given by Mirabel et al. (2000) may not be accurate. The multilayer of adsorbed water is believed to be ordered and relatively pure, i.e., free of ions. This interesting phenomenon is under investigation, i.e., pure water whose chemical potential is under unity (Peters and Ewing 1997a,b). The above formulation for the free energy of the system (Eqn. 26) can be revised for a multilayer interface involving the surface tensions of the water layer against the vapor and of the crystal against the water layer, as follows:

$$\begin{aligned} G_{vapor/adsorbed/crystal} &= \left(N_{total}^{H_2O} - N_{adsorbed}^{H_2O}\right)\mu_{vapor}^{H_2O} + \left(N_{adsorbed}^{H_2O}\mu_{adsorbed}^{H_2O} + \sigma^{vapor/adsorbed}A^{adsorbed}\right) \\ &\quad + \left(N_{solute}\mu_{solute,bulk}^{crystal} + \sigma^{adsorbed/crystal}A^{crystal}\right) \\ G_{vapor/aqueous} &= \left(\left(N_{total}^{H_2O} - N_{aqueous}^{H_2O}\right)\mu_{vapor}^{H_2O}\right) + \\ &\quad \left(N_{solute}\mu_{solute,bulk}^{aqueous}(x) + N_{aqueous}^{H_2O}\mu_{H_2O,bulk}^{aqueous}(x) + \sigma^{aqueous/vapor}(x)A^{aqueous}\right) \end{aligned} \tag{27}$$

A key point is that both $\sigma^{vapor/absorbed}$ and $\sigma^{absorbed/crystal}$ are less than $\sigma^{crystal/vapor}$. Recent work (L.M. Russell, unpublished results) has shown that for soluble salts that adsorb water below their deliquescence point (including NaCl), the DRH values increase with decreasing particle size for typical values of $\sigma^{vapor/absorbed}$ and $\sigma^{absorbed/crystal}$. This conclusion differs markedly from the formulation given in Equation (26) and shown in Figure 20, where the limiting value of DRH is zero at small particle sizes. These differences highlight the uncertainty and the need for further theoretical and laboratory work on the effect of particle size on DRH values.

Chemical reactivity

Chemical reactions involving atmospheric nanoparticles are potentially important for at least two reasons: (1) reactions might alter the chemical composition and thus the physicochemical properties of nanoparticles and (2) reactions might change the composition of the gas-phase. For example, reactions with O_3 make soot particles more hydrophilic, which should enhance their ability to act as cloud condensation nuclei (Kotzick et al. 1997). Although it has been speculated that nanoparticle reactions also alter gas-phase composition, this is probably quite uncommon except perhaps in aerosols with very high nanoparticle surface areas. Even so, in some specific cases there might be a unique role for nanoparticles in atmospheric chemistry. Such cases could occur, for example, if a nanoparticle reaction is fast compared to the equivalent reaction in the gas-phase or in/on other condensed phases such as cloud drops or larger aerosol particles. Nanoparticles could also significantly alter the composition of the gas phase if the nanoparticle reaction is unique in terms of the reactants destroyed or products produced. To the best of our knowledge, there are no identified instances of nanoparticle reactions in the atmosphere where either of these two criteria is satisfied. However, this field is largely unexplored at the time of this writing.

As a first step in assessing the potential importance of nanoparticle reactions, we compare the volume and surface areas of these particles with the same values from other condensed phases with known chemical effects. We first consider nanoparticle volumes. As an upper limit, we consider an urban air parcel containing 20-nm diameter nanoparticles at a number concentration of 10^5 cm^{-3}. Under this scenario, the nanoparticle volume is a small fraction (10^{-13}) of the total air parcel volume. Thus the nanoparticle reaction rate (in units of mol m^{-3}-air s^{-1}) would have to be ca. 10^{13} times as fast as the equivalent gas phase reaction (mol m^{-3}-air s^{-1}) to have a comparable overall rate in the air parcel. For comparison, clouds typically have liquid water contents of 10^{-7} to 10^{-6} (volume fraction) and can have significant effects upon atmospheric chemistry (Seinfeld and Pandis 1998). For simplicity of argument, if the medium of the cloud droplets and nanoparticles are assumed similar (e.g., dilute aqueous), then the fundamental rate constants in each medium are similar. Under this condition, reactant concentrations in the nanoparticles would need to be enhanced by 10^6, as compared to the cloud droplets, to have equal rates. Based on this analysis, it appears unlikely that reactions occurring in the bulk of nanoparticles could affect the composition of the gas phase.

The other possibility is a reaction at the surface of the nanoparticles. For the assumed urban scenario above, the nanoparticle surface area concentration is approximately 100 μm^2 per cm^3-air. At this surface area loading, nanoparticle reactions could plausibly affect the composition of the gas phase. For example, de Reus et al. (2000a) report that plumes of mineral dust particles (primarily in the accumulation mode), with surface area concentrations of ca. 100 μm^2 cm^{-3}, significantly reduce mixing ratios of ozone and nitric acid. While this comparison suggests nanoparticle surface reactions could be significant in urban atmospheres, the conclusion depends on a number of factors, including high particle number concentrations (which are unlikely outside of urban areas), a reactive particle surface with sufficiently high uptake coefficients for trace gases, and surface renewal during the course of the reaction.

In the remaining portion of this section we examine several nanoparticle types and reactions that could potentially be significant in the atmosphere. Of course, until the chemical composition, morphology, and reactivity of nanoparticles are better characterized, it is not possible to assess accurately the significance of these or other reactions.

Concentrated H_2SO_4. Sulfuric acid is an important participant in the nucleation and the growth of new particles in the troposphere. The mole fraction acidity (viz. H_2SO_4 content) of these particles is believed to be high. It is likely that the presence of this strong sulfate acidity influences the chemistry of nanoparticles because sulfuric acid promotes a number of organic reactions that otherwise either do not occur or occur much more slowly (Liler 1971). We will consider several of these reaction classes here with a focus on those involving organic species common in the atmosphere.

One class of organic reactions in sulfuric acid is hydrolysis, which affects functional groups such as ethers, esters, nitriles, halogenated organics, and converts amides to the corresponding carboxylic acid, as follows:

$$RC(O)NH_2 + H_2SO_4 + H_2O \rightarrow RC(O)OH + NH_4^+ + HSO_4^- \qquad (28)$$

In addition, concentrated sulfuric acid catalyzes the decarbonylation of aldehydes, aromatic acids, and aliphatic carboxylic acids, such as formic acid:

$$HC(O)OH + H_2SO_4 \rightarrow CO + H_2O + H_2SO_4 \qquad (29)$$

Other classes of reactions are the dehydrations and condensations, such as the conversion of acetone to protonated mesityl oxide, as follows:

$$2\ CH_3C(O)CH_3 + H_2SO_4 \rightarrow (CH_3)_2C{=}CHC(OH^+)CH_3 + H_2O + HSO_4^- \qquad (30)$$

Finally, a number of commonly occurring atmospheric trace species are converted to reactive electrophiles in concentrated sulfuric acid solutions. For example, nitric acid is converted to NO_2^+, which is a potent nitrating agent for aromatic compounds (Liler 1971), and hydrogen peroxide can be converted to peroxymonosulfuric acid ($HOOSO_2OH$), which is a strong oxidant (Mozurkewich 1995; Dalleska et al. 2000).

Studies of these reactions have typically been carried out in very concentrated sulfuric acid solutions (85 to 100 wt % H_2SO_4) (Liler 1971). More recent work examines whether similar types of reactions occur at lower sulfuric acid concentrations. Roberts and co-workers, for example, examine the condensation reactions of acetone in thin films of 70 to 96 wt % H_2SO_4 at temperatures found in the upper troposphere and lower stratosphere (180-220 K) (Duncan et al. 1998, 1999). According to this work, acetone is protonated in 70 wt % H_2SO_4, undergoes condensation to mesityl oxide (Eqn. 30) above 75 wt % H_2SO_4, and undergoes further condensation to trimethylbenzene above 85% H_2SO_4. Thus acetone condensation to larger carbon products requires quite high acidities. It has not yet been investigated how the rates and extents of the other reactions discussed above vary at lower sulfate acidities.

The weight percent of H_2SO_4 in stratospheric sulfuric acid particles is typically between 40 and 80% and stratospheric temperatures are generally between 210 and 240 K (Shen et al. 1995; Tabazadeh et al. 1997, 2000; Finlayson-Pitts and Pitts 2000). On the other hand, tropospheric sulfuric acid particles typically contain less than 75 wt % H_2SO_4 (this concentration is in equilibrium with a RH value of less than 1% at 298 K), and temperatures are typically above 230 K. For example, at 298 K pure sulfuric acid-water droplets (D_p > 100 nm) contain 42.5 wt % H_2SO_4 at a relative humidity of 50% (Seinfeld and Pandis 1998). Thus it is not known whether the reactions mentioned above occur to any appreciable extent in the troposphere or stratosphere. In the troposphere, nanoparticles of sulfuric acid may be much more acidic than larger sulfuric acid particles due to an increase in water vapor pressure with decreasing particle size (i.e., the Kelvin effect) (Fig. 21). For example, at 50% RH the weight percent of H_2SO_4 is 1.7 times

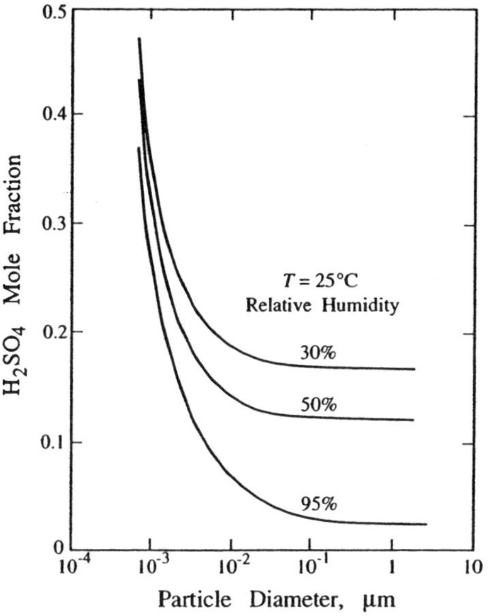

Figure 21. Mole fraction of H_2SO_4 ($x_{H_2SO_4}$) in a spherical drop of sulfuric acid and water as a function of relative humidity and drop diameter. The weight percent (wt %) of H_2SO_4 in solution is related to the mole fraction by: wt % = $\{MW_{H_2SO_4} / [(MW_{H_2O}(1/x_{H_2SO_4} - 1) + MW_{H_2SO_4}]\} \times 100\%$, where $MW_{H_2SO_4}$ and MW_{H_2O} are the molecular weights of sulfuric acid and water, respectively. From Seinfeld and Pandis (1998). Used by permission of Wiley-Interscience.

greater for a 1-nm particle (72 wt %) as compared to a 100-nm particle (43 wt %). Thus, organic reactions in sulfuric acid, such as those described above, are more likely to occur in nanoparticles and under conditions of low relative humidity. Because some reactions, such as the condensation of aldehydes (e.g., Eqn. 30), produce organic products that are of higher molecular weight and lower volatility than the initial organic compound, these chemical pathways could contribute to the growth of nanoparticles.

Soot. Combustion particles containing elemental carbon with variable amounts of organic carbon and trace metals (i.e., soot) consist of 30- to 40-nm diameter nanoparticles agglomerated in chains with typical lengths of a few hundred nanometers (Kennedy 1997; Seinfeld and Pandis 1998; Clague et al. 1999). In the past five years there has been growing interest in atmospheric reactions involving soot. Much of the attention has focused on reactions with O_3 and nitrogen oxides (i.e., NO and NO_2) because of the importance of these species as air pollutants.

The most significant soot reaction identified to date is the conversion of nitrogen dioxide (NO_2) to nitrous acid (HONO) and smaller amounts of nitric oxide (NO) (Ammann et al. 1998; Gerecke et al. 1998; Kalberer et al. 1999; Longfellow et al. 1999; Al-Abadleh and Grassian 2000; Alcala-Jornod et al. 2000; Stadler and Rossi 2000; Saathoff et al. 2001), as follows:

$$NO_2 + \text{soot} \rightarrow\rightarrow \text{HONO} + \text{NO} \qquad (31)$$

This reaction is of interest because current models underestimate the production of HONO, which is an important photolytic source of hydroxyl radical (Ammann et al. 1998; Finlayson-Pitts and Pitts 2000). This reaction could also be important as a reductive pathway for nitrogen species in the atmosphere, which is in contrast to most atmospheric reactions which tend to oxidize trace gases (e.g., the conversion of SO_2 to H_2SO_4). Molar yields of HONO via Equation (31) (relative to the moles of NO_2 lost) typically range from 50 to 100%, while yields of NO are 4 to 30% (Gerecke et al. 1998; Stadler and Rossi 2000). It appears that the formation of NO results from the secondary reaction of HONO at the surface of the soot (Stadler and Rossi 2000). The mechanism of this reaction is not known, but water vapor appears to play a significant role (Longfellow et al. 1999). There is also some evidence that the reaction involves organic compounds that can be solvent-extracted from the soot (Stadler and Rossi 2000). There is no consensus on whether this reaction is a significant source of HONO in the polluted troposphere (Ammann et al. 1999; Kalberer et al. 1999; Kirchner et al. 2000; Saathoff et al. 2001) because there is uncertainty whether reactive sites on soot are catalytic (Longfellow et al. 1999) or are deactivated during the reaction (Kalberer et al. 1999; Kirchner et al. 2000).

A number of other atmospheric gases also react with soot. For example, ozone is destroyed on soot to form O_2, CO, and CO_2 (Gao et al. 1998; Kamm et al. 1999), while HNO_3 is reduced either to NO or NO_2 on soot (Choi and Leu 1998; Disselkamp et al. 2000a) or is simply physically adsorbed (Longfellow et al. 2000). In any case, the uptake coefficients for O_3 and HNO_3 are small enough that these soot reactions are not important as sinks for these gases or as a source of NO and NO_2 (Choi and Leu 1998; Gao et al. 1998; Kamm et al. 1999; Disselkamp et al. 2000a, 2000b; Longfellow et al. 2000). There is some evidence that the soot uptake of hydroperoxyl radical ($HO_2^{\bullet}$), which is a key component of ozone (smog) chemistry in the troposphere, is large enough to reduce O_3 mixing ratios in heavily polluted areas (Saathoff et al. 2001).

Although these O_3 and HNO_3 reactions are not important as sinks for the gases, they are likely important as pathways that chemically transform the surface of soot. For example, exposure of soot to HNO_3 or NO_2 produces surface species such as R-C=O, R-NO_2, R-ONO_2, and R-ONO (Al-Abadleh and Grassian 2000; Kirchner et al. 2000), while exposure to O_3 produces oxygen-containing functional groups such as -OH, R-C=O, and R-C-O (Kotzick et al. 1997; Al-Abadleh and Grassian 2000; Kirchner et al. 2000). These reactions are important because they make the surface of soot particles more polar, which increase soot's efficiency as a cloud condensation nucleus (Kotzick et al. 1997). These reactions might also affect the toxicity of soot particles. Given that soot is also an important reservoir of atmospheric polycylic aromatic hydrocarbons (PAHs) (Dachs and Eisenreich 2000), reactions of trace gases such as O_3 with soot may also be important in the transformation and destruction of these toxic molecules (Finlayson-Pitts and Pitts 2000; Poschl et al. 2001).

Although a number of studies have used commercially available carbon black as a surrogate for soot, there are large differences in the composition, and therefore probably in the chemistry, between these two types of materials (Watson and Valberg 2001). Diesel soot, for example, is much more oxygenated, has a more polar surface, and contains higher amounts of trace species such as Fe, S, and Ca than carbon black (Clague et al. 1999). Similarly, even with authentic soots there can be large differences in chemistry, both as a result of different fuels as well as different fuel/air ratios during soot formation (Alcala-Jornod et al. 2000; Kirchner et al. 2000; Stadler and Rossi 2000).

Metal oxides. Although most published chemical studies with prepared nanoparticles involve reactants bearing little or no relevance to the atmosphere, there are a number of reports of reactions involving metal or metal oxide nanoparticles that might have analogues in the atmosphere. For example, Klabunde and co-workers report that halocarbons, which are a significant class of anthropogenic atmospheric pollutants, are destroyed by nanoparticles of calcium oxide and magnesium oxide to form products such as CO, CO_2, elemental carbon, and the corresponding metal chloride (Koper et al. 1993, 1997; Li et al. 1994; Koper and Klabunde 1997). However, these reactions are likely insignificant in the atmosphere because they require high temperatures (typically above 600 K). It has also been shown that gases such as NO, HCl, SO_2, and CO_2 adsorb on the surface of MgO and in some cases react at room temperature (Stark and Klabunde 1996;

Stark et al. 1996). However, for these CaO and MgO reactions, the metal oxides are stoichiometrically consumed during the process. Thus even if the reactions could stoichiometrically destroy pollutant gases at atmospheric temperatures, their importance in the atmosphere should be limited by the small available masses of nanoparticulate CaO and MgO.

This stoichiometric constraint on reactivity is eliminated for catalytic or photocatalytic reactions involving nanoparticles. Although no quantitative investigations have yet been performed in either laboratory experiments or modeling, a reasonable speculation is that metals or metal oxides embedded in individual atmospheric nanoparticles might serve as catalysts for oxidation and reduction reactions in the atmosphere, as suggested by José-Yacamán (1998). In support of this view, synthesized nanoparticles containing Fe, Cr, Cu, Ti and other metals are typically very reactive, often catalytically, in reduction and oxidation reactions involving atmospheric pollutants such as CO, CH_4, and larger organic gases (Park and Ledford 1997, 1998; Zhang et al. 1998; Perkas et al. 2001). Furthermore, the rates and the efficiencies of reactions of metal oxide nanoparticles are often enhanced relative to the bulk material due to different surface structures (Stark et al. 1996; Jefferson 2000). However, it is unclear whether the reactivities of nanoparticles used in these studies are comparable to metal and metal oxide nanoparticles in the atmosphere. Moreover, given their small amounts, it is unclear whether reactions involving metal and metal oxide nanoparticles are as significant as similar reactions that might occur on accumulation and coarse mode mineral dusts.

Electronic. Several minerals, including oxides of Fe, Si, Ti, or Zn, have been investigated as active heterogeneous photocatalysts in the atmosphere. Gas/solid reactions have been investigated, for example, as tropospheric sinks for chlorofluorocarbons on mineral dusts (Ausloos et al. 1977; Benzoni and Garbassi, 1984; Filby et al. 1981; Zakharenko 1997). Photochemical conversions of aqueous species at the surface of iron oxyhydroxides (e.g., hematite, goethite, lepicrodocite, or ferrihydrite) contained within cloud droplets have also been the focus of considerable study (Sulzberger et al. 1988; Behra and Sigg 1990; Pehkonen et al. 1993, 1995; Litter et al. 1994; Siefert et al. 1994). These reactions favor the formation of oxidized species, both inorganic and organic, within cloud droplets.

The photochemical mechanism is worked out most completely for hematite. This semiconducting mineral generates electron-hole pairs when irradiated with sunlight (Sulzberger 1990; Hoffmann et al. 1995). The valence band holes oxidize adsorbed organics such as oxalate to carbon dioxide and other reduced species such as bisulfite to sulfate. As a by-product, ferrous iron is released to the aqueous solution from the ferric oxide matrix of the mineral. Although the absorption spectrum of hematite is in the red, this color arises mostly from *d*-orbital transitions in the band structure (2.2 eV) (Finklea 1988). The photoaction spectrum does not show significant activity until 3 eV (Siffert and Sulzberger 1991), which is explained by the band structure. Transitions begin at 3 eV from the occupied O(2*p*) valence band to the empty conduction band, which is composed of Fe centered *d*-orbitals. The photochemistry is explained either by hole scavenging by adsorbates or by direct charge transfer from localized orbitals of the oxygen atoms of adsorbed ligands into the conduction band.

The general process of mineral dissolution by photochemical reduction, as well as by pH-dependent non-oxidative thermal dissolution of other minerals in cloud waters (Desboeufs et al. 1999), yields the formation of dissolved aqueous species. Upon cloud droplet evaporation, one reasonable fate we can suggest for these species is reoxidation by abundant atmospheric O_2 followed by recrystallization into nanoparticles (since oxidized ions in general have lower solubility (Stumm and Morgan 1996)). There are two assumptions inherent in this suggestion: (1) since acidic media dissolve minerals and thus disfavor precipitation of nanoparticles, the acidic cloud drop needs to be neutralized (e.g., by alkaline mineral constituents or ammonia) and (2) mineral precipitation forms new nanoparticles, possibly as accessory minerals, rather than growing material on preexisting surfaces of the same or similar minerals. Furthermore, Ostwald's rule of stages suggests amorphous phases with less developed band structures would be favored, at least initially, over more crystalline phases (Lasaga 1998). Under conditions favorable with respect to these assumptions, we offer the speculation that nanocrystalline mineral particles crystallize from cloud droplets.

Due to their small size, if these nanoparticles form they should possess altered electronic properties (Brus 1983), including a wider bandgap and thus a blue-shifted photoaction spectrum. Although this effect reduces the sunlight absorbed, the holes formed in the valence band have a higher thermodynamic driving force to promote chemical reactions. Novel chemical channels may then be opened or, alternatively, rates of chemical conversion may be increased for known channels. The idea of the formation of nanocrystalline particles with unique properties within cloud droplets is at best speculative at this time. Even so, it seems appropriate within the aims of this chapter to provide a perspective on possible properties and occurrence of atmospheric nanoparticles with electronic properties different from the bulk. Another occurrence may be in the stratosphere where 50% of the particles are estimated to contain 0.5 to 1 (w/w) iron oxide (Cziczo et al. 2001). It is unknown if the thermodynamics of highly acidic particles (60-80 wt % H_2SO_4) at extremely low temperatures (210-225 K) favors a dissolved or precipitated state for the iron. However, if crystalline, the iron should exhibit size-quantization effects in its band structure.

OUTLOOK

Many critical uncertainties must still be addressed to assess quantitatively the sources and sinks of atmospheric nanoparticles, their physical and chemical evolution during atmospheric aging, their direct and indirect impacts on atmospheric chemistry and physics, and their connections to adverse effects on human health. Several important and complex areas that need much additional study include:

1. New particle formation and growth. These processes are important sources of new particles and CCN, but field observations suggest that current theories relying on H_2SO_4 and H_2O are often inadequate to explain nucleation or growth. The role of NH_3 in a possible ternary nucleation mechanism needs to be examined quantitatively at more field sites. The identities of nonvolatile and low-volatile organic species involved in nucleation and growth should be further examined. To do so, highly sensitive and specific analytical techniques must be developed. In theoretical work, contemporary nonclassical theories of homogeneous nucleation are promising, but they are urgently in need of further data on the kinetics of cluster formation from laboratory studies.
2. Particle composition and morphology. Understanding the adverse health effects and possible chemical reactivity of atmospheric nanoparticles requires knowing their chemical composition and morphology. Quantitative measurements of the composition of ambient nanoparticles, as well as of nanoparticles from emission sources, are needed. Both traditional fine particle emission sources (e.g., combustion) as well as coarse particle emission sources (e.g., oceanic sea-salt and mineral dust) should be examined. Real-time single-particle compositional analysis, albeit still qualitative, has occurred through impressive instrumentation advances in the last few years. These advances highlight the need for commensurate progress in techniques sensitive not

only to overall particle morphology but also to internal structures such as water pockets, grain structures, surfactant layers, or radial compositional layering. In most cases, investigators must be cognizant of the overriding role of relative humidity in affecting morphology.

3. Chemical and photochemical reactions. Evaluating the potential effects of nanoparticles on atmospheric chemistry and composition will require much additional laboratory and modeling work. Experiments need to be performed to measure uptake coefficients, reaction rates, and products formed from the interaction of atmospheric trace gases with authentic or representative nanoparticles. Special attention should be given to important reactions in urban areas (e.g., ozone formation) where nanoparticle concentrations are typically greatest. An additional focus should be to examine reactions with important species that are destroyed only slowly in the gas-phase (e.g., CO or CH_4) since in these cases nanoparticle reactions might be fast enough to compete effectively with the gas-phase reactions. With these data it should then be possible to model atmospheric nanoparticle reactions and their significance.
4. Health effects. More epidemiological and clinical studies are needed to further examine the hypothesis that nano- and ultrafine particles are responsible for adverse human health effects. It will be especially valuable to perform experiments that allow elucidation of the particle properties that are responsible for toxicity, such as number concentration, surface area concentration, or specific chemical species.
5. Unique size-dependent properties. Although it is known that nanosized objects have properties diverging from the bulk, as highlighted in contemporary work on the electronic properties of semiconductors, unique chemical and physical properties in systems relevant to atmospheric nanoparticles are mostly speculative at this time and poorly informed by data. However, further intensive work is warranted based upon available data such as observations that heterogeneous nucleation properties depend on size and reach a critical regime at small size due to the statistics of small numbers, e.g., not every particle has an active site. Deliquescence relative humidities of salts depend critically on size at the nanoscale due to significant contributions of surface energies to the overall free energies. Electronic properties, including photochemistry, of metallic and semiconducting particles present in the atmosphere are strongly altered in nanosized objects. These objects may be spatially separate individual particles or they may instead be a single grain or inclusion embedded within a larger particle. These possible properties and their effects on atmospheric processes need to be investigated through additional laboratory work.

ACKNOWLEDGMENTS

This chapter is dedicated to the memory of Glen Cass and his many important contributions to atmospheric aerosol science.

We gratefully acknowledge assistance from the authors who provided us access to their unpublished manuscripts: V.H. Grassian, M.V. Johnston, L.M. Russell, and A.S. Wexler. We are also grateful to A.S. Wexler, A. Navrotsky, H.M. Hung, J. Schlenker, and A. Fillinger for valuable comments on the manuscript. C.A. appreciates the help of J. Chan, I. George, and Q. Zhang in completing this project, as well as funding from the National Science Foundation Atmospheric Chemistry Program through a CAREER award (ATM-9701995), the NIEHS through the University of California-Davis Superfund Program, and the NSF-funded NEAT (Nanophases in the Environment, Agriculture and Technology) IGERT at UC-Davis. S.M. is grateful for support received from the NSF Atmospheric Chemistry Program, a Presidential Early Career Award in Science and Engineering (PECASE), and the New York Community Trust Merck Fund.

REFERENCES

Abbatt J, Molina MJ (1993) Status of stratospheric ozone depletion. Ann Rev Energy Environ 18:1-29

Ackerman AS, Toon OB, Stevens DE, Heymsfield AJ, Ramanathan V, Welton EJ (2000) Reduction of tropical cloudiness by soot. Science 288:1042-1047

Al-Abadleh HA, Grassian VH (2000) Heterogeneous reaction of NO_2 on hexane soot: A Knudsen cell and FT-IR study. J Phys Chem A 104:11926-11933

Albrecht BA (1989) Aerosols, cloud microphysics, and fractional cloudiness. Science 245:1227-1230

Alcala-Jornod C, van den Bergh H, Rossi MJ (2000) Reactivity of NO_2 and H_2O on soot generated in the laboratory: A diffusion tube study at ambient temperature. Phys Chem Chem Phys 2:5584-5593

Ammann M, Kalberer M, Jost DT, Tobler L, Rossler E, Piguet D, Gaggeler HW, Baltensperger U (1998) Heterogeneous production of nitrous acid on soot in polluted air masses. Nature 395:157-160

Ammann M, Madronich S, Kalberer M, Baltensperger U, Hauglustaine D, Brocheton F (1999) On the NO_2 plus soot reaction in the atmosphere. J Geophys Res 104:1729-1736

Anderson JG (1995) Polar processes in ozone depletion. *In* Progress and Problems in Atmospheric Chemistry. Barker JR (ed) p 744-770. Singapore: World Scientific

Andreae MO, Crutzen PJ (1997) Atmospheric aerosols: Biogeochemical sources and role in atmospheric chemistry. Science 276:1052-1058

Arts JHE, Spoor SM, Muijser H (2000) Short-term inhalation exposure of healthy and compromised rats and mice to fine and ultrafine carbon particles. Inhal Toxicol 12 (Suppl 1):261-266

Ausloos P, Rebbert RE, Glasgow L (1977) Photo-decomposition of chloromethanes adsorbed on silica surfaces. J Res Nat Bur Stand 82:1-8

Bagley ST, Baumgard KJ, Gratz LG, Johnson JH, Leddy DG (1996) Characterization of fuel and aftertreatment device effects on diesel emissions. Research Report 76. Cambridge, Massachusetts: Health Effects Institute (HEI)

Ball SM, Hanson DR, Eisele FL, McMurry PH (1999) Laboratory studies of particle nucleation: Initial results for H_2SO_4, H_2O, and NH_3 vapors. J Geophys Res 104:23709-23718

Barnett TP, Hasselmann K, Chelliah M, Delworth T, Hegerl G, Jones P, Rasmusson E, Roeckner E, Ropelewski C, Santer B, Tett S (1999) Detection and attribution of recent climate change: A status report. Bull Am Meterol Soc 80:2631-2659

Barraclough PB, Hall PB (1974) Adsorption of water-vapor by lithium-fluoride, sodium-fluoride, and sodium-chloride. Surf Sci 46:393

Barthelmie RJ, Pryor SC (1999) A model mechanism to describe oxidation of monoterpenes leading to secondary organic aerosol: 1. Alpha-pinene and beta-pinene. J Geophys Res 104:23657-23669

Becker EJ, O'Dowd CD, Hoell C, Aalto P, Mäkelä JM, Kulmala M (1999) Organic contribution to sub-micron aerosol evolution over a boreal forest: A case study. Phys Chem Chem Phys 1:5511-5516

Behra P, Sigg L (1990) Evidence for redox cycling of iron in atmospheric water droplets. Nature 344: 419-421

Benzoni L, Garbassi F (1984) Reactivity of fluorochloromethanes with desert sands. Ber Bunsen Phys Chem 88:379-382

Berg OH, Swietlicki E, Krejci R (1998) Hygroscopic growth of aerosol particles in the marine boundary layer over the Pacific and Southern Oceans during the First Aerosol Characterization Experiment (ACE 1). J Geophys Res 103:16535-16545

Boer GJ, Flato G, Ramsden D (2000a) A transient climate change simulation with greenhouse gas and aerosol forcing: Projected climate to the twenty-first century. Clim Dynam 16:427-450

Boer GJ, Flato G, Reader MC, Ramsden D (2000b) A transient climate change simulation with greenhouse gas and aerosol forcing: Experimental design and comparison with the instrumental record for the twentieth century. Clim Dynam 16:405-425

Borensen C, Kirchner U, Scheer V, Vogt R, Zellner R (2000) Mechanism and kinetics of the reactions of NO_2 or HNO_3 with alumina as a mineral dust model compound. J Phys Chem A 104:5036-5045

Borys RD, Lowenthal DH, Mitchell DL (2000) The relationships among cloud microphysics, chemistry, and precipitation rate in cold mountain clouds. Atmos Environ 34:2593-2602

Bowles RK, McGraw R, Schaaf P, Senger B, Voegel JC, Reiss H. (2000) A molecular based derivation of the nucleation theorem. J Chem Phys 113:4524-4532

Boyd PW, Watson AJ, Law CS, Abraham ER, Trull T, Murdoch R, Bakker DCE, Bowie AR, Buesseler KO, Change H, Charette M, Croot P, Downing K, Frew R, Gall M, Hadfield M, Hall J, Harvey M, Jameson G, LaRoche J, Liddicoat M, Ling R, Maldonado MT, McKay RM, Nodder S, Pickmere S, Pridmore R, Rintoul S, Safi K, Sutton P, Strzepek R, Tanneberger K, Turner S, Waite A, Zeldis J (2000) A mesoscale phytoplankton bloom in the polar Southern Ocean stimulated by iron fertilization. Nature 407:695-702

Brock CA, Schröder F, Kärcher B, Petzold A, Busen R, Fiebig M (2000) Ultrafine particle size distributions measured in aircraft exhaust plumes. J Geophys Res 105:26555-26567

Brunekreef B (2000) SESSION 2: What properties of particulate matter are responsible for health effects? Inhal Toxicol 12 (Suppl 1):15-18

Brus LE (1983) A simple model for the ionization potential, electron affinity, and aqueous redox potentials of small semiconductor crystallites. J Chem Phys 79:5566

Burtscher H, Kunzel S, Huglin C (1998) Characterization of particles in combustion engine exhaust. J Aerosol Sci 29:389-396

Buseck PR, Pósfai M (1999) Airborne minerals and related aerosol particles: Effects on climate and the environment. Proc Nat Acad Sci USA 96:3372-3379

Buseck PR, Jacob DJ, Pósfai M, Li J, Anderson JR (2000) Minerals in the air: An environmental perspective. Int Geol Rev 42:577-593

Carson PG, Johnston MV, Wexler AS (1997) Laser desorption/ionization of ultrafine aerosol particles. Rapid Commun Mass Spectrom 11:993-996

Cass GR, Hughes LA, Bhave P, Kleeman MJ, Allen JO, Salmon LG (2000) The chemical composition of atmospheric ultrafine particles. Phil Trans Roy Soc Lond A 358:2581-2592

Chameides WL, Stelson AW (1992) Aqueous-phase chemical processes in deliquescent sea-salt aerosols—A mechanism that couples the atmospheric cycles of S and sea salt. J Geophys Res 97:20565-20580

Chameides WL, Lindsay RW, Richardson J, Kiang CS (1988) The role of biogenic hydrocarbons in urban photochemical smog: Atlanta as a case study. Science 241:1473-1475

Charlesworth DH, Marshall WR (1960) Evaporation from drops containing dissolved solids. Am Inst Chem Engin J 6:9-23

Charlson RJ, Heintzenberg J (eds) (1994) Aerosol Forcing of Climate. New York: Wiley

Charlson RJ, Schwartz SE, Hales JM, Cess RD, Coakley JA, Hansen JE, Hofmann DJ (1992) Climate forcing by anthropogenic aerosols. Science 255:423-430

Chen DR, Pui DYH (1997) Numerical modeling of the performance of differential mobility analyzers for nanometer aerosol measurements. J Aerosol Sci 28:985-1004

Chiapello I, Prospero JM, Herman JR, Hsu NC (1999) Detection of mineral dust over the north Atlantic ocean and Africa with the Nimbus 7 TOMS. J Geophys Res 104:9277-9291

Choi W, Leu MT (1998) Nitric acid uptake and decomposition on black carbon (soot) surfaces: Its implications for the upper troposphere and lower stratosphere. J Phys Chem A 102:7618-7630

Clague ADH, Donnet J, Wang TK, Peng JCM (1999) A comparison of diesel engine soot with carbon black. Carbon 37:1553-1565

Claquin T, Schulz M, Balkanski YJ (1999) Modeling the mineralogy of atmospheric dust sources. J Geophys Res 104:22243-22256

Clarke AD, Davis D, Kapustin VN, Eisele F, Chen G, Paluch I, Lenschow D, Bandy AR, Thornton D, Moore K, Mauldin L, Tanner D, Litchy M, Carroll MA, Collins J, Albercook G (1998) Particle nucleation in the tropical boundary layer and its coupling to marine sulfur sources. Science 282:89-92

Clarke AD, Kapustin VN, Eisele FL, Weber RJ, McMurry PH (1999) Particle production near marine clouds: Sulfuric acid and predictions from classical binary nucleation. Geophys Res Lett 26:2425-2428

Clegg SL, Brimblecombe P, Wexler AS (1998) Thermodynamic model of the system H^+-NH_4^+-Na^+-SO_4^{2-}-NO_3^--Cl^--H_2O at 298.15 K. J Phys Chem. A 102:2155-2171

Clement CF, Pirjola L, dal Maso M, Mäkelä JM, Kulmala M (2001) Analysis of particle formation bursts observed in Finland. J Aerosol Sci 32:217-236

Coffman DJ, Hegg DA (1995) A preliminary study of the effect of ammonia on particle nucleation in the marine boundary layer. J Geophys Res 100:7147-7160

Collings N, Graskow BR (2000) Particles from internal combustion engines—What we need to know. Phil Trans Roy Soc Lond A 358:2611-2622

Covert DS, Kapustin VN, Quinn PK, Bates TS (1992) New particle formation in the marine boundary-layer. J Geophys Res 97:20581-20589

Covert DS, Wiedensohler A, Aalto P, Heintzenberg J, McMurry PH, Leck C (1996) Aerosol number size distributions from 3 to 500-nm diameter in the Arctic marine boundary layer during summer and autumn. Tellus B-Chem Phys Meteorol 48:197-212

Craig A, McIntosh R (1952) The preparation of sodium chloride of large specific surface. Can J Chem 30:448

Crowley TJ (2000) Causes of climate change over the past 1000 years. Science 289:270-277

Cruz CN, Pandis SN (2000) Deliquescence and hygroscopic growth of mixed inorganic-organic atmospheric aerosol. Environ Sci Technol 34:4313-4319

Czizco DJ, Nowak JB, Hu JH, Abbatt JPD (1997) Infrared spectroscopy of model tropospheric aerosols as a function of relative humidity: Observation of deliquescence and crystallization. J Geophys Res 102:18843-18850
Cziczo DJ, Thomson DS, Murphy DM (2001) Ablation, flux, and atmospheric implications of meteors inferred from stratospheric aerosol. Science 291:1772-1775
Dachs J, Eisenreich SJ (2000) Adsorption onto aerosol soot carbon dominates gas-particle partitioning of polycyclic aromatic hydrocarbons. Environ Sci Technol 34:3690-3697
Dalleska NF, Colussi AJ, Hyldahl AM, Hoffmann MR (2000) Rates and mechanism of carbonyl sulfide oxidation by peroxides in concentrated sulfuric acid. J Phys Chem A 104:10794-10796
D'Almeida GA, Schütz L (1983) Number, mass, and volume distributions of mineral aerosol and soils of the Sahara. J Clim Appl Meteorol 22:233-243
De Haan DO, Brauers T, Oum K, Stutz J, Nordmeyer T, Finlayson-Pitts BJ (1999) Heterogeneous chemistry in the troposphere: Experimental approaches and applications to the chemistry of sea salt particles. Int'l Rev Phys Chem 18:343-385
de Reus M, Ström J, Kulmala M, Pirjola L, Lelieveld J, Schiller C, Zoger M (1998) Airborne aerosol measurements in the tropopause region and the dependence of new particle formation on preexisting particle number concentration. J Geophys Res 103:31255-31263
de Reus M, Dentener F, Thomas A, Borrmann S, Ström J, Lelieveld J (2000a) Airborne observations of dust aerosol over the North Atlantic ocean during ACE 2: Indications for heterogenous ozone destruction. J Geophys Res 105:15263-15275
de Reus M, Ström J, Curtius J, Pirjola L, Vignati E, Arnold F, Hansson HC, Kulmala M, Lelieveld J, Raes F (2000b) Aerosol production and growth in the upper free troposphere. J Geophys Res 105: 24751-24762
Dentener FJ, Carmichael GR, Zhang Y, Lelieveld J, Crutzen PJ (1996) Role of mineral aerosol as a reactive surface in the global atmosphere. J Geophys Res 101:22869-22889
Desboeufs KV, Losno R, Vimeux F, Cholbi S (1999) The pH-dependent dissolution of wind-transported Saharan dust. J Geophys Res 104:21287-21299
Disselkamp RS, Carpenter MA, Cowin JP (2000a) A chamber investigation of nitric acid-soot aerosol chemistry at 298 K. J Atmos Chem 37:113-123
Disselkamp RS, Carpenter MA, Cowin JP, Berkowitz CM, Chapman EG, Zaveri RA, Laulainen NS (2000b) Ozone loss in soot aerosols. J Geophys Res 105:9767-9771
Dixon JB, Weed SB (eds) (1989) Minerals in Soil Environments. Madison, Wisconsin: Soil Science Society of America
Dockery DW, Pope CA, Xu XP, Spengler JD, Ware JH, Fay ME, Ferris BG, Speizer FE (1993) An association between air pollution and mortality in 6 United States cities. N Engl J Med 329:1753-1759
Donaldson K, Li XY, MacNee W (1998) Ultrafine (nanometre) particle mediated lung injury. J Aerosol Sci 29:553-560
Donaldson K, Stone V, Gilmour PS, Brown DM, MacNee W (2000) Ultrafine particles: mechanisms of lung injury. Phil Trans Roy Soc Lond A 358:2741-2748
Donaldson K, Stone V, Clouter A, Renwick L, MacNee W (2001) Ultrafine particles. Occup Environ Med 58:211-216
Doremus RH (1985) Rates of Phase Transformations. New York: Academic Press
Dreher KL, Jaskot RH, Lehmann JR, Richards JH, McGee JK, Ghio AJ, Costa DL (1997) Soluble transition metals mediate residual oil fly ash induced acute lung injury. J Toxicol Environ Health 50:285-305
Duce RA (1969) On the source of gaseous chlorine in marine atmosphere. J Geophys Res 74:4597
Duce RA, Unni CK, Ray BJ, Prospero JM, Merrill JT (1980) Long-range atmospheric transport of soil dust from Asia to the tropical north Pacific: Temporal variability. Science 209:1522-1524
Duncan JL, Schindler LR, Roberts JT (1998) A new sulfate-mediated reaction: Conversion of acetone to trimethylbenzene in the presence of liquid sulfuric acid. Geophys Res Lett 25:631-634
Duncan JL, Schindler LR, Roberts JT (1999) Chemistry at and near the surface of liquid sulfuric acid: A kinetic, thermodynamic, and mechanistic analysis of heterogeneous reactions of acetone. J Phys Chem B 103:7247-7259
Eisele FL, Hanson DR (2000) First measurement of prenucleation molecular clusters. J Phys Chem A 104:830-836
Elder ACP, Gelein R, Finkelstein JN, Cox C, Oberdörster G (2000) Pulmonary inflammatory response to inhaled ultrafine particles is modified by age, ozone exposure, and bacterial toxin. Inhal Toxicol 12 (Suppl 1):227-246
Eldering A, Cass GR, Moon KC (1994) An air monitoring network using continuous particle size distribution monitors—Connecting pollutant properties to visibility via Mie scattering calculations. Atmos Environ 28:2733-2749

Ellison GB, Tuck AF, Vaida V (1999) Atmospheric processing of organic aerosols. J Geophys Res 104:11633-11641

Fahey DW, Gao RS, Carslaw KS, Kettleborough J, Popp PJ, Northway MJ, Holecek JC, Ciciora SC, McLaughlin RJ, Thompson TL, Winkler RH, Baumgardner DG, Gandrud B, Wennberg PO, Dhaniyala S, McKinney K, Peter T, Salawitch RJ, Bui TP, Elkins JW, Webster CR, Atlas EL, Jost H, Wilson JC, Herman RL, Kleinböhl A, von König M (2001) The detection of large HNO_3-containing particles in the winter Arctic stratosphere. Science 291:1026-1031

Fan CW, Zhang JJ (2001) Characterization of emissions from portable household combustion devices: Particle size distributions, emission rates and factors, and potential exposures. Atmos Environ 35:1281-1290

Faust BC (1994) A review of the photochemical redox reactions of iron(III) species in atmospheric, oceanic, and surface waters: Influences on geochemical cycles and oxidant formation. *In* Aquatic and Surface Photochemistry. RG Zepp, DG Crosby, GR Helz (eds) p 3-37. Boca Raton, Florida: Lewis Publishers

Fenter FF, Caloz F, Rossi MJ (1995) Experimental-evidence for the efficient dry deposition of nitric-acid on calcite. Atmos Environ 29:3365-3372

Filby WG, Mintas M, Gusten H (1981) Heterogeneous catalytic degradation of chlorofluoromethanes on zinc-oxide surfaces. Ber Bunsen Phys Chem 85:189-192

Finklea HO, 1988. Semiconductor Electrodes. New York: Elsevier

Finlayson-Pitts BJ, Hemminger JC (2000) Physical chemistry of airborne sea salt particles and their components. J Phys Chem A 104:11463-11477

Finlayson-Pitts BJ, Pitts JN (2000) Chemistry of the Upper and Lower Atmosphere: Theory, Experiments, and Applications. San Diego: Academic Press

Finlayson-Pitts BJ, Ezell MJ, Pitts JN (1989) Formation of chemically active chlorine compounds by reactions of atmospheric NaCl particles with gaseous N_2O_5 and $ClONO_2$. Nature 337:241-244

Fletcher NH (1969) Actives sites and ice crystal nucleation. J Atmos Sci 26:1266-1271

Forstner HJL, Flagan RC, Seinfeld JH (1997) Secondary organic aerosol from the photooxidation of aromatic hydrocarbons: Molecular composition. Environ Sci Technol 31:1345-1358

Foster KL, Plastridge RA, Bottenheim JW, Shepson PB, Finlayson-Pitts BJ, Spicer CW (2001) The role of Br_2 and BrCl in surface ozone destruction at polar sunrise. Science 291:471-474

Friedlander SK (2000) Smoke, Dust, and Haze: Fundamentals of Aerosol Dynamics. Topics in Chemical Engineering. New York: Oxford University Press

Gao RS, Kärcher B, Keim ER, Fahey DW (1998) Constraining the heterogeneous loss of O_3 on soot particles with observations in jet engine exhaust plumes. Geophys Res Lett 25:3323-3326

Ge Z, Wexler AS, Johnston MV (1996) Multicomponent aerosol crystallization. J Coll Interface Sci 183:68-77

Ge ZZ, Wexler AS, Johnston MV (1998) Laser desorption/ionization of single ultrafine multicomponent aerosols. Environ Sci Technol 32:3218-3223

Gerecke A, Thielmann A, Gutzwiller L, Rossi MJ (1998) The chemical kinetics of HONO formation resulting from heterogeneous interaction of NO_2 with flame soot. Geophys Res Lett 25:2453-2456

Ghio AJ, Stonehuerner J, Pritchard RJ, Piantadosi CA, Quigley DR, Dreher KL, Costa DL (1996) Humic-like substances in air pollution particulates correlate with concentrations of transition metals and oxidant generation. Inhal Toxicol 8:479-494

Gill PS, Graedel TE, Weschler CJ (1983) Organic films on atmospheric aerosol particles, fog droplets, cloud droplets, raindrops, and snowflakes. Rev Geophys Space Phys 21:903-920

Gong SL, Barrie LA, Blanchet JP (1997) Modeling sea-salt aerosols in the atmosphere. 1. Model development. J Geophys Res 102:3805-3818

Goodman AL, Underwood GM, Grassian VH (2000) A laboratory study of the heterogeneous reaction of nitric acid on calcium carbonate particles. J Geophys Res 105:29053-29064

Gorbunov BZ, Kakutkina NA (1982) Ice crystal formation on aerosol particles with a non-uniform surface. J Aerosol Sci 13:21-28

Graedel TE, Hawkins DT, Claxton LD (1986) Atmospheric Chemical Compounds: Sources, Occurrence, and Bioassay. Orlando, Florida: Academic Press

Griffin RJ, Cocker DR, Seinfeld JH, Dabdub D (1999) Estimate of global atmospheric organic aerosol from oxidation of biogenic hydrocarbons. Geophys Res Lett 26:2721-2724

Hameri K, Vakeva M, Hansson HC, Laaksonen A (2000) Hygroscopic growth of ultrafine ammonium sulphate aerosol measured using an ultrafine tandem differential mobility analyzer. J Geophys Res 105:22231-22242

Han J, Martin ST (1999) Heterogeneous nucleation of the efflorescence of $(NH_4)_2SO_4$ particles internally mixed with Al_2O_3, TiO_2, and ZrO_2. J Geophys Res 104:3543-3553

Han JH, Hung HM, Martin ST (2001) The size effect of hematite and corundum inclusions on the efflorescence relative humidities of aqueous ammonium nitrate particles. J Geophys Res (submitted)

Hanisch F, Crowley JN (2001) Heterogeneous reactivity of gaseous nitric acid on Al_2O_3, $CaCO_3$, and atmospheric dust samples: A Knudsen cell study. J Phys Chem A 105:3096-3106

Harrison RM, Shi JP, Xi SH, Khan A, Mark D, Kinnersley R, Yin JX (2000) Measurement of number, mass and size distribution of particles in the atmosphere. Phil Trans Roy Soc Lond A 358:2567-2579

Harvey LDD (2000) Constraining the aerosol radiative forcing and climate sensitivity. Clim Change 44:413-418

Hauser R, Godleski JJ, Hatch V, Christiani DC (2001) Ultrafine particles in human lung macrophages. Arch Environ Health 56:150-156

Hemminger JC (1999) Heterogeneous chemistry in the troposphere: A modern surface chemistry approach to the study of fundamental processes. Int'l Rev Phys Chem 18:387-417

Herman JR, Bhartia PK, Torres O, Hsu C, Seftor C, Celarier E (1997) Global distribution of uv-absorbing aerosols from Nimbus 7/TOMS data. J Geophys Res 102:16911-16922

Hoffmann MR, Martin ST, Choi W, Bahnemann DW (1995) Environmental applications of semiconductor photocatalysis. Chem Rev 95:69-96

Holland HD (1978) The Chemistry of Atmospheres and Oceans. New York: Wiley

Huffman GP, Huggins FE, Shah N, Huggins R, Linak WP, Miller CA, Pugmire RJ, Meuzelaar HLC, Seehra MS, Manivannan A (2000) Characterization of fine particulate matter produced by combustion of residual fuel oil. J Air Waste Manage Assoc 50:1106-1114

Hughes LS, Cass GR, Gone J, Ames M, Olmez I (1998) Physical and chemical characterization of atmospheric ultrafine particles in the Los Angeles area. Environ Sci Technol V32:1153-1161

Husar RB, Prospero JM, Stowe LL (1997) Characterization of tropospheric aerosols over the oceans with the NOAA advanced very high resolution radiometer optical thickness operational product. J Geophys Res 102:16889-16909

IPCC (Intergovernmental Panel on Climate Change) (2001) Climate Change 2001: The Scientific Basis. Houghton JT, Ding Y, Griggs DJ, Noguer M, van der Linden PJ, Xiaosu D (eds) Cambridge: Cambridge University Press

Jacobson MZ (2001) Strong radiative heating due to the mixing state of black carbon in atmospheric aerosols. Nature 409:695-697

Jaecker-Voirol A, Mirabel P (1988) Nucleation rate in a binary mixture of sulfuric-acid and water- vapor. J Phys Chem 92:3518-3521

Jaecker-Voirol A, Mirabel P, Reiss H (1987) Hydrates in supersaturated binary sulfuric acid-water vapor—A re-examination. J Chem Phys 87:4849-4852

Jaques PA, Kim CS (2000) Measurement of total lung deposition of inhaled ultrafine particles in healthy men and women. Inhal Toxicol 12 (Suppl 1):715-731

Jefferson DA (2000) The surface activity of ultrafine particles. Phil Trans Roy Soc Lond A 358:2683-2692

José-Yacamán M (1998) The role of nanosized particles. A frontier in modern materials science, from nanoelectronics to environmental problems. Metall Mater Trans A 29:713-725

Kalberer M, Ammann M, Arens F, Gaggeler HW, Baltensperger U (1999) Heterogeneous formation of nitrous acid (HONO) on soot aerosol particles. J Geophys Res 104:13825-13832

Kamm S, Mohler O, Naumann KH, Saathoff H, Schurath U (1999) The heterogeneous reaction of ozone with soot aerosol. Atmos Environ 33:4651-4661

Kane DB, Johnston MV (2000) Size and composition biases on the detection of individual ultrafine particles by aerosol mass spectrometry. Environ Sci Technol 34:4887-4893

Kärcher B, Turco RP, Yu F, Danilin MY, Weisenstein DK, Miake-Lye RC, Busen R (2000) A unified model for ultrafine aircraft particle emissions. J Geophys Res 105:29379-29386

Kashchiev D (1982) On the relation between nucleation work, nucleus size, and nucleation rate. J Chem Phys 76:5098-5102

Kavouras IG, Mihalopoulos N, Stephanou EG (1999) Secondary organic aerosol formation vs primary organic aerosol emission: *In situ* evidence for the chemical coupling between monoterpene acidic photooxidation products and new particle formation over forests. Environ Sci Technol 33:1028-1037

Keene WC, Sander R, Pszenny AAP, Vogt R, Crutzen PJ, Galloway JN (1998) Aerosol pH in the marine boundary layer: A review and model evaluation. J Aerosol Sci 29:339-356

Kennedy IM (1997) Models of soot formation and oxidation. Prog Energ Combust Sci 23:95-132

Kerminen VM (1999) Roles of SO_2 and secondary organics in the growth of nanometer particles in the lower atmosphere. J Aerosol Sci 30:1069-1078

Kerminen VM, Wexler AS, Potukuchi S (1997) Growth of freshly nucleated particles in the troposphere: Roles of NH_3, H_2SO_4, HNO_3, and HCl. J Geophys Res 102:3715-3724

Kerminen VM, Virkkula A, Hillamo R, Wexler AS, Kulmala M (2000) Secondary organics and atmospheric cloud condensation nuclei production. J Geophys Res 105:9255-9264

Keywood MD, Ayers GP, Gras JL, Gillett RW, Cohen DD (1999) Relationships between size-segregated mass concentration data and ultrafine particle number concentrations in urban areas. Atmos Environ 33:2907-2913

Kiehl JT, Schneider TL, Rasch PJ, Barth MC, Wong J (2000) Radiative forcing due to sulfate aerosols from simulations with the National Center for Atmospheric Research Community Climate Model, Version 3. J Geophys Res 105:1441-1457

Kim TO, Ishida T, Adachi M, Okuyama K, Seinfeld JH (1998) Nanometer-sized particle formation from $NH_3/SO_2/H_2O$/air mixtures by ionizing irradiation. Aerosol Sci Tech 29:111-125

Kirchner U, Scheer V, Vogt R (2000) FTIR spectroscopic investigation of the mechanism and kinetics of the heterogeneous reactions of NO_2 and HNO_3 with soot. J Phys Chem A 104:8908-8915

Kittelson DB (1998) Engines and nanoparticles: A review. J Aerosol Sci 29:575-588

Kleeman MJ, Schauer JJ, Cass GR (1999) Size and composition distribution of fine particulate matter emitted from wood burning, meat charbroiling, and cigarettes. Environ Sci Technol 33:3516-3523

Kleeman MJ, Schauer JJ, Cass GR (2000) Size and composition distribution of fine particulate matter emitted from motor vehicles. Environ Sci Technol 34:1132-1142

Kolb CE, Worsnop DR, Zahniser MS, Davidovits P, Keyser LF, Leu MT, Molina MR, Hanson DR, Ravishankara AR (1995) Laboratory studies of atmospheric heterogeneous chemistry. *In* Progress and Problems in Atmospheric Chemistry. Barker JR (ed) p 771-875. Singapore: World Scientific

Koper O, Klabunde KJ (1997) Destructive adsorption of chlorinated hydrocarbons on ultrafine (nanoscale) particles of calcium oxide.3. Chloroform, trichloroethene, and tetrachloroethene. Chem Mater 9: 2481-2485

Koper O, Li YX, Klabunde KJ (1993) Destructive adsorption of chlorinated hydrocarbons on ultrafine (nanoscale) particles of calcium oxide. Chem Mater 5:500-505

Koper O, Lagadic I, Klabunde KJ (1997) Destructive adsorption of chlorinated hydrocarbons on ultrafine (nanoscale) particles of calcium oxide. 2. Chem Mater 9:838-848

Korhonen P, Kulmala M, Laaksonen A, Viisanen Y, McGraw R, Seinfeld JH (1999) Ternary nucleation of H_2SO_4, NH_3, and H_2O in the atmosphere. J Geophys Res 104:26349-26353

Kotzick R, Panne U, Niessner R (1997) Changes in condensation properties of ultrafine carbon particles subjected to oxidation by ozone. J Aerosol Sci 28:725-735

Krämer L, Poschl U, Niessner R (2000) Microstructural rearrangement of sodium chloride condensation aerosol particles on interaction with water vapor. J Aerosol Sci 31:673-685

Kreidenweis SM, Tyndall GS, Barth MC, Dentener F, Lelieveld J, Mozurkewich M (1999) Aerosols and clouds. *In* Atmospheric Chemistry and Global Change. GP Brasseur, JJ Orlando, GS Tyndall (eds) p 117-155. New York: Oxford University Press

Kulmala M (1993) Condensational growth and evaporation in the transition regime: An analytical expression. Aerosol Sci Tech 19:381-388

Kulmala M, Vesala T, Wagner PE (1993) An analytical expression for the rate of binary condensational particle growth. Proc R Soc London A 441:589-605

Kulmala M, Kerminen VM, Laaksonen A (1995) Simulations on the effect of sulphuric acid formation on atmospheric aerosol concentrations. Atmos Environ 29:377-382

Kulmala M, Laaksonen A, Pirjola L (1998a) Parameterizations for sulfuric acid/water nucleation rates. J Geophys Res 103:8301-8307

Kulmala M, Toivonen A, Mäkelä JM, Laaksonen A (1998b) Analysis of the growth of nucleation mode particles observed in boreal forest. Tellus B 50:449-462

Kulmala M, Pirjola U, Mäkelä JM (2000) Stable sulphate clusters as a source of new atmospheric particles. Nature 404:66-69

Kusaka I, Wang ZG, Seinfeld JH (1998a) Direct evaluation of the equilibrium distribution of physical clusters by a grand canonical Monte Carlo simulation. J Chem Phys 108:3416-3423

Kusaka I, Wang ZG, Seinfeld JH (1998b) Binary nucleation of sulfuric acid-water: Monte Carlo simulation. J Chem Phys 108:6829-6848

Kusaka I, Oxtoby DW, Wang ZG (1999) On the direct evaluation of the equilibrium distribution of clusters by simulation. J Chem Phys 111:9958-9964

Laaksonen A, Talanquer V, Oxtoby DW (1995) Nucleation: Measurements, theory, and atmospheric applications. Ann Rev Phys Chem 46:489-524

Lasaga AC (1998) Kinetic Theory in the Earth Sciences. Princeton University Press, Princeton

Leinen M, Prospero JM, Arnold E, Blank M (1994) Mineralogy of aeolian dust reaching the North Pacific Ocean. 1. Sampling and analysis. J Geophys Res 99:21017-21023

Leong KH (1981) Morphology of aerosol-particles generated from the evaporation of solution drops. J Aerosol Sci 12:417-435

Li YX, Li H, Klabunde KJ (1994) Destructive adsorption of chlorinated benzenes on ultrafine (nanoscale) particles of magnesium oxide and calcium oxide. Environ Sci Technol 28:1248-1253

Li P, Perreau KA, Covington E, Song CH, Carmichael GR, Grassian VH, (2001) Heterogeneous reactions of volatile organic compounds on oxide particles of the most abundant crustal elements: Surface reactions of acetaldehyde, acetone, and propionaldehyde on SiO_2, Al_2O_3, Fe_2O_3, TiO_2, and CaO. J Geophys Res 106:5517-5529

Liler M (1971) Reaction Mechanisms in Sulphuric Acid and Other Strong Acid Solutions. London - New York: Academic Press

Linak WP, Miller CA (2000) Comparison of particle size distributions and elemental partitioning from the combustion of pulverized coal and residual fuel oil. J Air Waste Manage Assoc 50:1532-1544

Litter MI, Villegas M, Blesa MA (1994) Photodissolution of iron-oxides in malonic-acid. Can J Chem 72:2037-2043

Lohmann U, Feichter J (2001) Can the direct and semi-direct aerosol effect compete with the indirect effect on a global scale? Geophys Res Lett 28:159-161

Longfellow CA, Ravishankara AR, Hanson DR (1999) Reactive uptake on hydrocarbon soot: Focus on NO_2. J Geophys Res 104:13833-13840

Longfellow CA, Ravishankara AR, Hanson DR (2000) Reactive and nonreactive uptake on hydrocarbon soot: HNO_3, O_3, and N_2O_5. J Geophys Res 105:24345-24350

Loomis D (2000) Sizing up air pollution research. Epidemiology 11:2-4

MacTaylor RS, Castleman AW (2000) Cluster ion reactions: Insights into processes of atmospheric significance. J Atmos Chem 36:23-63

Mäkelä JM, Aalto P, Jokinen V, Pohja T, Nissinen A, Palmroth S, Markkanen T, Seitsonen K, Lihavainen H, Kulmala M (1997) Observations of ultrafine aerosol particle formation and growth in boreal forest. Geophys Res Lett 24:1219-1222

Marion GM, Farren RE (1999) Mineral solubilities in the Na-K-Mg-Ca-Cl-SO_4^{2-}-H_2O: A re-evaluation of the sulfate chemistry in the Spencer-Moller-Weare model. Geochim Cosmochim Acta 63:1305-1318

Marti JJ, Jefferson A, Cai XP, Richert C, McMurry PH, Eisele F (1997a) H_2SO_4 vapor pressure of sulfuric acid and ammonium sulfate solutions. J Geophys Res 102:3725-3735

Marti JJ, Weber RJ, McMurry PH, Eisele F, Tanner D, Jefferson A (1997b) New particle formation at a remote continental site: Assessing the contributions of SO_2 and organic precursors. J Geophys Res 102:6331-6339

Martin JH, Gordon RM (1988) Northeast Pacific iron distributions in relation to phytoplankton productivity. Deep-Sea Res 35:177-196

Martin ST (2000) Phase transitions of aqueous atmospheric particles. Chem Rev 100:3403-3453

Martin ST, Han JH, Hung HM (2001) The size effect of hematite and alumina inclusions on the efflorescence relative humidities of aqueous ammonium sulfate particles. Geophys Res Lett 28: 2601-2604

Mathai CV (1990) Visibility and fine particles. 4. A summary of the A&WMA EPA International Specialty Conference. J Air Waste Manage Assoc 40:1486-1494

Matteson MJ, Preiming O, Fox JF (1972) Density distribution of sodium-chloride aerosols formed by condensation. Nature-Phys Sci 238:61

Mattila T, Kulmala M, Vesala T (1997) On the condensational growth of a multicomponent droplet. J Aerosol Sci 28:553-564

McCaffrey MA, Lazar B, Holland HD (1987) The evaporation path of seawater and the coprecipitation of Br^- and K^+ with halite. J Sed Petrology 57:928-937

McInnes LM, Covert DS, Quinn PK, Germani MS (1994) Measurements of chloride depletion and sulfur enrichment in individual sea-salt particles collected from the remote marine boundary-layer. J Geophys Res 99:8257-8268

McMurry PH, Woo KS, Weber R, Chen DR, Pui DYH (2000) Size distributions of 3-10 nm atmospheric particles: Implications for nucleation mechanisms. Philos Trans Roy Soc Lond A 358:2625-2642

Merrill J, Arnold E, Leinen M, Weaver C (1994) Mineralogy of aeolian dust reaching the north pacific-ocean. 2. Relationship of mineral assemblages to atmospheric transport patterns. J Geophys Res 99:21025-21032

Miller RL, Tegen I (1998) Climate response to soil dust aerosols. J Climate 11:3247-3267

Mirabel P, Ponche JL (1991) Studies of gas-phase clustering of water on sulfuric-acid molecules. Chem Phys Lett 183:21-24

Mirabel P, Reiss H, Bowles RK (2000) A theory for the deliquescence of small particles. J Chem Phys 113:8200-8205

Mozurkewich M (1995) Mechanisms for the release of halogens from sea-salt particles by free radical reactions. J Geophys Res 100:14199-14207

Noble CA, Prather KA (1996) Real-time measurement of correlated size and composition profiles of individual atmospheric aerosol particles. Environ Sci Technol 30:2667-2680

Noble CA, Prather KA (2000) Real-time single particle mass spectrometry: A historical review of a quarter century of the chemical analysis of aerosols. Mass Spectrom Rev 19:248-274

Nolte CG, Schauer JJ, Cass GR, Simoneit BRT (2001) Highly polar organic compounds present in wood smoke and in the ambient atmosphere. Environ Sci Technol 35:1912-1919

Oberdörster G (2001) Pulmonary effects of inhaled ultrafine particles. Int Arch Occup Envir Health 74:1-8

O'Dowd C, McFiggans G, Creasey DJ, Pirjola L, Hoell C, Smith MH, Allan BJ, Plane JMC, Heard DE, Lee JD, Pilling MJ, Kulmala M (1999) On the photochemical production of new particles in the coastal boundary layer. Geophys Res Lett 26:1707-1710

Odum JR, Jungkamp TPW, Griffin RJ, Forstner HJL, Flagan RC, Seinfeld JH (1997) Aromatics, reformulated gasoline, and atmospheric organic aerosol formation. Environ Sci Technol 31:1890-1897

Oxtoby DW (1992) Homogeneous nucleation: Theory and experiment. J Phys: Condens Matter 4: 7627-7650

Oxtoby DW, Kashchiev D (1994) A general relation between the nucleation work and the size of the nucleus in multicomponent nucleation. J Chem Phys 100:7665-7671

Park PW, Ledford JS (1997) Characterization and CH_4 oxidation activity of Cr/Al_2O_3 catalysts. Langmuir 13:2726-2730

Park PW, Ledford JS (1998) Characterization and CO oxidation activity of $Cu/Cr/Al_2O_3$ catalysts. Indust Engin Chem Res 37:887-893

Pehkonen SO, Siefert R, Erel Y, Webb S, Hoffmann MR (1993) Photoreduction of iron oxyhydroxides in the presence of important atmospheric organic compounds. Environ Sci Technol 27:2056-2062

Pehkonen SO, Siefert RL, Hoffmann MR (1995) Photoreduction of iron oxyhydroxides and the photooxidation of halogenated acetic-acids. Environ Sci Technol 29:1215-1222

Pekkanen J, Timonen KL, Ruuskanen J, Reponen A, Mirme A (1997) Effects of ultrafine and fine particles in urban air on peak expiratory flow among children with asthmatic symptoms. Environ Res 74:24-33

Penner JE, Chuang CC, Grant K (1998) Climate forcing by carbonaceous and sulfate aerosols. Clim Dynam 14:839-851

Perkas N, Koltypin Y, Palchik O, Gedanken A, Chandrasekaran S (2001) Oxidation of cyclohexane with nanostructured amorphous catalysts under mild conditions. Appl Catal A 209:125-130

Peter T (1997) Microphysics and heterogeneous chemistry of polar stratospheric clouds. Ann Rev Phys Chem 48:785-822

Peter T (1999) Physico-chemistry of polar stratospheric clouds. *In* Ice Physics and the Natural Environment. Wettlaufer JS (ed) p 143-167. Berlin: Springer-Verlag

Peters SJ, Ewing GE (1997a) Thin film water on NaCl(100) under ambient conditions: An infrared study. Langmuir 13:6345-6348

Peters SJ, Ewing GE (1997b) Water on salt: An infrared study of adsorbed H_2O on NaCl(100) under ambient conditions. J Phys Chem B 101:10880-10886

Peters A, Wichmann HE, Tuch T, Heinrich J, Heyder J (1997) Respiratory effects are associated with the number of ultrafine particles. Am J Respir Crit Care Med 155:1376-1383

Phares DJ, Rhoades KP, Wexler AS (2001a) Performance of a single-ultrafine-particle mass spectrometer. Aerosol Sci Tech (accepted)

Phares DJ, Rhoades KP, Johnston MV, Wexler AS (2001b) Size-resolved ultrafine particle composition analysis. Part 2: Houston. J Geophys Res (submitted)

Pilinis C (1989) Numerical simulation of visibility degradation due to particulate matter: Model development and evaluation. J Geophys Res 94:9937-9946

Pirjola L, Kulmala M (1998) Modelling the formation of H_2SO_4-H_2O particles in rural, urban and marine conditions. Atmos Res 46:321-347

Pirjola L, Laaksonen A, Aalto P, Kulmala M (1998) Sulfate aerosol formation in the Arctic boundary layer. J Geophys Res 103:8309-8321

Pirjola L, Kulmala M, Wilck M, Bischoff A, Stratmann F, Otto E (1999) Formation of sulphuric acid aerosols and cloud condensation nuclei: An expression for significant nucleation and model comparison. J Aerosol Sci 30:1079-1094

Pirjola L, O'Dowd CD, Brooks IM, Kulmala M (2000) Can new particle formation occur in the clean marine boundary layer? J Geophys Res 105:26531-26546

Pitchford ML, McMurry PH (1994) Relationship between measured water vapor growth and chemistry of atmospheric aerosol for Grand Canyon, Arizona, in winter 1990. Atmos Environ 28:827-839

Pope CA (2000) What do epidemiologic findings tell us about health effects of environmental aerosols? J Aerosol Med 13:335-354

Pope CA, Dockery DW, Schwartz J (1995a) Review of epidemiological evidence of health effects of particulate air pollution. Inhal Toxicol 7:1-18

Pope CA, Thun MJ, Namboodiri MM, Dockery DW, Evans JS, Speizer FE, Heath CW (1995b) Particulate air pollution as a predictor of mortality in a prospective study of US adults. Am J Respir Crit Care Med V151:669-674

Poschl U, Letzel T, Schauer C, Niessner R (2001) Interaction of ozone and water vapor with spark discharge soot aerosol particles coated with benzo[a]pyrene: O_3 and H_2O adsorption, benzo[a]pyrene degradation, and atmospheric implications. J Phys Chem A 105:4029-4041

Prospero JM (1999a) Long-range transport of mineral dust in the global atmosphere: Impact of African dust on the environment of the southeastern United States. Proc Natl Acad Sci 96:3396-3403

Prospero JM (1999b) Long-term measurements of the transport of African mineral dust to the southeastern United States: Implications for regional air quality. J Geophys Res 104:15917-15927

Pruppacher HR, Klett JD (1997) Microphysics of Clouds and Precipitation. Dordrecht: Kluwer

Pszenny AAP, Keene WC, Jacob DJ, Fan S, Maben JR, Zetwo MP, Springer-Young M, Galloway JN (1993) Evidence of inorganic chlorine gases other than hydrogen chloride in marine surface air. Geophys Res Lett 20:699-702

Pryor SC, Barthelmie RJ (2000) REVEAL II: Seasonality and spatial variability of particle and visibility conditions in the Fraser Valley. Sci Total Environ 257:95-110

Pye K (1987) Aeolian Dust and Dust Deposits. San Diego: Academic Press

Rhoades KP, Phares DJ, Wexler AS, Johnston MV (2001) Size-resolved ultrafine particle composition analysis. Part 1: Atlanta. J Geophys Res (submitted)

Richter H, Howard JB (2000) Formation of polycyclic aromatic hydrocarbons and their growth to soot: A review of chemical reaction pathways. Prog Energ Combust Sci 26:565-608

Rogge WF, Hildemann LM, Mazurek MA, Cass GR, Simonelt BRT (1991) Sources of fine organic aerosol. 1. Charbroilers and meat cooking operations. Environ Sci Technol 25:1112-1125

Rogge WF, Hildemann LM, Mazurek MA, Cass GR (1994) Sources of fine organic aerosol. 6. Cigarette smoke in the urban atmosphere. Environ Sci Technol 28:1375-1388

Rogge WF, Hildemann LM, Mazurek MA, Cass GR, Simoneit BRT (1998) Sources of fine organic aerosol. 9. Pine, oak and synthetic log combustion in residential fireplaces. Environ Sci Technol 32:13-22

Rosenfeld D (2000) Suppression of rain and snow by urban and industrial air pollution. Science 287: 1793-1796

Saathoff H, Naumann KH, Riemer N, Kamm S, Mohler O, Schurath U, Vogel H, Vogel B (2001) The loss of NO_2, HNO_3, NO_3/N_2O_5, and $HO_2/HOONO_2$ on soot aerosol: A chamber and modeling study. Geophys Res Lett 28:1957-1960

Samet JM (2000) What properties of particulate matter are responsible for health effects? Inhal Toxicol 12 (Suppl 1):19-21

Saxena P, Hildemann LM, McMurry PH, Seinfeld JH (1995) Organics alter hygroscopic behavior of atmospheric particles. J Geophys Res 100:18755-18770

Schauer JJ, Kleeman MJ, Cass GR, Simoneit BRT (1999) Measurement of emissions from air pollution sources. 2. C-1 through C-30 organic compounds from medium duty diesel trucks. Environ Sci Technol 33:1578-1587

Schauer JJ, Kleeman MJ, Cass GR, Simoneit BRT (2001) Measurement of emissions from air pollution sources. 3. C-1-C-29 organic compounds from fireplace combustion of wood. Environ Sci Technol 35:1716-1728

Schlesinger RB (2000) Properties of ambient PM responsible for human health effects: Coherence between epidemiology and toxicology. Inhal Toxicol 12 (Suppl 1):23-25

Schütz L (1989) Atmospheric mineral dust—properties and source markers. *In* Paleoclimatology and Paleometeorology: Modern and Past Patterns of Global Atmospheric Transport. Leinen M, Sarnthein M (eds) p 359-384. Dordrecht: Kluwer

Schütz L (1997) Mineral dust and source-relevant data. *In* Proceedings of the Alfred-Wegener-Conference: Sediment and Aerosol. von Hoyningen-Huene W, Tetzlaff G (eds) March 10-12, Leipzig, Germany. Cologne: Terra Nostra

Schütz L, Sebert M (1987) Mineral aerosols and source identification. J Aerosol Sci 18:1

Schwartz SE, Buseck PR (2000) Atmospheric science—Absorbing phenomena. Science 288:989-990

Seinfeld JH, Pandis SN (1998) Atmospheric Chemistry and Physics: From Air Pollution to Climate Change. New York: Wiley

Senior CL, Helble JJ, Sarofim AF (2000) Emissions of mercury, trace elements, and fine particles from stationary combustion sources. Fuel Process Technol 65:263-288

Seto T, Nakamoto T, Okuyama K, Adachi M, Kuga Y, Takeuchi K (1997) Size distribution measurement of nanometer-sized aerosol particles using DMA under low-pressure conditions. J Aerosol Sci 28: 193-206

Shen T-L, Wooldridge PJ, Molina MJ (1995) Stratospheric pollution and ozone depletion. *In* Composition, Chemistry and Climate of the Atmosphere. HB Singh (ed) p 394-442. New York: Van Norstrand

Shi JP, Harrison RM (1999) Investigation of ultrafine particle formation during diesel exhaust dilution. Environ Sci Technol 33:3730-3736

Shi JP, Khan AA, Harrison RM (1999) Measurements of ultrafine particle concentration and size distribution in the urban atmosphere. Sci Total Environ 235:51-64

Siefert RL, Pehkonen SO, Erel Y, Hoffmann MR (1994) Iron photochemistry of aqueous suspensions of ambient aerosol with added organic-acids. Geochim Cosmochim Acta 58:3271-3279

Sievering H, Boatman J, Gorman E, Kim Y, Anderson L, Ennis G, Luria M, Pandis S (1992) Removal of sulphur from the marine boundary layer by ozone oxidation in sea-salt aerosols. Nature 360:571-573

Siffert C, Sulzberger B (1991) Light-induced dissolution of hematite in the presence of oxalate—A case-study. Langmuir 7:1627-1634

Simoneit BRT, Schauer JJ, Nolte CG, Oros DR, Elias VO, Fraser MP, Rogge WF, Cass GR (1999) Levoglucosan, a tracer for cellulose in biomass burning and atmospheric particles. Atmos Environ 33:173-182

Sloane CS, Watson J, Chow J, Pritchett L, Richards LW (1991) Size-segregated fine particle measurements by chemical species and their impact on visibility impairment in Denver. Atmos Environ A 25: 1013-1024

Smith KR, Aust AE (1997) Mobilization of iron from urban particulates leads to generation of reactive oxygen species in vitro and induction of ferritin synthesis in human lung epithelial cells. Chem Res Toxicol 10:828-834

Song CH, Carmichael GR (1999) The aging process of naturally emitted aerosol (sea-salt and mineral aerosol) during long range transport. Atmos Environ 33:2203-2218

Stadler D, Rossi MJ (2000) The reactivity of NO_2 and HONO on flame soot at ambient temperature: The influence of combustion conditions. Phys Chem Chem Phys 2:5420-5429

Stark JV, Klabunde KJ (1996) Nanoscale metal oxide particles/clusters as chemical reagents—Adsorption of hydrogen halides, nitric oxide, and sulfur trioxide on magnesium oxide nanocrystals and compared with microcrystals. Chem Mater 8:1913-1918

Stark JV, Park DG, Lagadic I, Klabunde KJ (1996) Nanoscale metal oxide particles/clusters as chemical reagents—Unique surface chemistry on magnesium oxide as shown by enhanced adsorption of acid gases (sulfur dioxide and carbon dioxide) and pressure dependence. Chem Mater 8:1904-1912

Stolzenburg MR, McMurry PH (1991) An ultrafine aerosol condensation nucleus counter. Aerosol Sci Tech 14:48-65

Stumm W, Morgan JJ (1996) Aquatic Chemistry. New York: Wiley

Sulzberger B (1990) Photoredox reactions at hydrous metal oxide surfaces: A surface coordination chemistry approach. *In* Aquatic Chemical Kinetics. Stumm W (ed) p 401-430. New York: Wiley

Sulzberger B, Siffert C, Stumm W (1988) Surface coordination chemistry and redox processes—Photoinduced iron (III) oxide dissolution. Chimia 42:257-261

Swap R, Garstang M, Greco S, Talbot R, Kallberg P (1992) Saharan dust in the Amazon basin. Tellus B 44:133-149

Tabazadeh A, Toon OB, Clegg SL, Hamill P (1997) A new parameterization of H_2SO_4/H_2O aerosol composition: Atmospheric implications. Geophys Res Lett 24:1931-1934

Tabazadeh A, Martin ST, Lin JS (2000) The effect of particle size and nitric acid uptake on the homogeneous freezing of sulfate aerosols. Geophys Res Lett 27:1111-1114

Tang IN, Tridico AC, Fung KH (1997) Thermodynamic and optical properties of sea salt aerosols. J Geophys Res 102:23269-23275

Tegen I, Fung I (1994) Modeling of mineral dust in the atmosphere—Sources, transport, and optical-thickness. J Geophys Res 99:22897-22914

Tegen I, Fung I (1995) Contribution to the atmospheric mineral aerosol load from land surface modification. J Geophys Res 100:18707-18726

Tegen I, Lacis A, Fung I (1996) The influence on climate forcing of mineral aerosols from disturbed soils. Nature 380:419-422

Tobias HJ, Ziemann PJ (2000) Thermal desorption mass spectrometric analysis of organic aerosol formed from reactions of 1-tetradecene and O_3 in the presence of alcohols and carboxylic acids. Environ Sci Technol 34:2105-2115

Tobias HJ, Beving DE, Ziemann PJ, Sakurai H, Zuk M, McMurry PH, Zarling D, Waytulonis R, Kittelson DB (2001) Chemical analysis of diesel engine nanoparticles using a nano-DMA/thermal desorption particle beam mass spectrometer. Environ Sci Technol 35:2233-2243

Tolbert MA, Toon OB (2001) Atmospheric science—Solving the PSC mystery. Science 292:61-63

Twomey S (1977) The influence of pollution on the shortwave albedo of clouds. J Atmos Sci 34:1149-1152

Underwood GM, Li P, Usher CR, Grassian VH (2000) Determining accurate kinetic parameters of potentially important heterogeneous atmospheric reactions on solid particle surfaces with a Knudsen cell reactor. J Phys Chem A 104:819-829

Underwood GM, Song CH, Phadnis M, Carmichael GR, Grassian VH (2001) Heterogeneous reactions of NO_2 and HNO_3 on oxides and mineral dust: A combined laboratory and modeling study. J Geophys Res (in press)

Utell MJ, Frampton MW (2000) SESSION 5: Who is susceptible to particulate matter and why? Inhal Toxicol 12 (Suppl 1):37-40

van Maanen JMS, Borm PJA, Knaapen A, van Herwijnen M, Schilderman PAEL, Smith KR, Aust AE, Tomatis M, Fubini B (1999) *In vitro* effects of coal fly ashes: Hydroxyl radical generation, iron release, and DNA damage and toxicity in rat lung epithelial cells. Inhal Toxicol 11:1123-1141

Vesala T, Kulmala M, Rudolf R, Vrtala A, Wagner PE (1997) Models for condensational growth and evaporation of binary aerosol particles. J Aerosol Sci 28:565-598

Vogt R, Crutzen PJ, Sander R (1996) A mechanism for halogen release from sea-salt aerosol in the remote marine boundary layer. Nature 383:327-330

Watson AY, Valberg PA (2001) Carbon black and soot: Two different substances. Am Indust Hyg Assoc J 62:218-228

Weber RJ, McMurry PH, Eisele FL, Tanner DJ (1995) Measurement of expected nucleation precursor species and 3-500 nm diameter particles at Mauna Loa Observatory, Hawaii. J Atmos Sci 52: 2242-2257

Weber RJ, Marti JJ, McMurry PH, Eisele FL, Tanner DJ, Jefferson A (1996) Measured atmospheric new particle formation rates: Implications for nucleation mechanisms. Chem Engin Commun 151:53-64

Weber RJ, Marti JJ, McMurry PH, Eisele FL, Tanner DJ, Jefferson A (1997) Measurements of new particle formation and ultrafine particle growth rates at a clean continental site. J Geophys Res 102:4375-4385

Weber RJ, McMurry PH, Mauldin L, Tanner DJ, Eisele FL, Brechtel FJ, Kreidenweis SM, Kok GL, Schillawski RD, Baumgardner D (1998a) A study of new particle formation and growth involving biogenic and trace gas species measured during ACE-1. J Geophys Res 103:16385-16396

Weber RJ, Stolzenburg MR, Pandis SN, McMurry PH (1998b) Inversion of ultrafine condensation nucleus counter pulse height distributions to obtain nanoparticle (~3-10 nm) size distributions. J Aerosol Sci 29:601-615

Weber RJ, McMurry PH, Mauldin RL, Tanner DJ, Eisele FL, Clarke AD, Kapustin VN (1999) New particle formation in the remote troposphere: A comparison of observations at various sites. Geophys Res Lett 26:307-310

Weis DD, Ewing GE (1999) Water content and morphology of sodium chloride aerosol particles. J Geophys Res 104:21275-21285

Wexler AS, Lurmann FW, Seinfeld JH (1994) Modelling urban and regional aerosols—I. Model development. Atmos Environ 28:531-546

Wiedensohler A, Aalto P, Covert D, Heintzenberg J, McMurry PH (1994) Intercomparison of 4 methods to determine size distributions of low-concentration (~100 cm^{-3}), ultrafine aerosols ($3 < D(p) < 10$ nm) with illustrative data from the Arctic. Aerosol Sci Tech 21:95-109

Wiedensohler A, Hansson HC, Orsini D, Wendisch M, Wagner F, Bower KN, Chourlarton TW, Wells M, Parkin M, Acker K, Wieprecht W, Facchini MC, Lind JA, Fuzzi S, Arends BG, Kulmala M (1997) Night-time formation and occurrence of new particles associated with orographic clouds. Atmos Environ 31:2545-2559

Woo KS, Chen DR, Pui DYH, McMurry PH (2001) Measurement of Atlanta aerosol size distributions: Observations of ultrafine particle events. Aerosol Sci Tech 34:75-87

Wooldridge MS (1998) Gas-phase combustion synthesis of particles. Prog Energ Combust Sci 24:63-87

Wooldridge MS, Danczyk SA, Wu JF (1999) Demonstration of gas-phase combustion synthesis of nanosized particles using a hybrid burner. Nanostruct Mater 11:955-964

Wurzler S, Reisin TG, Levin Z (2000) Modification of mineral dust particles by cloud processing and subsequent effects on drop size distributions. J Geophys Res 105:4501-4512

Xie JY, Marlow WH (1997) Water vapor pressure over complex particles. 1. Sulfuric acid solution effect. Aerosol Sci Technol 27:591-603

Xiong Y, Kodas TT (1993) Droplet evaporation and solute precipitation during spray-pyrolysis. J Aerosol Sci. 24:893-908

Xiong JQ, Zhong M, Fang C, Chen LC, Lippmann M (1998) Influence of organic films on the hygroscopicity of ultrafine sulfuric acid aerosol. Environ Sci Technol 32:3536-3541

Yu JZ, Griffin RJ, Cocker DR, Flagan RC, Seinfeld JH, Blanchard P (1999) Observation of gaseous and particulate products of monoterpene oxidation in forest atmospheres. Geophys Res Lett 26:1145-1148

Zakharenko VS (1997) Photoadsorption and photocatalytic oxidation on the metal oxides—Components of tropospheric solid aerosols under the earth's atmosphere conditions. Catalysis Today 39:243-249

Zhang Y, Carmichael GR (1999) The role of mineral aerosol in tropospheric chemistry in East Asia—A model study. J Appl Meteorol 38:353-366

Zhang WX, Wang CB, Lien HL (1998) Treatment of chlorinated organic contaminants with nanoscale bimetallic particles. Catalysis Today 40:387-395

Zhuang Y, Biswas P (2001) Submicrometer particle formation and control in a bench-scale pulverized coal combustor. Energy & Fuels 15:510-516

Zondlo MA, Hudson PK, Prenni AJ, Tolbert MA (2000) Chemistry and microphysics of polar stratospheric clouds and cirrus clouds. Ann Rev Phys Chem 51:473-499